Bone Marrow Stem Cell Therapy for Stroke

Kunlin Jin • Xunming Ji • Qichuan Zhuge
Editors

Bone Marrow Stem Cell Therapy for Stroke

Springer

Editors
Kunlin Jin
Department of Pharmacology Neuroscience
University of North Texas Health Science Center
Fort Worth, TX, USA

Xunming Ji
Department of Neurosurgery, Xuanwu Hospital
Capital Medical University
Beijing, Beijing, China

Qichuan Zhuge
Department of Neurosurgery
First Affiliated Hospital
Wenzhou Medical University
Wenzhou, China

Zhejiang Provincial Key Laboratory of Aging and Neurological Disorder Research
First Affiliated Hospital
Wenzhou Medical University
Wenzhou, China

ISBN 978-981-10-2928-8 ISBN 978-981-10-2929-5 (eBook)
DOI 10.1007/978-981-10-2929-5

Library of Congress Control Number: 2016959532

Printed on acid-free paper

This Springer imprint is published by Springer Nature
The registered company is Springer Science+Business Media Singapore Pte Ltd.
The registered company address is: 152 Beach Road, #22-06/08 Gateway East, Singapore 189721, Singapore

To our families

Preface

The book *Bone Marrow Stem Cell Therapy for Stroke* has been created to be the primary resource for scientists, clinicians, teachers, students, and the public at large in the area of cell therapy for stroke. I am honored to have had the opportunity to edit the book with our coeditors, Qichuan Zhuge and Xunming Ji. The contributions of the editors and editorial broad cannot be overestimated.

We know that stroke remains the fifth leading cause of death, with an incidence of ~700,000 events per year and a prevalence of about 4.8 million individuals, of which about 80 % are ischemic and 20 % are hemorrhagic. About one-third of these people fail to survive the event, and of the survivors, 90 % suffer permanent deficits. The most recent major advance in treatment, the use of thrombolytic agents to dissolve clots, appears to be effective only within about the first 3 hours after the initial onset of symptoms. Widely effective treatments for stroke remain elusive, and no effective treatment is available for chronic stroke except a lengthy program of rehabilitation, all the more emphasizing the need for new therapeutic developments. Stem cell transplantation offers an exciting new therapeutic avenue for stroke, because stem cell not only prevent damage, but also actually repair the injured brain.

Many studies have demonstrated favorable results in animal models with various stem cells including bone marrow-derived stem cells (BMSCs), which consequently have resulted in several early Phase I and II clinical trials with promising outcomes. In addition, BMSCs are easy to obtain and expand in culture; using the patient's own BMSCs would eliminate the risk of rejection, and BMSCs have the capacity to migrate to the injury site, permitting systemic administration. This book reviews recent advances in all aspects of BMSC-mediated stroke treatment in animal models of stroke and clinical trials.

The book is organized into 14 major areas, starting with an introduction of stroke pathophysiology and cell therapy, followed by BMSCs characterization, isolation, culture, and identification. The third and forth chapters cover the topics of mobilization and homing of bone marrow stem cells and interaction of BMSCs with other cells. The next three chapters deal with the various types of BMSCs in stroke treatment and co-transplantation strategies and combination therapies for stroke.

The tenth section reviews the clinical studies of BMSC therapy in stroke patients. The 11th chapter is dedicated to the area of the optimal condition of transplantation for stroke treatment. The 12th and 13th chapters summarize the potential mechanisms of transplanted cell-mediated recovery after stroke. The last section of the book is dedicated to imaging and tracking transplanted BMSCs after stroke.

I thank our colleagues, who have contributed their expert advice to the preparation of the first book to focus on BMSC therapy for stroke. I am especially indebted to the staff at Springer Publishing, who have been enormously helpful in moving this book through editing and production. I hope our efforts will help to demystify stem cell therapy for stroke for researchers and clinicians and contribute to provide stroke patients better and more effective treatment.

Fort Worth, TX, USA
September 8, 2016

Kunlin Jin

Contents

Contributors

Ana-Maria Buga Department of Psychiatry, University of Medicine Rostock, Rostock, Germany

Department of Functional Sciences, Center of Clinical and Experimental Medicine, University of Medicine and Pharmacy of Craiova, Craiova, Romania

Jieli Chen Department of Neurology, Henry Ford Hospital, Detroit, MI, USA

Lefu Chen Zhejiang Provincial Key Laboratory of Aging and Neurological Disorder Research, First Affiliated Hospital, Wenzhou Medical University, Wenzhou City, People's Republic of China

Zhiguo Chen Cell Therapy Center, Xuanwu Hospital, Capital Medical University, Beijing, China

Ovidiu Ciobanu Vivantes Humboldt-Klinikum Psyhiatrie, Psychotherapie und Psychosomatik Zentrum für affektive Erkrankungen, Berlin, Germany

Irene Escudero Stroke Unit, Neurology Department, Instituto de Biomedicina de Sevilla-IBiS, Hospital Universitario Virgen del Rocío, Seville, Spain

Yunqian Guan Cell Therapy Center, Xuanwu Hospital Capital Medical University, Beijing, China

Hongbin Han Department of Radiology, Peking University Third Hospital, Beijing, China

Beijing Key Laboratory of Magnetic Resonance Imaging Device and Technique, Beijing, China

Rongrong Han Institute of Hypoxia Medicine, Xuanwu Hospital, Capital Medical University, Beijing, China

Jiangnan Hu Zhejiang Provincial Key Laboratory of Aging and Neurological Disorder Research, First Affiliated Hospital, Wenzhou Medical University, Wenzhou, China

Center for Neuroscience Discovery, Institute for Healthy Aging, University of North Texas Health Science Center, Fort Worth, TX, USA

Lijie Huang Department of Neurosurgery, First Affiliated Hospital, Wenzhou Medical University, Wenzhou, China

Zhejiang Provincial Key Laboratory of Aging and Neurological Disorder Research, First Affiliated Hospital, Wenzhou Medical University, Wenzhou, China

Shengwei Huang Zhejiang Provincial Key Laboratory of Aging and Neurological Disorder Research, First Affiliated Hospital, Wenzhou Medical University, Wenzhou, China

Center for Neuroscience Discovery, Institute for Healthy Aging, University of North Texas Health Science Center, Fort Worth, TX, USA

Andrew Hui Texas College of Osteopathic Medicine, University of North Texas Health Science Center, Fort Worth, TX, USA

Xunming Ji Department of Neurosurgery, Xuanwu Hospital, Capital Medical University, Beijing, China

Institute of Hypoxia Medicine, Xuanwu Hospital, Capital Medical University, Beijing, China

China-America Joint Institute for Neuroscience, Xuanwu Hospital, Capital Medical University, Beijing, China

Michael Jiang Department of Anesthesiology, Emory University School of Medicine, Atlanta, GA, USA

Kunlin Jin Department of Pharmacology Neuroscience, University of North Texas Health Science Center, Fort Worth, TX, USA

Yunqian Li Medical and Health Analysis Center, Peking University, Beijing, China

Siyang Lin Zhejiang Provincial Key Laboratory of Aging and Neurological Disorder Research, Department of Neurology, First Affiliated Hospital, Wenzhou Medical University, Wenzhou, Zhejiang, China

Fernando Mancha Instituto de Biomedicina de Sevilla-IBiS, Hospital Universitario Virgen del Rocío, Seville, Spain

Francisco Moniche Stroke Unit, Neurology Department, Instituto de Biomedicina de Sevilla-IBiS, Hospital Universitario Virgen del Rocío, Seville, Spain

Instituto de Biomedicina de Sevilla-IBiS, Hospital Universitario Virgen del Rocío, Seville, Spain

Joan Montaner Instituto de Biomedicina de Sevilla-IBiS, Hospital Universitario Virgen del Rocío, Seville, Spain

Mark Nyanzu Department of Neurosurgery, First Affiliated Hospital, Wenzhou Medical University, Wenzhou, China

Zhejiang Provincial Key Laboratory of Aging and Neurological Disorder Research, First Affiliated Hospital, Wenzhou Medical University, Wenzhou, China

Mengxiong Pan Zhejiang Provincial Key Laboratory of Aging and Neurological Disorder Research, Department of Neurology, First Affiliated Hospital, Wenzhou Medical University, Wenzhou, Zhejiang, China

Blanca Pardo Stroke Unit, Neurology Department, Instituto de Biomedicina de Sevilla-IBiS, Hospital Universitario Virgen del Rocío, Seville, Spain

Suning Ping Department of Neurosurgery, State University of New York, Upstate Medical University, Syracuse, NY, USA

Aurel Popa-Wagner Department of Psychiatry, University of Medicine Rostock, Rostock, Germany

Changhong Ren Institute of Hypoxia Medicine, Xuanwu Hospital, Capital Medical University, Beijing, China

Linhui Ruan Zhejiang Provincial Key Laboratory of Aging and Neurological Disorder Research, First Affiliated Hospital, Wenzhou Medical University, Wenzhou City, People's Republic of China

Raluca Elena Sandu Department of Functional Sciences, Center of Clinical and Experimental Medicine, University of Medicine and Pharmacy of Craiova, Craiova, Romania

Chunyan Shi Department of Radiology, Beijing An Zhen Hospital, Capital Medical University, Beijing Institute of Heart, Lung and Blood Vessel Disease, Beijing, China

Jingfei Shi China-America Joint Institute for Neuroscience, Xuanwu Hospital, Capital Medical University, Beijing, China

Felix Siaw-Debrah Department of Neurosurgery, First Affiliated Hospital, Wenzhou Medical University, Wenzhou, China

Zhejiang Provincial Key Laboratory of Aging and Neurological Disorder Research, First Affiliated Hospital, Wenzhou Medical University, Wenzhou, China

Yaying Song Department of Neurology, Shanghai Ruijin Hospital, School of Medicine, Shanghai Jiao Tong University, Shanghai, China

Neuroscience and Neuroengineering Center, Med-X Research Institute and School, Biomedical Engineering, Shanghai Jiao Tong University, Shanghai, China

Yaohui Tang Department of Orthopaedic Surgery and Bioengineering, School of Medicine, Stanford University, Stanford, CA, USA

Ángela Vega-Salvatierra Instituto de Biomedicina de Sevilla-IBiS, Hospital Universitario Virgen del Rocío, Seville, Spain

Brian Wang Center for Neuroscience Discovery, Institute for Healthy Aging, University of North Texas Health Science Center, Fort Worth, TX, USA

Wei Wang Department of Radiology, Peking University Third Hospital, Beijing, China

Beijing Key Laboratory of Magnetic Resonance Imaging Device and Technique, Beijing, China

Weikai Wang Zhejiang Provincial Key Laboratory of Aging and Neurological Disorder Research, First Affiliated Hospital, Wenzhou Medical University, Wenzhou City, People's Republic of China

Ling Wei Department of Anesthesiology, Emory University School of Medicine, Atlanta, GA, USA

Junhao Yan Department of Anatomy and Embryology, School of Basic Medical Sciences, Peking University, Beijing, China

Guo-Yuan Yang Department of Neurology, Shanghai Ruijin Hospital, School of Medicine, Shanghai Jiao Tong University, Shanghai, China

Neuroscience and Neuroengineering Center, Med-X Research Institute and School of Biomedical Engineering, Shanghai Jiao Tong University, Shanghai, China

Jianjing Yang Department of Neurosurgery, First Affiliated Hospital, Wenzhou Medical University, Wenzhou, China

Zhejiang Provincial Key Laboratory of Aging and Neurological Disorder Research, First Affiliated Hospital, Wenzhou Medical University, Wenzhou, China

Lan Yuan Medical and Health Analysis Center, Peking University, Beijing, China

Elena Zapata Stroke Unit, Neurology Department, Instituto de Biomedicina de Sevilla-IBiS, Hospital Universitario Virgen del Rocío, Seville, Spain

Hongxia Zhang Zhejiang Provincial Key Laboratory of Aging and Neurological Disorder Research, First Affiliated Hospital, Wenzhou Medical University, Wenzhou, China

Center for Neuroscience Discovery, Institute for Healthy Aging, University of North Texas Health Science Center, Fort Worth, TX, USA

Weifang Zhang Nuclear Medicine Department, Peking University Third Hospital, Beijing, China

Y. Alex Zhang Cell Therapy Center, Xuanwu Hospital, Capital Medical University, Beijing, China

Zhijun Zhang Department of Neurology, Shanghai Ruijin Hospital, School of Medicine, Shanghai Jiao Tong University, Shanghai, China

Li-Ru Zhao Department of Neurosurgery, State University of New York, Upstate Medical University, Syracuse, NY, USA

Chengcai Zheng Zhejiang Provincial Key Laboratory of Aging and Neurological Disorder Research, Department of Neurology, First Affiliated Hospital, Wenzhou Medical University, Wenzhou, China

Yulei Zhou Zhejiang Provincial Key Laboratory of Aging and Neurological Disorder Research, Department of Neurology, First Affiliated Hospital, Wenzhou Medical University, Wenzhou, Zhejiang, China

Qichuan Zhuge Department of Neurosurgery, First Affiliated Hospital, Wenzhou Medical University, Wenzhou, China

Zhejiang Provincial Key Laboratory of Aging and Neurological Disorder Research, First Affiliated Hospital, Wenzhou Medical University, Wenzhou, China

Chapter 1
Ischemic Stroke Pathophysiology and Cell Therapy

Changhong Ren, Rongrong Han, Jingfei Shi, and Xunming Ji

Abstract Current evidence shows great promise for stem cell transplantation as a new therapeutic strategy for stroke. However, stem cell transplantation for stroke is still in its infancy, with many issues that need to be addressed in order to achieve the full potential of stem cell therapy for stroke. Among the major hurdles for successful clinical translation is determining the therapeutic time window, stem cell type selection, delivery route, and underlying cellular and molecular mechanisms. In this chapter, we attempt to review the basic knowledge of pathophysiology and summarize the different stem cells for stroke treatment.

Keywords Stroke • Ischemia • Transplantation

1.1 Introduction

Stroke remains a worldwide health burden, causing high morbidity, mortality, and costs to health care [51]. It is the foremost cause of long-term disability in the USA and has a high cost to society totaling $38.6 billion in 2009 [65]. Among the various types of strokes, patients are most commonly seen with ischemic stroke, with about

C. Ren • R. Han
Institute of Hypoxia Medicine, Xuanwu Hospital, Capital Medical University, Chang Chun Road 45, Beijing 100053, China

J. Shi
China-America Joint Institute for Neuroscience, Xuanwu Hospital, Capital medical University, Chang Chun Road 45, Beijing 100053, China

X. Ji (✉)
Department of Neurosurgery, Xuanwu Hospital, Capital Medical University, Beijing, China

Institute of Hypoxia Medicine, Xuanwu Hospital, Capital Medical University, Beijing, China

China-America Joint Institute for Neuroscience, Xuanwu Hospital, Capital Medical University, Beijing, China
e-mail: jixm@ccmu.edu.cn

K. Jin et al. (eds.), *Bone Marrow Stem Cell Therapy for Stroke*,
DOI 10.1007/978-981-10-2929-5_1

80 % and over being stricken with this specific disease. Presently, the only FDA-approved treatment is thrombolysis by way of tPA administration within a narrow 4.5 h therapeutic window after stroke onset [40]. Because of its short treatment window and the accompanying concern of hemorrhage when given outside this window, thrombolytic therapy is thus not widely used [123]. About 4 % of stroke patients have been seen to benefit from reperfusion therapies, but only a dismal 10 % of stroke survivors could return to living on their own. In this regard, a huge challenge is posed to researchers and clinical investigators to develop other restorative therapies for the treatment of ischemic stroke. Mechanistic studies of neuronal cell death have resulted in several treatment plans that seek to impede secondary biochemical changes, which consequently decrease the size of brain damage stemming from cerebral ischemia. Incidentally, these mechanisms underlying disproportionate neuronal death in ischemia could be relevant to other neurodegenerative diseases facing the same issue.

This chapter seeks to summarize several mechanisms that have been shown to worsen neuronal death in cerebral ischemia as well as provide a succinct review of current transplanted cell types and their efficacies in the clinic.

1.2 Pathophysiology of Ischemic Stroke

Although there are many etiologic mechanisms, the common pathway of ischemic stroke is lack of sufficient blood flow to perfuse cerebral tissue. Interruption of forward blood flow at any point can lead to irreversible neuronal damage. Ischemic stroke is caused by either of the three main mechanisms, namely, thrombosis, embolism, or global ischemia. Of course, there is no one mechanism that can account for all ischemic strokes. In actuality, the number of factors responsible for uncommon stroke syndromes are numerous. However, vasospastic strokes and a form of arterial inflammation are the more prominent infrequent causes of stroke.

1.2.1 Thrombosis

In situ thrombosis is the formation of a clot in an artery that persists long enough to cause ischemic insult to the cerebral tissue supplied by the affected vessel. Thrombosis is often triggered by pathology in the local endothelium [62]. Atherosclerosis is the most common pathological feature of vascular obstruction resulting in thrombotic stroke. Atherosclerotic plaques can undergo pathological changes such as ulcerations, thrombosis, calcifications, and intra-plaque hemorrhage [80, 126]. In addition to atherosclerosis, other pathological conditions that cause thrombotic occlusion of a vessel include clot formation due to hypercoagulable state, fibromuscular dysplasia, arteritis, and dissection of a vessel wall [182].

1.2.2 Embolism

Embolic stroke (ES) can result from embolization of an artery in the central circulation from a variety of sources. Although the heart is the most common source of a thromboembolus, several types of material can be carried to the brain through the cerebral circulation and lodge in a vessel, leading to stroke, including clot, fibrin, pieces of atheromatous plaque, fat, air, tumor or metastasis, bacterial clumps, and foreign bodies.

1.2.3 Systemic Hypoperfusion

A third mechanism of ischemic stroke is systemic hypoperfusion due to a generalized loss of arterial pressure. Several processes can lead to systemic hypoperfusion, the most widely recognized and studied being cardiac arrest due to myocardial infarction and/or arrhythmia. Global ischemia has been shown to cause the largest damage to areas between the major cerebral and cerebellar arteries widely known as the "boundary zone" or "watershed area." The parietal-temporal-occipital triangle located at the junction of the posterior, middle, and anterior cerebral arteries is very frequently affected. Watershed infarction in this area results in a clinical syndrome involving sensory loss in the arm predominantly as well as paralysis; speech and facial muscles are surprisingly spared. These infarcts contribute to about 10 % of total ischemic strokes, and approximately 40 % occur in patients exhibiting carotid stenosis or occlusion.

1.2.4 Cellular Pathophysiology

The brain accounts for 2 % of body weight but 20 % of total oxygen consumption. Approximately 70 % of the metabolic demand in the brain is due to the Na+/K+-ATPase pump that maintains the ion gradient responsible for neuronal membrane potential. Under ischemic conditions, mitochondrial production of ATP ceases and intracellular ATP stores deplete within 2 min. Cell membranes depolarize, leading to a large influx of calcium and sodium and an efflux of potassium. Cells in the infarct core are rapidly and irreversibly destroyed by lipolysis, proteolysis, and disaggregation of microtubules due to metabolic failure. The ischemic penumbra—the zone of tissue between the infarct core and normal brain—experiences diminished blood flow but preserved cellular metabolism. The goal of acute stroke therapies is to normalize perfusion and intervene in the cascade of biochemical dysfunction to preserve the maximal amount of penumbral tissue [42, 46].

At a molecular level, the development of hypoxic–ischemic neuronal injury is greatly influenced by "overreaction" of certain neurotransmitters, primarily glutamate and aspartate. This process called "excitotoxicity" is triggered by depletion of cellular energy stores. Glutamate, which is normally stored inside the synaptic terminals, is cleared from the extracellular space by an energy-dependent process. The greatly increased concentration of glutamate (and aspartate) in the extracellular space in a depleted energy state results in the opening of calcium channels associated with N-methyl-D-aspartate (NMDA) and alpha-amino-3-hydroxy-5-methyl-4-isoxanole propionate (AMPA) receptors [175]. Persistent membrane depolarization causes influx of calcium, sodium, and chloride ions and efflux of potassium ions [108]. Intracellular calcium is responsible for activation of a series of destructive enzymes such as proteases, lipases, and endonucleases that allow release of cytokines and other mediators, resulting in the loss of cellular integrity [186]. Inflammatory response to tissue injury is initiated by the rapid production of many different inflammatory mediators, tumor necrosis factor being one of the key agents. Leukocyte recruitment to the ischemic areas occurs as early as 30 min after ischemia and reperfusion [39].

1.2.5 Cell Death After Stroke

Within the center or core of the ischemic territory, blood flow deficits, low ATP levels and energy stores, ionic disruption, and metabolic failure are severe, and cell death progresses in minutes [126]. Infarction is synonymous with necrosis (i.e., cytoplasmic swelling, dissolution of organelles and plasma membranes, and inflammation are present). Necrotic cell death is characterized by energy failure, which results in inhibition of protein synthesis. Therefore, new gene products may not be expressed. In the permanent middle cerebral artery occlusion model in the rat, loss of glucose utilization is rapid and complete within a few hours [184], with little time or energy available for the synthesis of new gene products. Thus, one would surmise that necrosis is the primary mode of cell death in this model. Histologic characteristics of necrotic cell death are cytoplasmic and nuclear swelling, loss of integrity of cell organelles, rupture of the cell membrane, and dissolution of all cell structures. In vivo, necrotic cell death is often accompanied by intense inflammation with recruitment of inflammatory cells. This inflammatory response can injure adjacent normal cells. Although the ischemic core is usually characterized by necrosis, when the ischemia is transient or less severe (penumbra), changes consistent with apoptosis have been described [66]. Apoptosis refers to the morphologic changes that occur after programmed cell death (PCD) [66, 91]. Programmed cell death has several key characteristics: (a) the death process is active, and the expression of new proteins is often involved. (b) Cellular energy stores are normal until the final stages of cellular death; therefore, energy failure is a late, secondary event in programmed cell death. (c) The activation of endonucleases results in numerous double-stranded DNA breaks at the boundaries between histosomes. (d) Morphologic changes characteristic of apoptosis, including cytoplasmic and nuclear budding ("apoptotic bodies"), are present [91]. In contrast to necrosis, programmed cell death results in neuronal death with little or no accompanying inflammation. Thus, "collateral damage" to neighboring cells is avoided.

1.2.6 Mechanisms of Necrotic Cell Death

There are at least three fundamental mechanisms leading to cell death during ischemic brain injury: excitotoxicity, ionic imbalance, and oxidative/nitrosative stress. These mechanisms demonstrate overlapping and redundant features. They mediate injury within neurons, glia, and vascular elements, and at the subcellular level, they impact the function of mitochondria, nuclei, cell membranes, endoplasmic reticula, and lysosomes [126].

After stroke onset, the loss of energy stores results in ionic imbalance, neurotransmitter release, and inhibition of reuptake (e.g., of glutamate, the major excitatory transmitter in the mammalian brain). Subsequently, binding of glutamate to ionotropic NMDA (N-methyl-D-aspartate) and AMPA (alpha-amino-3-hydroxy-5-methyl-4-isoxanole propionate receptor) receptors promotes excessive calcium influx, which triggers an array of downstream phospholipases and proteases that degrade membranes and proteins that are essential for cellular integrity. Mitochondria have also been implicated in toxicity, because of oxygen radical generation and the release of death-inducing factors. In addition, ionotropic glutamate receptors (GluRs) promote an excessive influx of sodium with concomitant cell swelling and edema [126]. Besides calcium, imbalances in other ions are important after ischemia. Large amounts of zinc are stored in vesicles of excitatory neurons and are co-released upon depolarization [21], and loss of zinc from presynaptic terminals correlates with zinc translocation into cell bodies and subsequent neuronal death after focal cerebral ischemia [194]. Recently, imbalances in potassium have also been implicated in ischemic cell death. Neurons express a class of calcium-sensitive high-conductance potassium channels, and compounds that selectively modulate these channels protect the brain against stroke in animal models [67].

The reactive oxygen radical is a key mediator of tissue damage after reperfusion in many organs including the heart, kidney, and brain. Mitochondria are strongly implicated, and this might be due to excessive superoxide production during electron transport and inhibition of mitochondrial electron transport mechanisms by free radicals, leading to even more oxygen radical generation [54]. High calcium, sodium, and ADP levels in ischemic cells stimulate excessive mitochondrial oxygen radical production. Oxygen radicals are also produced during enzymatic conversions, such as the cyclooxygenase-dependent conversion of arachidonic acid to prostanoids and the degradation of hypoxanthine, especially upon reperfusion. Furthermore, free radicals are also generated during the inflammatory response after ischemia. Then, oxidative stress, excitotoxicity, energy failure, and ionic imbalances are inextricably linked and contribute to ischemic cell death. After ischemia and particularly reperfusion, production of reactive oxygen species, including superoxide and hydroxyl radicals, overwhelms endogenous scavenging mechanisms and directly damages lipids, proteins, nucleic acids, and carbohydrates [54]. Importantly, oxygen radicals and oxidative stress facilitate mitochondrial transition pore (MTP) formation. MTP dissipates the proton motive force that is required for oxidative phosphorylation and ATP generation, and, as a result, mitochondria release their constituents—including apoptosis-related proteins—within the inner and outer mitochondrial membranes [109]. Upon reperfusion and renewed tissue oxygenation, dysfunctional mitochondria might generate oxidative stress and MTP formation [10]. Oxidative and nitrosative stresses are modulated by enzyme systems such as SOD and the NOS family. Mice deficient in expression of the neuronal NOS isoform or the inducible isoform show less tissue damage compared with their wild-type counterparts after cerebral ischemia [84, 86]. Similarly, the generation of nitric oxide and oxidative stress is linked to DNA damage and activation of poly-(ADP-ribose) polymerase (PARP1), a nuclear enzyme that facilitates DNA repair and regulates transcription [38]. In response to DNA strand breaks, PARP1 activity becomes excessive and depletes the cell of NAD+ and possibly ATP. Ischemic cell death by necrotic and apoptotic mechanisms is suppressed by inhibiting PARP1 activity or by deleting the parp1 gene [47].

1.2.7 Mechanisms of Programmed Cell Death

There are at least three sites where PCD can be triggered: the mitochondria, the cell membrane receptors, and the chromosomal DNA [66]. Once PCD is triggered, there are at least three major pathways by which it might be initiated: (1) the intrinsic caspase pathway is activated when cytochrome c is released from the mitochondria and activates caspases; (2) the extrinsic caspase pathway is activated when cell membrane receptor systems, namely, Fas and TNF-α, activate caspases; and (3) the third pathway is activated when AIF initiates apoptosis by caspase-independent mechanisms [66]. Oncogenes Bcl-2 and Bax inhibit and promote PCD, respectively. Caspase-3 acts with Bax and promotes PCD. Treatments aimed at blocking the caspase-dependent and caspase-independent pathways of PCD (novel protease inhibitors) could potentially decrease ischemic brain damage [91].

Just as calcium entry into the neuron is a key step in excitotoxicity, the release of cytochrome c from the mitochondria is a key event in initiating apoptosis in many cell types. Cytosolic cytochrome c complexes with APAF-1 and procaspase-9 [115]. As a result, procaspase-9 is cleaved into its active form, caspase-9. Caspase-9 then cleaves and activates other caspases, including caspase-3. There is a large body of evidence that brain ischemia can cause activation of caspases. Upregulation and activation of caspase-3 was found to precede death of neurons, especially in the hippocampus and caudate–putamen, in models of transient focal and global brain ischemia [128]. Luo et al. found that deoxyribonuclease activity resulting from transient focal ischemia in the rat could be prevented by inhibitors of caspase-3-like activity [129].

Evidence indicates that many of the mechanisms that initiate programmed cell death are activated in ischemic neurons under certain conditions. The mRNA of the Fas ligand is induced by forebrain ischemia [136]. Expression of the Fas ligand and associated proteins and infarction volumes was smaller in LPR mice that expressed a dysfunctional Fas ligand than in wild-type controls. The Fas receptor is also upregulated after cerebral ischemia in rat brain [52]. TNF-αmRNA transcription is induced as an early response after cerebral ischemia [228]. Expression of the TNF receptor is also increased after cerebral ischemia. TNF-binding protein, a protein that binds and inhibits TNF, reduced infarction volume after middle cerebral artery occlusion in rats [148]. However, ischemic injury was exacerbated in TNF-α -receptor null mice, which suggests that TNF signaling pathways may instead have beneficial effects in ischemic injury under some circumstances. Caspase-8, which is activated by both the Fas and TNF receptors, is expressed and activated after cerebral ischemia [15].

The third pathway by which PCD may occur is caspase-independent. A key factor in this pathway is AIF, a novel proapoptotic molecule that is involved in the final execution of apoptosis and that has been identified and partially characterized [66]. The AIF cDNA codes for a protein of 612 and 613 amino acids in mouse and human, respectively [199]. The AIF protein contains an amino-terminal mitochondrial localization sequence that confines AIF to residing exclusively in the mitochondria in healthy cells [127]. During apoptosis, AIF is released from mitochondria, loses its mitochondrial localization sequence domain (mature AIF), and expresses its apoptogenic effects [199]. When microinjected into the cytoplasm of normal cells, recombinant AIF is sufficient to cause the following four apoptotic hallmarks: (1) the exposure of phosphatidylserine on the plasma membrane surface, (2) the condensation of nuclear chromatin (stage I), (3) large-scale DNA fragmentation, and (4) the dissipation of the mitochondrial transmembrane potential and release of cytochrome c [53]. These apoptogenic effects of ectopic (extramitochondrial) AIF are independent of the action of caspases and are not affected by overexpression of the antiapoptotic protein, Bcl-2 [199].

1.3 Types of Stem Cells Used in Experimental Ischemic Stroke Therapy

Many types of stem cells have been tested and evaluated for their therapeutic potentials in the treatment of ischemic stroke, including mesenchymal stem cells (MSCs), neural stem cells (NSCs), vascular progenitor cells (VPCs), endothelial progenitor cells (EPCs), embryonic stem cells (ESCs), and induced pluripotent stem cell (iPS). The majority of published studies explored the efficacy of transplantation of single type of stem cells [206]. Recently, there are also several studies that investigated the efficacy of transplantation of a combination of different stem cells.

1.3.1 Embryonic Stem Cells (ESCs)

Embryonic stem cells (ESCs) are derived from totipotent cells of the early mammalian embryo [49, 135], which is capable to differentiate not only into all three embryonic germ layers (endoderm, mesoderm, and ectoderm) but also in the embryonic annexes (e.g., placenta, amniotic membranes, etc.), eventually leading to the possibility to obtain any of the more than 220 cell types found in the human organism [68].

The advantage of ESCs is based on its capability of unlimited expansion in vitro to meet the needed amount of cells. In addition, ESCs can be induced to differentiate into neural lineage under specific culturing condition in vitro [8, 153, 172, 243, 258]. Hence, ESCs has been initially considered as an ideal source of transplanted cell for the treatment of neural disorders. Murine ESCs implanted into the contralateral hemisphere following transient cerebral ischemia migrated along the corpus callosum to the ventricular walls, massively populating the border zone of the damaged brain tissue [79], and correlated with improvements in histological and behavioral outcomes [145, 240]. Grafted mouse ESCs also form synaptic connection in the recipient brain [200]. After transplantation of mouse ESCs into rat cortex with a severe focal ischemia, ESC-derived cells expressing cell surface markers of neurons, astrocytes, oligodendrocytes, and endothelial cells could be found in the lesion cavity, and improved structural repair and functional recovery has been demonstrated [230] Intrastriatal transplantation of mouse ESCs or ESC-derived neuron-like cells improved the dopaminergic function and subsequently recovered behavioral dysfunction in focal ischemic rats subjected to middle cerebral artery occlusion (MCAO) [240]. Intracerebral transplantation of mouse ESCs could improve the motor and sensory function of rat with MCAO and reduce the infarct size [200].

Regarding the therapeutic use of hESCs, theoretically they can be differentiated into any type of cell that forms the organism; the undifferentiated ESCs tend to generate teratomas or even highly malignant teratocarcinomas in intact animals [173, 209] and in the stroke rodents [48, 145]. However, xenotransplantation (i.e., stem cells derived from a different species) appears to suppress tumorigenic formation compared to homologous grafting [48] but raises many issues concerning graft rejection. One possible approach is to use in vitro pre-differentiated ESCs that become postmitotic to minimize their tumorigenic potential. Although one group found teratoma formation was independent of pre-differentiation [48], several other groups have found efficacy of pre-differentiated grafts without evidence of tumor formation. ESC-derived neural progenitor cells [59, 75, 202], vascular progenitor cells [160], and mesenchymal stem cells (MSCs) [125] have been shown to exert salutary effects after stroke without noticeable tumorigenesis.

Many studies have explored the effect of ESC-derived neural stem/progenitor cells (NSPC) in animal models of stroke [16, 37, 75, 77, 105]. Most results showed improved behavioral deficit, reduced infarct area, and increased differentiation into neurons after cell transplantation, despite different transplanted cell sources, different stroke animal models, and different infusion routes. However, several studies found that the grafted human ESC-derived neural cells also have the risk of teratomas formation [14, 192]. Culturing condition might reduce tumorigenesis risk of transplanted ESC-derived neural cells. For example, neural cells derived from human ESCs under defined inductive culturing condition (named SD56) did not show chromosome abnormalities after differentiation and tumor formation after implantation into ischemic rat brains and naive nude rat brains and flanks [37]. Malignant transformation of ESC-derived neural cells has been demonstrated to be related to postischemic environment probably by the stimulation of various local cytokine [181]. It is widely acknowledged that higher cerebral blood vessel density results in less possibility and later occurrence of patients suffering from stroke. Any therapeutic measure aimed at promoting angiogenesis would play a pivotal role in function recovery of stroke patients. Intra-arterial transplantation of human ESC-derived endothelial cells and mural cells significantly increased cerebral blood vessel and vascular density in the ischemic striatum, followed by reduction of the infarct volume and of apoptosis as well as acceleration of neurological recovery in mice with transient MCAO [160].

Currently there are no clinical studies using ESCs for stroke treatment. Several drawbacks of ESCs limit their potential for clinical translation, including ethical concerns, immunological response, limited availability, and heterogeneity of donor cells. These issues need to be resolved to encourage more extensive investigations on ESCs.

1.3.2 Mesenchymal Stem Cells (MSCs)

Mesenchymal stem cells (MSCs) are spindle-shaped plastic-adherent cells consisting of a heterogeneous collection of mesenchymal stem and progenitor cells and were first defined as a population of plastic-adherent fibroblastic cells isolated by Percoll density centrifugation [64]. MSCs constitute a population of nonhematopoietic cells in the bone marrow from which these were identified for the first time [58]. MSCs as adult multipotent stem cells can be isolated not only from bone marrow stroma [140] but also from other tissues such as the adipose tissue [185], neural tissue [19], olfactory mucosa [96], heart tissue [138], skin [139], gingiva [56], and many others. Classically, MSCs exhibit CD105, CD73, and CD90 as specific cell surface markers and lack expression of CD45, CD34, CD14 or CD11b, CD79alpha or CD19, and HLA-DR surface molecules. Also, MSCs have the ability to differentiate into adipocytes, osteoblasts, and chondroblasts [45]. Moreover, studies show that they can also differentiate into non-mesenchymal cells such as pancreatic islets [139], neuron-like cells [96], and hepatocytes [197].

Mounting evidence indicated that these MSCs are a favorable cell type for autologous cell transplantations in the scenario of stroke [81]. Though MSCs from different sources are not entirely the same [179], their therapeutic difference for stroke treatment is still unclear. We will briefly review MSCs derived from the bone marrow, adipose tissue, and placenta.

1.3.3 Bone Marrow-Derived MSCs (BMSCs)

In animal models, post-stroke human BMSC transplantation improved sensorimotor function [22, 25, 83, 260], enhanced synaptogenesis, stimulated nerve regeneration [212], decreased tissue plasminogen activator (tPA)-induced brain damage [121], and mediated immunomodulatory effects [245]. Transplanted adult BMSCs migrate to damaged tissue in the brain and decrease post-stroke functional deficits [116, 118]. Migration may be aided by the disruption of the blood brain barrier (BBB) to allow selective entry of BMSCs into the ischemic brain compared to normal cerebral tissue. BMSCs have been reported to stimulate tPA secretion by astrocytes through downregulation of tPA inhibitor plasminogen activator inhibitor-1 [236]. In fact, endogenous tPA could promote BMSC-induced outgrowth of neurites [183] to result in increased synaptic plasticity that correlates with improved functional outcome [236].

In focal ischemic models, BMSC therapy could decrease apoptosis and damage to neurons by inhibiting the downregulation of Bcl-2 and surviving [154]. Survivin-transduced BMSCs saw reduced infarction volume, increased transplanted cell survival, improved functional recovery, and increased protective cytokine expressions such as VEGF and bFGF in a rat model of stroke [122]. Additional evidence of BMSCs protecting against stroke was shown through the i.v. administration of male BMSCs into female rats after MCAO. This type of systemic grafting actually led to decreased apoptosis, increased endogenous neurogenesis, and bFGF expression in the internal border zone; the totality of these processes led to improved neurobehavioral recovery [23]. In a model of chronic stroke, BMSC transplantation improved white matter survival, which may contribute to improved cognitive outcomes [159].

As there are fewer ethical hurdles to overcome when compared to the use of fetal cells, BMSCs may likely be the best alternative for cell therapy at the present moment. Nonetheless, the time required to obtain BMSCs from the individual himself could be a potential limiting factor for the application in acute stroke patients. Yet, BMSC transplantation seems to be an efficacious and practical therapy as seen from clinical studies [9, 114, 198]. The first of many reports originated from Korea. Bang's group infused autologous BMSCs through i.v. into five chronic stroke patients (while 25 received placebo) and analyzed their neurobehavioral deficits after 1 year. Results showed no adverse transplanted BMSC-related effects, and they concluded that i.v. infusion of autologous BMSCs could enhance neurological recovery [9]. They further performed a long-term study of 5 years to evaluate BMSC transplantation safety and efficacy in 52 patients; 16 patients received BMSC treatment. Similarly, they noted no significant side effects following BMSC infusion. In another similar study conducted in Spain, this phase I/II clinical trial transfused autologous bone marrow mononuclear cells via i.a. at 5 and 9 days after stroke in MCA stroke patients and found improved neurobehavioral outcomes as well as feasibility and safety of the transplantation. At the follow-up after 6 months, they reported the absence of stroke recurrence, adverse effects, tumor formation, and deaths except for two individuals with partial seizures that occurred at 3 months [143].

1.3.4 Adipose-Derived Mesenchymal Stem Cells (AD-MSCs)

An abundant number of fully functional MSCs were discovered in the adipose tissue (AD-MSCs) of all adult body. AD-MSCs are easy to obtain without invasive surgery [211], through procedures such as liposuction or abdominoplasty. AD-MSCs can be routinely isolated by collagenase digestion of adipose tissue and cultured [71]. The isolated cells display surface antigens and phenotype similar (but not identical) to BMSCs [63]. There is evidence that AD-MSCs are as effective as BMSCs following permanent middle cerebral artery occlusion (MCAO) in obtaining improved functional recovery in rats [72, 88]. Their lack of expression of MHC-II facilitates their allogeneic administration and possibly allows MSCs from healthy donors to be stored in biobanks for the treatment of stroke patients during the acute phase of the disease [70].

1.3.5 Placenta-Derived Mesenchymal Stem Cells (PD-MSCs)

MSCs are a part of the adult human bone marrow and are seemingly sparse in number, representing less than 0.01 % of the total peripheral blood cells found in the human body. Moreover, as we age, their numbers and function decrease significantly [195, 196]. Nevertheless, MSCs can be readily collected from the chorion, amnion, and villous stroma of the human placenta, independent of the stage of pregnancy [162], which makes it relatively easy to generate a generous number of MSCs in culture.

PD-MSCs have been shown to expand in vitro without showing phenotypic or karyotypic changes. Those derived from allogeneic sources do not require histocompatible tissue matching and therefore are relatively convenient to use vs. bone marrow- or adipose-derived MSCs [168]. Placental cells including PD-MSCs have been found to express MHC class I chain-related proteins A and B that are ligands that can bind to the NK cell receptor NKG2D to downregulate its actions, thereby conferring a type of immune escape strategy [76]. Taken together, PD-MSCs can be considered feasible for application in the clinic. Similar to BMSCs, PD-MSCs possess multilineage differentiation potential with respect to gene expression patterns, morphology, and cell surface antigen expression [61, 87]. Furthermore, PD-MSCs could differentiate into many cell types including neuronal cells [167], e.g., differentiation of PD-MSCs into dopamine neurons was shown to rescue locomotor activity in a hypoxic-ischemia animal model. PD-MSCs have been shown to orient, migrate, and survive in the IBZ when transplanted via intracerebral [180] or i.v. injection [242] after stroke. In fact, PD-MSCs administered through i.v. just 4 h after MCAO demonstrated reduced infarct volume and improved functional outcome [26]. On the other hand, i.v. infusions of PD-MSCs at 8 and 24 h after MCAO proved more efficacious than a single infusion at 24 h [107]. More importantly, compared with fetal-derived MSCs, PD-MSCs showed more robust protective effects [107]. Additionally, PD-MSCs have proangiogenic effects seen in mice such as increased VEGF expression and generation of new blood vessels in ischemic limbs [149].

The mechanisms of action of MSCs have been explored on two levels: a peripheral level accounting for reduction of inflammation and immunomodulation and a central level expressed by the effects on neurogenesis, astrocytes, oligondendrocytes, axons, and angiogenesis.

Although MSCs have been demonstrated to be capable of differentiating into cells of neural lineage in vitro and express neuronal or glial markers in ischemic brain of animal models [41, 89, 110, 260], the survival number of grafted and differentiated cells was small. Also there is a controversy about the function of neuronal cells derived from MSCs [57, 90, 165, 219, 232, 251]. Hence, cell replacement might not be mainly responsible for the beneficial effect of MSCs on ischemic brain injury in vivo [73]. MSCs are stimulated to secrete various neurotrophic factors including cytokines, chemokines, and extracellular matrix protein by damaged surrounding environment [216]. The paracrine effect hypothesis has been strengthened by recent evidence that stem cells release extracellular vesicles which elicit similar biological activity to the stem cells themselves [20, 111, 235]. These released extracellular lipid vesicles provide a novel means of intercellular communication [60, 171, 215, 250], and among them a particular importance seems to have exosomes. New data show that MSCs release large amounts of exosomes which mediate the communication of MSCs with other cells [33, 113, 223, 234]. Exosomes are complex "living" structures generated by many cell types containing a multitude of cell surface receptors [241], encapsulating proteins, trophic factors, miRNAs, and RNAs [31, 170, 213]. Secretion of trophic factors by MSCs and/or MSC-stimulated resident cerebral cells has been considered to contribute to the beneficial effects mentioned above. MSCs constitutively express BDNF, which was significantly increased when MSCs were transplanted into MCAO model. MSCs overexpressing BDNF showed stronger therapeutic effects than original MSCs alone [110]. Other neurotrophic factors, such as HGF, VEGF, NGF, bFGF, FGF-2, and IGF-1, have been demonstrated to be implicated in endogenous repair mechanisms mediated by MSCs [23, 24, 117, 221, 229, 252]. The trophic factors might play critical roles in neuroprotection, angiogenesis, synaptogenesis, endogenous neurogenesis, and inflammatory and immune response.

1.3.6 Induced Pluripotent Stem Cells (iPSCs)

Induced pluripotent stem cells (iPSCs) are a type of pluripotent stem cell that can be reprogrammed from somatic cells with defined factors. In 2006, Shinya Yamanaka et al. first convert mouse embryonic or adult fibroblasts into pluripotent stem cells by introducing four factors, Oct3/4, Sox2, c-Myc, and Klf4 [204]. In 2007 three groups used second generation method to successful reprogrammed mouse fibroblasts into iPSCs [133, 156, 231]. Unlike the first generation of iPSC, these cells could produce viable chimeric mice and could contribute to the germ line, which is the "gold standard" for pluripotent stem cells. The researchers used a different marker Nanog, a gene that is functionally important in ESCs, to select for pluripotent cells. By using this different strategy, the researchers were able to create iPSC that were more similar to ESCs than the first generation of iPSC. In 2007, two independent research groups had successfully induced human fibroblasts into pluripotent stem cells using the pivotal genes [203, 246] Recently, researchers can generate iPSC from human renal epithelial cells in urine and peripheral blood cells [147, 262]. These methods of obtaining donor cells are comparatively less invasive and simple. Because some of the reprogramming factors are oncogenes that bring on a potential tumor risk, the clinical application is not for practice [55, 134]. However, Scientist reported that the generation of iPSC is possible without any genetic alteration of the adult cell. These protocols for the establishment and maintenance of iPSC are both safer and more effective [146, 157, 158]. In 2013, the first trial of stem cells produced from a patient's own body has been approved. Researchers in Japan will use autologous hiPSC-derived retinal pigment epithelium [237] cells to attempt to treat wet age-related macular degeneration.

1.3.6.1 Cell Therapy Using iPSCs for Stroke

iPSCs share similar features compared to ESCs in the morphology, growth properties, high reproduction ability, and pluripotency to differentiate into various types of cells [1, 204]. However, comparing with ESCs, iPSCs could be obtained from cells of any part of an adult patient, avoiding both ethical concerns and reducing the need for immunosuppression. The programed iPSCs cells could specifically differentiate into glutamatergic neurons, motor neurons, and GABAergic neurons [17, 35]. These cells hold great promises for the treatment of various neurological diseases [188, 206]. Wang et al. compared the functionality of the transplanted iPSCs and ESCs in a rat model of cerebral ischemia with use of (18)F-FDG PET imaging [222]. It was shown that both of the cells demonstrated comparable effects in neurological functional and metabolic recovery. Direct injection of mouse embryonic fibroblasts generated iPSCs into damaged areas of rat cortex was shown to significantly decrease the infarct size and improved the motor function in rats with MCAO [30]. Intracerebral transplanted human iPSCs were shown to migrate to the ischemic region and differentiate into neuronal cells in rats with MCAO [93]. After 4–16 days of iPSCs grafting, sensorimotor function of rats has been improved significantly. These findings indicated that iPSC therapy in stroke was effective [263]. The neurorestorative effects of iPSC grafts were comparable with those reported using NSCs [249].

Safety is one major issue in the preclinical cell therapy for stroke. Despite the lack of an ethical problem, depending on the methods used, reprogramming of adult cells to obtain iPSCs may pose the high tumorigenicity of grafted iPSCs [106] that clearly needs to be overcome [233]. The tumorigenicity of pluripotent stem cells (including ESCs and iPSCs) has been well documented in vivo [142, 263]. In a side-by-side comparison between the teratoma composition of human ESCs and iPSCs, human iPSCs develop teratoma more efficiently and faster than human ESC regardless the site of injection [69]. Kawai et al. reported that iPSCs transplanted into ipsilateral striatum and cortex expanded and formed much larger tumors in mice postischemic brain than in sham-operated brain [103]. Klf4, c-Myc, Oct4 and Sox2, the four reprogramming factors exhibit high expression in iPSC-derived tumors, indicating their highly relevant role in tumorigenicity [103, 263]. Subsequently research demonstrated that increased expression of MMP-9 and activated VEGFR2 in the tumors might be involved in promoting the teratoma formation [239]. Inactivation or deletion of the tumor suppressor p53, which is the master regulator of cancer, significantly increases reprogramming efficiency [134]. Thus, there seems to be a trade-off between reprogramming efficiency and tumor generation [263].

Recent studies focused mostly on direct injection of iPSCs pre-differentiated neural cells for stroke therapy [206]. Somatic cells are reprogrammed to iPSC and then differentiated into the cells required for transplantation. A series of studies showed that iPS-derived neural stem cells (NSCs), neural progenitor cells (NPCs), or neuroepithelial-like stem (NES) were safer and more effective than iPSC [21, 92, 166]. Yuan et al. [248] reported that NSCs induced from human iPSCs using retinoic acid and serum-free medium showed stable neural phenotype. After acute transplantation into the ischemic stroke model, these cells survived, migrated into the ischemic penumbra, differentiated into mature neural cells, and showed beneficial effects on functional recovery. Tornero et al. used human iPSC-derived long-term expandable NES cells for stroke therapy [214]. After intracortical implantation into the stroke-damaged striatum and cortex, the cells integrated into stroke-injured host brain and promoted functional recovery. Transplanted iPSC-derived NPC improved sensorimotor functional 14 day after stroke by providing trophic factors, increasing angiogenesis and neurogenesis, and providing new cells for tissue repair [21]. After translation of human iPSC-derived NPC into the postischemic striatum, the graft cells received host tyrosine hydroxylase-positive afferents and contained developing interneurons and homotopic GABAergic medium spiny neurons that, with time, sent axons to the host substantia nigra [166]. Oki et al. [155] reported that human iPSCS-generated NES, which translated into the stroke-damaged mouse and rat striatum or cortex, could survive without forming tumors for at least 4 months. These studies provide evidence that transplantation of human iPSC-derived cells is a safe and efficient approach to promote recovery after stroke [206].

1.3.6.2 Mechanism of iPSC Therapy in Stroke

One major feature of ischemic stroke caused by large vascular occlusion is the significant reduction of cerebral blood flow in the infarct core and surrounding area. The ischemic penumbra has been defined as a hypoperfused area around the ischemic core that is potentially reversible with a timely intervention [120]. Murine NSCs derived from fetal central nervous tissues were reported to release elevated levels of proangiogenic factors such as vascular endothelial growth factor [74] and nitric oxide and to promote angiogenesis with increased cerebral blood flow in the ischemic boundary region. Therefore, one reasonable hypothesis is that human iPSC-NSCs might be able to restore cerebral blood flow in the ischemic penumbra through angiogenesis [120]. In addition, human iPSC-NSCs might also exert other effects on the penumbra tissues, such as rescue of apoptotic cells, modulation of stroke-induced inflammation, and enhancement of intrinsic repair mechanism. Further studies are urgently needed to understand the possible mechanisms behind the functional recovery and to provide reliable and strong evidence for translating iPSC-based cell therapy into the clinical setting.

1.3.7 Endothelial Progenitor Cells (EPCs)

Endothelial progenitor cells (EPCs) are immature endothelial cells which circulate in peripheral blood [176]. In 1997, Asahara first isolated Flk-1^{+}/CD34^{+} cells from human peripheral blood and defined as bone marrow (BM)-derived immature cells with the ability to differentiate into mature endothelial cells [7]. They are halfway in their maturation process to become endothelial cells. Therefore, EPCs possess functional and structural characteristics of both stem cells and mature endothelial cells [176]. During their development, EPCs gradually lose stem cell characteristics and progressively gain endothelial cell characteristics [176]. Therefore, EPC are usually defined as cells expressing both stem cell markers and endothelial cell markers [131]. Most of EPCs quiescently lodge in a microenvironment within the BM, termed the stem cell niche [112]. Upon tissue ischemia, EPCs are mobilized from BM to circulation and migrate toward injured blood vessels and ischemic tissue [97]. The mobilized and recruited EPC participate in endothelial repair and contribute to postnatal angiogenesis and repairing function [101].

In 1999, Takahashi et al. first reported that EPCs are mobilized endogenously in response to hindlimb ischemia [205]. In recent years, more and more evidences show that EPCs are present after ischemic stroke [206]. Clinical studies show that acute stroke induces a transient increase of EPCs [98]. It has been found that level of circulating EPCs is independently predictive of prognosis after ischemic stroke [244]. Taguchi et al. reported that EPC levels gradually increased to the highest level by day 7 after stroke onset and remained significantly above the pre-stroke baseline on days 7 and 14, returning to baseline levels by day 30 [201]. In 48 stroke patients, Sobrino et al. [46] demonstrated that an observed increase of EPC cluster numbers 7 and 90 days after a stroke also related to a good functional outcome [191]. The level of circulated EPCs is also shown to be reduced in various stroke risk factors such as hypertension, hypercholesterolemia, diabetes, and atherosclerosis [29, 217, 261]. Another observational study stated that an increase in EPC levels (measured as EPC-CFUs) after acute ischemic stroke correlated with good functional outcome and reduced infarct growth [191]. More evidences demonstrate that

EPCs not only serve as biomarker but also might offer a new therapeutic strategy for ischemic stroke [201, 261].

EPCs are usually reduced in number and dysfunctional in disease conditions, importantly, the level of EPCs in vivo is limited, comprising just less than 0.0001–0.01 % of the peripheral circulating mononuclear cells, therefore, the transfusion of exogenous EPCs could accelerate the repairing processes after stroke [119]. EPC used for transplantation, in these studies were derived from different sources including bone marrow, cord blood and peripheral blood [112, 261]. Zhang et al. first reported that transplantation of EPC improved cerebral neovascularization after focal cerebral ischemia in the adult mouse [259]. Lately, report showed that injection of EPC isolated from human cord blood decrease cell apoptosis, promote angiogenesis and neurogenesis, and improve functional recovery in rat focal cerebral ischemia model [144]. Transarterial administration of EPC derived from bone marrow reduced infarct volume and neurological deficits in acute focal brain ischemia rat [152]. Fan et al. [50] showed that labeled human EPCs were found around microvessels in the cerebral ischemic boundary 24 h after EPC transplantation, and promote neurovascular repair and improves long-term outcome in mice. All these studies indicate that EPCs could serve as a cellular reservoir for the replacement/repair of dysfunctional ECs in stroke and are promising EPC for the treatment of ischemic stroke [261]. In order to enhance the therapeutic effect, EPC modifications such as gene transfection, ischemia preconditioning and pretreatment have been investigated [247, 261]. In a hind limb ischemic model, combination of intravenous infusion of EPCs overexpressing VEGF with local SDF-1 application showed to be more efficient in improving local blood supply than either of them used alone [247].

Currently, several clinical trials (clinicaltrials.gov identifier: NCT01468064, NCT00535197, NCT01289795) are undergoing to evaluate the feasibility and safety of autologous EPC and CD34+ stem cell transplantation for the treatment of patients with ischemic stroke [206, 261].

The safe aspects of EPC transfusion have been explored in recent years. Although many studies have shown that EPCs have a therapeutic effect on ischemic stroke, still, the release of inflammatory factors may compromise the therapeutic efficacy. It was shown that EPC could produce inflammatory factors such as interleukin-6 receptor, interleukin-8, thrombin, monocyte chemotactic protein-1, and recruit monocytes [218, 261]. These factors can effect on the level, mobilization, and survival of EPCs. Several studies put forward that pharmacological or gene modulation of EPCs before and after transplantation may benefit EPCs with regard to both function and survival [261]. In addition, Bone marrow-derived EPCs are a major determinant of nascent tumor neovascularization [150]. The level of cEPCs has been reported higher in patients with lung, hepatocellular, breast, and colorectal cancers [44]. The evidence indicates that EPCs participate in the neovascularization of tumors and that EPC transfusion to patients with tumors should be avoided.

1.3.7.1 Mechanism of EPC Therapy in Stroke

It is well known that a major pathophysiological event of ischemic stroke is vascular endothelial damage that is induced by various high-risk factors such as hypertension, hyperlipidemia, diabetes, and many more. Evidentially, the development of therapeutic approaches to repair damaged ECs has become an attractive topic of research. EPCs have an ability to differentiate into mature ECs and secrete different protective cytokines and growth factors, such as VEGF, FGF-b, and PDGF-bb, to play a significant role in endothelial homeostasis and repair [238]. While on the other side, they may mediate neighboring-injured ECs with normal structure and function extending into injured sites and performing the function of repair.

Neovascularization is necessary for blood vessel reconstruction and collateral circulation establishment, which are important to deliver nutrients and protectants to the injured tissue for repair. Zhang et al. reported that intravenous bone marrow-derived EPC not only increased the angiogenesis but also increased vasculogenesis at the border of the infarct after focal cerebral ischemia in the adult mouse [259]. EPCs are involved in angiogenesis by secreting an array of growth factors and cytokines, such as VEGF, SDF-1, IGF-1, and G-CSF, which can enhance EC proliferation, reduce cell apoptosis, and recruit endogenous progenitor cells [174]. Several studies have proven that endogenous EPCs participate in the neovascularization via C-X-C chemokine receptor type (CXCR) 4/SDF-1 axis after permanent middle cerebral artery occlusion (MCAO) in rats. Bone marrow-derived EPCs have been shown to account for up to 26 % of all ECs in neovascularization. In a way, these findings confirm the idea that EPCs promote the repair and regeneration of injured vessels simultaneously, which refers to the combined action of angiogenesis and vasculogenesis after ischemic stroke.

1.3.8 Neural Stem Cells

Neural stem cells (NSCs) are self-renewing, multipotent cells that generate the main phenotype of the nervous system [190]. There are two basic types of stem cell: adult stem cells, which are limited in their ability to differentiate, and embryonic stem cells (ESCs), which are pluripotent. ESCs are not limited to a particular cell fate; rather they have the capability to differentiate into any cell type [32] ESCs are derived from the inner cell mass of the blastocyst with the potential to self-replicate [3].

NSCs are considered adult stem cells because they are limited in their capability to differentiate. NSCs are generated throughout an adult's life via the process of neurogenesis [164]. Since neurons do not divide within the central nervous system (CNS), NSCs can be differentiated to replace lost or injured neurons or in many cases even glial cells [3]. NSCs are differentiated into new neurons within the SVZ of lateral ventricles, a remnant of the embryonic germinal neuroepithelium, as well as the dentate gyrus of the hippocampus [164].

Adult NSCs were first isolated from mouse striatum in the early 1990s. They are capable of forming multipotent neurospheres when cultured in vitro. Neurospheres can produce self-renewing and proliferating specialized cells. These neurospheres can differentiate to form the specified neurons, glial cells, and oligodendrocytes [164]. In previous studies, cultured neurospheres have been transplanted into the brains of immunodeficient neonatal mice and have shown engraftment, proliferation, and neural differentiation [164].

NSCs are stimulated to begin differentiation via exogenous cues from the microenvironment or stem cell niche. This capability of the NSCs to replace lost or damaged neural cells is called neurogenesis [3]. Some neural cells are migrated from the SVZ along the rostral migratory stream which contains a marrow-like structure with ependymal cells and astrocytes when stimulated. The ependymal cells and astrocytes form glial tubes used by migrating neuroblasts. The astrocytes in the tubes provide support for the migrating cells as well as insulation from electrical and chemical signals released from surrounding cells. The astrocytes are the primary precursors for rapid cell amplification. The neuroblasts form tight chains and migrate toward the specified site of cell damage to repair or replace neural cells. One example is a neuroblast migrating toward the olfactory bulb to differentiate into periglomerular or granule neurons which have a radial migration pattern rather than a tangential one [177].

On the other hand, the dentate gyrus neural stem cells produce excitatory granule neurons which are involved in learning and memory. One example of learning and memory is pattern separation, a cognitive process used to distinguish similar inputs [3].

1.3.8.1 The Attempts to Treat Stroke with NSCs

Two different types of insults can cause ischemic damage to the brain. Cardiac arrest or coronary artery occlusion, which leads to abrupt and near-total interruption of cerebral blood flow, causes selective neuronal death to certain vulnerable neuronal populations such as the pyramidal neurons of hippocampal CA1. In contrast, occlusion of a cerebral artery—that is, stroke—gives rise to irreversible damage in a core region and a partially reversible injury in the surrounding penumbra zone. In animals, so-called global ischemia models mimic the effects of cardiac arrest or coronary artery occlusion, whereas focal ischemia models replicate the consequences of stroke. These models are useful to explore various restorative strategies [11].

Can cell transplants reconstruct neural circuits that have been damaged by ischemic insults and thereby lead to functional recovery? In the case of global ischemia, this has been achieved at least to some extent by transplantation of fetal hippocampal tissue into the damaged hippocampal CA1 area in rats [78]. Significant improvement in this model requires homotypic replacement of the degenerated CA1 cells and establishment of reciprocal graft–host connectivity [78]. In animals subjected to focal ischemia, fetal cortical grafts placed in the infarcted cortical area receive afferent connections from cortex, thalamus, and subcortical nuclei of the host, whereas efferent projections from the graft to the host brain are sparse [193]. In this model, the grafts were able to promote functional recovery only if the rats were housed in an enriched environment, for reasons that are unclear [137].

To avoid the use of human embryonic tissue, other sources of cells have been tested in ischemia models. In rats with focal ischemia, functional improvement was reported after intrastriatal implantation of neurons derived from a human teratocarcinoma cell line [12, 13]. However, although cells expressing neuronal markers are detected within the grafts, there is no evidence that these cells develop into appropriate striatal neurons in this model. Functional recovery was also observed after global ischemia in rats, following implantation of a mouse hippocampal neuroepithelial cell line within the damaged hippocampus [187]. In this case, grafted cells were demonstrated in the CA1 region, and a minority of them were identified either as astrocytes or neurons, mostly with a pyramidal-like morphology.

Human ES cell-derived NSCs, grafted into the ischemic boundary zone in rats subjected to stroke, have been shown to migrate toward the lesion and improve forelimb performance [37]. Electrophysiological recordings showed functional neuronal properties in the grafted cells and synaptic input from host neurons [36], as has been observed for mouse ES cell-derived precursors implanted in stroke-damaged rat brain [16]. Transplanted human fetal NSCs have also given rise to neurons that migrate toward the ischemic lesion in rodents [104], while human NSCs isolated from embryonic striatum and cortex [100] have generated morphologically mature neurons after transplantation into stroke-damaged rat striatum [38].

Taken together, these findings provide evidence that replacement of functional neurons using stem cell grafts is possible in the stroke-damaged brain.

1.3.8.2 Underlying Mechanisms

The subventricular zone (SVZ) of the lateral ventricle and the dentate gyrus of the hippocampus of adult rodent brains contain neural stem cells that produce neuroblasts [5, 43]. Under physiological conditions, neuroblasts in the SVZ travel via the rostral migratory stream to the olfactory bulb where they differentiate into granule and periglomerular neurons throughout adult life [130]. In the SVZ of adult human brains, neural stem cells are present in a band of astrocytes separated from the ependyma [34, 169, 178]. In experimental stroke, focal cerebral ischemia increases neurogenesis in the ipsilateral SVZ, and neuroblasts emigrate from the SVZ to the ischemic boundary regions of the striatum and cortex where they have the phenotypes of mature neurons [6, 94, 161, 208, 210, 254, 257]. Stroke-induced neurogenesis also takes place in the SVZ and ischemic boundary of adult human brains, even in elderly patients aged 60–87 years [95, 132, 141].

Neurogenesis induced by stroke involves proliferation of neural stem and progenitor cells, differentiation of neural progenitor cells, and migration of neuroblasts to the ischemic boundary where neuroblasts mature into resident neurons and integrate into the parenchymal tissue [207]. In adult mice, gene profile analysis of neural progenitor cells from the SVZ that were isolated by laser-capture microdissection has shown that these cells share more than 70 % of all expressed genes with embryonic cortical neural progenitor cells [2]. In murine neural progenitor cells from the SVZ, stroke activates many genes involved in neurogenesis during embryonic development. The most upregulated genes after stroke are those in the transforming growth factor β superfamily, such as bone morphogenetic protein 8, bone morphogenetic protein type I receptors, and growth differentiation factor 2 [124]. After stroke, adult neural progenitor cells seem to recapture embryonic molecular signals, which probably mediate neuroblast migration and stroke-induced proliferation and differentiation of neural progenitor cells.

In vivo analysis of the cytokinetics of neural progenitor cells has suggested that stroke might trigger actively proliferating neural progenitor cells from the SVZ in adult rodents to repeat the cell cycle kinetics of the embryonic form of these cells [255]. During cortical neurogenesis, cell cycle length is associated with progression of neural progenitor cells from proliferation to neurogenic division, and lengthening of the G1 phase of the neuroepithelial cell cycle activates neuronal differentiation [18, 85]. In rats, studies done in vivo that used cumulative and single S-phase labeling with 5-bromo-2′-deoxyuridine (BrdU) [151] showed that dynamic changes in cell cycle kinetics of neural progenitor cells correlated with the proportion of daughter cells that remained within and left the cell cycle over a period of 2–14 days after stroke [256]. Decreasing the length of the G1 phase of the cell cycle at 2–4 days after stroke was associated with an increase in dividing daughter cells that remained within the cell cycle to expand the SVZ progenitor pool rapidly. By contrast, lengthening the G1 phase at 4–14 days after stroke was accompanied by an increased number of daughter cells that left the cell cycle to differentiate into neurons. These data indicate that stroke triggers dynamic changes in the G1 phase of the actively dividing SVZ cell cycle, resulting in early expansion of a neural progenitor pool and subsequent neuronal differentiation, which leads to increased neurogenesis [256]. Neuroblasts in the ischemic boundary have the phenotypes of mature neurons; [6] by use of the patch clamp technique, new neurons in the ischemic boundary were shown to have the electrophysiological characteristics of mature neurons. These findings suggest that neuroblasts mature into resident neurons and integrate into local neuronal circuitry [102]. However, neurogenesis is diminished after stroke and many newly formed neurons die [6].

Cell-based and pharmacological therapies increase neurogenesis in the ischemic brain. These therapies activate the phosphatidylinositol 3-kinase (PI3K)-Akt signaling pathway in neural progenitor cells [28, 225]. The PI3K-Akt pathway affects several cellular functions such as cell survival, proliferation, differentiation, and migration [102, 220]. Akt regulates proliferation of neural stem cells and neuronal differentiation in embryonic mice [189, 220] and blockage of Akt activation with a selective PI3K inhibitor decreases proliferation of neural progenitor cells [253]. Therefore, the PI3K-Akt signaling pathway seems to be important in the regulation of neurogenesis enhanced by restorative therapies. However, initiation of the PI3K-Akt signaling pathway could differ with individual therapies. Treatment with bone marrow mesenchymal cells stimulates brain parenchymal cells to secrete an array of neurotrophic factors, including basic fibroblast growth factor and brain-derived neurotrophic factor, which are known to activate Akt [4]. Erythropoietin activates the PI3K-Akt pathway by interaction with its receptor in neural progenitor cells, whereas phosphodiesterase five inhibitors and statins are thought to activate Akt via increased concentrations of cGMP [27, 224]. Mammalian achaete-scute homolog 1 (Mash1) and neurogenin 1 (Neurog1, also known as Ngn1) are pro-neuronal basic helix-loop-helix (bHLH) transcription factors that mediate differentiation of neural progenitor cells into neurons [99, 163]. Akt regulates the assembly and activity of bHLH-coactivator complexes to promote this differentiation [220]. Inhibition of the PI3K-Akt pathway in neural progenitor cells suppresses expression of Mash1 and Ngn1. As a result, neuronal differentiation induced by erythropoietin and statins is prevented [227]. Small interfering RNA in neural progenitor cells also attenuates expression of endogenous Mash1 and Ngn1, which further minimizes the rise in the neuronal population caused by erythropoietin and statins [226, 227]. These findings indicate that the PI3K-Akt signaling pathway activated by these restorative therapies can trigger pro-neuronal bHLH transcription factors in neural progenitor cells, leading to neuronal, but not astrocytic, differentiation [82, 226, 227].

References

1. Abe K, Yamashita T, Takizawa S, Kuroda S, Kinouchi H, Kawahara N. Stem cell therapy for cerebral ischemia: from basic science to clinical applications. J Cereb Blood Flow Metab Off J Int Soc Cereb Blood Flow Metab. 2012;32(7):1317–31. doi:10.1038/jcbfm.2011.187.
2. Abramova N, Charniga C, Goderie SK, Temple S. Stage-specific changes in gene expression in acutely isolated mouse CNS progenitor cells. Dev Biol. 2005;283(2):269–81.
3. Alenzi F, Bahkali A. Stem cells: biology and clinical potential. Afr J Biotechnol. 2011;10(86):19929–40.
4. Alessi DR, Andjelkovic M, Caudwell B, Cron P, Morrice N, Cohen P, et al. Mechanism of activation of protein kinase B by insulin and IGF-1. Embo J. 1996;15(23):6541–51.
5. Alvarez-Buylla A, Garcia-Verdugo JM. Neurogenesis in adult subventricular zone. J Neurosci. 2002;22(3):629–34.
6. Arvidsson A, Collin T, Kirik D, Kokaia Z, Lindvall O. Neuronal replacement from endogenous precursors in the adult brain after stroke. Nat Med. 2002;8(9):963–70.
7. Asahara T, Murohara T, Sullivan A, Silver M, van der Zee R, Li T, et al. Isolation of putative progenitor endothelial cells for angiogenesis. Science (New York, NY). 1997;275(5302):964–7.
8. Bain G, Kitchens D, Yao M, Huettner JE, Gottlieb DI. Embryonic stem cells express neuronal properties in vitro. Dev Biol. 1995;168(2):342–57. doi:10.1006/dbio.1995.1085. S0012-1606(85)71085-8 [pii].
9. Bang OY, Lee JS, Lee PH, Lee G. Autologous mesenchymal stem cell transplantation in stroke patients. Ann Neurol. 2005;57(6):874–82. doi:10.1002/ana.20501.
10. Bernardi P, Petronilli V, Di Lisa F, Forte M. A mitochondrial perspective on cell death. Trends Biochem Sci. 2001;26(2):112–7.
11. Bjorklund A, Lindvall O. Cell replacement therapies for central nervous system disorders. Nat Neurosci. 2000;3(6):537–44.

12. Borlongan CV, Tajima Y, Trojanowski JQ, Lee VM, Sanberg PR. Cerebral ischemia and CNS transplantation: differential effects of grafted fetal rat striatal cells and human neurons derived from a clonal cell line. Neuroreport. 1998;9(16):3703–9.
13. Borlongan CV, Tajima Y, Trojanowski JQ, Lee VM, Sanberg PR. Transplantation of cryopreserved human embryonal carcinoma-derived neurons (NT2N cells) promotes functional recovery in ischemic rats. Exp Neurol. 1998;149(2):310–21.
14. Brederlau A, Correia AS, Anisimov SV, Elmi M, Paul G, Roybon L, et al. Transplantation of human embryonic stem cell-derived cells to a rat model of Parkinson's disease: effect of in vitro differentiation on graft survival and teratoma formation. Stem Cells. 2006;24(6):1433–40. doi:10.1634/stemcells.2005-0393. 2005-0393 [pii].
15. Bruce AJ, Boling W, Kindy MS, Peschon J, Kraemer PJ, Carpenter MK, et al. Altered neuronal and microglial responses to excitotoxic and ischemic brain injury in mice lacking TNF receptors. Nat Med. 1996;2(7):788–94.
16. Buhnemann C, Scholz A, Bernreuther C, Malik CY, Braun H, Schachner M, et al. Neuronal differentiation of transplanted embryonic stem cell-derived precursors in stroke lesions of adult rats. Brain. 2006;129(Pt 12):3238–48.
17. Cai J, Yang M, Poremsky E, Kidd S, Schneider JS, Iacovitti L. Dopaminergic neurons derived from human induced pluripotent stem cells survive and integrate into 6-OHDA-lesioned rats. Stem Cells Dev. 2010;19(7):1017–23. doi:10.1089/scd.2009.0319.
18. Calegari F, Haubensak W, Haffner C, Huttner WB. Selective lengthening of the cell cycle in the neurogenic subpopulation of neural progenitor cells during mouse brain development. J Neurosci. 2005;25(28):6533–8.
19. Calzolari F, Michel J, Baumgart EV, Theis F, Gotz M, Ninkovic J. Fast clonal expansion and limited neural stem cell self-renewal in the adult subependymal zone. Nat Neurosci. 2015;18(4):490–2. doi:10.1038/nn.3963. nn.3963 [pii].
20. Camussi G, Deregibus MC, Cantaluppi V. Role of stem-cell-derived microvesicles in the paracrine action of stem cells. Biochem Soc Trans. 2013;41(1):283–7. doi:10.1042/BST20120192. BST20120192 [pii].
21. Chau MJ, Deveau TC, Song M, Gu X, Chen D, Wei L. iPSC transplantation increases regeneration and functional recovery after ischemic stroke in neonatal rats. Stem Cells (Dayton, Ohio). 2014;32(12):3075–87. doi:10.1002/stem.1802.
22. Chen J, Chopp M. Neurorestorative treatment of stroke: cell and pharmacological approaches. NeuroRx. 2006;3(4):466–73. doi:10.1016/j.nurx.2006.07.007. S1545-5343(06)00129-5 [pii].
23. Chen J, Li Y, Katakowski M, Chen X, Wang L, Lu D, et al. Intravenous bone marrow stromal cell therapy reduces apoptosis and promotes endogenous cell proliferation after stroke in female rat. J Neurosci Res. 2003;73(6):778–86. doi:10.1002/jnr.10691.
24. Chen J, Li Y, Wang L, Lu M, Chopp M. Caspase inhibition by Z-VAD increases the survival of grafted bone marrow cells and improves functional outcome after MCAo in rats. J Neurol Sci. 2002;199(1–2):17–24. doi:S0022510X02000758 [pii].
25. Chen J, Li Y, Wang L, Zhang Z, Lu D, Lu M, et al. Therapeutic benefit of intravenous administration of bone marrow stromal cells after cerebral ischemia in rats. Stroke. 2001;32(4):1005–11.
26. Chen J, Shehadah A, Pal A, Zacharek A, Cui X, Cui Y, et al. Neuroprotective effect of human placenta-derived cell treatment of stroke in rats. Cell Transplant. 2013;22(5):871–9. doi:10.3727/096368911X637380. ct0613chen [pii].
27. Chen J, Zhang C, Jiang H, Li Y, Zhang L, Robin A, et al. Atorvastatin induction of VEGF and BDNF promotes brain plasticity after stroke in mice. J Cereb Blood Flow Metab. 2005;25(2):281–90.
28. Chen J, Zhang ZG, Li Y, Wang Y, Wang L, Jiang H, et al. Statins induce angiogenesis, neurogenesis, and synaptogenesis after stroke. Ann Neurol. 2003;53(6):743–51.

29. Chen JZ, Zhang FR, Tao QM, Wang XX, Zhu JH, Zhu JH. Number and activity of endothelial progenitor cells from peripheral blood in patients with hypercholesterolaemia. Clin Sci (London, England: 1979). 2004;107(3):273–80. doi:10.1042/cs20030389.
30. Chen SJ, Chang CM, Tsai SK, Chang YL, Chou SJ, Huang SS, et al. Functional improvement of focal cerebral ischemia injury by subdural transplantation of induced pluripotent stem cells with fibrin glue. Stem Cells Dev. 2010;19(11):1757–67. doi:10.1089/scd.2009.0452.
31. Chen TS, Lim SK. Measurement of precursor miRNA in exosomes from human ESC-derived mesenchymal stem cells. Methods Mol Biol. 2013;1024:69–86. doi:10.1007/978-1-62703-453-1_6.
32. Clarke DL, Johansson CB, Wilbertz J, Veress B, Nilsson E, Karlstrom H, et al. Generalized potential of adult neural stem cells. Science. 2000;288(5471):1660–3.
33. Collino F, Deregibus MC, Bruno S, Sterpone L, Aghemo G, Viltono L, et al. Microvesicles derived from adult human bone marrow and tissue specific mesenchymal stem cells shuttle selected pattern of miRNAs. PLoS ONE. 2010;5(7), e11803. doi:10.1371/journal.pone.0011803.
34. Curtis MA, Kam M, Nannmark U, Anderson MF, Axell MZ, Wikkelso C, et al. Human neuroblasts migrate to the olfactory bulb via a lateral ventricular extension. Science. 2007;315(5816):1243–9.
35. Czepiel M, Balasubramaniyan V, Schaafsma W, Stancic M, Mikkers H, Huisman C, et al. Differentiation of induced pluripotent stem cells into functional oligodendrocytes. Glia. 2011;59(6):882–92. doi:10.1002/glia.21159.
36. Daadi MM, Li Z, Arac A, Grueter BA, Sofilos M, Malenka RC, et al. Molecular and magnetic resonance imaging of human embryonic stem cell-derived neural stem cell grafts in ischemic rat brain. Mol Ther. 2009;17(7):1282–91.
37. Daadi MM, Maag AL, Steinberg GK. Adherent self-renewable human embryonic stem cell-derived neural stem cell line: functional engraftment in experimental stroke model. PLoS ONE. 2008;3(2), e1644.
38. Darsalia V, Kallur T, Kokaia Z. Survival, migration and neuronal differentiation of human fetal striatal and cortical neural stem cells grafted in stroke-damaged rat striatum. Eur J Neurosci. 2007;26(3):605–14.
39. DeGraba TJ. The role of inflammation after acute stroke: utility of pursuing anti-adhesion molecule therapy. Neurology. 1998;51(3 Suppl 3):S62–8.
40. Del Zoppo GJ, Saver JL, Jauch EC, Adams Jr HP. Expansion of the time window for treatment of acute ischemic stroke with intravenous tissue plasminogen activator: a science advisory from the American Heart Association/American Stroke Association. Stroke J Cereb Circ. 2009;40(8):2945–8. doi:10.1161/strokeaha.109.192535.
41. Ding DC, Shyu WC, Chiang MF, Lin SZ, Chang YC, Wang HJ, et al. Enhancement of neuroplasticity through upregulation of beta1-integrin in human umbilical cord-derived stromal cell implanted stroke model. Neurobiol Dis. 2007;27(3):339–53. doi:10.1016/j.nbd.2007.06.010. S0969-9961(07)00114-3 [pii].
42. Dirnagl U, Iadecola C, Moskowitz MA. Pathobiology of ischaemic stroke: an integrated view. Trends Neurosci. 1999;22(9):391–7.
43. Doetsch F, Garcia-Verdugo JM, Alvarez-Buylla A. Cellular composition and three-dimensional organization of the subventricular germinal zone in the adult mammalian brain. J Neurosci. 1997;17(13):5046–61.
44. Dome B, Timar J, Ladanyi A, Paku S, Renyi-Vamos F, Klepetko W, et al. Circulating endothelial cells, bone marrow-derived endothelial progenitor cells and proangiogenic hematopoietic cells in cancer: from biology to therapy. Crit Rev Oncol Hematol. 2009;69(2):108–24. doi:10.1016/j.critrevonc.2008.06.009.
45. Dominici M, Le Blanc K, Mueller I, Slaper-Cortenbach I, Marini F, Krause D, et al. Minimal criteria for defining multipotent mesenchymal stromal cells. The International Society for Cellular Therapy position statement. Cytotherapy. 2006;8(4):315–7. doi:10.1080/14653240600855905. Q2183N8UT042W62H [pii].

46. Doyle KP, Simon RP, Stenzel-Poore MP. Mechanisms of ischemic brain damage. Neuropharmacology. 2008;55(3):310–8. doi:10.1016/j.neuropharm.2008.01.005.
47. Endres M, Wang ZQ, Namura S, Waeber C, Moskowitz MA. Ischemic brain injury is mediated by the activation of poly(ADP-ribose)polymerase. J Cereb Blood Flow Metab Off J Int Soc Cereb Blood Flow Metab. 1997;17(11):1143–51. doi:10.1097/00004647-199711000-00002.
48. Erdo F, Buhrle C, Blunk J, Hoehn M, Xia Y, Fleischmann B, et al. Host-dependent tumorigenesis of embryonic stem cell transplantation in experimental stroke. J Cereb Blood Flow Metab. 2003;23(7):780–5. doi:10.1097/01.WCB.0000071886.63724.FB.
49. Evans MJ, Kaufman MH. Establishment in culture of pluripotential cells from mouse embryos. Nature. 1981;292(5819):154–6.
50. Fan Y, Shen F, Frenzel T, Zhu W, Ye J, Liu J, et al. Endothelial progenitor cell transplantation improves long-term stroke outcome in mice. Ann Neurol. 2010;67(4):488–97. doi:10.1002/ana.21919.
51. Feigin VL, Lawes CM, Bennett DA, Barker-Collo SL, Parag V. Worldwide stroke incidence and early case fatality reported in 56 population-based studies: a systematic review. Lancet Neurol. 2009;8(4):355–69. doi:10.1016/s1474-4422(09)70025-0.
52. Felderhoff-Mueser U, Taylor DL, Greenwood K, Kozma M, Stibenz D, Joashi UC, et al. Fas/CD95/APO-1 can function as a death receptor for neuronal cells in vitro and in vivo and is upregulated following cerebral hypoxic-ischemic injury to the developing rat brain. Brain Pathol (Zurich, Switzerland). 2000;10(1):17–29.
53. Ferri KF, Jacotot E, Blanco J, Este JA, Zamzami N, Susin SA, et al. Apoptosis control in syncytia induced by the HIV type 1-envelope glycoprotein complex: role of mitochondria and caspases. J Exp Med. 2000;192(8):1081–92.
54. Fiskum G, Murphy AN, Beal MF. Mitochondria in neurodegeneration: acute ischemia and chronic neurodegenerative diseases. J Cereb Blood Flow Metab Off J Int Soc Cereb Blood Flow Metab. 1999;19(4):351–69. doi:10.1097/00004647-199904000-00001.
55. Fong CY, Gauthaman K, Bongso A. Teratomas from pluripotent stem cells: a clinical hurdle. J Cell Biochem. 2010;111(4):769–81. doi:10.1002/jcb.22775.
56. Fournier BP, Larjava H, Hakkinen L. Gingiva as a source of stem cells with therapeutic potential. Stem Cells Dev. 2013;22(24):3157–77. doi:10.1089/scd.2013.0015.
57. Fox LE, Shen J, Ma K, Liu Q, Shi G, Pappas GD, et al. Membrane properties of neuron-like cells generated from adult human bone-marrow-derived mesenchymal stem cells. Stem Cells Dev. 2010;19(12):1831–41. doi:10.1089/scd.2010.0089.
58. Friedenstein AJ, Piatetzky II S, Petrakova KV. Osteogenesis in transplants of bone marrow cells. J Embryol Exp Morphol. 1966;16(3):381–90.
59. Fujimoto M, Hayashi H, Takagi Y, Hayase M, Marumo T, Gomi M, et al. Transplantation of telencephalic neural progenitors induced from embryonic stem cells into subacute phase of focal cerebral ischemia. Lab Invest. 2012;92(4):522–31. doi:10.1038/labinvest.2012.1. labinvest20121 [pii].
60. Fujita Y, Yoshioka Y, Ito S, Araya J, Kuwano K, Ochiya T. Intercellular communication by extracellular vesicles and their microRNAs in asthma. Clin Ther. 2014;36(6):873–81. doi:10.1016/j.clinthera.2014.05.006. S0149-2918(14)00263-X [pii].
61. Fukuchi Y, Nakajima H, Sugiyama D, Hirose I, Kitamura T, Tsuji K. Human placenta-derived cells have mesenchymal stem/progenitor cell potential. Stem Cells. 2004;22(5):649–58. doi:10.1634/stemcells.22-5-649. 22/5/649 [pii].
62. Furie B, Furie BC. Mechanisms of thrombus formation. N Engl J Med. 2008;359(9):938–49. doi:10.1056/NEJMra0801082.
63. Gimble JM, Bunnell BA, Guilak F. Human adipose-derived cells: an update on the transition to clinical translation. Regen Med. 2012;7(2):225–35. doi:10.2217/rme.11.119.
64. Glover LE, Tajiri N, Weinbren NL, Ishikawa H, Shinozuka K, Kaneko Y, et al. A step-up approach for cell therapy in stroke: translational hurdles of bone marrow-derived stem cells. Transl Stroke Res. 2012;3(1):90–8. doi:10.1007/s12975-011-0127-8. 127 [pii].

65. Go AS, Mozaffarian D, Roger VL, Benjamin EJ, Berry JD, Borden WB, et al. Heart disease and stroke statistics – 2013 update: a report from the American Heart Association. Circulation. 2013;127(1):e6–245. doi:10.1161/CIR.0b013e31828124ad.
66. Graham SH, Chen J. Programmed cell death in cerebral ischemia. J Cereb Blood Flow Metab Off J Int Soc Cereb Blood Flow Metab. 2001;21(2):99–109. doi:10.1097/00004647-200102000-00001.
67. Gribkoff VK, Starrett Jr JE, Dworetzky SI, Hewawasam P, Boissard CG, Cook DA, et al. Targeting acute ischemic stroke with a calcium-sensitive opener of maxi-K potassium channels. Nat Med. 2001;7(4):471–7. doi:10.1038/86546.
68. Gucciardo L, Lories R, Ochsenbein-Kolble N, Done E, Zwijsen A, Deprest J. Fetal mesenchymal stem cells: isolation, properties and potential use in perinatology and regenerative medicine. BJOG. 2009;116(2):166–72. doi:10.1111/j.1471-0528.2008.02005.x. BJO2005 [pii].
69. Gutierrez-Aranda I, Ramos-Mejia V, Bueno C, Munoz-Lopez M, Real PJ, Macia A, et al. Human induced pluripotent stem cells develop teratoma more efficiently and faster than human embryonic stem cells regardless the site of injection. Stem Cells (Dayton, Ohio). 2010;28(9):1568–70. doi:10.1002/stem.471.
70. Gutierrez-Fernandez M, Fuentes B, Rodriguez-Frutos B, Ramos-Cejudo J, Vallejo-Cremades MT, Diez-Tejedor E. Trophic factors and cell therapy to stimulate brain repair after ischaemic stroke. J Cell Mol Med. 2012;16(10):2280–90. doi:10.1111/j.1582-4934.2012.01575.x.
71. Gutierrez-Fernandez M, Rodriguez-Frutos B, Otero-Ortega L, Ramos-Cejudo J, Fuentes B, Diez-Tejedor E. Adipose tissue-derived stem cells in stroke treatment: from bench to bedside. Discov Med. 2013;16(86):37–43.
72. Gutierrez-Fernandez M, Rodriguez-Frutos B, Ramos-Cejudo J, Teresa Vallejo-Cremades M, Fuentes B, Cerdan S, et al. Effects of intravenous administration of allogenic bone marrow- and adipose tissue-derived mesenchymal stem cells on functional recovery and brain repair markers in experimental ischemic stroke. Stem Cell Res Ther. 2013;4(1):11. doi:10.1186/scrt159. scrt159 [pii].
73. Hao L, Zou Z, Tian H, Zhang Y, Zhou H, Liu L. Stem cell-based therapies for ischemic stroke. Biomed Res Int. 2014;2014:468748. doi:10.1155/2014/468748.
74. Harms KM, Li L, Cunningham LA. Murine neural stem/progenitor cells protect neurons against ischemia by HIF-1alpha-regulated VEGF signaling. PLoS ONE. 2010;5(3), e9767. doi:10.1371/journal.pone.0009767.
75. Hayashi J, Takagi Y, Fukuda H, Imazato T, Nishimura M, Fujimoto M, et al. Primate embryonic stem cell-derived neuronal progenitors transplanted into ischemic brain. J Cereb Blood Flow Metab. 2006;26(7):906–14. doi:10.1038/sj.jcbfm.9600247. 9600247 [pii].
76. Hedlund M, Stenqvist AC, Nagaeva O, Kjellberg L, Wulff M, Baranov V, et al. Human placenta expresses and secretes NKG2D ligands via exosomes that down-modulate the cognate receptor expression: evidence for immunosuppressive function. J Immunol. 2009;183(1):340–51. doi:10.4049/jimmunol.0803477. 183/1/340 [pii].
77. Hicks AU, Lappalainen RS, Narkilahti S, Suuronen R, Corbett D, Sivenius J, et al. Transplantation of human embryonic stem cell-derived neural precursor cells and enriched environment after cortical stroke in rats: cell survival and functional recovery. Eur J Neurosci. 2009;29(3):562–74. doi:10.1111/j.1460-9568.2008.06599.x. EJN6599 [pii].
78. Hodges H, Sowinski P, Fleming P, Kershaw TR, Sinden JD, Meldrum BS, et al. Contrasting effects of fetal CA1 and CA3 hippocampal grafts on deficits in spatial learning and working memory induced by global cerebral ischaemia in rats. Neuroscience. 1996;72(4):959–88.
79. Hoehn M, Kustermann E, Blunk J, Wiedermann D, Trapp T, Wecker S, et al. Monitoring of implanted stem cell migration in vivo: a highly resolved in vivo magnetic resonance imaging investigation of experimental stroke in rat. Proc Natl Acad Sci U S A. 2002;99(25):16267–72. doi:10.1073/pnas.242435499. 242435499 [pii].
80. Holvoet P, Collen D. Thrombosis and atherosclerosis. Curr Opin Lipidol. 1997;8(5):320–8.

81. Honmou O, Onodera R, Sasaki M, Waxman SG, Kocsis JD. Mesenchymal stem cells: therapeutic outlook for stroke. Trends Mol Med. 2012;18(5):292–7. doi:10.1016/j.molmed.2012.02.003. S1471-4914(12)00034-2 [pii].
82. Hou SW, Wang YQ, Xu M, Shen DH, Wang JJ, Huang F, et al. Functional integration of newly generated neurons into striatum after cerebral ischemia in the adult rat brain. Stroke. 2008;39(10):2837–44.
83. Huang W, Mo X, Qin C, Zheng J, Liang Z, Zhang C. Transplantation of differentiated bone marrow stromal cells promotes motor functional recovery in rats with stroke. Neurol Res. 2013;35(3):320–8. doi:10.1179/1743132812Y.0000000151.
84. Huang Z, Huang PL, Panahian N, Dalkara T, Fishman MC, Moskowitz MA. Effects of cerebral ischemia in mice deficient in neuronal nitric oxide synthase. Science (New York, NY). 1994;265(5180):1883–5.
85. Huttner WB, Kosodo Y. Symmetric versus asymmetric cell division during neurogenesis in the developing vertebrate central nervous system. Curr Opin Cell Biol. 2005;17(6):648–57.
86. Iadecola C, Zhang F, Casey R, Nagayama M, Ross ME. Delayed reduction of ischemic brain injury and neurological deficits in mice lacking the inducible nitric oxide synthase gene. J Neurosci Off J Soc Neurosci. 1997;17(23):9157–64.
87. Igura K, Zhang X, Takahashi K, Mitsuru A, Yamaguchi S, Takashi TA. Isolation and characterization of mesenchymal progenitor cells from chorionic villi of human placenta. Cytotherapy. 2004;6(6):543–53.
88. Ikegame Y, Yamashita K, Hayashi S, Mizuno H, Tawada M, You F, et al. Comparison of mesenchymal stem cells from adipose tissue and bone marrow for ischemic stroke therapy. Cytotherapy. 2011;13(6):675–85. doi:10.3109/14653249.2010.549122.
89. Ikegame Y, Yamashita K, Nakashima S, Nomura Y, Yonezawa S, Asano Y, et al. Fate of graft cells: what should be clarified for development of mesenchymal stem cell therapy for ischemic stroke? Front Cell Neurosci. 2014;8:322. doi:10.3389/fncel.2014.00322.
90. Isik S, Zaim M, Yildiz MT, Negis Y, Kunduraci T, Karakas N, et al. DNA topoisomerase IIbeta as a molecular switch in neural differentiation of mesenchymal stem cells. Ann Hematol. 2015;94(2):307–18. doi:10.1007/s00277-014-2209-7.
91. Janardhan V, Qureshi AI. Mechanisms of ischemic brain injury. Curr Cardiol Rep. 2004;6(2):117–23.
92. Jensen MB, Yan H, Krishnaney-Davison R, Al Sawaf A, Zhang SC. Survival and differentiation of transplanted neural stem cells derived from human induced pluripotent stem cells in a rat stroke model. J Stroke Cerebrovasc Dis Off J Natl Stroke Assoc. 2013;22(4):304–8. doi:10.1016/j.jstrokecerebrovasdis.2011.09.008.
93. Jiang M, Lv L, Ji H, Yang X, Zhu W, Cai L, et al. Induction of pluripotent stem cells transplantation therapy for ischemic stroke. Mol Cell Biochem. 2011;354(1–2):67–75. doi:10.1007/s11010-011-0806-5.
94. Jin K, Minami M, Lan JQ, Mao XO, Batteur S, Simon RP, et al. Neurogenesis in dentate subgranular zone and rostral subventricular zone after focal cerebral ischemia in the rat. Proc Natl Acad Sci U S A. 2001;98(8):4710–5.
95. Jin K, Wang X, Xie L, Mao XO, Zhu W, Wang Y, et al. Evidence for stroke-induced neurogenesis in the human brain. Proc Natl Acad Sci U S A. 2006;103(35):13198–202.
96. Johnstone SA, Liley M, Dalby MJ, Barnett SC. Comparison of human olfactory and skeletal MSCs using osteogenic nanotopography to demonstrate bone-specific bioactivity of the surfaces. Acta Biomater. 2015;13:266–76. doi:10.1016/j.actbio.2014.11.027. S1742-7061(14)00521-2 [pii].
97. Jung KH, Chu K, Lee ST, Park HK, Kim DH, Kim JH, et al. Circulating endothelial progenitor cells as a pathogenetic marker of moyamoya disease. J Cereb Blood Flow Metab Off J Int Soc Cereb Blood Flow Metab. 2008;28(11):1795–803. doi:10.1038/jcbfm.2008.67.
98. Jung KH, Roh JK. Circulating endothelial progenitor cells in cerebrovascular disease. J Clin Neurol (Seoul, Korea). 2008;4(4):139–47. doi:10.3988/jcn.2008.4.4.139.

99. Kageyama R, Ohtsuka T, Hatakeyama J, Ohsawa R. Roles of bHLH genes in neural stem cell differentiation. Exp Cell Res. 2005;306(2):343–8.
100. Kallur T, Darsalia V, Lindvall O, Kokaia Z. Human fetal cortical and striatal neural stem cells generate region-specific neurons in vitro and differentiate extensively to neurons after intrastriatal transplantation in neonatal rats. J Neurosci Res. 2006;84(8):1630–44.
101. Kaneko Y, Tajiri N, Shinozuka K, Glover LE, Weinbren NL, Cortes L, et al. Cell therapy for stroke: emphasis on optimizing safety and efficacy profile of endothelial progenitor cells. Curr Pharm Des. 2012;18(25):3731–4.
102. Katakowski M, Zhang ZG, Chen J, Zhang R, Wang Y, Jiang H, et al. Phosphoinositide 3-kinase promotes adult subventricular neuroblast migration after stroke. J Neurosci Res. 2003;74(4):494–501.
103. Kawai H, Yamashita T, Ohta Y, Deguchi K, Nagotani S, Zhang X, et al. Tridermal tumorigenesis of induced pluripotent stem cells transplanted in ischemic brain. J Cereb Blood Flow Metab Off J Int Soc Cereb Blood Flow Metab. 2010;30(8):1487–93. doi:10.1038/jcbfm.2010.32.
104. Kelly S, Bliss TM, Shah AK, Sun GH, Ma M, Foo WC, et al. Transplanted human fetal neural stem cells survive, migrate, and differentiate in ischemic rat cerebral cortex. Proc Natl Acad Sci U S A. 2004;101(32):11839–44.
105. Kim DY, Park SH, Lee SU, Choi DH, Park HW, Paek SH, et al. Effect of human embryonic stem cell-derived neuronal precursor cell transplantation into the cerebral infarct model of rat with exercise. Neurosci Res. 2007;58(2):164–75. doi:10.1016/j.neures.2007.02.016. S0168-0102(07)00087-9 [pii].
106. Knoepfler PS. Deconstructing stem cell tumorigenicity: a roadmap to safe regenerative medicine. Stem Cells (Dayton, Ohio). 2009;27(5):1050–6. doi:10.1002/stem.37.
107. Kranz A, Wagner DC, Kamprad M, Scholz M, Schmidt UR, Nitzsche F, et al. Transplantation of placenta-derived mesenchymal stromal cells upon experimental stroke in rats. Brain Res. 2010;1315:128–36. doi:10.1016/j.brainres.2009.12.001. S0006-8993(09)02602-X [pii].
108. Kroemer G, Petit P, Zamzami N, Vayssiere JL, Mignotte B. The biochemistry of programmed cell death. FASEB J Off Publ Fed Am Soc Exp Biol. 1995;9(13):1277–87.
109. Kroemer G, Reed JC. Mitochondrial control of cell death. Nat Med. 2000;6(5):513–9. doi:10.1038/74994.
110. Kurozumi K, Nakamura K, Tamiya T, Kawano Y, Kobune M, Hirai S, et al. BDNF gene-modified mesenchymal stem cells promote functional recovery and reduce infarct size in the rat middle cerebral artery occlusion model. Mol Ther. 2004;9(2):189–97. doi:10.1016/j.ymthe.2003.10.012. S152500160300354X [pii].
111. Lai RC, Chen TS, Lim SK. Mesenchymal stem cell exosome: a novel stem cell-based therapy for cardiovascular disease. Regen Med. 2011;6(4):481–92. doi:10.2217/rme.11.35.
112. Lapergue B, Mohammad A, Shuaib A. Endothelial progenitor cells and cerebrovascular diseases. Prog Neurobiol. 2007;83(6):349–62. doi:10.1016/j.pneurobio.2007.08.001.
113. Lee JK, Park SR, Jung BK, Jeon YK, Lee YS, Kim MK, et al. Exosomes derived from mesenchymal stem cells suppress angiogenesis by down-regulating VEGF expression in breast cancer cells. PLoS ONE. 2013;8(12), e84256. doi:10.1371/journal.pone.0084256. PONE-D-13-14677 [pii].
114. Lee JS, Hong JM, Moon GJ, Lee PH, Ahn YH, Bang OY. A long-term follow-up study of intravenous autologous mesenchymal stem cell transplantation in patients with ischemic stroke. Stem Cells. 2010;28(6):1099–106. doi:10.1002/stem.430.
115. Li P, Nijhawan D, Budihardjo I, Srinivasula SM, Ahmad M, Alnemri ES, et al. Cytochrome c and dATP-dependent formation of Apaf-1/caspase-9 complex initiates an apoptotic protease cascade. Cell. 1997;91(4):479–89.
116. Li Y, Chen J, Chopp M. Adult bone marrow transplantation after stroke in adult rats. Cell Transplant. 2001;10(1):31–40.
117. Li Y, Chen J, Wang L, Lu M, Chopp M. Treatment of stroke in rat with intracarotid administration of marrow stromal cells. Neurology. 2001;56(12):1666–72.

118. Li Y, Chopp M, Chen J, Wang L, Gautam SC, Xu YX, et al. Intrastriatal transplantation of bone marrow nonhematopoietic cells improves functional recovery after stroke in adult mice. J Cereb Blood Flow Metab. 2000;20(9):1311–9. doi:10.1097/00004647-200009000-00006.
119. Li YF, Ren LN, Guo G, Cannella LA, Chernaya V, Samuel S, et al. Endothelial progenitor cells in ischemic stroke: an exploration from hypothesis to therapy. J Hematol Oncol. 2015;8:33. doi:10.1186/s13045-015-0130-8.
120. Liu J. Induced pluripotent stem cell-derived neural stem cells: new hope for stroke? Stem Cell Res Ther. 2013;4(5):115. doi:10.1186/scrt326.
121. Liu N, Deguchi K, Yamashita T, Liu W, Ikeda Y, Abe K. Intracerebral transplantation of bone marrow stromal cells ameliorates tissue plasminogen activator-induced brain damage after cerebral ischemia in mice detected by in vivo and ex vivo optical imaging. J Neurosci Res. 2012;90(11):2086–93. doi:10.1002/jnr.23104.
122. Liu N, Zhang Y, Fan L, Yuan M, Du H, Cheng R, et al. Effects of transplantation with bone marrow-derived mesenchymal stem cells modified by survivin on experimental stroke in rats. J Transl Med. 2011;9:105. doi:10.1186/1479-5876-9-105. 1479-5876-9-105 [pii].
123. Liu X. Beyond the time window of intravenous thrombolysis: standing by or by stenting? Interv Neurol. 2012;1(1):3–15. doi:10.1159/000338389.
124. Liu XS, Zhang ZG, Zhang RL, Gregg S, Morris DC, Wang Y, et al. Stroke induces gene profile changes associated with neurogenesis and angiogenesis in adult subventricular zone progenitor cells. J Cereb Blood Flow Metab. 2007;27(3):564–74.
125. Liu YP, Seckin H, Izci Y, Du ZW, Yan YP, Baskaya MK. Neuroprotective effects of mesenchymal stem cells derived from human embryonic stem cells in transient focal cerebral ischemia in rats. J Cereb Blood Flow Metab. 2009;29(4):780–91. doi:10.1038/jcbfm.2009.1. jcbfm20091 [pii].
126. Lo EH, Dalkara T, Moskowitz MA. Mechanisms, challenges and opportunities in stroke. Nat Rev Neurosci. 2003;4(5):399–415. doi:10.1038/nrn1106.
127. Lorenzo HK, Susin SA, Penninger J, Kroemer G. Apoptosis inducing factor (AIF): a phylogenetically old, caspase-independent effector of cell death. Cell Death Differ. 1999;6(6):516–24. doi:10.1038/sj.cdd.4400527.
128. Love S. Apoptosis and brain ischaemia. Prog Neuro-Psychopharmacol Biol Psychiatry. 2003;27(2):267–82. doi:10.1016/s0278-5846(03)00022-8.
129. Luo Y, Cao G, Pei W, O'Horo C, Graham SH, Chen J. Induction of caspase-activated deoxyribonuclease activity after focal cerebral ischemia and reperfusion. J Cereb Blood Flow Metab Off J Int Soc Cereb Blood Flow Metab. 2002;22(1):15–20. doi:10.1097/00004647-200201000-00002.
130. Luskin MB, Zigova T, Soteres BJ, Stewart RR. Neuronal progenitor cells derived from the anterior subventricular zone of the neonatal rat forebrain continue to proliferate in vitro and express a neuronal phenotype. Mol Cell Neurosci. 1997;8(5):351–66.
131. Ma F, Morancho A, Montaner J, Rosell A. Endothelial progenitor cells and revascularization following stroke. Brain Res. 1623;2015:150–9. doi:10.1016/j.brainres.2015.02.010.
132. Macas J, Nern C, Plate KH, Momma S. Increased generation of neuronal progenitors after ischemic injury in the aged adult human forebrain. J Neurosci. 2006;26(50):13114–9.
133. Maherali N, Sridharan R, Xie W, Utikal J, Eminli S, Arnold K, et al. Directly reprogrammed fibroblasts show global epigenetic remodeling and widespread tissue contribution. Cell Stem Cell. 2007;1(1):55–70. doi:10.1016/j.stem.2007.05.014.
134. Marion RM, Strati K, Li H, Murga M, Blanco R, Ortega S, et al. A p53-mediated DNA damage response limits reprogramming to ensure iPS cell genomic integrity. Nature. 2009;460(7259):1149–53. doi:10.1038/nature08287.
135. Martin GR. Isolation of a pluripotent cell line from early mouse embryos cultured in medium conditioned by teratocarcinoma stem cells. Proc Natl Acad Sci U S A. 1981;78(12):7634–8.
136. Matsuyama T, Hata R, Yamamoto Y, Tagaya M, Akita H, Uno H, et al. Localization of Fas antigen mRNA induced in postischemic murine forebrain by in situ hybridization. Brain Res Mol Brain Res. 1995;34(1):166–72.

137. Mattsson B, Sorensen JC, Zimmer J, Johansson BB. Neural grafting to experimental neocortical infarcts improves behavioral outcome and reduces thalamic atrophy in rats housed in enriched but not in standard environments. Stroke. 1997;28(6):1225–31. discussion 31–2.
138. Mayfield AE, Tilokee EL, Davis DR. Resident cardiac stem cells and their role in stem cell therapies for myocardial repair. Can J Cardiol. 2014;30(11):1288–98. doi:10.1016/j.cjca.2014.03.018. S0828-282X(14)00164-0 [pii].
139. Mehrabi M, Mansouri K, Hosseinkhani S, Yarani R, Yari K, Bakhtiari M, et al. Differentiation of human skin-derived precursor cells into functional islet-like insulin-producing cell clusters. In Vitro Cell Dev Biol Anim. 2015;51(6):595–603. doi:10.1007/s11626-015-9866-2.
140. Mendez-Ferrer S, Michurina TV, Ferraro F, Mazloom AR, Macarthur BD, Lira SA, et al. Mesenchymal and haematopoietic stem cells form a unique bone marrow niche. Nature. 2010;466(7308):829–34. doi:10.1038/nature09262. nature09262 [pii].
141. Minger SL, Ekonomou A, Carta EM, Chinoy A, Perry RH, Ballard CG. Endogenous neurogenesis in the human brain following cerebral infarction. Regen Med. 2007;2(1):69–74.
142. Miura K, Okada Y, Aoi T, Okada A, Takahashi K, Okita K, et al. Variation in the safety of induced pluripotent stem cell lines. Nat Biotechnol. 2009;27(8):743–5. doi:10.1038/nbt.1554.
143. Moniche F, Gonzalez A, Gonzalez-Marcos JR, Carmona M, Pinero P, Espigado I, et al. Intra-arterial bone marrow mononuclear cells in ischemic stroke: a pilot clinical trial. Stroke. 2012;43(8):2242–4. doi:10.1161/STROKEAHA.112.659409. STROKEAHA.112.659409 [pii].
144. Moubarik C, Guillet B, Youssef B, Codaccioni JL, Piercecchi MD, Sabatier F, et al. Transplanted late outgrowth endothelial progenitor cells as cell therapy product for stroke. Stem Cell Rev. 2011;7(1):208–20. doi:10.1007/s12015-010-9157-y.
145. Nagai N, Kawao N, Okada K, Okumoto K, Teramura T, Ueshima S, et al. Systemic transplantation of embryonic stem cells accelerates brain lesion decrease and angiogenesis. Neuroreport. 2010;21(8):575–9. doi:10.1097/WNR.0b013e32833a7d2c.
146. Nakagawa M, Koyanagi M, Tanabe K, Takahashi K, Ichisaka T, Aoi T, et al. Generation of induced pluripotent stem cells without Myc from mouse and human fibroblasts. Nat Biotechnol. 2008;26(1):101–6. doi:10.1038/nbt1374.
147. Nakagawa M, Taniguchi Y, Senda S, Takizawa N, Ichisaka T, Asano K, et al. A novel efficient feeder-free culture system for the derivation of human induced pluripotent stem cells. Sci Rep. 2014;4:3594. doi:10.1038/srep03594.
148. Nawashiro H, Martin D, Hallenbeck JM. Neuroprotective effects of TNF binding protein in focal cerebral ischemia. Brain Res. 1997;778(2):265–71.
149. Nishishita T, Ouchi K, Zhang X, Inoue M, Inazawa T, Yoshiura K, et al. A potential pro-angiogenic cell therapy with human placenta-derived mesenchymal cells. Biochem Biophys Res Commun. 2004;325(1):24–31. doi:10.1016/j.bbrc.2004.10.003. S0006-291X(04)02289-2 [pii].
150. Nolan DJ, Ciarrocchi A, Mellick AS, Jaggi JS, Bambino K, Gupta S, et al. Bone marrow-derived endothelial progenitor cells are a major determinant of nascent tumor neovascularization. Genes Dev. 2007;21(12):1546–58. doi:10.1101/gad.436307.
151. Nowakowski RS, Lewin SB, Miller MW. Bromodeoxyuridine immunohistochemical determination of the lengths of the cell cycle and the DNA-synthetic phase for an anatomically defined population. J Neurocytol. 1989;18(3):311–8.
152. Ohta T, Kikuta K, Imamura H, Takagi Y, Nishimura M, Arakawa Y, et al. Administration of ex vivo-expanded bone marrow-derived endothelial progenitor cells attenuates focal cerebral ischemia-reperfusion injury in rats. Neurosurgery. 2006;59(3):679–86. doi:10.1227/01.neu.0000229058.08706.88. discussion -86.
153. Okabe S, Forsberg-Nilsson K, Spiro AC, Segal M, McKay RD. Development of neuronal precursor cells and functional postmitotic neurons from embryonic stem cells in vitro. Mech Dev. 1996;59(1):89–102. doi:0925477396005722 [pii].
154. Okazaki T, Magaki T, Takeda M, Kajiwara Y, Hanaya R, Sugiyama K, et al. Intravenous administration of bone marrow stromal cells increases survivin and Bcl-2 protein expression

and improves sensorimotor function following ischemia in rats. Neurosci Lett. 2008;430(2):109–14. doi:10.1016/j.neulet.2007.10.046. S0304-3940(07)01140-8 [pii].
155. Oki K, Tatarishvili J, Wood J, Koch P, Wattananit S, Mine Y, et al. Human-induced pluripotent stem cells form functional neurons and improve recovery after grafting in stroke-damaged brain. Stem Cells (Dayton, Ohio). 2012;30(6):1120–33. doi:10.1002/stem.1104.
156. Okita K, Ichisaka T, Yamanaka S. Generation of germline-competent induced pluripotent stem cells. Nature. 2007;448(7151):313–7. doi:10.1038/nature05934.
157. Okita K, Matsumura Y, Sato Y, Okada A, Morizane A, Okamoto S, et al. A more efficient method to generate integration-free human iPS cells. Nat Methods. 2011;8(5):409–12. doi:10.1038/nmeth.1591.
158. Okita K, Nakagawa M, Hyenjong H, Ichisaka T, Yamanaka S. Generation of mouse induced pluripotent stem cells without viral vectors. Science (New York, NY). 2008;322(5903):949–53. doi:10.1126/science.1164270.
159. Onda T, Honmou O, Harada K, Houkin K, Hamada H, Kocsis JD. Therapeutic benefits by human mesenchymal stem cells (hMSCs) and Ang-1 gene-modified hMSCs after cerebral ischemia. J Cereb Blood Flow Metab. 2008;28(2):329–40. doi:10.1038/sj.jcbfm.9600527. 9600527 [pii].
160. Oyamada N, Itoh H, Sone M, Yamahara K, Miyashita K, Park K, et al. Transplantation of vascular cells derived from human embryonic stem cells contributes to vascular regeneration after stroke in mice. J Transl Med. 2008;6:54. doi:10.1186/1479-5876-6-54. 1479-5876-6-54 [pii].
161. Parent JM, Vexler ZS, Gong C, Derugin N, Ferriero DM. Rat forebrain neurogenesis and striatal neuron replacement after focal stroke. Ann Neurol. 2002;52(6):802–13.
162. Park S, Koh SE, Maeng S, Lee WD, Lim J, Lee YJ. Neural progenitors generated from the mesenchymal stem cells of first-trimester human placenta matured in the hypoxic-ischemic rat brain and mediated restoration of locomotor activity. Placenta. 2011;32(3):269–76. doi:10.1016/j.placenta.2010.12.027. S0143-4004(11)00011-7 [pii].
163. Parras CM, Galli R, Britz O, Soares S, Galichet C, Battiste J, et al. Mash1 specifies neurons and oligodendrocytes in the postnatal brain. Embo J. 2004;23(22):4495–505.
164. Paspala SA, Murthy TV, Mahaboob VS, Habeeb MA. Pluripotent stem cells – a review of the current status in neural regeneration. Neurol India. 2011;59(4):558–65.
165. Paul G, Ozen I, Christophersen NS, Reinbothe T, Bengzon J, Visse E, et al. The adult human brain harbors multipotent perivascular mesenchymal stem cells. PLoS ONE. 2012;7(4), e35577. doi:10.1371/journal.pone.0035577. PONE-D-11-11140 [pii].
166. Polentes J, Jendelova P, Cailleret M, Braun H, Romanyuk N, Tropel P, et al. Human induced pluripotent stem cells improve stroke outcome and reduce secondary degeneration in the recipient brain. Cell Transplant. 2012;21(12):2587–602. doi:10.3727/096368912x653228.
167. Portmann-Lanz CB, Schoeberlein A, Huber A, Sager R, Malek A, Holzgreve W, et al. Placental mesenchymal stem cells as potential autologous graft for pre- and perinatal neuroregeneration. Am J Obstet Gynecol. 2006;194(3):664–73. doi:10.1016/j.ajog.2006.01.101. S0002-9378(06)00175-X [pii].
168. Prather WR, Toren A, Meiron M, Ofir R, Tschope C, Horwitz EM. The role of placental-derived adherent stromal cell (PLX-PAD) in the treatment of critical limb ischemia. Cytotherapy. 2009;11(4):427–34. doi:10.1080/14653240902849762. 912390024 [pii].
169. Quinones-Hinojosa A, Sanai N, Soriano-Navarro M, Gonzalez-Perez O, Mirzadeh Z, Gil-Perotin S, et al. Cellular composition and cytoarchitecture of the adult human subventricular zone: a niche of neural stem cells. J Comp Neurol. 2006;494(3):415–34.
170. Record M, Carayon K, Poirot M, Silvente-Poirot S. Exosomes as new vesicular lipid transporters involved in cell-cell communication and various pathophysiologies. Biochim Biophys Acta. 2014;1841(1):108–20. doi:10.1016/j.bbalip.2013.10.004. S1388-1981(13)00219-9 [pii].

171. Record M, Subra C, Silvente-Poirot S, Poirot M. Exosomes as intercellular signalosomes and pharmacological effectors. Biochem Pharmacol. 2011;81(10):1171–82. doi:10.1016/j.bcp.2011.02.011. S0006-2952(11)00115-8 [pii].
172. Reubinoff BE, Itsykson P, Turetsky T, Pera MF, Reinhartz E, Itzik A, et al. Neural progenitors from human embryonic stem cells. Nat Biotechnol. 2001;19(12):1134–40. doi:10.1038/nbt1201-1134. nbt1201-1134 [pii].
173. Reubinoff BE, Pera MF, Fong CY, Trounson A, Bongso A. Embryonic stem cell lines from human blastocysts: somatic differentiation in vitro. Nat Biotechnol. 2000;18(4):399–404. doi:10.1038/74447.
174. Rosell A, Morancho A, Navarro-Sobrino M, Martinez-Saez E, Hernandez-Guillamon M, Lope-Piedrafita S, et al. Factors secreted by endothelial progenitor cells enhance neurorepair responses after cerebral ischemia in mice. PLoS ONE. 2013;8(9), e73244. doi:10.1371/journal.pone.0073244.
175. Rothman SM, Olney JW. Excitotoxicity and the NMDA receptor – still lethal after eight years. Trends Neurosci. 1995;18(2):57–8.
176. Rouhl RP, van Oostenbrugge RJ, Damoiseaux J, Tervaert JW, Lodder J. Endothelial progenitor cell research in stroke: a potential shift in pathophysiological and therapeutical concepts. Stroke J Cereb Circ. 2008;39(7):2158–65. doi:10.1161/strokeaha.107.507251.
177. Sakaguchi M, Okano H. Neural stem cells, adult neurogenesis, and galectin-1: from bench to bedside. Dev Neurobiol. 2012;72(7):1059–67.
178. Sanai N, Tramontin AD, Quinones-Hinojosa A, Barbaro NM, Gupta N, Kunwar S, et al. Unique astrocyte ribbon in adult human brain contains neural stem cells but lacks chain migration. Nature. 2004;427(6976):740–4.
179. Sanberg PR, Eve DJ, Metcalf C, Borlongan CV. Advantages and challenges of alternative sources of adult-derived stem cells for brain repair in stroke. Prog Brain Res. 2012;201:99–117. doi:10.1016/B978-0-444-59544-7.00006-8. B978-0-444-59544-7.00006-8 [pii].
180. Schoeberlein A, Mueller M, Reinhart U, Sager R, Messerli M, Surbek DV. Homing of placenta-derived mesenchymal stem cells after perinatal intracerebral transplantation in a rat model. Am J Obstet Gynecol. 2011;205(3):277 e1–6. doi:10.1016/j.ajog.2011.06.044. S0002-9378(11)00772-1 [pii].
181. Seminatore C, Polentes J, Ellman D, Kozubenko N, Itier V, Tine S, et al. The postischemic environment differentially impacts teratoma or tumor formation after transplantation of human embryonic stem cell-derived neural progenitors. Stroke. 2010;41(1):153–9. doi:10.1161/STROKEAHA.109.563015. STROKEAHA.109.563015 [pii].
182. Shah S, Stroke Pathophysiology. Foundation for education and research in neurological emergencies. www.uicedu/com/ferne/pdf/pathophys0501pdf,2000. 2000.
183. Shen LH, Xin H, Li Y, Zhang RL, Cui Y, Zhang L, et al. Endogenous tissue plasminogen activator mediates bone marrow stromal cell-induced neurite remodeling after stroke in mice. Stroke. 2011;42(2):459–64. doi:10.1161/STROKEAHA.110.593863. STROKEAHA.110.593863 [pii].
184. Shiraishi K, Sharp FR, Simon RP. Sequential metabolic changes in rat brain following middle cerebral artery occlusion: a 2-deoxyglucose study. J Cereb Blood Flow Metab Off J Int Soc Cereb Blood Flow Metab. 1989;9(6):765–73. doi:10.1038/jcbfm.1989.110.
185. Shiratsuki S, Terai S, Murata Y, Takami T, Yamamoto N, Fujisawa K, et al. Enhanced survival of mice infused with bone marrow-derived as compared with adipose-derived mesenchymal stem cells. Hepatol Res. 2015;45(13):1353–9. doi:10.1111/hepr.12507.
186. Siesjo BK. Cell damage in the brain: a speculative synthesis. J Cereb Blood Flow Metab Off J Int Soc Cereb Blood Flow Metab. 1981;1(2):155–85. doi:10.1038/jcbfm.1981.18.
187. Sinden JD, Rashid-Doubell F, Kershaw TR, Nelson A, Chadwick A, Jat PS, et al. Recovery of spatial learning by grafts of a conditionally immortalized hippocampal neuroepithelial cell line into the ischaemia-lesioned hippocampus. Neuroscience. 1997;81(3):599–608.
188. Singh VK, Kalsan M, Kumar N, Saini A, Chandra R. Induced pluripotent stem cells: applications in regenerative medicine, disease modeling, and drug discovery. Front Cell Dev Biol. 2015;3:2. doi:10.3389/fcell.2015.00002.

189. Sinor AD, Lillien L. Akt-1 expression level regulates CNS precursors. J Neurosci. 2004;24(39):8531–41.
190. Snyder EY, Deitcher DL, Walsh C, Arnold-Aldea S, Hartwieg EA, Cepko CL. Multipotent neural cell lines can engraft and participate in development of mouse cerebellum. Cell. 1992;68(1):33–51.
191. Sobrino T, Hurtado O, Moro MA, Rodriguez-Yanez M, Castellanos M, Brea D, et al. The increase of circulating endothelial progenitor cells after acute ischemic stroke is associated with good outcome. Stroke J Cereb Circ. 2007;38(10):2759–64. doi:10.1161/strokeaha.107.484386.
192. Sonntag KC, Pruszak J, Yoshizaki T, van Arensbergen J, Sanchez-Pernaute R, Isacson O. Enhanced yield of neuroepithelial precursors and midbrain-like dopaminergic neurons from human embryonic stem cells using the bone morphogenic protein antagonist noggin. Stem Cells. 2007;25(2):411–8. doi:10.1634/stemcells.2006-0380. 2006-0380 [pii].
193. Sorensen JC, Grabowski M, Zimmer J, Johansson BB. Fetal neocortical tissue blocks implanted in brain infarcts of adult rats interconnect with the host brain. Exp Neurol. 1996;138(2):227–35.
194. Sorensen JC, Mattsson B, Andreasen A, Johansson BB. Rapid disappearance of zinc positive terminals in focal brain ischemia. Brain Res. 1998;812(1–2):265–9.
195. Stenderup K, Justesen J, Clausen C, Kassem M. Aging is associated with decreased maximal life span and accelerated senescence of bone marrow stromal cells. Bone. 2003;33(6):919–26. doi:S8756328203002679 [pii].
196. Stephan RP, Reilly CR, Witte PL. Impaired ability of bone marrow stromal cells to support B-lymphopoiesis with age. Blood. 1998;91(1):75–88.
197. Stock P, Bruckner S, Winkler S, Dollinger MM, Christ B. Human bone marrow mesenchymal stem cell-derived hepatocytes improve the mouse liver after acute acetaminophen intoxication by preventing progress of injury. Int J Mol Sci. 2014;15(4):7004–28. doi:10.3390/ijms15047004. ijms15047004 [pii].
198. Suarez-Monteagudo C, Hernandez-Ramirez P, Alvarez-Gonzalez L, Garcia-Maeso I, de la Cuetara-Bernal K, Castillo-Diaz L, et al. Autologous bone marrow stem cell neurotransplantation in stroke patients. An open study. Restor Neurol Neurosci. 2009;27(3):151–61. doi:10.3233/RNN-2009-0483. 085W5360088471V2 [pii].
199. Susin SA, Lorenzo HK, Zamzami N, Marzo I, Snow BE, Brothers GM, et al. Molecular characterization of mitochondrial apoptosis-inducing factor. Nature. 1999;397(6718):441–6. doi:10.1038/17135.
200. Tae-Hoon L, Yoon-Seok L. Transplantation of mouse embryonic stem cell after middle cerebral artery occlusion. Acta Cir Bras. 2012;27(4):333–9. doi:S0102-86502012000400009 [pii].
201. Taguchi A, Matsuyama T, Moriwaki H, Hayashi T, Hayashida K, Nagatsuka K, et al. Circulating CD34-positive cells provide an index of cerebrovascular function. Circulation. 2004;109(24):2972–5. doi:10.1161/01.cir.0000133311.25587.de.
202. Takagi Y, Nishimura M, Morizane A, Takahashi J, Nozaki K, Hayashi J, et al. Survival and differentiation of neural progenitor cells derived from embryonic stem cells and transplanted into ischemic brain. J Neurosurg. 2005;103(2):304–10. doi:10.3171/jns.2005.103.2.0304.
203. Takahashi K, Tanabe K, Ohnuki M, Narita M, Ichisaka T, Tomoda K, et al. Induction of pluripotent stem cells from adult human fibroblasts by defined factors. Cell. 2007;131(5):861–72. doi:10.1016/j.cell.2007.11.019.
204. Takahashi K, Yamanaka S. Induction of pluripotent stem cells from mouse embryonic and adult fibroblast cultures by defined factors. Cell. 2006;126(4):663–76. doi:10.1016/j.cell.2006.07.024.
205. Takahashi T, Kalka C, Masuda H, Chen D, Silver M, Kearney M, et al. Ischemia- and cytokine-induced mobilization of bone marrow-derived endothelial progenitor cells for neovascularization. Nat Med. 1999;5(4):434–8. doi:10.1038/7434.

206. Tang YH, Ma YY, Zhang ZJ, Wang YT, Yang GY. Opportunities and challenges: stem cell-based therapy for the treatment of ischemic stroke. CNS Neurosci Ther. 2015;21(4):337–47. doi:10.1111/cns.12386.
207. Taupin P, Gage FH. Adult neurogenesis and neural stem cells of the central nervous system in mammals. J Neurosci Res. 2002;69(6):745–9.
208. Temple S. Division and differentiation of isolated CNS blast cells in microculture. Nature. 1989;340(6233):471–3.
209. Thomson JA, Itskovitz-Eldor J, Shapiro SS, Waknitz MA, Swiergiel JJ, Marshall VS, et al. Embryonic stem cell lines derived from human blastocysts. Science. 1998;282(5391):1145–7.
210. Thored P, Arvidsson A, Cacci E, Ahlenius H, Kallur T, Darsalia V, et al. Persistent production of neurons from adult brain stem cells during recovery after stroke. Stem Cells. 2006;24(3):739–47.
211. Tobita M, Orbay H, Mizuno H. Adipose-derived stem cells: current findings and future perspectives. Discov Med. 2011;11(57):160–70.
212. Tohill M, Mantovani C, Wiberg M, Terenghi G. Rat bone marrow mesenchymal stem cells express glial markers and stimulate nerve regeneration. Neurosci Lett. 2004;362(3):200–3. doi:10.1016/j.neulet.2004.03.077. S0304394004002861 [pii].
213. Tomasoni S, Longaretti L, Rota C, Morigi M, Conti S, Gotti E, et al. Transfer of growth factor receptor mRNA via exosomes unravels the regenerative effect of mesenchymal stem cells. Stem Cells Dev. 2013;22(5):772–80. doi:10.1089/scd.2012.0266.
214. Tornero D, Wattananit S, Gronning Madsen M, Koch P, Wood J, Tatarishvili J, et al. Human induced pluripotent stem cell-derived cortical neurons integrate in stroke-injured cortex and improve functional recovery. Brain J Neurol. 2013;136(Pt 12):3561–77. doi:10.1093/brain/awt278.
215. Turturici G, Tinnirello R, Sconzo G, Geraci F. Extracellular membrane vesicles as a mechanism of cell-to-cell communication: advantages and disadvantages. Am J Physiol Cell Physiol. 2014;306(7):C621–33. doi:10.1152/ajpcell.00228.2013. ajpcell.00228.2013 [pii].
216. Uccelli A, Laroni A, Freedman MS. Mesenchymal stem cells for the treatment of multiple sclerosis and other neurological diseases. Lancet Neurol. 2011;10(7):649–56. doi:10.1016/S1474-4422(11)70121-1. S1474-4422(11)70121-1 [pii].
217. Umemura T, Soga J, Hidaka T, Takemoto H, Nakamura S, Jitsuiki D, et al. Aging and hypertension are independent risk factors for reduced number of circulating endothelial progenitor cells. Am J Hypertens. 2008;21(11):1203–9. doi:10.1038/ajh.2008.278.
218. van der Strate BW, Popa ER, Schipper M, Brouwer LA, Hendriks M, Harmsen MC, et al. Circulating human CD34+ progenitor cells modulate neovascularization and inflammation in a nude mouse model. J Mol Cell Cardiol. 2007;42(6):1086–97. doi:10.1016/j.yjmcc.2007.03.907.
219. Varga G, Gerber G. Mesenchymal stem cells of dental origin as promising tools for neuroregeneration. Stem Cell Res Ther. 2014;5(2):61. doi:10.1186/scrt450. scrt450 [pii].
220. Vojtek AB, Taylor J, DeRuiter SL, Yu JY, Figueroa C, Kwok RP, et al. Akt regulates basic helix-loop-helix transcription factor-coactivator complex formation and activity during neuronal differentiation. Mol Cell Biol. 2003;23(13):4417–27.
221. Wakabayashi K, Nagai A, Sheikh AM, Shiota Y, Narantuya D, Watanabe T, et al. Transplantation of human mesenchymal stem cells promotes functional improvement and increased expression of neurotrophic factors in a rat focal cerebral ischemia model. J Neurosci Res. 2010;88(5):1017–25. doi:10.1002/jnr.22279.
222. Wang J, Chao F, Han F, Zhang G, Xi Q, Li J, et al. PET demonstrates functional recovery after transplantation of induced pluripotent stem cells in a rat model of cerebral ischemic injury. J Nucl Med Off Publ Soc Nucl Med. 2013;54(5):785–92. doi:10.2967/jnumed.112.111112.
223. Wang J, Hendrix A, Hernot S, Lemaire M, De Bruyne E, Van Valckenborgh E, et al. Bone marrow stromal cell-derived exosomes as communicators in drug resistance in multiple

myeloma cells. Blood. 2014;124(4):555–66. doi:10.1182/blood-2014-03-562439. blood-2014-03-562439 [pii].
224. Wang L, Gang Zhang Z, Lan Zhang R, Chopp M. Activation of the PI3-K/Akt pathway mediates cGMP enhanced-neurogenesis in the adult progenitor cells derived from the subventricular zone. J Cereb Blood Flow Metab. 2005;25(9):1150–8.
225. Wang L, Zhang Z, Wang Y, Zhang R, Chopp M. Treatment of stroke with erythropoietin enhances neurogenesis and angiogenesis and improves neurological function in rats. Stroke. 2004;35(7):1732–7.
226. Wang L, Zhang ZG, Gregg SR, Zhang RL, Jiao Z, LeTourneau Y, et al. The sonic hedgehog pathway mediates carbamylated erythropoietin-enhanced proliferation and differentiation of adult neural progenitor cells. J Biol Chem. 2007;282(44):32462–70.
227. Wang L, Zhang ZG, Zhang RL, Jiao ZX, Wang Y, Pourabdollah-Nejad DS, et al. Neurogenin 1 mediates erythropoietin enhanced differentiation of adult neural progenitor cells. J Cereb Blood Flow Metab. 2006;26(4):556–64.
228. Wang X, Yue TL, Barone FC, White RF, Gagnon RC, Feuerstein GZ. Concomitant cortical expression of TNF-alpha and IL-1 beta mRNAs follows early response gene expression in transient focal ischemia. Mol Chem Neuropathol Sponsored Int Soc Neurochem World Fed Neurol Res Group Neurochem Cerebrospinal Fluid. 1994;23(2–3):103–14.
229. Warrier S, Haridas N, Bhonde R. Inherent propensity of amnion-derived mesenchymal stem cells towards endothelial lineage: vascularization from an avascular tissue. Placenta. 2012;33(10):850–8. doi:10.1016/j.placenta.2012.07.001. S0143-4004(12)00260-3 [pii].
230. Wei L, Cui L, Snider BJ, Rivkin M, Yu SS, Lee CS, et al. Transplantation of embryonic stem cells overexpressing Bcl-2 promotes functional recovery after transient cerebral ischemia. Neurobiol Dis. 2005;19(1–2):183–93. doi:10.1016/j.nbd.2004.12.016. S0969-9961(04)00319-5 [pii].
231. Wernig M, Meissner A, Foreman R, Brambrink T, Ku M, Hochedlinger K, et al. In vitro reprogramming of fibroblasts into a pluripotent ES-cell-like state. Nature. 2007;448(7151):318–24. doi:10.1038/nature05944.
232. Wislet-Gendebien S, Hans G, Leprince P, Rigo JM, Moonen G, Rogister B. Plasticity of cultured mesenchymal stem cells: switch from nestin-positive to excitable neuron-like phenotype. Stem Cells. 2005;23(3):392–402. doi:10.1634/stemcells.2004-0149. doi:23/3/392 [pii].
233. Xie N, Tang B. The application of human iPSCs in neurological diseases: from bench to bedside. Stem Cells Int. 2016;2016:6484713. doi:10.1155/2016/6484713.
234. Xin H, Li Y, Buller B, Katakowski M, Zhang Y, Wang X, et al. Exosome-mediated transfer of miR-133b from multipotent mesenchymal stromal cells to neural cells contributes to neurite outgrowth. Stem Cells. 2012;30(7):1556–64. doi:10.1002/stem.1129.
235. Xin H, Li Y, Chopp M. Exosomes/miRNAs as mediating cell-based therapy of stroke. Front Cell Neurosci. 2014;8:377. doi:10.3389/fncel.2014.00377.
236. Xin H, Li Y, Shen LH, Liu X, Wang X, Zhang J, et al. Increasing tPA activity in astrocytes induced by multipotent mesenchymal stromal cells facilitate neurite outgrowth after stroke in the mouse. PLoS ONE. 2010;5(2), e9027. doi:10.1371/journal.pone.0009027.
237. Xu C, Inokuma MS, Denham J, Golds K, Kundu P, Gold JD, et al. Feeder-free growth of undifferentiated human embryonic stem cells. Nat Biotechnol. 2001;19(10):971–4. doi:10.1038/nbt1001-971. nbt1001-971 [pii].
238. Yamashita T, Abe K. Mechanisms of endogenous endothelial repair in stroke. Curr Pharm Des. 2012;18(25):3649–52.
239. Yamashita T, Kawai H, Tian F, Ohta Y, Abe K. Tumorigenic development of induced pluripotent stem cells in ischemic mouse brain. Cell Transplant. 2011;20(6):883–91. doi:10.3727/096368910x539092.
240. Yanagisawa D, Qi M, Kim DH, Kitamura Y, Inden M, Tsuchiya D, et al. Improvement of focal ischemia-induced rat dopaminergic dysfunction by striatal transplantation of mouse

embryonic stem cells. Neurosci Lett. 2006;407(1):74–9. doi:10.1016/j.neulet.2006.08.007. S0304-3940(06)00799-3 [pii].

241. Yang JM, Gould SJ. The cis-acting signals that target proteins to exosomes and microvesicles. Biochem Soc Trans. 2013;41(1):277–82. doi:10.1042/BST20120275. BST20120275 [pii].
242. Yarygin KN, Kholodenko IV, Konieva AA, Burunova VV, Tairova RT, Gubsky LV, et al. Mechanisms of positive effects of transplantation of human placental mesenchymal stem cells on recovery of rats after experimental ischemic stroke. Bull Exp Biol Med. 2009;148(6):862–8.
243. Ying QL, Stavridis M, Griffiths D, Li M, Smith A. Conversion of embryonic stem cells into neuroectodermal precursors in adherent monoculture. Nat Biotechnol. 2003;21(2):183–6. doi:10.1038/nbt780. nbt780 [pii].
244. Yip HK, Chang LT, Chang WN, Lu CH, Liou CW, Lan MY, et al. Level and value of circulating endothelial progenitor cells in patients after acute ischemic stroke. Stroke J Cereb Circ. 2008;39(1):69–74. doi:10.1161/strokeaha.107.489401.
245. Yoo KH, Jang IK, Lee MW, Kim HE, Yang MS, Eom Y, et al. Comparison of immunomodulatory properties of mesenchymal stem cells derived from adult human tissues. Cell Immunol. 2009;259(2):150–6. doi:10.1016/j.cellimm.2009.06.010. S0008-8749(09)00115-4 [pii].
246. Yu J, Vodyanik MA, Smuga-Otto K, Antosiewicz-Bourget J, Frane JL, Tian S, et al. Induced pluripotent stem cell lines derived from human somatic cells. Science (New York, NY). 2007;318(5858):1917–20. doi:10.1126/science.1151526.
247. Yu JX, Huang XF, Lv WM, Ye CS, Peng XZ, Zhang H, et al. Combination of stromal-derived factor-1alpha and vascular endothelial growth factor gene-modified endothelial progenitor cells is more effective for ischemic neovascularization. J Vasc Surg. 2009;50(3):608–16. doi:10.1016/j.jvs.2009.05.049.
248. Yuan T, Liao W, Feng NH, Lou YL, Niu X, Zhang AJ, et al. Human induced pluripotent stem cell-derived neural stem cells survive, migrate, differentiate, and improve neurologic function in a rat model of middle cerebral artery occlusion. Stem Cell Res Ther. 2013;4(3):73. doi:10.1186/scrt224.
249. Zhang H, Song F, Xu C, Liu H, Wang Z, Li J, et al. Spatiotemporal PET imaging of dynamic metabolic changes after therapeutic approaches of induced pluripotent stem cells, neuronal stem cells, and a Chinese patent medicine in stroke. J Nucl Med Off Publ Soc Nucl Med. 2015;56(11):1774–9. doi:10.2967/jnumed.115.163170.
250. Zhang HG, Grizzle WE. Exosomes: a novel pathway of local and distant intercellular communication that facilitates the growth and metastasis of neoplastic lesions. Am J Pathol. 2014;184(1):28–41. doi:10.1016/j.ajpath.2013.09.027. S0002-9440(13)00738-4 [pii].
251. Zhang HT, Liu ZL, Yao XQ, Yang ZJ, Xu RX. Neural differentiation ability of mesenchymal stromal cells from bone marrow and adipose tissue: a comparative study. Cytotherapy. 2012;14(10):1203–14. doi:10.3109/14653249.2012.711470.
252. Zhang J, Li Y, Chen J, Yang M, Katakowski M, Lu M, et al. Expression of insulin-like growth factor 1 and receptor in ischemic rats treated with human marrow stromal cells. Brain Res. 2004;1030(1):19–27. doi:10.1016/j.brainres.2004.09.061. S0006-8993(04)01571-9 [pii].
253. Zhang R, Zhang Z, Tsang W, Wang L, Chopp M. Down-regulation of p27kip1 increases proliferation of progenitor cells in adult rats. Neuroreport. 2004;15(11):1797–800.
254. Zhang R, Zhang Z, Wang L, Wang Y, Gousev A, Zhang L, et al. Activated neural stem cells contribute to stroke-induced neurogenesis and neuroblast migration toward the infarct boundary in adult rats. J Cereb Blood Flow Metab. 2004;24(4):441–8.
255. Zhang RL, Zhang ZG, Chopp M. Ischemic stroke and neurogenesis in the subventricular zone. Neuropharmacology. 2008;55(3):345–52.
256. Zhang RL, Zhang ZG, Roberts C, LeTourneau Y, Lu M, Zhang L, et al. Lengthening the G(1) phase of neural progenitor cells is concurrent with an increase of symmetric neuron generating division after stroke. J Cereb Blood Flow Metab. 2008;28(3):602–11.

257. Zhang RL, Zhang ZG, Zhang L, Chopp M. Proliferation and differentiation of progenitor cells in the cortex and the subventricular zone in the adult rat after focal cerebral ischemia. Neuroscience. 2001;105(1):33–41.
258. Zhang SC, Wernig M, Duncan ID, Brustle O, Thomson JA. In vitro differentiation of transplantable neural precursors from human embryonic stem cells. Nat Biotechnol. 2001;19(12):1129–33. doi:10.1038/nbt1201-1129. nbt1201-1129 [pii].
259. Zhang ZG, Zhang L, Jiang Q, Chopp M. Bone marrow-derived endothelial progenitor cells participate in cerebral neovascularization after focal cerebral ischemia in the adult mouse. Circ Res. 2002;90(3):284–8.
260. Zhao LR, Duan WM, Reyes M, Keene CD, Verfaillie CM, Low WC. Human bone marrow stem cells exhibit neural phenotypes and ameliorate neurological deficits after grafting into the ischemic brain of rats. Exp Neurol. 2002;174(1):11–20. doi:10.1006/exnr.2001.7853. S0014488601978537 [pii].
261. Zhao YH, Yuan B, Chen J, Feng DH, Zhao B, Qin C, et al. Endothelial progenitor cells: therapeutic perspective for ischemic stroke. CNS Neurosci Ther. 2013;19(2):67–75. doi:10.1111/cns.12040.
262. Zhou T, Benda C, Dunzinger S, Huang Y, Ho JC, Yang J, et al. Generation of human induced pluripotent stem cells from urine samples. Nat Protoc. 2012;7(12):2080–9. doi:10.1038/nprot.2012.115.
263. Zhu Y, Wan S, Zhan RY. Inducible pluripotent stem cells for the treatment of ischemic stroke: current status and problems. Rev Neurosci. 2012;23(4):393–402. doi:10.1515/revneuro-2012-0042.

Chapter 2
Bone Marrow Stem Cells: Source, Characterization, Isolation, Culture, and Identification

Chengcai Zheng, Yulei Zhou, Mengxiong Pan, Siyang Lin, Brian Wang, and Kunlin Jin

Abstract Stem cell transplantation offers an exciting new therapeutic avenue for stroke, as many studies have demonstrated favorable results in animal models with various cell types. Among them, the bone marrow stem cells (BMSCs) and mesenchymal stem cells (MSCs) in particular, may have enormous therapeutic potential because they can be harvested from the patients themselves without posing ethical or immunological difficulties. More importantly, BMSCs represent an important stem cell population with multipotent functions, including migration and transport functions to sites of local injuries or tissue damage to support appropriate cell and tissue renewal to replace the damaged areas, which are extremely useful for clinical applications, particularly in regenerative medicine. In this chapter, we summarize the source, characterization, isolation, culture, and identification of BMSCs.

Keywords Bone marrow stem cells • Characterization • Isolation • Culture • Identification

C. Zheng • Y. Zhou • M. Pan • S. Lin
Zhejiang Provincial Key Laboratory of Aging and Neurological Disorder Research, Department of Neurology, First Affiliated Hospital, Wenzhou Medical University, Wenzhou, Zhejiang 325000, China

B. Wang
Center for Neuroscience Discovery, Institute for Healthy Aging, University of North Texas Health Science Center, Fort Worth, TX 76107, USA

K. Jin (✉)
Department of Pharmacology Neuroscience, University of North Texas Health Science Center, Fort Worth, TX, USA
e-mail: kunlin.jin@unthsc.edu

K. Jin et al. (eds.), *Bone Marrow Stem Cell Therapy for Stroke*,
DOI 10.1007/978-981-10-2929-5_2

2.1 Introduction

Many studies have demonstrated favorable results in animal models with various cell types [1–4], which consequently have resulted in several early Phase I and II clinical trials with promising outcomes [5–7]. Therefore, the potential therapeutic impact of stem cell transplantation on regeneration of damaged brain tissue opens up numerous possibilities. A variety of cell types, including embryonic stem (ES) cells, neural stem cells (NSCs), inducible pluripotent (iPS) cells, and bone marrow stromal cells (BMSCs), have been studied as sources for use in transplantation into animal models of stroke. Initially, stem cells seemed to work by a "cell replacement" mechanism. NSCs are favored as donor cells for transplantation, since NSCs can differentiate into neural lineages. However, it is now thought that cell therapy may work mostly by providing trophic support to the injured tissue, fostering neurogenesis and angiogenesis [7]. Importantly, BMSCs may have the largest therapeutic potential among them because they can be harvested from the patients themselves without posing ethical or immunological difficulties [8, 9]. In other words, BSMCs are easy to obtain and expand in culture; using the patient's own BMSCs would eliminate the risk of rejection, and BMSCs have the capacity to migrate to the injury site thereby allowing for systemic administration. There is increasing evidence that transplanted BMSCs enhance functional recovery by differentiating into neural cells and/or by producing various kinds of cytokines or growth factors that can rescue the host neurons [10, 11].

2.2 Source

The bone marrow (BM) consists of a heterogeneous population of stem and progenitor cells including hematopoietic stem cells (HSCs) and mesenchymal stem cells (MSCs), two of the most studied BM-derived stem cells. Additionally, endothelial progenitor cells (EPCs) have also been isolated from the BM [12].

2.2.1 Mesenchymal Stem Cells

The terms mesenchymal stem cells (MSCs) and marrow stromal cells are used interchangeably, referring to a population of plastic adherent fibroblastic cells isolated by Percoll density centrifugation [12]. MSCs have the potential to differentiate into mesodermal cell lineages such as adipocytes, chondroblasts, fibroblasts, osteoblasts, and skeletal myoblasts both in vitro and in vivo. Human MSCs (hMSCs) lack telomerase activity and can only undergo about 18 population doublings (PDs) [13].

MSCs are present in a variety of tissues during development but are frequently seen in the bone marrow of adults. They reside in the bone marrow around blood vessels (as pericytes), in the fat, skin, and muscle [13, 14]. From these readily available sources, MSCs can be isolated, expanded in culture, and stimulated to differentiate into bone, cartilage, muscle, marrow stroma, tendon, fat, and a variety of other connective tissues.

2.2.2 *Hematopoietic Stem Cells*

Hematopoietic stem cells (HSCs) are defined as self-renewing cells with the capacity to differentiate into any distinct blood-cell lineages [15]. HSCs are mainly produced by the bone marrow in adults. They could also arise from embryonic stem cells. The generation of blood cells from embryonic stem cells (ESCs) and the manipulation of HSCs continue to provide insights into other stem cell systems.

The most successful HSC production method involves the conditional expression of the HoxB4 gene. Previous studies documented that HoxB4 could enhance the potential of hematopoietic cells. The retroviral-mediated overexpression of HoxB4 enhances hematopoietic repopulation through the enhancement of self-renewing activity without inducing leukemia or abnormal differentiation. In addition, the constitutive expression of HoxB4 enhanced the formation of immature mixed hematopoietic colonies [16].

HoxB4 expression transformed ES cell-derived hematopoietic progenitors into HSCs, and, surprisingly, the transient expression of HoxB4 for several weeks was sufficient for the conversion. That is, the overexpression of HoxB4 for several weeks in hematopoietic progenitors seemed to reprogram the differentiation program from hematopoietic progenitors to immature transplantable hematopoietic stem cells. The other possibility is that HoxB4 expanded the ES-induced HSCs, which have not been detected yet because of the low cell numbers [15].

2.2.3 *Endothelial Progenitor Cells*

Endothelial progenitor cells, initially described by Asahara [17], are immature endothelial cells that circulate in peripheral blood (PB). EPCs mature as endothelial cells and are important components of the vascular system. In their pioneering study, transplanted EPCs, isolated from human umbilical cord blood (UCB), were found in the endothelium of newly formed vessels in ischemic regions, indicating that a discrete cell population within the human blood participates in the formation of new vessels after ischemia [18, 19].

EPCs circulate in the adult blood, so they could be derived from peripheral blood [20]. A population of cells with similar characteristics can also be derived from the human UCB. Several lines of investigations from both animal and human studies indicate that EPCs principally participate in re-endothelialization during the neovascularization of ischemic organs, suggesting that EPC modulation could be directed toward the treatment of cerebrovascular diseases [12].

Progress in the field has been hampered by the lack of agreed parameters to define EPCs or even to distinguish them from adult differentiated endothelial cells. The definition of what constitutes an EPC remains controversial; monocytic-myeloid cells can transiently express cell surface markers typically found on endothelial cells and thus mimic an endothelial phenotype without necessarily meeting other criteria used to define a progenitor cell such as the ability of a single progenitor cell to give rise to a proliferative colony of cells. Many studies have attempted to identify cell surface markers that are unique to EPCs and distinguish them from mature adult endothelial cells as well as from myeloid-monocytic cells, but these attempts have been fairly futile [20].

Indeed, EPCs have been isolated from such seemingly diverse locations such as the bone marrow, circulating blood, heart, skeletal muscle, adipose tissue, spleen, small intestine, and vascular adventitia. However, until recently, the strongest links have unquestionably been documented with the hematopoietic system, where EPCs have been shown to reside within populations enriched with primitive hematopoietic stem cells, more-committed hematopoietic precursors, and even cells within the myeloid pool. When combined with evidence that cells administered at bone marrow transplantation integrate within normal vasculature, atherosclerotic arteries, and tumor neovessels, it is not surprising that an informal consensus has gradually emerged supposing the bone marrow to be the primary source of EPCs in humans [21].

2.3 Characterization and Identification

2.3.1 MSCs

2.3.1.1 Morphology and Cell Cycle Study

MSCs have morphological properties such as small cell bodies with few long and thin cell processes. Their cell bodies have large round nuclei containing prominent nucleoli surrounded by chromatin particles that are dispersed sparingly to give the nuclei a noticeable appearance [22]. Only a small proportion of MSCs are actively undergoing proliferation (~10 %), while the rest of them are at the G0/G1 phase [23]. Although the check points and length of each phase of the cell cycle have not been determined, the high percentage of G0/G1 cells suggests a high competence of MSC to differentiate. Moreover, the G0/G1 MSC population includes a minor and variable subset of resting quiescent cells, as evidenced by RNA and DNA content [24] or by FACS (Fluorescence activated Cell Sorting) analysis of size and granularity [25].

2.3.1.2 Differentiation Capacity

In early studies in vitro [26], MSCs were characterized as a cell population capable of differentiating into osteoblasts [27], adipocytes [28], and chondrocytes [29]. Shortly after, findings were showing the ability of MSCs to differentiate into other tissues of mesodermal origin such as the tendon and ligament [30], cardiomyocytes, and muscle [31]. In parallel, an increasing number of studies reported a wider ecto- and endodermal differentiation potential of MSCs, which includes the skin [26], retinal pigment epithelium [32], lung [33], hepatocytes [34], renal tubular cells [35], pancreatic islets [35], sebaceous duct cells [36], and neural cells [37, 38]. These studies showed that the phenotypic potential of MSCs was wider than anticipated. MSCs indeed have the capacity to differentiate toward all three lineages, i.e., the ecto-, meso-, and endodermal tissues. It has long been known that MSCs are multipotent cells that can give rise to a wide range of cell types upon their differentiation that ends with a terminal cell [39]. Several in vitro studies have been conducted to assess the differentiation potential of MSCs as well as setting up culture conditions, differentiation stimuli, and methods for the identification of each terminal differentiated phenotype. Minguell and coworkers have summarized these information succinctly [40].

2.3.1.3 Function

As mentioned earlier, MSCs possess the ability to secrete bioactive molecules. Bone marrow MSCs produce growth factors and cytokines that contribute to the formation and function of the stromal microenvironment, which produces inductive/regulatory signals not only for MSCs but also for the development of hematopoietic progenitors and other non-mesenchymal stromal cells present in the bone marrow. This trophic effect is important in the maintenance and regulation of the local microenvironment [41]. Nonetheless, the hematopoietic compartment is highly vascularized. Hematopoiesis takes place around the specialized sinusoids that drain into the central vein. Mature cells translocate from the site of their growth and mature through the wall of the sinusoids by active trans-endothelial migration. The sinusoids are lined with specialized endothelial cells and subendothelial pericytes [42] that have very active phagocytosis and are able to produce growth factors (mainly hematopoietic cytokines). In the hematopoietic niche, stromal cells are involved in the maintenance and regulation of the microenvironment. These effects can be direct or indirect: (1) secreted biomolecules activate intracellular signaling pathways, while (2) signaling is initiated in a neighboring cell resulting in the release of bioactive molecules (paracrine signaling). Examples include G-CSF, M-CSF [43], or the RANK ligand [44]. Furthermore, they are also involved in inflammatory processes. Therefore, it seems that MSCs may play specific roles like taking on an immunomodulatory role in several processes such as transplantation tolerance, autoimmunity, tumor evasion, and even fetal maternal tolerance in the case of a pregnancy.

2.3.1.4 Phenotype and Identification

Monoclonal antibodies that recognize MSC antigens [27, 45] have been instrumental in the characterization of BMSCs [46–48]. MSC markers are not unique as they consist of antigenic phenotypes borrowed from endothelial, muscle, epithelial, and mesenchymal cells. In fact, MSCs do not possess typical hematopoietic markers such as CD45, CD14, and CD34 [23, 49]. An extended cytokine expression profile has been described for MSCs. MSCs produce several hematopoietic and non-hematopoietic growth factors, interleukins and chemokines. While many of these cytokines are constitutively produced, others are only expressed after stimulation [43]. In addition, MSCs express several cytokines and growth factor receptors. There are studies that have demonstrated MSCs to be able to produce a vast array of matrix molecules including fibronectin, laminin, collagen, and proteoglycans [50, 51] as well as the expression of several counter-receptors associated with matrix- and cell-to-cell adhesion interactions. Of particular relevance is the strong expression of CD44 [23, 52], a receptor for various ligands like hyaluronan and osteopontin, which plays a central role in the organization of the extracellular matrix in the marrow and bone, respectively [53, 54].

2.3.2 HSCs

2.3.2.1 Multipotency

HSCs are highly capable of multidirectional differentiation; they can give rise to both the myeloid and lymphoid lineages of blood cells [55]. Myeloid cells include monocytes, macrophages, neutrophils, basophils, eosinophils, erythrocytes, dendritic cells, and megakaryocytes or platelets, while lymphoid cells include T cells, B cells, and natural killer cells [56]. HSCs differentiate into more mature hematopoietic progenitor cells (HPCs) that cease to retain self-renewal capabilities but are still able to give rise to colonies that contain functionally active hematopoietic cells. The earliest HPC for the myeloid and lymphoid lineages are the common myeloid progenitor and common lymphoid progenitor, respectively. The HPCs in more differentiated states awaiting to commit to a particular lineage usually grow into mature cells of the myeloid lineage i.e., monocytes, megakaryocytes, granulocytes, and erythroblasts, or the lymphoid lineage e.g., B cells, T cells, and natural killer cells [57]. A study by MacKey showed that to produce a mature circulating blood cell, the original HSC have had undergone about 17–19.5 divisions [58].

2.3.2.2 Self-Renewal

Differentiation is associated with a loss of self-renewing capacity, but HSCs as a population requires self-renewing to maintain the HSC pool. Stem cell self-renewal is thought to occur in the stem cell niche in the bone marrow, and it is reasonable to assume that key signals present in this niche will be important for self-renewal.

New gene-expression analytical tools presently allow researchers to examine developmental changes in telomeres and telomerase activity. The enzyme telomerase is known for extending the regions of DNA at the end of chromosomes known as telomeres. Importantly, it has been found that telomerase activity is critical for cellular proliferation however, as its enzymatic activity decreases with age, it results in shortened telomeres, which is a strong hypothesis for decreased stem cell renewal. To put this in context, telomerase activity in HSCs is correlated with their self-renewing capabilities [59].

As the process of self-renewal is infrequent and quite impossible to induce in cell culture, researchers still do not have a definite clue as to what regulates HSC self-renewal. HSCs that have been injected into an anemic patient were shown to migrate to the bone marrow and actively divide to supply all types of blood cells and give rise to more HSCs with self-renewing capabilities; this is a mystery that remains to be solved with the use of HSC cultures. Two recent studies cultured mouse HSCs and found a number of important factors such as the stem cell factor and thrombopoietin as well as the signaling molecule gp130, which are critical for the renewal of the HSC pool in culture [60, 61].

More clues on how to increase the numbers of stem cells may arise from examining other animals and various developmental stages. During early developmental stages in the fetal liver, HSCs may undergo more active cell division to increase their numbers, but later in life, they divide far less frequently [62]. Culturing HSCs from 10- to 11-day-old mouse embryos gave rise to the possibility of obtaining a 15-fold increase in HSCs within the first 2 or 3 days after removal of the aorta-gonad-mesonephros (AGM) from the embryos [63]. This finding may be viewed as important as the increase could be attributed to the downstream actions of the number of divisions that were directed by the specific embryonic microenvironment. Hence, determining the cellular compositions of adult-derived HSCs and their microenvironments in the AGM region could be critical for HSC expansion and manipulation for clinical purposes.

In another study, mouse HSCs were maintained for 4–7 weeks when they were grown on a clonal line of cells (AFT024) derived from the stroma, the other major cellular constituent of the bone marrow [64]. However, no one knows which specific factors secreted by the stromal cells maintains the stem cells. If it is true that recreating the stromal environment could expand the number of HSCs, then these specific factors, however complicated, could be worth examining [23].

2.3.2.3 Markers and Identification

During embryogenesis, HSCs migrate from one anatomical site to another. This developmental journey begins at the primitive streak/yolk sac and the AGM region and continues through the fetal liver to their final destination in the bone marrow and in rodents also, the spleen [57, 65, 66]. At different stages of development, the human HSCs show diverse surface markers. HSC development has been characterized in minute detail in the mouse, which serves as a model for understanding human hematopoiesis. For obvious ethical reasons, more functional assays are available for studying murine HSCs than human HSCs. Human HSCs are studied in vivo by using surrogate heterotransplantation models into immunodeficient mice or intrauterine transplants into sheep. It is, however, clear that there are some considerable differences in the expression of surface markers between mouse and human HSCs, which have been summarize in [67].

2.3.3 EPCs

2.3.3.1 Regenerative Potential of EPCs

After thrombotic microangiopathy [68, 69], EPCs seem to participate in the repair of the injured endothelium. It is not known which physiological or pathological factors influence the homing and differentiation of EPCs, and little is known about the signals that direct circulating EPCs to the sites of injured vessels. Recent studies addressing the integration of EPCs into the mature endothelium found that a small fraction of these cells can also transdifferentiate into smooth muscle cells in vitro [70]. This process seems to be dependent on the presence of the transforming growth factor-β1 and cell-to-cell contact. If this were the case, strategically located EPCs within the vascular endothelium could be an emerging standby tool for the regeneration of surrounding endothelial cells (ECs) or smooth muscle cells subsequent to an injury. Furthermore, EPCs may be involved in the regeneration of the ischemic myocardium by modulation of angiogenesis and myogenesis, cardiomyocyte apoptosis, and remodeling in the ischemic cardiac tissue [71–73]. EPCs have also been reported to participate in cerebral neovascularization after ischemic stroke [74]. Thus, EPCs derived from the hematopoietic tissue of the postnatal bone marrow may possess high regenerative potentials and some characteristics of embryonic stem cells. How these cells, if migrating from the bone marrow to the periphery, remain restricted in the circulation and what signals cause their homing to sites of injured endothelium or extravascular tissue, remains an enigma.

2.3.3.2 Phenotype and Identification

In 1997, the isolation of endothelial progenitors for angiogenesis was first described. Cells that were positive for the marker CD34 were obtained from human peripheral blood with the use of magnetic microbeads. By plating these on a fibronectin-coated surface, the cells differentiated and seem to take on endothelial characteristics [17]. Three markers were then discovered some time later characterizing early EPCs i.e., CD133, CD34, and the vascular endothelial growth factor receptor-2 (VEGFR-2) also known as Flk-1 [75, 76]. CD133, originally known as AC133, is a 120-kDa transmembrane polypeptide expressed in early hematopoietic stem and progenitor cells of the human bone marrow as well as peripheral blood and the fetal liver. Cells possessing the markers CD133, CD34, and VEGFR-2 but not vascular endothelial (VE)-cadherin and the von Willebrand factor (vWF) are predominantly seen in the bone marrow [17]. Mature EPCs in the adult peripheral circulation, however, have been reported to lose CD133 expression but not CD34 and VEGFR-2. In fact, mature ECs are characterized by high levels of VEGFR-2, VE-cadherin, and vWF [77]. Thus, it is highly suggestive that the loss of CD133 expression could indicate the maturation of circulating EPCs, but the point in which EPCs begin to lose this expression is unclear; it may occur either at the time they are transmigrating from the bone marrow into the systemic circulation or later on in the peripheral circulation. Circulating EPCs possess a wide range of markers with varying levels of expressions, typical of the endothelial lineage. Some of the markers include CD31, CD146, VE-cadherin, vWF, and E-selectin [17, 78, 79]. Taken together, we see at least two types of EPCs that are present in the peripheral blood at the same time wherein they possess different sets of markers, thereby aiding in identification.

2.4 Isolation

2.4.1 MSCs

MSCs have been isolated and characterized from many species including the rat, murine, dog, pig, goat, and horse [80]. Classical methods include differentially adhering to tissue culture substrates [81] and density separation [23].

MSCs were first isolated from bone marrow mononuclear cells based on adherence to tissue culture plastic. Guillot's paper described that the resultant cells were allowed to adhere to a standard tissue culture flask for 72 h in MSC medium. Afterwards, they were washed and passaged with TrypLE-Select upon confluence, cultured under humidified conditions in 5 % CO_2, and routinely cryopreserved in 90 % FCS and 10 % dimethyl sulfoxide (DMSO) [82]. Nonetheless, the whole marrow sample adhesion method can result in the nonspecific isolation of mononuclear cells and activation and/or potential loss of target cells. Density separation methods are complicated and toxic substances can easily contaminate the cultures and affect cell viability.

The density gradient centrifugation method is based on BMSC components that are in the 1.05–1.08 density range separate from other cell density separation method. Commonly used reagents are Percoll and Ficoll. It was suggested that the Percoll gradient density separation protocol was the best in terms of MSC yield and self-renewal potential of the MSCs retrieved, while MSCs retrieved with the Ficoll protocol had the lowest self-renewal capacity [83]. Density-gradient centrifugation requires frequent manual interventions and manipulation, which may lead to a higher risk of microbial contamination. Another disadvantage of density fractionation is that it is more difficult to standardize and may be influenced by the operator's skills.

RBC Lysis with Ammonium Chloride It was reported that the highest BMSCs yields were obtained with RBC lysis, compared with density gradient centrifugation [84]. Horn and colleagues [85] compared RBC lysis with Ficoll density fractionation and untreated whole BM adherent cultures, and found that BMSCs can be efficiently isolated by RBC lysis. This technique is faster and can be standardized more easily for clinical applications of MSCs [86]. In addition, based on the principles of hypotonic lysis and natural sedimentation, a heterogeneous mixture of bone marrow cells can be isolated [80]. Untreated whole BM adherent cultures are best for rabbit BMSC isolation. Pierini et al. isolated MSCs efficiently by the BD Vacutainer® Cell Preparation Tube™ (CPT) peripheral blood. Surprisingly, a fibroblast-colony-forming unit (CFU-F) assay indicated that with CPT, the number of MSC progenitors was 1.8 times higher compared with the Ficoll gradient separation method [87].

2.4.2 HSCs

A major obstacle to studying HSC biology is that the cells are extremely rare and can only be directly assayed using xenograft assays. Only 1 in 106 cells in the human BM is regarded as transplantable HSC. For more than 30 years, researchers have been uncovering novel markers that when used in combination, significantly enhance the purification of HSCs from the murine and human bone marrows. Nearly 20 years now, the separation methods for HSCs include flow cytometry, immunomagnetic beads, and affinity adsorption, among others.

The advent of flow cytometry has proven to be critical in the identification of mature hematopoietic cells due to the heterogeneous and undefined liquidic nature of the hematopoietic organ. A range of CD34 selection techniques have been developed to purify peripheral blood progenitor cells or the bone marrow prior to transplantation [89]. One successful approach that identified cells with hematopoietic stem and progenitor cell (HSPC) characteristics was the use of fluorescently tagged monoclonal antibodies that bind to specific cell surface proteins. Based on the differential binding of the antibodies to the surface of the cell (high, low, negative), researchers selectively isolated cells bearing a specific cell surface immunophenotype FACS and functionally characterized them through in vivo transplant assays. The second approach that was being employed simultaneously was to select BM cells using supravital dyes that were nontoxic to cells. Using a similar approach to immunophenotypic methods, cells were isolated based on the same high, low, and negative staining criteria and then functionally characterized (see Comprehensive Hematopoietic Stem Cell Isolation Methods).

2.4.3 EPCs

Various isolation procedures of EPCs from different sources could be performed using adherence culture or magnetic microbeads. There are some that find that floating cells exhibited therapeutic potential in a mouse myocardial infarction model showing specific local recruitment to the ischemic border zone [90].

Fluorescence-Activated Cell Sorting (FACS) System Isolation of EPCs from BM or peripheral blood is required for this cell-based therapeutic approach. Among a variety of EPC isolation methods, specifically in the case of experimental animal models, a recent mouse-cultured EPC isolation protocol was developed, which is feasible for obtaining a sufficient number of viable cells for both in vitro and in vivo experiments. The protocol for human CD34+ cell (EPC-rich cell population) isolation from peripheral blood can be characterized by FACS [91].

Immunomagnetic Beads A study by Peichev [75] using CD34 magnetic microbeads for isolation showed rather low amounts of CD133/CD34/VEGFR-2 cells (0.002 % of total PMC), corresponding to a number of approximately 70–210 cells per milliliter of blood [88]. However, there appear to be differences between progenitor cells isolated from the bone marrow or from peripheral blood because multipotent progenitor cells from the bone marrow may represent a more undifferentiated cell type with higher plasticity as compared with the peripheral blood-derived cells.

2.5 Culture

2.5.1 MSCs

Many modern culture techniques still utilize a colony-forming unit-fibroblasts (CFU-F) approach wherein the unpurified bone marrow or Ficoll-purified mononuclear cells of the bone marrow are directly plated onto culture plates [92]. In a span of 24–48 h, MSCs could adhere to tissue culture plastic, but not RBCs or hematopoietic progenitors. One report, however, characterized a population of non-adherent MSCs that were not achieved through direct plating [93]. Flow cytometry-based techniques allow the identification of bone marrow cells using unique surface markers such as STRO-1 [94]. STRO-1^+ cells are observed to be more homogeneous and possess greater adherent and proliferative rates. The differences between MSCs and STRO-1^+ cells, however, are unclear [95]. Immunodepletion using MACS have also been reported for negatively selecting MSCs [96].

2.5.2 EPCs

Isolated cells are cultured in medium containing special factors (e.g., VEGF, EGF, and bovine brain extract) to enhance the proliferation of cells with endothelial-like properties. After the initial adhesion, EPCs start to differentiate and lose progenitor characteristics. They form monolayers with endothelial-like appearances in about 3–4 weeks [78, 97]. EPCs can also form capillary tubes given the right conditions [17, 98, 99]. To differentiate adult human marrow-derived progenitor cells toward the endothelial lineage, one can seed the cells at high density in serum-free or low-serum medium and with the addition of VEGF. To differentiate into other cell types such as osteoblasts, adipocytes, and chondroblasts, one should use a medium containing more than 10% FCS. EPCs show exponential proliferative rates after remaining 30–60 days in culture, contrasting with the early outgrowth of vessel wall-derived endothelial cells that only have limited proliferative capacities.

2.6 Conclusion

Mesenchymal stem cells, hematopoietic stem cells, and endothelial progenitor cells are the major components of bone marrow stem cells. Bone marrow-derived mesenchymal stem cells come from the early development of the mesoderm and ectoderm. They possess small cell bodies with few cell processes that are long and thin, and the vast majority of cells are at the G0/G1 phase of the cell cycle. The ability of MSCs to differentiate into other tissues of mesodermal origin includes the tendon and ligament, cardiomyocytes, and muscle, among others. In addition, they can secrete bioactive molecules. BMSCs produce growth factors and cytokines that contribute to the formation and function of the stromal microenvironment. Furthermore, they are involved in inflammatory processes. The antigenic phenotype of MSC is not unique in that it borrows features of mesenchymal, endothelial, epithelial, and muscle cells. MSCs do not express the typical hematopoietic antigens, CD45, CD34, and CD14. Isolation methods include differentially adhering to tissue culture substrates, density separation, RBC lysis with ammonium chloride, and immunomagnetic beads.

Hematopoietic stem cells come from embryonic stem cells. Its developmental journey begins in the primitive streak/yolk sac and the aorta-gonad-mesonephros (AGM) region and continues through the fetal liver to their final destination in the bone marrow. HSCs are highly capable of multidirectional differentiation and self-renewal. At different stages of development, HSCs show diverse surface markers; thus the use of flow cytometry and supravital dyes can aid in their isolation.

EPCs can be isolated from the human umbilical cord blood. The bone marrow is the primary source of EPCs in humans [21]. EPCs seem to participate in the repair of the injured endothelium. These also possess high regenerative potentials and some characteristics of embryonic stem cells. CD133, CD34, and the vascular endothelial growth factor receptor-2 were discovered in the early EPC. By adherence culture or magnetic microbeads, one can isolate EPCs. After isolation, the cells are cultured in a medium with specific growth factors (e.g., VEGF, bovine brain extract, and epidermal growth factor) to facilitate the growth of endothelial-like cells.

References

1. Lindvall O, Kokaia Z. Stem cells for the treatment of neurological disorders. Nature. 2006;441(7097):1094–6.
2. Emsley JG, Mitchell BD, Magavi SS, Arlotta P, Macklis JD. The repair of complex neuronal circuitry by transplanted and endogenous precursors. NeuroRx. 2004;1(4):452–71.
3. Kerr DA, Llado J, Shamblott MJ, Maragakis NJ, Irani DN, Crawford TO, Krishnan C, Dike S, Gearhart JD, Rothstein JD. Human embryonic germ cell derivatives facilitate motor recovery of rats with diffuse motor neuron injury. J Neurosci. 2003;23(12):5131–40.
4. Klein S, Svendsen CN. Stem cells in the injured spinal cord: reducing the pain and increasing the gain. Nat Neurosci. 2005;8(3):259–60.
5. Kalladka D, Muir KW. Stem cell therapy in stroke: designing clinical trials. Neurochem Int. 2011;59:367.
6. Locatelli F, Bersano A, Ballabio E, Lanfranconi S, Papadimitriou D, Strazzer S, Bresolin N, Comi GP, Corti S. Stem cell therapy in stroke. Cell Mol Life Sci. 2009;66(5):757–72.
7. Bersano A, Ballabio E, Lanfranconi S, Boncoraglio GB, Corti S, Locatelli F, Baron P, Bresolin N, Parati E, Candelise L. Clinical studies in stem cells transplantation for stroke: a review. Curr Vasc Pharmacol. 2010;8(1):29–34.
8. Bliss T, Guzman R, Daadi M, Steinberg GK. Cell transplantation therapy for stroke. Stroke. 2007;38(2 Suppl):817–26.
9. Parr AM, Tator CH, Keating A. Bone marrow-derived mesenchymal stromal cells for the repair of central nervous system injury. Bone Marrow Transplant. 2007;40(7):609–19.
10. Hokari M, Kuroda S, Shichinohe H, Yano S, Hida K, Iwasaki Y. Bone marrow stromal cells protect and repair damaged neurons through multiple mechanisms. J Neurosci Res. 2008;86(5):1024–35.
11. Prockop DJ, Gregory CA, Spees JL. One strategy for cell and gene therapy: harnessing the power of adult stem cells to repair tissues. Proc Natl Acad Sci U S A. 2003;100 Suppl 1:11917–23.
12. Borlongan CV, Glover LE, Tajiri N, Kaneko Y, Freeman TB. The great migration of bone marrow-derived stem cells toward the ischemic brain: therapeutic implications for stroke and other neurological disorders. Prog Neurobiol. 2011;95(2):213–28.
13. Munoz-Elias G, Woodbury D, Black IB. Marrow stromal cells, mitosis, and neuronal differentiation: stem cell and precursor functions. Stem Cells. 2003;21(4):437–48.
14. Caplan AI, Bruder SP. Mesenchymal stem cells: building blocks for molecular medicine in the 21st century. Trends Mol Med. 2001;7(6):259–64.
15. Nakano T. Hematopoietic stem cells: generation and manipulation. Trends Immunol. 2003;24(11):589–94.
16. Kyba M, Perlingeiro RC, Daley GQ. HoxB4 confers definitive lymphoid-myeloid engraftment potential on embryonic stem cell and yolk sac hematopoietic progenitors. Cell. 2002;109(1):29–37.
17. Asahara T, Murohara T, Sullivan A, Silver M, van der Zee R, Li T, Witzenbichler B, Schatteman G, Isner JM. Isolation of putative progenitor endothelial cells for angiogenesis. Science. 1997;275(5302):964–7.
18. Zhao SP, Li YF. Downregulation of PPARgamma expression in peripheral blood monocytes correlated with adhesion molecules in acute coronary syndrome. Clin Chim Acta. 2003;336(1-2):19–25.
19. Nishikawa SI. A complex linkage in the developmental pathway of endothelial and hematopoietic cells. Curr Opin Cell Biol. 2001;13(6):673–8.
20. Zhang B, Pu WT. Notching up vascular regeneration. Cell Res. 2014;24(7):777–8.
21. Doyle B, Caplice N. A new source of endothelial progenitor cells – vascular biology redefined? Trends Biotechnol. 2005;23(9):444–6.
22. Brighton CT, Hunt RM. Early histological and ultrastructural changes in medullary fracture callus. J Bone Joint Surg Am. 1991;73(6):832–47.

23. Pittenger MF, Mackay AM, Beck SC, Jaiswal RK, Douglas R, Mosca JD, Moorman MA, Simonetti DW, Craig S, Marshak DR. Multilineage potential of adult human mesenchymal stem cells. Science. 1999;284(5411):143–7.
24. Conget PA, Minguell JJ. Adenoviral-mediated gene transfer into ex vivo expanded human bone marrow mesenchymal progenitor cells. Exp Hematol. 2000;28(4):382–90.
25. Colter DC, Class R, DiGirolamo CM, Prockop DJ. Rapid expansion of recycling stem cells in cultures of plastic-adherent cells from human bone marrow. Proc Natl Acad Sci U S A. 2000;97(7):3213–8.
26. Dominici M, Le Blanc K, Mueller I, Slaper-Cortenbach I, Marini F, Krause D, Deans R, Keating A, Prockop D, Horwitz E. Minimal criteria for defining multipotent mesenchymal stromal cells. The International Society for Cellular Therapy position statement. Cytotherapy. 2006;8(4):315–7.
27. Haynesworth SE, Baber MA, Caplan AI. Cell surface antigens on human marrow-derived mesenchymal cells are detected by monoclonal antibodies. Bone. 1992;13(1):69–80.
28. Dennis JE, Merriam A, Awadallah A, Yoo JU, Johnstone B, Caplan AI. A quadripotential mesenchymal progenitor cell isolated from the marrow of an adult mouse. J Bone Miner Res. 1999;14(5):700–9.
29. Yoo JU, Johnstone B. The role of osteochondral progenitor cells in fracture repair. Clin Orthop Relat Res. 1998;355(Suppl):S73–81.
30. Kahn CJ, Vaquette C, Rahouadj R, Wang X. A novel bioreactor for ligament tissue engineering. Biomed Mater Eng. 2008;18(4-5):283–7.
31. Wakitani S, Saito T, Caplan AI. Myogenic cells derived from rat bone marrow mesenchymal stem cells exposed to 5-azacytidine. Muscle Nerve. 1995;18(12):1417–26.
32. Arnhold S, Absenger Y, Klein H, Addicks K, Schraermeyer U. Transplantation of bone marrow-derived mesenchymal stem cells rescue photoreceptor cells in the dystrophic retina of the rhodopsin knockout mouse. Graefes Arch Clin Exp Ophthalmol. 2007;245(3):414–22.
33. Gavino AC, Chung JS, Sato K, Ariizumi K, Cruz Jr PD. Identification and expression profiling of a human C-type lectin, structurally homologous to mouse dectin-2. Exp Dermatol. 2005;14(4):281–8.
34. Jiang Y, Jahagirdar BN, Reinhardt RL, Schwartz RE, Keene CD, Ortiz-Gonzalez XR, Reyes M, Lenvik T, Lund T, Blackstad M, et al. Pluripotency of mesenchymal stem cells derived from adult marrow. Nature. 2002;418(6893):41–9.
35. Morigi M, Imberti B, Zoja C, Corna D, Tomasoni S, Abbate M, Rottoli D, Angioletti S, Benigni A, Perico N, et al. Mesenchymal stem cells are renotropic, helping to repair the kidney and improve function in acute renal failure. J Am Soc Nephrol. 2004;15(7):1794–804.
36. Fu YS, Cheng YC, Lin MY, Cheng H, Chu PM, Chou SC, Shih YH, Ko MH, Sung MS. Conversion of human umbilical cord mesenchymal stem cells in Wharton's jelly to dopaminergic neurons in vitro: potential therapeutic application for Parkinsonism. Stem Cells. 2006;24(1):115–24.
37. Kopen GC, Prockop DJ, Phinney DG. Marrow stromal cells migrate throughout forebrain and cerebellum, and they differentiate into astrocytes after injection into neonatal mouse brains. Proc Natl Acad Sci U S A. 1999;96(19):10711–6.
38. Munoz-Elias G, Marcus AJ, Coyne TM, Woodbury D, Black IB. Adult bone marrow stromal cells in the embryonic brain: engraftment, migration, differentiation, and long-term survival. J Neurosci. 2004;24(19):4585–95.
39. Caplan AI. Adult mesenchymal stem cells for tissue engineering versus regenerative medicine. J Cell Physiol. 2007;213(2):341–7.
40. Minguell JJ, Erices A, Conget P. Mesenchymal stem cells. Exp Biol Med (Maywood). 2001;226(6):507–20.
41. Caplan AI, Dennis JE. Mesenchymal stem cells as trophic mediators. J Cell Biochem. 2006;98(5):1076–84.

42. Andreeva ER, Pugach IM, Gordon D, Orekhov AN. Continuous subendothelial network formed by pericyte-like cells in human vascular bed. Tissue Cell. 1998;30(1):127–35.
43. Haynesworth SE, Baber MA, Caplan AI. Cytokine expression by human marrow-derived mesenchymal progenitor cells in vitro: effects of dexamethasone and IL-1 alpha. J Cell Physiol. 1996;166(3):585–92.
44. Takahashi M, Iwashita T, Santoro M, Lyonnet S, Lenoir GM, Billaud M. Co-segregation of MEN2 and Hirschsprung's disease: the same mutation of RET with both gain and loss-of-function? Hum Mutat. 1999;13(4):331–6.
45. Bruder SP, Jaiswal N, Haynesworth SE. Growth kinetics, self-renewal, and the osteogenic potential of purified human mesenchymal stem cells during extensive subcultivation and following cryopreservation. J Cell Biochem. 1997;64(2):278–94.
46. Galmiche MC, Koteliansky VE, Briere J, Herve P, Charbord P. Stromal cells from human long-term marrow cultures are mesenchymal cells that differentiate following a vascular smooth muscle differentiation pathway. Blood. 1993;82(1):66–76.
47. Simmons PJ, Torok-Storb B. Identification of stromal cell precursors in human bone marrow by a novel monoclonal antibody, STRO-1. Blood. 1991;78(1):55–62.
48. Gronthos S, Simmons PJ. The growth factor requirements of STRO-1-positive human bone marrow stromal precursors under serum-deprived conditions in vitro. Blood. 1995;85(4):929–40.
49. Muraglia A, Cancedda R, Quarto R. Clonal mesenchymal progenitors from human bone marrow differentiate in vitro according to a hierarchical model. J Cell Sci. 2000;113(Pt 7):1161–6.
50. Prockop DJ. Marrow stromal cells as stem cells for nonhematopoietic tissues. Science. 1997;276(5309):71–4.
51. Chichester CO, Fernandez M, Minguell JJ. Extracellular matrix gene expression by human bone marrow stroma and by marrow fibroblasts. Cell Adhes Commun. 1993;1(2):93–9.
52. Conget PA, Minguell JJ. Phenotypical and functional properties of human bone marrow mesenchymal progenitor cells. J Cell Physiol. 1999;181(1):67–73.
53. Minguell JJ. Is hyaluronic acid the "organizer" of the extracellular matrix in marrow stroma? Exp Hematol. 1993;21(1):7–8.
54. Yamazaki A, Homma T, Uchiyama S, Katsumi Y, Okumura H. Morphologic limitations of posterior decompression by midsagittal splitting method for myelopathy caused by ossification of the posterior longitudinal ligament in the cervical spine. Spine (Phila Pa 1976). 1999;24(1):32–4.
55. Iwasaki H, Akashi K. Myeloid lineage commitment from the hematopoietic stem cell. Immunity. 2007;26(6):726–40.
56. Schmeisser A, Garlichs CD, Zhang H, Eskafi S, Graffy C, Ludwig J, Strasser RH, Daniel WG. Monocytes coexpress endothelial and macrophagocytic lineage markers and form cord-like structures in Matrigel under angiogenic conditions. Cardiovasc Res. 2001;49(3):671–80.
57. Mikkola HK, Orkin SH. The journey of developing hematopoietic stem cells. Development. 2006;133(19):3733–44.
58. MacKey MC. Cell kinetic status of haematopoietic stem cells. Cell Prolif. 2001;34(2):71–83.
59. Morrison SJ, Prowse KR, Ho P, Weissman IL. Telomerase activity in hematopoietic cells is associated with self-renewal potential. Immunity. 1996;5(3):207–16.
60. Audet J, Miller CL, Rose-John S, Piret JM, Eaves CJ. Distinct role of gp130 activation in promoting self-renewal divisions by mitogenically stimulated murine hematopoietic stem cells. Proc Natl Acad Sci U S A. 2001;98(4):1757–62.
61. Ema H, Takano H, Sudo K, Nakauchi H. In vitro self-renewal division of hematopoietic stem cells. J Exp Med. 2000;192(9):1281–8.
62. Lansdorp PM, Dragowska W, Mayani H. Ontogeny-related changes in proliferative potential of human hematopoietic cells. J Exp Med. 1993;178(3):787–91.

63. Medvinsky A, Dzierzak E. Definitive hematopoiesis is autonomously initiated by the AGM region. Cell. 1996;86(6):897–906.
64. Melrose J, Ghosh P, Taylor TK, Latham J, Moore R. Topographical variation in the catabolism of aggrecan in an ovine annular lesion model of experimental disc degeneration. J Spinal Disord. 1997;10(1):55–67.
65. Godin I, Cumano A. Hematopoietic stem cells: where do they come from at last? Med Sci (Paris). 2007;23(8–9):681–4.
66. Purton LE, Scadden DT. Limiting factors in murine hematopoietic stem cell assays. Cell Stem Cell. 2007;1(3):263–70.
67. Ratajczak MZ, Wysoczynski M, Reca R, Wan W, Zuba-Surma EK, Kucia M, Ratajczak J. A pivotal role of activation of complement cascade (CC) in mobilization of hematopoietic stem/progenitor cells (HSPC). Adv Exp Med Biol. 2008;632:47–60.
68. Rookmaaker MB, Tolboom H, Goldschmeding R, Zwaginga JJ, Rabelink TJ, Verhaar MC. Bone-marrow-derived cells contribute to endothelial repair after thrombotic microangiopathy. Blood. 2002;99(3):1095.
69. Walter DH, Dimmeler S. Endothelial progenitor cells: regulation and contribution to adult neovascularization. Herz. 2002;27(7):579–88.
70. Frid MG, Kale VA, Stenmark KR. Mature vascular endothelium can give rise to smooth muscle cells via endothelial-mesenchymal transdifferentiation: in vitro analysis. Circ Res. 2002;90(11):1189–96.
71. Kocher AA, Schuster MD, Szabolcs MJ, Takuma S, Burkhoff D, Wang J, Homma S, Edwards NM, Itescu S. Neovascularization of ischemic myocardium by human bone-marrow-derived angioblasts prevents cardiomyocyte apoptosis, reduces remodeling and improves cardiac function. Nat Med. 2001;7(4):430–6.
72. Kawata H, Yoshida K, Kawamoto A, Kurioka H, Takase E, Sasaki Y, Hatanaka K, Kobayashi M, Ueyama T, Hashimoto T, et al. Ischemic preconditioning upregulates vascular endothelial growth factor mRNA expression and neovascularization via nuclear translocation of protein kinase C epsilon in the rat ischemic myocardium. Circ Res. 2001;88(7):696–704.
73. Kamihata H, Matsubara H, Nishiue T, Fujiyama S, Tsutsumi Y, Ozono R, Masaki H, Mori Y, Iba O, Tateishi E, et al. Implantation of bone marrow mononuclear cells into ischemic myocardium enhances collateral perfusion and regional function via side supply of angioblasts, angiogenic ligands, and cytokines. Circulation. 2001;104(9):1046–52.
74. Zhang YH, Hume K, Cadonic R, Thompson C, Hakim A, Staines W, Sabourin LA. Expression of the Ste20-like kinase SLK during embryonic development and in the murine adult central nervous system. Brain Res Dev Brain Res. 2002;139(2):205–15.
75. Peichev M, Naiyer AJ, Pereira D, Zhu Z, Lane WJ, Williams M, Oz MC, Hicklin DJ, Witte L, Moore MA, et al. Expression of VEGFR-2 and AC133 by circulating human CD34(+) cells identifies a population of functional endothelial precursors. Blood. 2000;95(3):952–8.
76. Gehling UM, Ergun S, Schumacher U, Wagener C, Pantel K, Otte M, Schuch G, Schafhausen P, Mende T, Kilic N, et al. In vitro differentiation of endothelial cells from AC133-positive progenitor cells. Blood. 2000;95(10):3106–12.
77. Harraz M, Jiao C, Hanlon HD, Hartley RS, Schatteman GC. CD34- blood-derived human endothelial cell progenitors. Stem Cells. 2001;19(4):304–12.
78. Quirici N, Soligo D, Caneva L, Servida F, Bossolasco P, Deliliers GL. Differentiation and expansion of endothelial cells from human bone marrow CD133(+) cells. Br J Haematol. 2001;115(1):186–94.
79. Kaushal GP, Kaushal V, Hong X, Shah SV. Role and regulation of activation of caspases in cisplatin-induced injury to renal tubular epithelial cells. Kidney Int. 2001;60(5):1726–36.
80. Lin C, Shen M, Chen W, Li X, Luo D, Cai J, Yang Y. Isolation and purification of rabbit mesenchymal stem cells using an optimized protocol. In Vitro Cell Dev Biol Anim. 2015;51(10):1102–8.
81. Friedenstein AJ, Chailakhyan RK, Gerasimov UV. Bone marrow osteogenic stem cells: in vitro cultivation and transplantation in diffusion chambers. Cell Tissue Kinet. 1987;20(3):263–72.

82. Guillot PV, Gotherstrom C, Chan J, Kurata H, Fisk NM. Human first-trimester fetal MSC express pluripotency markers and grow faster and have longer telomeres than adult MSC. Stem Cells. 2007;25(3):646–54.
83. Bourzac C, Smith LC, Vincent P, Beauchamp G, Lavoie JP, Laverty S. Isolation of equine bone marrow-derived mesenchymal stem cells: a comparison between three protocols. Equine Vet J. 2010;42(6):519–27.
84. Connolly JF, Guse R, Tiedeman J, Dehne R. Autologous marrow injection as a substitute for operative grafting of tibial nonunions. Clin Orthop Relat Res. 1991;266:259–70.
85. Healey KM, Pavletic SZ, Al-Omaishi J, Leuschen MP, Pirruccello SJ, Filipi ML, Enke C, Ursick MM, Hahn F, Bowen JD, et al. Discordant functional and inflammatory parameters in multiple sclerosis patients after autologous haematopoietic stem cell transplantation. Mult Scler. 2004;10(3):284–9.
86. Horn P, Bork S, Diehlmann A, Walenda T, Eckstein V, Ho AD, Wagner W. Isolation of human mesenchymal stromal cells is more efficient by red blood cell lysis. Cytotherapy. 2008;10(7):676–85.
87. Pierini M, Dozza B, Lucarelli E, Tazzari PL, Ricci F, Remondini D, di Bella C, Giannini S, Donati D. Efficient isolation and enrichment of mesenchymal stem cells from bone marrow. Cytotherapy. 2012;14(6):686–93.
88. Hancock JP, Goulden NJ, Oakhill A, Steward CG. Quantitative analysis of chimerism after allogeneic bone marrow transplantation using immunomagnetic selection and fluorescent microsatellite PCR. Leukemia. 2003;17(1):247–51.
89. Holyoake TL, Alcorn MJ, Franklin IM. The CD34 antigen: potential clinical advantages of CD34 selection. Clin Oncol (R Coll Radiol). 1996;8(4):214–21.
90. Sekiguchi H, Ii M, Jujo K, Yokoyama A, Hagiwara N, Asahara T. Improved culture-based isolation of differentiating endothelial progenitor cells from mouse bone marrow mononuclear cells. PLoS One. 2011;6(12):e28639.
91. Ii M. Bone marrow-derived endothelial progenitor cells: isolation and characterization for myocardial repair. Methods Mol Biol. 2010;660:9–27.
92. Iudicone P, Fioravanti D, Bonanno G, Miceli M, Lavorino C, Totta P, Frati L, Nuti M, Pierelli L. Pathogen-free, plasma-poor platelet lysate and expansion of human mesenchymal stem cells. J Transl Med. 2014;12:28.
93. Chen X, Xu H, Wan C, McCaigue M, Li G. Bioreactor expansion of human adult bone marrow-derived mesenchymal stem cells. Stem Cells. 2006;24(9):2052–9.
94. Gronthos S, Graves SE, Ohta S, Simmons PJ. The STRO-1+ fraction of adult human bone marrow contains the osteogenic precursors. Blood. 1994;84(12):4164–73.
95. Oyajobi BO, Lomri A, Hott M, Marie PJ. Isolation and characterization of human clonogenic osteoblast progenitors immunoselected from fetal bone marrow stroma using STRO-1 monoclonal antibody. J Bone Miner Res. 1999;14(3):351–61.
96. Tondreau T, Lagneaux L, Dejeneffe M, Delforge A, Massy M, Mortier C, Bron D. Isolation of BM mesenchymal stem cells by plastic adhesion or negative selection: phenotype, proliferation kinetics and differentiation potential. Cytotherapy. 2004;6(4):372–9.
97. Murohara T. Therapeutic vasculogenesis using human cord blood-derived endothelial progenitors. Trends Cardiovasc Med. 2001;11(8):303–7.
98. Kaushal S, Amiel GE, Guleserian KJ, Shapira OM, Perry T, Sutherland FW, Rabkin E, Moran AM, Schoen FJ, Atala A, et al. Functional small-diameter neovessels created using endothelial progenitor cells expanded ex vivo. Nat Med. 2001;7(9):1035–40.
99. Murohara T, Ikeda H, Duan J, Shintani S, Sasaki K, Eguchi H, Onitsuka I, Matsui K, Imaizumi T. Transplanted cord blood-derived endothelial precursor cells augment postnatal neovascularization. J Clin Invest. 2000;105(11):1527–36.

Chapter 3
Mobilization and Homing of Bone Marrow Stem Cells After Stroke

Weikai Wang, Lefu Chen, Linhui Ruan, Kunlin Jin, and Qichuan Zhuge

Abstract Generally, bone marrow stem cells (BMSCs) reside in the bone marrow, where the microenvironments maintain a dynamic balance between self-renewal and differentiation. However, BMSCs can also be forced into the blood, a process termed mobilization, which is clinically used to harvest large number of cells for transplantation. On the other side, stroke-induced local and systemic pathological responses also lead to the mobilization of BMSCs to peripheral blood and then "homing" to the damaged regions, which is considered as an important regenerative process. In this chapter, we summarize current understanding of the physiological and pathological mechanisms that guide BMSC mobilization and homing to the damaged brain. The underlying cellular and molecular mechanisms, which largely depend on an interplay between chemokines, chemokine receptors, intracellular signaling, adhesion molecules, and proteases, are also discussed. Increasing the number of BMSC mobilization and homing is critical for the promotion of stroke cell-based therapies.

Keywords Bone marrow • Stem cells • Stroke • Characterization • Mobilization • Homing • Trophic factor

W. Wang • L. Chen • L. Ruan
Zhejiang Provincial Key Laboratory of Aging and Neurological Disorder Research, First Affiliated Hospital, Wenzhou Medical University, Wenzhou City, Zhejiang Province 325000, People's Republic of China

K. Jin
Department of Pharmacology Neuroscience, University of North Texas Health Science Center, Fort Worth, TX, USA

Q. Zhuge (✉)
Department of Neurosurgery, First Affiliated Hospital, Wenzhou Medical University, Wenzhou, China

Zhejiang Provincial Key Laboratory of Aging and Neurological Disorder Research, First Affiliated Hospital, Wenzhou Medical University, Wenzhou, China
e-mail: zhugeqichuan@vip.163.com

K. Jin et al. (eds.), *Bone Marrow Stem Cell Therapy for Stroke*,
DOI 10.1007/978-981-10-2929-5_3

3.1 Introduction

Stroke remains a leading cause of death in the USA and industrial countries. Due to the sudden occlusion of cerebral blood vessels, brain tissues undergo hypoxia, inflammatory cascades, and ultimately cell necrosis, which is associated with neurological dysfunctions. Over the past decade, stem cell therapy has shown great promise in experimental stroke models [1–3]. In the clinical setting, intravascular route transplantation with a minimal invasion is thought to be practical. However, this peripheral route of cell injection requires mobilization of the cells before the cells secrete their products to the site of injury in order to afford brain plasticity and remodeling. Bone marrow (BM) consists of a heterogeneous group of stem and progenitor cells [4], including hematopoietic stem cells (HSCs) and mesenchymal stem cells (MSCs), endothelial progenitor cells (EPCs) and very small embryonic-like stem cells (VSELs). Previous investigations have demonstrated the differentiation of BMSCs into neural lineage cells [5], and their secretion of growth/trophic factors for neuronal survivals [6–8]. This forms the basis of BMSC transplantation or mobilization for therapy in neurological disorders including stroke.

BMSCs mainly resides within BM under multiple regulations of bone marrow niche components, with only a fraction of these cells circulating in the peripheral blood (PB) [9]. Stroke evokes the release of inflammatory cytokines/chemokines and also causes the excitation of sympathetic nervous system to modulate the micro-environments in the bone marrow niches. These mechanisms mobilize and shift BMSCs into the PB. Injured brain also produces factors, which eventually guide the migration of BMSCs to the brain lesions. In this chapter, we will discuss the mechanisms of BMSC mobilization and homing, especially under stroke condition. Introduction of the clinical and experiment applications of reagents to enhance mobilization and homing for better therapeutic effects on stroke would also be highlighted.

3.2 Bone Mesenchymal Stem Cells

3.2.1 The Characterization of BMSCs

BMSCs are multipotent, self-renewing cells that can differentiate into many different types of tissues, such as bones, fat, cartilage, neurons and glial cells. According to the criteria proposed by the International Society for Cellular Therapy (ISCT), BMSCs must express CD105, CD90, and CD73 without the expression of pan-leukocyte (CD45), primitive hematopoietic (CD34), monocytic (CD14 or CD11b), or B-cell (CD79a or CD19) marker. Steady-state BMSCs should also lack the expression of HLA-DR [10].

Among the three positive expressed surface antigens, only CD105 can be found in the uncultured (fresh) human MSPCs [11]. However, those fresh MSCs expressing CD105 are not capable of differentiating into other cells or self-renewing.

It is commonly acknowledged that adhesion protein CD44 is highly expressed in vitro in MSPCs [12, 13]. However, recent study has suggested that, the CD44 fraction contains almost all the primary functional mesenchymal cells in both mice and humans, and these cells acquired CD44 during in vitro culture. This may explain the previous misunderstanding of CD44 expression [14]. Changing of expression of surface markers may coincide with the cell differentiation in vitro culture.

Although, tons of researches have revealed lots of facts concerning characterization of BMSCs, still some detailed and underlying truth about BMSCs remains unknown. Further researches may give us the answer.

3.2.2 BMSC Mobilization After Stroke

The mobilization of MSCs establishes a rich circulating stem cell pool in the peripheral blood, which provides a convenient approach to harvest the cell for autogenic or allogenic transplantation, as well as promotes the cell migration (homing) to injured tissue. Studies have attempted to mobilize MSC into the peripheral blood (PB) using hematopoietic growth factors such as granulocyte-colony stimulating factor (G-CSF). However, most of the investigations brought about disappointing results [15–18], and little evidence showed feasible and efficacious strategies for MSCs mobilization [17–20].

To date, the mobilization of MSC is largely observed in animals undergoing hypoxia. This makes hypoxic condition is an important factor [21, 22]. For example, hypoxic-ischemic brain damaged mice show increased MSCs in the circulation [23]. Although the precise molecular mechanisms underlying hypoxia-induced MSC mobilization are not completely understood, to the best of our knowledge, hypoxia-inducible factor-1 (HIF-1) signaling is an essential mediator [21]. HIF-1 is a transcription factor in response to hypoxia in an organism and mediates homeostatic regulation under low oxygen conditions [24]. HIF-1 is a heterodimer composed of α and β subunits. Under normoxic conditions, the unstable α subunit is prone to hydroxylation, which leads to polyubiquitination and degradation by proteasome; in contrast, hypoxic conditions prevent the α subunits from hydroxylation, maintaining the active function of HIF-1 in the nucleus [25, 26]. Further studies have revealed that HIF-1-induced synthesis of vascular endothelial growth factor (VEGF) might be the direct regulatory molecule for mobilization. VEGF promotes the vasodilation, vascular permeability [27, 28], and angiogenesis [29, 30], which facilitates MSCs drifting from the BM into the PB. However, as VEGF was upregulated during day 2 to day 7 of hypoxia, there is no elevation of VEGF in chronic hypoxic animals, indicating that MSC mobilization in the chronic phase is probably controlled by other factors [21]. In addition to VEGF, stromal cell-derived factor-1 (SDF-1) is also induced by HIF-1 [21, 31]. SDF-1 (or termed CXCL12) is a potent ligand that binds to CXC chemokine receptor 4 (CXCR4) which is expressed on MSCs, and the interaction of SDF-1/CXCR4 plays an important role in MSC apoptosis, migration, and cytokine secretion [32]. The SDF-1α level in the PB was increased on day 2 of hypoxia, which induced the MSC egress to the circulation [21]. On the basis of the function of serum SDF-1α, this chemokine has been used as a pharmacological agent for MSC mobilization in mice [23]. SDF-1 was demonstrated to increase in the injured brain tissue and serum after stroke [33–38], but little evidence supported the stroke-induced MSC mobilization by this molecule.

3.2.3 The Potential Signaling Pathway/Mechanisms Underlying BMSC Homing

After mobilization of the BMSCs, how to lead these stem cells to the damaged lesion becomes the main issue. Several years of hard work bring us some possible signaling pathways underlying the BMSC homing. Among them, SDF-1 and CXCR-4 come first. Scientists found that the interaction between stromal cell-derived factor-1 (SDF-1) and CXCR-4 may lead to the BMSC homing to the ischemic penumbra.

SDF-1, also called chemokine CXC ligand 12, can be found in all types of cells presented in the CNS. It works as a conjugation spot with CXCR-4, a G-protein coupled receptor, which is expressed on the surface of many different cells including neurons, astrocytes, BMSCs, and some other stem cells [39]. Recent study revealed SDF-1 is highly upregulated in the ischemic region, especially along with the ischemic boundary zone, peaked by 3–7 days and maintained at least 14 days. Meanwhile, CXCR-4 expression is dramatically increased on the surface of BMSCs under the circumstance of hypoxia. On the other hand, deficiency of CXCR-4 significantly decreases the migration of BMSCs toward the ischemic region, indicating that SDF-1/CXCR-4 plays an important role in regulating the homing of BMSCs [40]. Some studies suggested that these two molecules may also regulate the inflammatory response and angiogenesis thereby enhancing the neurorestorative effect after stroke.

3.2.3.1 MCP-1

Monocyte chemoattractant protein-1 (MCP-1), known as a chemoattractant factor, was suggested to contribute to the homing of the BMSCs. Wang et al. evaluated the amount of MCP-1 from the ischemic brain. Result showed the level of MCP-1 increased significantly at 6 h after MCAO and peaked at 48 h with exponential MSCs migration, indicating MCP-1 plays an important role in BMSC homing [41].

3.2.3.2 CD44 and Selectin

Scientists have verified the significant role selecting has played in the process of BMSC binding to endothelial cells. When the endothelial cells were treated with inhibitor of P-selectin, this binding process was suppressed. Meanwhile, the rolling of BMSC is enhanced with immobilized P-selectin [42]. CD44 has also been proven to induce BMSC migration [43]. Gokhan Yilmaz et al. have verified the function of selectin and CD44 by monitoring the cell recruitment. The MCAO/R mice were treated with antibody against P- or E-selectin. They also used the CD44-deficient mice to test the role CD44 played in cell migration. The results showed attenuated homing was observed in both antagonists treated and CD44-deficient mice, suggesting that cerebral vascular endothelium provides a pro-adhesive phenotype after ischemic stroke to facilitate the migration of BMSCs, which use both P-selectin and E-selectin to home the cells to the infarct zone. CD44 may contribute as a ligand during the BMSC recruitment [44].

The potential signaling molecules and underlying mechanisms concerning homing of BMSCs are complicated and comprehensive. It is not exactly clear whether increasing the target migration of BMSCs is crucial or not in neurological function recovery. Scientists found significant functional improvement with few targeted BMSCs, indicating that cell homing may not be mandatory for stroke treatment [45]. More studies, relating to the migration of BMSCs and poststroke treatment, need to be done.

3.2.4 BMSCs Induced Neurotrophic Factor Secretion and Restoration of Neural Function

BMSC's transplantation can promote sensorimotor function and neural regeneration, decrease tPA-induced brain damage, regulate immune response, and reduce inflammation after ischemic infarction.

Studies have shown that BMSCs can significantly ameliorate the neurological deficits by neurotrophic factor secretion, which drastically decrease the cell apoptosis and promote the proliferation of endogenous cells in the SVZ [46]. These cytokines can also facilitate functional recovery by inducing angiogenesis, regeneration of dendrites and axons as well as rebuilding the synapse.

BMSCs express various cytokines including Ang 1, basic fibroblast growth factor-2 (bFGF2), insulin-like growth factor (IGF), vascular endothelial growth factor (VEGF), and brain-derived neurotrophic factor (BDNF), as well as glial cell line-derived neurotrophic factor (GDNF) [47–51]. These factors have both paracrine and autocrine activities [52], which can regulate cell survival, proliferation, and differentiation.

Intrastriatal infusion of GDNF can promote cell proliferation in SVZ in MCAO rat models. Recruitment of neuroblasts and the improved survival of neurons are also detected [51]. These findings suggest the significant role GDNF played in enhancing neurogenesis, promoting the proliferation and migration of neuroblasts, and inhibiting cell apoptosis.

Wakabayashi K and his group found out that human insulin-like growth factor 1 (IGF-1) was highly increased after transplanting MSC into MCAO rat model.

Therefore, it leads to endogenous expression of vascular endothelial growth factor (VEGF), epidermal growth factor (EGF) and basic fibroblast growth factor (bFGF) [53].

Recently, the novel treatment strategy for ischemic stroke using exosome-mediated miRNAs attracted scientists' attention. Exosomes are nanoparticles released from cells enhancing intercellular communication. miRNAs are noncoding RNAs that can inhibit translation of mRNA. Studies have shown that BMSCs can secrete exosome-enriched miRNAs which facilitate bidirectional material exchanging between BMSCs and injured cells [54], thus contributing to neurorestorative effect by regulating neurogenesis and mediating transplanted cells differentiating to functional neurons [55].

Recent research has revealed that the microRNA-133b level of BMSCs was dramatically increased when treated with ipsilateral tissue extracted from MCAO rats. Additionally, these miRNAs could transfer to neurons and astrocytes via exosomes which resulted in increased miRNAs in these cells, indicating the intercellular communication between BMSCs and parenchymal cells. As a result, miRNAs could promote neurite outgrowth and remodeling by gene expression regulation, thus to facilitate neurological functional recovery after stroke [56, 57]. This novel strategy paved a promising new way for poststroke therapy and future clinical treatment.

3.3 Bone Marrow-Derived Hematopoietic Stem Cells

3.3.1 The Characterization of HSCs and the Niches

3.3.1.1 The Characterization of HSCs

Hematopoietic stem cells (HSCs) are one of the typical subsets of bone marrow (BM) stem cells, possessing the capacity for self-renewal and differentiation into all blood cells of myeloid and lymphoid lineages [58]. HSCs are mainly distributed in adult bone marrow, umbilical cord blood, as well as in peripheral blood in small numbers [59–61]. After harvesting HSCs from these sources, it is an indispensable step to assess the cell before scientific research and clinical applications. The most commonly used approach is detecting markers on cell surface. HSCs are negative for the markers that label lineage commitment, and it is termed as Lin^{-}. However, the markers may vary between species. For example, human HSCs present $CD34^{+}$, $CD59^{+}$, $Thy1/CD90^{+}$, $CD38^{lo/-}$, $C\text{-}kit/CD117^{+}$, and Lin-, while mouse HSCs display $CD34^{+}$, $CD59^{+}$, $Thy1/CD90^{+}$, $CD38^{lo/-}$, $C\text{-}kit/CD117^{+}$, and Lin^{-}. Actually, even in humans, there are hematopoietic stem cells presenting $CD34^{-}/CD38^{-}$ [62, 63]. One advance in defining human HSCs surface marker is the introduction of CD133, for both $CD34^{+}$ and $CD34^{-}$ HSCs share this molecule [64].

3.3.1.2 The HSC Niches

In the adult bone marrow, HSCs exist in contact with osteoblasts and vascular endothelial cells, which are considered as two important microenvironments named osteoblastic niches (or endosteal niches) [65, 66] and vascular niches [67]. Osteoblastic niches, comprised of osteoclasts and other mesenchymal-derived stromal cells, including reticular cells, fibroblasts, and adipocytes, create a hypoxic environment ideal for quiescent and homeostatic HSCs; by contrast, vascular niches, contain sinusoidal endothelial cells, reticular cells, and megakaryocytes, lacking osteoblastic cells, play an essential role in supporting injury-mediated HSCs mobilization [68–72].

The activation of HSCs is regulated by the niches through a combination of adhesion molecules, growth factors (GFs), and chemokines. Some essential molecules expressed by the endosteal niches, such as N-cadherin [73–75], osteopontin (OPN) [76, 77], and angiopoietin-1 (Ang-1) [78], contribute to anchoring HSCs to the niches and maintain the cells in a quiescent state. The interplay of vascular cell adhesion molecule 1 (VCAM-1) expressed by vascular niches and its receptor integrin very late antigen 4 (VLA-4) expressed on HSCs is proposed as a mechanism of homing, lodgment, and mobilization of HSCs [79, 80]. HSCs were found adjacent to the stromal cells with a high-level expression of CXCL12 (also called SDF-1) and sinusoidal endothelial cells. These cells were named CXCL12-abundant reticular (CAR) cells [81]. SDF-1, with its receptor CXCR4 on HSPCs, plays a pivotal role in maintaining the quiescent state of the cells allowing the stem cells to preserve a sustained hematopoietic pool [81–84]. Another important factor is transforming growth factor (TGF)-β which controls proliferation, differentiation, apoptosis [85], and cell cycle entry of HSCs [82, 86]. As a result, the quiescent or motile state of HSCs is switchable by the physiological and pathological changes of these GF- or cytokine-mediated signaling within the BM niches. Another important notion is that except for the dysfunctions of BM per se [87–91], the niches are equally influenced by diseases of other systems [92, 93] via neural and/or humoral regulations on these important molecules. Mounting evidence has shown that the changing of these factors caused by pathological states exert effects on the mobilization of HSCs and homing to affected tissues and organs [36, 94, 95].

3.3.2 HSC Mobilization After Stroke

During homeostasis, HSPCs are continuously released into the peripheral blood (PB) but predominantly reside in the bone marrow in a state of quiescence, owing to the anchoring via adhesion interactions to the specialized niches, which inhibit their motility and proliferation [9]. In response to injury or after administration of pharmacologic agents, hematopoietic stem/progenitor cells (HSPCs) quickly become motile into the blood circulation [96–99]. Researches have shown the mobilization of HPSCs from BM into the PB after the onset of some ischemic diseases, including acute myocardial infarction [100–102] and ischemic stroke [34, 103–105], and also demonstrated the repair effects of these mobilized cells on the diseases. However, clinical HSPC mobilization has not significantly progressed over the past decades, due to the complicity of the mechanisms by which HSPCs egress the BM [106]. Therefore, it is necessary for neural scientific researchers and clinicians to understand how ischemic stroke mobilizes HSPCs into the BM and the current approaches to enhance this process. The mechanism of HSPCs mobilization holds great promise for further exploration and provision of novel stroke therapies.

3.3.2.1 Mechanisms of HSPC Mobilization After Stroke

Although the definite or complete mechanisms of stroke-induced HSPC mobilization are not fully understood, scientists have so far presented explanations in two major aspects: neuronal (sympathetic) and humoral (chemokines, cytokines, growth factors, and the others) regulations [106]. All the related mechanisms are based on the interactions between the stroke-mediated systemic or local responses and the BM niches. Actually, mounting evidence has presented two major patterns in HSC mobilization: (1) increase the SDF-1 level in the circulation [34] and (2) reduce the SDF-1/CXCR4 interaction in BM (such as inhibiting SDF-1 expression or antagonizing CXCR4) [99, 107–109].

3.3.2.1.1 The SDF-1/CXCR4 Interaction

SDF-1, constitutively produced by BM niches, is a potent chemoattractant for HSCs which express its major receptor CXCR4 [110]. In adult life, the SDF-1/CXCR4 axis serves as a key player for retention, seeding, migration, and mobilization of HSPCs [111]. Evidence has indicated that the balance between adhesion and mobilization of hematopoietic cells depends on SDF-1 concentrations in the PB and the BM niches [112]. On this ground, expression and release of SDF-1 by remote site tissues and organs can probably initiate HSC mobilization. Ischemic stroke in both animal models and patients has shown an increased SDF-1 level in the injured brain tissue or the serum [33–38]. Serum SDF-1 change is positively correlated with infarct volume and severity of stroke in patients [36]. This change, combined with the increased numbers of HSCs in the PB after brain ischemia [103, 104], suggests that the stem cells are mobilized by stroke-induced SDF-1 production. A study conducted by William D. Hill et al. further revealed that SDF-1 is primarily secreted by astrocytes and endothelial cells after stroke [35].

However, the mechanisms of stroke-induced SDF-1 overexpression are still ambiguous, while the relationship between ischemia-induced hypoxia and SDF-1 production is being revealed in the further studies. In homeostatic state, high SDF-1 expression is sustained by the hypoxic condition in the BM niche, which leads to the retention of most HSCs. In a similar way, local hypoxia caused by ischemia may also lead to overexpression of SDF-1. Jean-Jacques Lataillade et al. have demonstrated that SDF-1 expression is regulated by the transcription factor hypoxia-inducible factor-1 (HIF-1) in endothelial cells under a hypoxic condition [113]. This HIF-1 induced SDF-1 expression was later proven in the ischemic stroke [114]. Actually, timely administration of tissue plasminogen activator (tPA) after acute cerebral ischemia, effectively alleviates the vessel occlusion, but cannot promote CD34+ increase in the PB [104], suggesting that SDF-1 might be induced by hypoxic condition.

3.3.2.1.2 Interplays of the Sympathetic Nervous System with SDF-1/CXCR4 and G-CSF

The BM is highly innervated with nerve fibers which extend into the parenchyma and along the blood vessels. Therefore, sympathetic neurotransmitter catecholamines (CAs) can regulate the BM resident cells, either directly by nervous endings or via circulation transference [9, 115]. Stroke-induced organism stress shows a high activation of the sympathetic nervous system (SNS) [116]. The increased CAs act on adrenergic ß-receptors on osteocytes in the niches, leading to the degradation of transcription factor SP1 which inhibit SDF-1 synthesis. Research has shown that HSC mobilization was reduced by sympathectomy or β-adrenergic receptor blockade [106]. Further evidence has shown that in both wild-type and norepinephrine-deficient mice, the administration of ß-adrenergic agonist could promote mobilization [117].

Granulocyte-colony stimulating factor (G-CSF), which was discovered two decades ago as a hematopoietic growth factor for its function in promoting the survival and growth of HSPCs [118], has been applied as an effective HSC mobilizer in clinical trials for stroke therapy [119–121]. Although the exact mechanisms of G-CSF-induced HSC mobilization are not completed revealed, evidence has shown three potential pathways in general: 1) proteases activation, 2) suppression of adhesion molecules, and 3) attenuation of SDF-1/CXCR4 function [122, 123]. For example, researches have shown that G-CSF interrupts HSC anchoring by suppressing SDF-1 expression and activating the protease CD26 which cleaves the amino-terminal dipeptide of SDF-1, thereby inhibiting the SDF-1/CXCR4 interaction [124–127]. Recent studies have found that the mobilization mechanism by SNS is related with this factor. Peripheral noradrenergic neurons can be activated by G-CSF to induce mobilization of HSCs through depression of osteoblasts and downregulation of SDF-1 expression [117, 128]. Other researches on chimeric NOD/SCID mice have demonstrated that repeated stimulation with G-CSF helps to promote catecholaminergic receptor expression on CD34+ progenitor cells, and these receptors increase the cell motility [129, 130]. In summary, G-CSF-mediated HSC mobilization includes direct act on BM stroma and indirect pathway trough SNS.

3.3.2.2 Agents That Enhance the Mobilization of HSCs for Stroke

In order to enhance the outcome of HSPCs on stroke therapy, pharmacologic agents have been used in preclinical experiments and clinical trials, such as G-CSF, stem cell factor (SCF), and granulocyte-macrophage colony-stimulating factor

(GM-CSF). Clinical mobilization of HSPCs is performed by G-CSF administration with or without other agents, which may develop into one of the bases for noninvasive therapy of autologous HSC for brain ischemia. It is observed that both mature and immature HSCs egress from the BM after several days of G-CSF stimulation [106]. Clinical trials have been widely conducted to mobilize CD34+ HSCS as a stroke therapy [120, 121, 131, 132], which actually improve the PB stem cell level. Its administration at an effective dose has been proven safe and tolerated for most patients [131]. Unfortunately, G-CSF is found to cause bone pain experiences and loss of body weight in patients or animals. In addition, G-CSF may fail in sufficient mobilization of HSCs and requires multiple daily subcutaneous injections for several days [133]. To overcome these drawbacks, a salvage treatment has been proposed, that is, a combination of G-CSF and other mobilizing cytokines (GM-CSF and SCF) [97, 134], that relies on a synergistic effect to enhance mobilization [135].

Recently, numerous types of novel agents have been developed to enhance HSC motility, most of which are also regulators for the BM niche (especially CXCR4 antagonists), not acting on HSCs per se. These experimentally studied agents might hold promise in enhancing human HSCs in clinical stroke treatment. AMD3100 (plerixafor), a reversible CXCR4 antagonist, was shown to rapidly mobilize human and murine HSPCs, and it greatly promoted G-CSF-induced mobilization of HSPCs [129, 136]. This synergic effect is also involved with immune regulations, such as complement cascade [137]. Other CXCR4 antagonists or inhibitors include GRO-β [20], GRO-p [138, 139], AMD3465 [140], T-140 (4F-benzoyl-TN14003) [141], POL6326 [142], and VLA-4 inhibitor BIO5192 [143], which are verified effective agents for HSC mobilization in animal models.

3.3.3 HSC Homing to the Ischemic Brain

HSC homing is a process whereby cells in circulation migrate to target tissues, including their origin the BM and other peripheral tissues. Molecular pathways of the formal process are a mirror image of mobilization [106]. Increasing evidence has demonstrated the capability of HSCs to home the injured tissue/organ after mobilization from the BM to the PB [144]. Herein, we mainly discuss the clinical and experimental observations and underlying mechanisms of HSC homing to the brain after stroke and highlight the applications of these mechanisms to promote homing in stroke therapy.

3.3.3.1 Mechanisms of HSC Homing After Stroke

Proper homing of HSCs to ischemic brain provides a means for the restoration of the injured neural tissues and improved functional outcomes. The main speculated mechanism underlying HSC migration to ischemic brain is considered to be the overexpression and secretion of SDF-1 by ischemic tissues, which develops a gradient in circulation and guides CXCR4+ HSC recruitment from PB to the injured site [144]. The splicing of the SDF-1 mRNA gives rise to the two isoforms SDF-1α and SDF-1β, which are identical except for four C-terminal amino acids [145]. Murine brain ischemia induced SDF-1β overexpression in penumbra endothelial cells and downregulation in non-injured brain areas, which presented a co-occurrence of infiltration of CXCR4-expressing cells from PB, suggesting that SDF-1β plays a more important role in the homing process [33]. In addition, SDF-1 overexpression was found to persist in ischemic brain tissue for at least 4 months after rat middle cerebral artery occlusion (MCAO) [38], which indicts this chemokine might continually guide HSC migration for a long time. The detailed mechanisms are related to several critical steps in the homing process, generally including (1) interaction with vascular endothelium, (2) trans-endothelial migration, and (3) migration to the brain. The whole process involves the interactions of multiple cells, adhesion molecules, cytokines, chemotactic factors, and extracellular matrix (ECM) degrading proteases, among which, stroke-induced cytokine secretions play an important role in directing HSCs into the brain parenchyma [144]. The first step is the recognition and interaction between HSCs and vascular endothelium. Similar to leukocytes in response to inflammatory signals, HSC rolls along the vessels, which is mainly mediated by E-selectin and P-selectin. E-selectin deficiency has been found to cause decreased homing capacity of endothelial progenitor cells in mice, and this effect is reversible after administrating soluble E-selectin (sE-selectin) [146], because sE-selectin upregulates the adhesion molecules (ICAM-1 and VCAM-1) on endothelial cells. Likewise, the vascularization can be improved by the HSCs adhesion to P-selectin endothelial cells [147]. The attachment is subsequently made firm by integrins, such as the VCAM/VLA-4 [79] and ICAM-l/LFA-1 [148]. Researches have shown that stroke brain released increased SDF-1 to the serum [36], and SDF-1 is a promoter of the adhesion of LFA-1 and VLA-4 to the endothelial ligands [149]. Trans-endothelial migration of HSCs requires basement membrane degradation, which is mediated by the matrix-degrading enzymes, such as matrix metalloproteinases (MMPs). Serum MMP2 and MMP9 level are also significantly increased in acute ischemic stroke patients compared to the healthy individuals [150–153], which indicates it might be a potential mechanism for HSC homing after stroke. However, MMP-2 was shown to suppress the activity and expression of SDF-1 and CXCR4 [154] which are crucial for the migration of HSC to the injured brain.

3.3.3.2 Methods of Enhancing HSC Homing to Ischemic Brain

Mobilized HSCs provide a rich stem cell pool in the circulation for the subsequent recruitment into the ischemic brain. Thereby, any method to augment HSC motility is indirectly amplify its homing capacity, and regular mobilizers, especially hematopoietic factors, are most widely employed agents to enhance motility [133]. A novel approach is on the basis of the chemotactic effect of SDF-1 on CXCR4+ HSCs, that is, to increase the local SDF-1 concentration in the ischemic brain by stereotaxic cerebral injection [155]. Some studies are attempting to prevent the degradation of SDF-1. For example, a protease-resistant form of SDF-1 substitute, termed S-SDF-1 (S4V), has been used to treat myocardial infarction [156]. However, intracerebral injection may cause iatrogenic damage to the injured brain. Another approach is found in an experiment by Woei-Cherng Shyu et al. that intracerebrally transplanted HSCs to stimulate endogenous HSC migration to the brain after cerebral stroke [157], which might be a potential therapy for clinical application. The underlying mechanism proposed partly related - the increased SDF-1 expression in the brain through the interaction between the implanted cells and the host brain tissue as a contributing factor. Actually, directing autogenous HSCs to the brain is a more ideal therapy for stroke compared to allogenic HSC transplantation (HSCT) systemically or intracerebrally, as it avoids some adverse events, such as the occurrence of graft-versus-host disease (GVHD) [158, 159] and side effects on nontarget organs which the HSC may also migrate into. Another adverse event caused by HSCT was reported that 2 days after HSCT for acute lymphoblastic leukemia, acute cerebral ischemia occurred to the patient, which was likely related to the elevated Factor VIII Level [160]. Therefore, novel efficient and safe methods are highly required to guide the HSC to ischemic brain tissue.

3.3.4 Regulations of HSC on Neural Restoration and Neural Function After Stroke

A sudden arterial occlusion and secondary inflammation in cerebral ischemia lead to the necrosis of neurons and glial cells, which results in functional deficits in movement, sensation, and cognition [161, 162]. Emerging evidence has shown that mobilized or transplanted HSCs migrate to the ischemic brain and improve the neurological function [114, 157, 163]. The underlying therapeutic mechanisms of HSCs include paracrine effects and cell differentiation, which contributes to inflammatory response modulation and cell replacement [157, 164]. In as much as allogenic HSCT could provide stroke patients with large numbers of cells, the number of mobilized autogenic HSCs after stroke may not sufficiently improve the patients' condition. Therefore, it is necessary to apply efficient pharmacological agents to mobilize ample HSCs for stroke, additionally, these hematopoietic growth factors, such as G-SCF and SCF could themselves, present other benefits such as neuroprotection and neuro-regeneration after ischemic stroke [110, 134, 165].

The activities of HSCs, including survival, proliferation, mobilization, differentiation, and migration, are under the regulations of numerous cytokines, growth factors and chemokines secreted in BM niches, other tissues or exogenous administrations [166–172]. Evidence has shown that HSPCs can also secrete the growth/trophic factors, such as GM-CSF, vascular endothelial growth factor (VEGF), basic fibroblast growth factor (bFGF), and insulin-like growth factor-1 (IGF-1) [173], some of which possess the capacity of neural development and restoration for the brain [174–178]. On the ground of the present literature, it is proposed that anti-inflammation/apoptosis is the major protective mechanism in the acute phase of cerebral ischemia [179–182], which helps to reduce the infarct volume and retain more neural functions. Systemic injection of HSCs into stroke mice model showed significant decrease in neuron apoptosis, infarct volumes, and the infiltration of T cells and macrophages in the ischemic hemisphere [164]. Blood flow supply is highly required for the attenuation of infarction and the reconstruction of neurovasculature. The function of HSC in angiogenesis for ischemic stroke is demonstrated by a research that systemically administrated human CD34+ cells to stroke mice, which created a permissive microenvironment for neuro-regeneration [183]. This angiogenesis is probably induced by angiogenic growth factors, such as VEGF and IGF-1 [173]. However, after the cerebral ischemic rat was treated with PB CD34+ cells, another angiogenic growth factor SDF-1 was significantly upregulated in the brain [157]. Since SDF-1 is not produced by HSCs [173], the result suggested that the angiogenesis is more likely the outcome of interactions between HSCs and brain parenchymal cells, instead of the unilateral regulation through factors secreted by HSCs [157]. Aside from promoting novel microvasculature, CD34+ cell-treated neonatal stroke mice showed an augmented diameter of the cerebral blood vessel in the peri-infarct area, which indicated nitric oxide or other undefined vasodilative factors are responsible for this regulation [184]. Neurogenesis is a pivotal process for the improvement of functional outcome after stroke, and HSCs are reported to improve the trophic milieu for neurogenesis after stroke. For example, HSC-treated ischemic brain presented an upregulation of neurotrophic factors, including SDF-1, brain-derived neurotrophic factor (BDNF), nerve growth factor (NGF), glial cell line-derived neurotrophic factor (GDNF), and transforming growth factor beta (TGF-β), which are essential molecules for neurogenesis [157]. Another important restoration mechanism is that the microenvironment provided by ischemic brain directs HSCs to differentiate to neural linage cells, including neurons and glial cells. As a consequence, the tissue defect caused by necrotic brain cells can be supplemented, which is particularly beneficial for chronic stroke with permanent cavities. In vitro, treatment of adult human HSCs with conditioned culture medium induced neuron morphology with extending processes and express specific markers of neurons, astrocytes, and neural stem cell [185, 186], which implicated the neural linage differentiation potential of HSCs under proper microenvironment. Woei-Cherng Shyu et al. intracerebrally implanted $CD34^+$ HSCs to rat ischemic brain, and then followed by immunohistochemistry which showed that the cells differentiated into neurons ($NeuN^+$, $Nestin^+$, $MAP\text{-}2^+$), glial cells ($GFAP^+$), and vascular endothelial cells (vWF^+), which enhanced - neuroplastic effects after stroke with subsequent significantly improved neurological functions [157]. Another investigation conducted by the same team showed that sufficient HSCs can be mobilized by G-CSF and home to rat ischemic brain, which promoted the recovery of neural functions. The mechanism is considered to be partly due to differentiation of endogenous HSCs into neurons, glial cells, and vascular endothelial cells, which are essential components of novel neurovasculature units [163]. This pre-clinic experiment has great promise for a noninvasive autologous HSC therapy for cerebral ischemic patients. However, it is not well established whether the differentiated HSCs play a critical role in neuro-restoration by integrating to the host neural circuits, or because of increased SDF-1/CXCR4 interaction after G-CSF administration [163] since other pathways by G-CSF [187] may also promote neural restoration.

In conclusion, all current investigations lend support to a notion that HSCs play outstanding regulatory roles in the repair of ischemic injured brains. The cross talk between the therapeutic effects of HSCs and the central nervous system (CNS) further stresses on the possibility of its application for clinical treatment strategies for ischemic stroke or even other neurological diseases. The autologous HSCs mobilized by hematopoietic factors may even have a potential applicability, due to their efficient, feasible, and safe properties.

3.4 Bone Marrow-Derived Endothelial Progenitor Cells

3.4.1 Characteristics of EPCs

Endothelial progenitor cells(EPCs) were first isolated from human peripheral blood in 1997 [188]. EPCs express both endothelial cell markers and stem cell markers [189], and also possess the ability to differentiate into mature endothelial cells which maintain vascular homeostasis and promote vasculogenesis. It is well-known that EPCs can be isolated from three different sources including peripheral blood (PB), bone marrow (BM), and umbilical cord blood (UCB). However, bone marrow-derived progenitor cells are considered as the main source of EPCs, although EPCs remain extremely rare in adult PB [190]. The immunological surface markers used for isolating EPCs include both endothelial cell markers such as CD31, vWF, Tie2, KDR, and VE-cadherin and hematopoietic stem cell marker CD34 or CD133 [191, 192]. However, the identification of EPCs remains controversial, and EPCs that originate from different sources express different cell surface markers. Two methods of functional measurements are used to define the phenotypic feature of EPCs. The first one is performed by counting the number of EPC colonies formed after 7 days of culture. These colonies exhibit many endothelial features including expression of CD31, Tie-2, and VEGFR2 [193]. Another approach is to measure the uptake of Dil-labeled acetylated low-density lipoprotein or binding to specific lectins.

At least two different subsets of EPCs have been identified based on culture characters. Early EPCs appear after short-term culture with spindle-shaped morphology and display peak growth at 2–3 weeks. Late EPCs appear after long-term culture with cobblestone-shaped morphology and display proliferative and tubulogenic potential [194]. Molecular analysis of endothelial progenitor cell (EPC) including genome-wide transcriptional profiling, 2D protein electrophoresis, and electron microscopy showed that early and late EPCs have different gene expression signatures [195]. The same research revealed that early EPCs showed a molecular phenotype linked to monocytes, and late EPCs expressed high level of vascular development and angiogenesis-related signaling pathways (Tie2, eNOS, Ephrins). Several studies also indicated that early EPCs produces large amount of growth factors, which as a result enhances angiogenesis and neovascularization [196, 197], whereas the late EPCs highly expressed KDR and VE-cadherin which contribute to vascular regeneration [198].

3.4.1.1 EPCs and Angiogenesis

Angiogenesis is defined as the formation of new blood vessels due to proliferation and migration of preexisting endothelial cells [199]. The first isolation of endothelial progenitor cells by Asahara T et al. has opened a new era in angiogenesis research [188]. Bone marrow-derived EPCs can home to neovascularization sites and have the capacity to differentiate into endothelial cells [200]. Accumulating evidence have established that EPCs play a key role in the process of angiogenesis. Early EPCs contribute to angiogenesis by producing a variety of growth factors including IGF-1, G-CSF, VEGF, and SDF-1, which can promote endothelial cell proliferation and recruit endogenous progenitor cells [196]. Another research reported that late EPCs also produce several soluble factors and contribute to angiogenesis [201]. Furthermore, late EPCs also have the ability to differentiate into endothelial cells and finally lead to neovascularization [192]. The contribution of EPCs in angiogenesis has also been established in the recovery processes of animal model of ischemic stroke [202, 203].

3.4.2 EPC Mobilization and Homing After Stroke

Mounting evidence suggests that the level of circulating EPCs correlates with the severity of ischemic stroke. A human study by T. Bogoslovsky et al. revealed that high levels of circulating EPCs were indicative of smaller volumes of acute lesion and less lesion growth. They further indicated that the level of circulating EPCs could serve as a marker of acute phase stroke severity [204]. Moreover, the high level of circulating EPC after acute ischemic stroke is associated with good functional outcome [205]. Another research by Joan MF also found that circulating EPC counts peaked at 7 days after acute stroke, and higher counts were related to better outcome at 3 months after stroke [206].

EPCs can be mobilized from bone marrow into the peripheral circulation and are further able to migrate to injured region. The mechanism underlying this process is still not clear. Howbeit, several chemokines and signaling pathways have been reported to play a role in EPC mobilization and homing. Stromal cell-derived factor-1 (SDF-1) and receptor CXCR4 have been established to play a key role in EPC mobilization and homing. The basal level of SDF-1 in bone marrow and in circulation is comparatively low. In ischemic tissue however, the expression of transcription factor hypoxia-inducible factor-1 (HIF-1) is significantly increased due to reduced oxygen tension. Increased HIF-1 expression will in turn activate chemokine SDF-1 [113]. Finally, EPCs are mobilized from bone marrow to peripheral circulation and then to ischemic region following SDF-1 chemotaxis. Cell surface receptor CXCR4 is also reported to regulate mobilization and homing of endothelial progenitors in response to chemokine SDF-1 [207]. In a recent study, researchers administrated human EPCs intravenously to adult nude mice after ischemic stroke and found EPCs protect the brain from ischemic injury and lead to improvement in neurobehavioral outcomes. Their results also suggested that hEPC homing is mediated by stromal cell-derived factor-1 (SDF-1)/CXCR4 signaling [208]. Granulocyte-colony stimulating factor (G-CSF) has also been reported to mobilize EPCs in patients with coronary artery disease [209]. Another study by Jean-Pierre found that hematopoietic progenitor cell mobilization after administration of G-CSF is due to the proteolytic cleavage of VCAM-1 by neutrophil proteases in the bone marrow [210].

3.5 Conclusion

Stem cell-based tissue restoration has become the major impetus for regenerative medicine in neurological disease including stroke. Bone marrow-derived MSCs, HSCs, and EPCs have been implicated as promising sources of adult stem cells for CNS regeneration. As is reported in mounting investigations, BMSCs promote neural function recovery via several important mechanisms, of which the ability of MSCs to give rise to neuronal cells is one [211]. However, this trans-differentiation process was questioned since in vitro contamination of cell culture media [212, 213] might alter the morphology of MSCs which influences therapeutic effect. Other underlying mechanisms explaining BMSC therapeutic effects include the secretion of growth factors or cytokines [53, 173, 174, 214], anti-inflammatory/anti-apoptotic modulations [179, 180], and EPC-induced angiogenesis [192, 202, 203]. More recently, a keen interest in repairing the BBB after stroke via EPC transplantation has been taken [215].

Stroke-induced BMSC mobilization to the peripheral blood and migration to the injured brain region potentially provides a self-repair process for the patients, with therapeutic outcome dependent on the number of stem cells aggregating in the brain. Although the mechanisms of these processes are not completely understood, the interaction of SDF-1/CXCR4 [21, 36, 38, 113, 144, 154], the stimulation of G-CSF [17–20, 119–121], and the regulation by sympathetic nervous system [106, 116, 117] after stroke have explicated part of the picture. Based on this fundamental understanding, reagents that augment mobilization and in experimental stroke modeling and clinical translation in patients.

References

1. Borlongan CV, et al. Potential of stem/progenitor cells in treating stroke: the missing steps in translating cell therapy from laboratory to clinic. Regen Med. 2008;3(3):249–50.

2. Chopp M, et al. Who's in favor of translational cell therapy for stroke: STEPS forward please? Cell Transplant. 2009;18(7):691–3.
3. Stem Cell Therapies as an Emerging Paradigm in Stroke, P. Stem Cell Therapies as an Emerging Paradigm in Stroke (STEPS): bridging basic and clinical science for cellular and neurogenic factor therapy in treating stroke. Stroke. 2009;40(2):510–5.
4. Herzog EL, Chai L, Krause DS. Plasticity of marrow-derived stem cells. Blood. 2003;102(10):3483–93.
5. Munoz-Elias G, Woodbury D, Black IB. Marrow stromal cells, mitosis, and neuronal differentiation: stem cell and precursor functions. Stem Cells. 2003;21(4):437–48.
6. Hara K, et al. Neural progenitor NT2N cell lines from teratocarcinoma for transplantation therapy in stroke. Prog Neurobiol. 2008;85(3):318–34.
7. Hess DC, Borlongan CV. Cell-based therapy in ischemic stroke. Expert Rev Neurother. 2008;8(8):1193–201.
8. Hess DC, Borlongan CV. Stem cells and neurological diseases. Cell Prolif. 2008;41 Suppl 1:94–114.
9. Kalinkovich A, et al. Blood-forming stem cells are nervous: direct and indirect regulation of immature human CD34+ cells by the nervous system. Brain Behav Immun. 2009;23(8):1059–65.
10. Dominici M, et al. Minimal criteria for defining multipotent mesenchymal stromal cells. The International Society for Cellular Therapy position statement. Cytotherapy. 2006;8(4):315–7.
11. Majumdar MK, et al. Isolation, characterization, and chondrogenic potential of human bone marrow-derived multipotential stromal cells. J Cell Physiol. 2000;185(1):98–106.
12. Caplan AI. The mesengenic process. Clin Plast Surg. 1994;21(3):429–35.
13. Prockop DJ. Marrow stromal cells as stem cells for nonhematopoietic tissues. Science. 1997;276(5309):71–4.
14. Qian H, Le Blanc K, Sigvardsson M. Primary mesenchymal stem and progenitor cells from bone marrow lack expression of CD44 protein. J Biol Chem. 2012;287(31):25795–807.
15. Wexler SA, et al. Adult bone marrow is a rich source of human mesenchymal 'stem' cells but umbilical cord and mobilized adult blood are not. Br J Haematol. 2003;121(2):368–74.
16. Hayashi Y, et al. Bannayan-Zonana syndrome associated with lipomas, hemangiomas, and lymphangiomas. J Pediatr Surg. 1992;27(6):722–3.
17. Lazarus HM, et al. Human bone marrow-derived mesenchymal (stromal) progenitor cells (MPCs) cannot be recovered from peripheral blood progenitor cell collections. J Hematother. 1997;6(5):447–55.
18. Pitchford SC, et al. Troubleshooting: quantification of mobilization of progenitor cell subsets from bone marrow in vivo. J Pharmacol Toxicol Methods. 2010;61(2):113–21.
19. Pitchford SC, et al. Differential mobilization of subsets of progenitor cells from the bone marrow. Cell Stem Cell. 2009;4(1):62–72.
20. Pelus LM. Peripheral blood stem cell mobilization: new regimens, new cells, where do we stand. Curr Opin Hematol. 2008;15(4):285–92.
21. Liu L, et al. Hypoxia-inducible factor-1alpha is essential for hypoxia-induced mesenchymal stem cell mobilization into the peripheral blood. Stem Cells Dev. 2011;20(11):1961–71.
22. Rochefort GY, et al. Multipotential mesenchymal stem cells are mobilized into peripheral blood by hypoxia. Stem Cells. 2006;24(10):2202–8.
23. Yu Q, et al. Stromal cell-derived factor-1 alpha alleviates hypoxic-ischemic brain damage in mice. Biochem Biophys Res Commun. 2015;464(2):447–52.
24. Semenza GL. HIF-1: mediator of physiological and pathophysiological responses to hypoxia. J Appl Physiol (1985). 2000;88(4):1474–80.
25. Semenza GL. HIF-1, O(2), and the 3 PHDs: how animal cells signal hypoxia to the nucleus. Cell. 2001;107(1):1–3.
26. Ivan M, et al. HIFalpha targeted for VHL-mediated destruction by proline hydroxylation: implications for O2 sensing. Science. 2001;292(5516):464–8.

27. Dafni H, et al. MRI and fluorescence microscopy of the acute vascular response to VEGF165: vasodilation, hyper-permeability and lymphatic uptake, followed by rapid inactivation of the growth factor. NMR Biomed. 2002;15(2):120–31.
28. Senger DR, et al. Vascular permeability factor (VPF, VEGF) in tumor biology. Cancer Metastasis Rev. 1993;12(3–4):303–24.
29. Kopp HG, et al. Tie2 activation contributes to hemangiogenic regeneration after myelosuppression. Blood. 2005;106(2):505–13.
30. Hattori K, et al. Vascular endothelial growth factor and angiopoietin-1 stimulate postnatal hematopoiesis by recruitment of vasculogenic and hematopoietic stem cells. J Exp Med. 2001;193(9):1005–14.
31. Schioppa T, et al. Regulation of the chemokine receptor CXCR4 by hypoxia. J Exp Med. 2003;198(9):1391–402.
32. Liu X, et al. SDF-1/CXCR4 axis modulates bone marrow mesenchymal stem cell apoptosis, migration and cytokine secretion. Protein Cell. 2011;2(10):845–54.
33. Stumm RK, et al. A dual role for the SDF-1/CXCR4 chemokine receptor system in adult brain: isoform-selective regulation of SDF-1 expression modulates CXCR4-dependent neuronal plasticity and cerebral leukocyte recruitment after focal ischemia. J Neurosci. 2002;22(14):5865–78.
34. Mocco J, et al. SDF1-a facilitates Lin-/Sca1+ cell homing following murine experimental cerebral ischemia. PLoS One. 2014;9(1):e85615.
35. Hill WD, et al. SDF-1 (CXCL12) is upregulated in the ischemic penumbra following stroke: association with bone marrow cell homing to injury. J Neuropathol Exp Neurol. 2004;63(1):84–96.
36. Liu P, Xiang JW, Jin SX. Serum CXCL12 levels are associated with stroke severity and lesion volumes in stroke patients. Neurol Res. 2015;37(10):853–8.
37. Chang YC, et al. Regenerative therapy for stroke. Cell Transplant. 2007;16(2):171–81.
38. Shen LH, et al. Therapeutic benefit of bone marrow stromal cells administered 1 month after stroke. J Cereb Blood Flow Metab. 2007;27(1):6–13.
39. Wang Y, et al. Roles of chemokine CXCL12 and its receptors in ischemic stroke. Curr Drug Targets. 2012;13(2):166–72.
40. Wang Y, Deng Y, Zhou GQ. SDF-1alpha/CXCR4-mediated migration of systemically transplanted bone marrow stromal cells towards ischemic brain lesion in a rat model. Brain Res. 2008;1195:104–12.
41. Wang L, et al. Ischemic cerebral tissue and MCP-1 enhance rat bone marrow stromal cell migration in interface culture. Exp Hematol. 2002;30(7):831–6.
42. Ruster B, et al. Mesenchymal stem cells display coordinated rolling and adhesion behavior on endothelial cells. Blood. 2006;108(12):3938–44.
43. Sackstein R, et al. Ex vivo glycan engineering of CD44 programs human multipotent mesenchymal stromal cell trafficking to bone. Nat Med. 2008;14(2):181–7.
44. Yilmaz G, et al. Selectin-mediated recruitment of bone marrow stromal cells in the postischemic cerebral microvasculature. Stroke. 2011;42(3):806–11.
45. Guzman R, et al. Intravascular cell replacement therapy for stroke. Neurosurg Focus. 2008;24(3–4):E15.
46. Li Y, et al. Human marrow stromal cell therapy for stroke in rat: neurotrophins and functional recovery. Neurology. 2002;59(4):514–23.
47. Chen J, et al. Intravenous bone marrow stromal cell therapy reduces apoptosis and promotes endogenous cell proliferation after stroke in female rat. J Neurosci Res. 2003;73(6):778–86.
48. Zacharek A, et al. Angiopoietin1/Tie2 and VEGF/Flk1 induced by MSC treatment amplifies angiogenesis and vascular stabilization after stroke. J Cereb Blood Flow Metab. 2007;27(10):1684–91.
49. Chen X, et al. Human bone marrow stromal cell cultures conditioned by traumatic brain tissue extracts: growth factor production. J Neurosci Res. 2002;69(5):687–91.
50. Chen X, et al. Ischemic rat brain extracts induce human marrow stromal cell growth factor production. Neuropathology. 2002;22(4):275–9.

51. Kobayashi T, et al. Intracerebral infusion of glial cell line-derived neurotrophic factor promotes striatal neurogenesis after stroke in adult rats. Stroke. 2006;37(9):2361–7.
52. Matsuda-Hashii Y, et al. Hepatocyte growth factor plays roles in the induction and autocrine maintenance of bone marrow stromal cell IL-11, SDF-1 alpha, and stem cell factor. Exp Hematol. 2004;32(10):955–61.
53. Wakabayashi K, et al. Transplantation of human mesenchymal stem cells promotes functional improvement and increased expression of neurotrophic factors in a rat focal cerebral ischemia model. J Neurosci Res. 2010;88(5):1017–25.
54. Chen TS, et al. Mesenchymal stem cell secretes microparticles enriched in pre-microRNAs. Nucleic Acids Res. 2010;38(1):215–24.
55. Lim PK, et al. Neurogenesis: role for microRNAs and mesenchymal stem cells in pathological states. Curr Med Chem. 2010;17(20):2159–67.
56. Xin H, et al. Exosome-mediated transfer of miR-133b from multipotent mesenchymal stromal cells to neural cells contributes to neurite outgrowth. Stem Cells. 2012;30(7):1556–64.
57. Xin H, et al. MiR-133b promotes neural plasticity and functional recovery after treatment of stroke with multipotent mesenchymal stromal cells in rats via transfer of exosome-enriched extracellular particles. Stem Cells. 2013;31(12):2737–46.
58. Spangrude GJ, Heimfeld S, Weissman IL. Purification and characterization of mouse hematopoietic stem cells. Science. 1988;241(4861):58–62.
59. Holtick U, et al. Comparison of bone marrow versus peripheral blood allogeneic hematopoietic stem cell transplantation for hematological malignancies in adults – a systematic review and meta-analysis. Crit Rev Oncol Hematol. 2015;94(2):179–88.
60. Deotare U, et al. G-CSF-primed bone marrow as a source of stem cells for allografting: revisiting the concept. Bone Marrow Transplant. 2015;50(9):1150–6.
61. Pineault N, Abu-Khader A. Advances in umbilical cord blood stem cell expansion and clinical translation. Exp Hematol. 2015;43(7):498–513.
62. Bhatia M, et al. A newly discovered class of human hematopoietic cells with SCID-repopulating activity. Nat Med. 1998;4(9):1038–45.
63. Guo Y, Lubbert M, Engelhardt M. CD34- hematopoietic stem cells: current concepts and controversies. Stem Cells. 2003;21(1):15–20.
64. Wognum AW, Eaves AC, Thomas TE. Identification and isolation of hematopoietic stem cells. Arch Med Res. 2003;34(6):461–75.
65. Calvi LM, et al. Osteoblastic cells regulate the haematopoietic stem cell niche. Nature. 2003;425(6960):841–6.
66. Zhang J, et al. Identification of the haematopoietic stem cell niche and control of the niche size. Nature. 2003;425(6960):836–41.
67. Kiel MJ, et al. SLAM family receptors distinguish hematopoietic stem and progenitor cells and reveal endothelial niches for stem cells. Cell. 2005;121(7):1109–21.
68. Jin DK, et al. Cytokine-mediated deployment of SDF-1 induces revascularization through recruitment of CXCR4+ hemangiocytes. Nat Med. 2006;12(5):557–67.
69. Abkowitz JL, et al. Mobilization of hematopoietic stem cells during homeostasis and after cytokine exposure. Blood. 2003;102(4):1249–53.
70. Dar A, et al. Chemokine receptor CXCR4-dependent internalization and resecretion of functional chemokine SDF-1 by bone marrow endothelial and stromal cells. Nat Immunol. 2005;6(10):1038–46.
71. Heissig B, et al. A role for niches in hematopoietic cell development. Hematology. 2005;10(3):247–53.
72. Yin T, Li L. The stem cell niches in bone. J Clin Invest. 2006;116(5):1195–201.
73. Kiel MJ, et al. Hematopoietic stem cells do not depend on N-cadherin to regulate their maintenance. Cell Stem Cell. 2009;4(2):170–9.
74. Kiel MJ, Radice GL, Morrison SJ. Lack of evidence that hematopoietic stem cells depend on N-cadherin-mediated adhesion to osteoblasts for their maintenance. Cell Stem Cell. 2007;1(2):204–17.

75. Wilson A, Trumpp A. Bone-marrow haematopoietic-stem-cell niches. Nat Rev Immunol. 2006;6(2):93–106.
76. Nilsson SK, et al. Osteopontin, a key component of the hematopoietic stem cell niche and regulator of primitive hematopoietic progenitor cells. Blood. 2005;106(4):1232–9.
77. Stier S, et al. Osteopontin is a hematopoietic stem cell niche component that negatively regulates stem cell pool size. J Exp Med. 2005;201(11):1781–91.
78. Arai F, et al. Tie2/angiopoietin-1 signaling regulates hematopoietic stem cell quiescence in the bone marrow niche. Cell. 2004;118(2):149–61.
79. Papayannopoulou T, et al. The VLA4/VCAM-1 adhesion pathway defines contrasting mechanisms of lodgement of transplanted murine hemopoietic progenitors between bone marrow and spleen. Proc Natl Acad Sci U S A. 1995;92(21):9647–51.
80. Avecilla ST, et al. Chemokine-mediated interaction of hematopoietic progenitors with the bone marrow vascular niche is required for thrombopoiesis. Nat Med. 2004;10(1):64–71.
81. Sugiyama T, et al. Maintenance of the hematopoietic stem cell pool by CXCL12-CXCR4 chemokine signaling in bone marrow stromal cell niches. Immunity. 2006;25(6):977–88.
82. Fortunel NO, et al. Control of hematopoietic stem/progenitor cell fate by transforming growth factor-beta. Oncol Res. 2003;13(6–10):445–53.
83. Nie Y, Han YC, Zou YR. CXCR4 is required for the quiescence of primitive hematopoietic cells. J Exp Med. 2008;205(4):777–83.
84. van Velthoven CT, et al. Regeneration of the ischemic brain by engineered stem cells: fuelling endogenous repair processes. Brain Res Rev. 2009;61(1):1–13.
85. Ruscetti FW, Akel S, Bartelmez SH. Autocrine transforming growth factor-beta regulation of hematopoiesis: many outcomes that depend on the context. Oncogene. 2005;24(37):5751–63.
86. Yamazaki S, et al. TGF-beta as a candidate bone marrow niche signal to induce hematopoietic stem cell hibernation. Blood. 2009;113(6):1250–6.
87. Bernasconi P, et al. Therapeutically targeting self-reinforcing leukemic niches in acute myeloid leukemia (AML): a worthy endeavour? Am J Hematol. 2016.
88. Cogle CR, et al. Bone marrow niche in the myelodysplastic syndromes. Leuk Res. 2015;39(10):1020–7.
89. Abe-Suzuki S, et al. CXCL12+ stromal cells as bone marrow niche for CD34+ hematopoietic cells and their association with disease progression in myelodysplastic syndromes. Lab Invest. 2014;94(11):1212–23.
90. Lim M, et al. Altered mesenchymal niche cells impede generation of normal hematopoietic progenitor cells in leukemic bone marrow. Leukemia. 2016;30(1):154–62.
91. Schepers K, Campbell TB, Passegue E. Normal and leukemic stem cell niches: insights and therapeutic opportunities. Cell Stem Cell. 2015;16(3):254–67.
92. Togel FE, Westenfelder C. Role of SDF-1 as a regulatory chemokine in renal regeneration after acute kidney injury. Kidney Int Suppl. 2011;1(3):87–9.
93. Cheng M, et al. A critical role of Src family kinase in SDF-1/CXCR4-mediated bone-marrow progenitor cell recruitment to the ischemic heart. J Mol Cell Cardiol. 2015;81:49–53.
94. Macarthur Jr JW, et al. Preclinical evaluation of the engineered stem cell chemokine stromal cell-derived factor 1alpha analog in a translational ovine myocardial infarction model. Circ Res. 2014;114(4):650–9.
95. Wu CH, et al. Stem cell mobilizers targeting chemokine receptor CXCR4: renoprotective application in acute kidney injury. J Med Chem. 2015;58(5):2315–25.
96. Lapidot T, Dar A, Kollet O. How do stem cells find their way home? Blood. 2005;106(6):1901–10.
97. Nervi B, Link DC, DiPersio JF. Cytokines and hematopoietic stem cell mobilization. J Cell Biochem. 2006;99(3):690–705.
98. Papayannopoulou T, Scadden DT. Stem-cell ecology and stem cells in motion. Blood. 2008;111(8):3923–30.
99. Christopher MJ, et al. Suppression of CXCL12 production by bone marrow osteoblasts is a common and critical pathway for cytokine-induced mobilization. Blood. 2009;114(7):1331–9.

100. Wojakowski W, Tendera M. Mobilization of bone marrow-derived progenitor cells in acute coronary syndromes. Folia Histochem Cytobiol. 2005;43(4):229–32.
101. Nagareddy PR, et al. A novel role for bioactive lipids in stem cell mobilization during cardiac ischemia: new paradigms in thrombosis: novel mediators and biomarkers. J Thromb Thrombolysis. 2014;37(1):24–31.
102. Wojakowski W, et al. Mobilization of stem and progenitor cells in cardiovascular diseases. Leukemia. 2012;26(1):23–33.
103. Paczkowska E, et al. Human hematopoietic stem/progenitor-enriched CD34(+) cells are mobilized into peripheral blood during stress related to ischemic stroke or acute myocardial infarction. Eur J Haematol. 2005;75(6):461–7.
104. Hennemann B, et al. Mobilization of CD34+ hematopoietic cells, colony-forming cells and long-term culture-initiating cells into the peripheral blood of patients with an acute cerebral ischemic insult. Cytotherapy. 2008;10(3):303–11.
105. Courties G, et al. Ischemic stroke activates hematopoietic bone marrow stem cells. Circ Res. 2015;116(3):407–17.
106. Bonig H, Papayannopoulou T. Hematopoietic stem cell mobilization: updated conceptual renditions. Leukemia. 2013;27(1):24–31.
107. Levesque JP, et al. Disruption of the CXCR4/CXCL12 chemotactic interaction during hematopoietic stem cell mobilization induced by GCSF or cyclophosphamide. J Clin Invest. 2003;111(2):187–96.
108. Petit I, et al. G-CSF induces stem cell mobilization by decreasing bone marrow SDF-1 and up-regulating CXCR4. Nat Immunol. 2002;3(7):687–94.
109. Semerad CL, et al. G-CSF potently inhibits osteoblast activity and CXCL12 mRNA expression in the bone marrow. Blood. 2005;106(9):3020–7.
110. Lanfranconi S, et al. Growth factors in ischemic stroke. J Cell Mol Med. 2011;15(8):1645–87.
111. Dar A, Kollet O, Lapidot T. Mutual, reciprocal SDF-1/CXCR4 interactions between hematopoietic and bone marrow stromal cells regulate human stem cell migration and development in NOD/SCID chimeric mice. Exp Hematol. 2006;34(8):967–75.
112. Dar A, et al. Rapid mobilization of hematopoietic progenitors by AMD3100 and catecholamines is mediated by CXCR4-dependent SDF-1 release from bone marrow stromal cells. Leukemia. 2011;25(8):1286–96.
113. Ceradini DJ, et al. Progenitor cell trafficking is regulated by hypoxic gradients through HIF-1 induction of SDF-1. Nat Med. 2004;10(8):858–64.
114. Lai TW, et al. HIF-1alpha binding to the Epac1 promoter recruits hematopoietic stem cells to the ischemic brain following stroke. J Mol Cell Biol. 2012;4(3):184–7.
115. Spiegel A, et al. Stem cell regulation via dynamic interactions of the nervous and immune systems with the microenvironment. Cell Stem Cell. 2008;3(5):484–92.
116. Winklewski PJ, Radkowski M, Demkow U. Cross-talk between the inflammatory response, sympathetic activation and pulmonary infection in the ischemic stroke. J Neuroinflammation. 2014;11:213.
117. Katayama Y, et al. Signals from the sympathetic nervous system regulate hematopoietic stem cell egress from bone marrow. Cell. 2006;124(2):407–21.
118. Welte K, et al. Purification and biochemical characterization of human pluripotent hematopoietic colony-stimulating factor. Proc Natl Acad Sci U S A. 1985;82(5):1526–30.
119. Sprigg N, et al. Granulocyte-colony-stimulating factor mobilizes bone marrow stem cells in patients with subacute ischemic stroke: the Stem cell Trial of recovery EnhanceMent after Stroke (STEMS) pilot randomized, controlled trial (ISRCTN 16784092). Stroke. 2006;37(12):2979–83.
120. Hochsmann B, et al. Hemispheric stroke does not mobilize cd34+ hematopoietic stem cells into the peripheral blood. Neurology. 2009;72(14):1277–8.
121. Boy S, et al. Mobilisation of hematopoietic CD34+ precursor cells in patients with acute stroke is safe – results of an open-labeled non randomized phase I/II trial. PLoS One. 2011;6(8):e23099.

122. Papayannopoulou T. Current mechanistic scenarios in hematopoietic stem/progenitor cell mobilization. Blood. 2004;103(5):1580–5.
123. Levesque JP, et al. Mobilization of bone marrow-derived progenitors. Handb Exp Pharmacol. 2007;(180):3–36.
124. Larsson J, Scadden D. Nervous activity in a stem cell niche. Cell. 2006;124(2):253–5.
125. Christopherson 2nd KW, et al. G-CSF- and GM-CSF-induced upregulation of CD26 peptidase downregulates the functional chemotactic response of CD34+CD38- human cord blood hematopoietic cells. Exp Hematol. 2006;34(8):1060–8.
126. Campbell TB, et al. Inhibition of CD26 in human cord blood CD34+ cells enhances their engraftment of nonobese diabetic/severe combined immunodeficiency mice. Stem Cells Dev. 2007;16(3):347–54.
127. Christopherson 2nd KW, et al. Modulation of hematopoietic stem cell homing and engraftment by CD26. Science. 2004;305(5686):1000–3.
128. Rankin SM. Chemokines and adult bone marrow stem cells. Immunol Lett. 2012;145(1–2):47–54.
129. Lapidot T, Kollet O. The brain-bone-blood triad: traffic lights for stem-cell homing and mobilization. Hematology Am Soc Hematol Educ Program. 2010;2010:1–6.
130. Spiegel A, et al. Catecholaminergic neurotransmitters regulate migration and repopulation of immature human CD34+ cells through Wnt signaling. Nat Immunol. 2007;8(10):1123–31.
131. England TJ, et al. Granulocyte-colony stimulating factor for mobilizing bone marrow stem cells in subacute stroke: the stem cell trial of recovery enhancement after stroke 2 randomized controlled trial. Stroke. 2012;43(2):405–11.
132. Nomani F, Kamal AK. Granulocyte-colony stimulating factor for mobilizing bone marrow stem cells in the sub acute stroke. How safe is the use of granulocyte-colony stimulating factor in sub-acute stroke? Is this stem cell trial of recovery enhancement beneficial? J Pak Med Assoc. 2013;63(12):1558–9.
133. Alvarez P, et al. Regulatory systems in bone marrow for hematopoietic stem/progenitor cells mobilization and homing. Biomed Res Int. 2013;2013:312656.
134. Zhao LR, et al. The role of stem cell factor and granulocyte-colony stimulating factor in treatment of stroke. Recent Pat CNS Drug Discov. 2013;8(1):2–12.
135. Duarte RF, Franf DA. The synergy between stem cell factor (SCF) and granulocyte colony-stimulating factor (G-CSF): molecular basis and clinical relevance. Leuk Lymphoma. 2002;43(6):1179–87.
136. Matthys P, et al. AMD3100, a potent and specific antagonist of the stromal cell-derived factor-1 chemokine receptor CXCR4, inhibits autoimmune joint inflammation in IFN-gamma receptor-deficient mice. J Immunol. 2001;167(8):4686–92.
137. Lee HM, et al. Mobilization studies in complement-deficient mice reveal that optimal AMD3100 mobilization of hematopoietic stem cells depends on complement cascade activation by AMD3100-stimulated granulocytes. Leukemia. 2010;24(3):573–82.
138. Fukuda S, et al. The chemokine GRObeta mobilizes early hematopoietic stem cells characterized by enhanced homing and engraftment. Blood. 2007;110(3):860–9.
139. Pelus LM, Fukuda S. Peripheral blood stem cell mobilization: the CXCR2 ligand GRObeta rapidly mobilizes hematopoietic stem cells with enhanced engraftment properties. Exp Hematol. 2006;34(8):1010–20.
140. Bodart V, et al. Pharmacology of AMD3465: a small molecule antagonist of the chemokine receptor CXCR4. Biochem Pharmacol. 2009;78(8):993–1000.
141. Abraham M, et al. Enhanced unique pattern of hematopoietic cell mobilization induced by the CXCR4 antagonist 4F-benzoyl-TN14003. Stem Cells. 2007;25(9):2158–66.
142. DeMarco SJ, et al. Discovery of novel, highly potent and selective beta-hairpin mimetic CXCR4 inhibitors with excellent anti-HIV activity and pharmacokinetic profiles. Bioorg Med Chem. 2006;14(24):8396–404.
143. Ramirez P, et al. BIO5192, a small molecule inhibitor of VLA-4, mobilizes hematopoietic stem and progenitor cells. Blood. 2009;114(7):1340–3.

144. Borlongan CV, et al. The great migration of bone marrow-derived stem cells toward the ischemic brain: therapeutic implications for stroke and other neurological disorders. Prog Neurobiol. 2011;95(2):213–28.
145. Bleul CC, et al. The lymphocyte chemoattractant SDF-1 is a ligand for LESTR/fusin and blocks HIV-1 entry. Nature. 1996;382(6594):829–33.
146. Oh IY, et al. Involvement of E-selectin in recruitment of endothelial progenitor cells and angiogenesis in ischemic muscle. Blood. 2007;110(12):3891–9.
147. Tauxe C, et al. P-selectin glycoprotein ligand-1 decameric repeats regulate selectin-dependent rolling under flow conditions. J Biol Chem. 2008;283(42):28536–45.
148. Turner ML, et al. Differential expression of cell adhesion molecules by human hematopoietic progenitor cells from bone marrow and mobilized adult peripheral blood. Stem Cells. 1995;13(3):311–6.
149. Peled A, et al. The chemokine SDF-1 activates the integrins LFA-1, VLA-4, and VLA-5 on immature human CD34(+) cells: role in transendothelial/stromal migration and engraftment of NOD/SCID mice. Blood. 2000;95(11):3289–96.
150. Heo JH, et al. Increase in plasma matrix metalloproteinase-9 in acute stroke patients with thrombolysis failure. Stroke. 2003;34(6):e48–50.
151. Castellanos M, et al. Plasma metalloproteinase-9 concentration predicts hemorrhagic transformation in acute ischemic stroke. Stroke. 2003;34(1):40–6.
152. Horstmann S, et al. Profiles of matrix metalloproteinases, their inhibitors, and laminin in stroke patients: influence of different therapies. Stroke. 2003;34(9):2165–70.
153. Montaner J, et al. Safety profile of tissue plasminogen activator treatment among stroke patients carrying a common polymorphism (C-1562T) in the promoter region of the matrix metalloproteinase-9 gene. Stroke. 2003;34(12):2851–5.
154. McQuibban GA, et al. Matrix metalloproteinase activity inactivates the CXC chemokine stromal cell-derived factor-1. J Biol Chem. 2001;276(47):43503–8.
155. Shyu WC, et al. Stromal cell-derived factor-1 alpha promotes neuroprotection, angiogenesis, and mobilization/homing of bone marrow-derived cells in stroke rats. J Pharmacol Exp Ther. 2008;324(2):834–49.
156. Segers VF, et al. Local delivery of protease-resistant stromal cell derived factor-1 for stem cell recruitment after myocardial infarction. Circulation. 2007;116(15):1683–92.
157. Shyu WC, et al. Intracerebral peripheral blood stem cell (CD34+) implantation induces neuroplasticity by enhancing beta1 integrin-mediated angiogenesis in chronic stroke rats. J Neurosci. 2006;26(13):3444–53.
158. Blazar BR, Murphy WJ, Abedi M. Advances in graft-versus-host disease biology and therapy. Nat Rev Immunol. 2012;12(6):443–58.
159. Ferrara JL, et al. Graft-versus-host disease. Lancet. 2009;373(9674):1550–61.
160. Hsiao HH, et al. Acute cerebral infarct with elevated factor VIII level during the thrombocytopenic stage after hematopoietic stem cell transplant. Exp Clin Transplant. 2014;12(2):171–2.
161. Moshayedi P, Carmichael ST. *Hyaluronan, neural stem cells and tissue reconstruction after acute ischemic stroke.* Biomatter. 2013;3(1).
162. Macrez R, et al. Stroke and the immune system: from pathophysiology to new therapeutic strategies. Lancet Neurol. 2011;10(5):471–80.
163. Shyu WC, et al. Functional recovery of stroke rats induced by granulocyte colony-stimulating factor-stimulated stem cells. Circulation. 2004;110(13):1847–54.
164. Schwarting S, et al. Hematopoietic stem cells reduce postischemic inflammation and ameliorate ischemic brain injury. Stroke. 2008;39(10):2867–75.
165. Fisher M. New approaches to neuroprotective drug development. Stroke. 2011;42(1 Suppl):S24–7.
166. Kim S, et al. Angiotensin II regulation of proliferation, differentiation, and engraftment of hematopoietic stem cells. Hypertension. 2016;67:574.
167. Lui WC, et al. Cytokine combinations on the potential for ex vivo expansion of murine hematopoietic stem cells. Cytokine. 2014;68(2):127–32.

168. Bendall LJ, Bradstock KF. G-CSF: from granulopoietic stimulant to bone marrow stem cell mobilizing agent. Cytokine Growth Factor Rev. 2014;25(4):355–67.
169. Metcalf D. The granulocyte-macrophage colony stimulating factors. Cell. 1985;43(1):5–6.
170. Bijou F, et al. Hematopoietic stem cells mobilization: state of the art in 2011 and perspectives. Transfus Clin Biol. 2011;18(5–6):503–15.
171. Metcalf D. The molecular control of cell division, differentiation commitment and maturation in haemopoietic cells. Nature. 1989;339(6219):27–30.
172. Fukuda S, et al. Survivin modulates genes with divergent molecular functions and regulates proliferation of hematopoietic stem cells through Evi-1. Leukemia. 2015;29(2):433–40.
173. Majka M, et al. Numerous growth factors, cytokines, and chemokines are secreted by human CD34(+) cells, myeloblasts, erythroblasts, and megakaryoblasts and regulate normal hematopoiesis in an autocrine/paracrine manner. Blood. 2001;97(10):3075–85.
174. Chmielnicki E, et al. Adenovirally expressed noggin and brain-derived neurotrophic factor cooperate to induce new medium spiny neurons from resident progenitor cells in the adult striatal ventricular zone. J Neurosci. 2004;24(9):2133–42.
175. Gustafsson E, et al. Anterograde delivery of brain-derived neurotrophic factor to striatum via nigral transduction of recombinant adeno-associated virus increases neuronal death but promotes neurogenic response following stroke. Eur J Neurosci. 2003;17(12):2667–78.
176. Schanzer A, et al. Direct stimulation of adult neural stem cells in vitro and neurogenesis in vivo by vascular endothelial growth factor. Brain Pathol. 2004;14(3):237–48.
177. Kuhn HG, et al. Epidermal growth factor and fibroblast growth factor-2 have different effects on neural progenitors in the adult rat brain. J Neurosci. 1997;17(15):5820–9.
178. Jin K, et al. Vascular endothelial growth factor (VEGF) stimulates neurogenesis in vitro and in vivo. Proc Natl Acad Sci U S A. 2002;99(18):11946–50.
179. Sun J, et al. Clostridium butyricum pretreatment attenuates cerebral ischemia/reperfusion injury in mice via anti-oxidation and anti-apoptosis. Neurosci Lett. 2016;613:30–5.
180. Gu Y, Chen J, Shen J. Herbal medicines for ischemic stroke: combating inflammation as therapeutic targets. J Neuroimmune Pharmacol. 2014;9(3):313–39.
181. Tobin MK, et al. Neurogenesis and inflammation after ischemic stroke: what is known and where we go from here. J Cereb Blood Flow Metab. 2014;34(10):1573–84.
182. Bu Q, et al. w007B protects brain against ischemia-reperfusion injury in rats through inhibiting inflammation, apoptosis and autophagy. Brain Res. 2014;1558:100–8.
183. Taguchi A, et al. Administration of CD34+ cells after stroke enhances neurogenesis via angiogenesis in a mouse model. J Clin Invest. 2004;114(3):330–8.
184. Tsuji M, et al. Effects of intravenous administration of umbilical cord blood CD34(+) cells in a mouse model of neonatal stroke. Neuroscience. 2014;263:148–58.
185. Reali C, et al. Differentiation of human adult CD34+ stem cells into cells with a neural phenotype: role of astrocytes. Exp Neurol. 2006;197(2):399–406.
186. Hao HN, et al. Fetal human hematopoietic stem cells can differentiate sequentially into neural stem cells and then astrocytes in vitro. J Hematother Stem Cell Res. 2003;12(1):23–32.
187. Lu CZ, Xiao BG. Neuroprotection of G-CSF in cerebral ischemia. Front Biosci. 2007;12:2869–75.
188. Asahara T, et al. Isolation of putative progenitor endothelial cells for angiogenesis. Science. 1997;275(5302):964–7.
189. Quirici N, et al. Differentiation and expansion of endothelial cells from human bone marrow CD133(+) cells. Br J Haematol. 2001;115(1):186–94.
190. Gehling UM, et al. In vitro differentiation of endothelial cells from AC133-positive progenitor cells. Blood. 2000;95(10):3106–12.
191. Miraglia S, et al. A novel five-transmembrane hematopoietic stem cell antigen: isolation, characterization, and molecular cloning. Blood. 1997;90(12):5013–21.
192. Fadini GP, Losordo D, Dimmeler S. Critical reevaluation of endothelial progenitor cell phenotypes for therapeutic and diagnostic use. Circ Res. 2012;110(4):624–37.
193. Hill JM, et al. Circulating endothelial progenitor cells, vascular function, and cardiovascular risk. N Engl J Med. 2003;348(7):593–600.

194. Hur J, et al. Characterization of two types of endothelial progenitor cells and their different contributions to neovasculogenesis. Arterioscler Thromb Vasc Biol. 2004;24(2):288–93.
195. Medina RJ, et al. Molecular analysis of endothelial progenitor cell (EPC) subtypes reveals two distinct cell populations with different identities. BMC Med Genomics. 2010;3:18.
196. Urbich C, et al. Soluble factors released by endothelial progenitor cells promote migration of endothelial cells and cardiac resident progenitor cells. J Mol Cell Cardiol. 2005;39(5):733–42.
197. Rehman J, et al. Peripheral blood "endothelial progenitor cells" are derived from monocyte/macrophages and secrete angiogenic growth factors. Circulation. 2003;107(8):1164–9.
198. Yoder MC, et al. Redefining endothelial progenitor cells via clonal analysis and hematopoietic stem/progenitor cell principals. Blood. 2007;109(5):1801–9.
199. Carmeliet P. Angiogenesis in life, disease and medicine. Nature. 2005;438(7070):932–6.
200. Shi Q, et al. Evidence for circulating bone marrow-derived endothelial cells. Blood. 1998;92(2):362–7.
201. Moubarik C, et al. Transplanted late outgrowth endothelial progenitor cells as cell therapy product for stroke. Stem Cell Rev. 2011;7(1):208–20.
202. Zhang ZG, et al. Bone marrow-derived endothelial progenitor cells participate in cerebral neovascularization after focal cerebral ischemia in the adult mouse. Circ Res. 2002;90(3):284–8.
203. Ohta T, et al. Administration of ex vivo-expanded bone marrow-derived endothelial progenitor cells attenuates focal cerebral ischemia-reperfusion injury in rats. Neurosurgery. 2006;59(3):679–86. discussion 679–86.
204. Bogoslovsky T, et al. Endothelial progenitor cells correlate with lesion volume and growth in acute stroke. Neurology. 2010;75(23):2059–62.
205. Sobrino T, et al. The increase of circulating endothelial progenitor cells after acute ischemic stroke is associated with good outcome. Stroke. 2007;38(10):2759–64.
206. Marti-Fabregas J, et al. Endothelial progenitor cells in acute ischemic stroke. Brain Behav. 2013;3(6):649–55.
207. Wu Q, et al. Extracellular calcium increases CXCR4 expression on bone marrow-derived cells and enhances pro-angiogenesis therapy. J Cell Mol Med. 2009;13(9B):3764–73.
208. Fan Y, et al. Endothelial progenitor cell transplantation improves long-term stroke outcome in mice. Ann Neurol. 2010;67(4):488–97.
209. Powell TM, et al. Granulocyte colony-stimulating factor mobilizes functional endothelial progenitor cells in patients with coronary artery disease. Arterioscler Thromb Vasc Biol. 2005;25(2):296–301.
210. Levesque JP, et al. Vascular cell adhesion molecule-1 (CD106) is cleaved by neutrophil proteases in the bone marrow following hematopoietic progenitor cell mobilization by granulocyte colony-stimulating factor. Blood. 2001;98(5):1289–97.
211. Jeong JA, et al. Rapid neural differentiation of human cord blood-derived mesenchymal stem cells. Neuroreport. 2004;15(11):1731–4.
212. Krabbe C, Zimmer J, Meyer M. Neural transdifferentiation of mesenchymal stem cells--a critical review. APMIS. 2005;113(11–12):831–44.
213. Lu P, Blesch A, Tuszynski MH. Induction of bone marrow stromal cells to neurons: differentiation, transdifferentiation, or artifact? J Neurosci Res. 2004;77(2):174–91.
214. Sun Y, et al. Vascular endothelial growth factor-B (VEGFB) stimulates neurogenesis: evidence from knockout mice and growth factor administration. Dev Biol. 2006;289(2):329–35.
215. Borlongan CV. Bone marrow stem cell mobilization in stroke: a 'bonehead' may be good after all! Leukemia. 2011;25(11):1674–86.

Chapter 4
Interaction of Bone Marrow Stem Cells with Other Cells

Hongxia Zhang, Shengwei Huang, Jiangnan Hu, Brian Wang, and Kunlin Jin

Abstract Transplantation of bone marrow stem cells (BMSCs) as a regenerative cell replacement therapy for stroke holds great promise. However, the mechanisms underlying functional recovery after stroke remain unclear. Clearly, BMSCs benefit the neurological dysfunction in either direct or indirect methods through the interaction with other cells. In this chapter, we review the direct and indirect interactions of transplanted BMSCs with HSCs, immune cells, neural stem cells (NSCs), neurons, astrocytes, and endothelial cells for better understanding of the mechanisms underlying BMSC-mediated neurological function improvement after stroke.

Keywords Transplantation • Bone marrow stem cells • Immune cells • Interaction

4.1 Introduction

Mesenchymal stem cells and hematopoietic stem cells (HSCs) are two major stem cells coexisting in the mammal bone marrow [13]. Bone marrow stem cells (BMSCs) are generally referred to as bone marrow-derived mesenchymal stem cells. Traditionally, these stem cells are harvested based on Friedenstein's procedure [1–4], which are so-called bone marrow stromal cells, but do not strictly meet the criteria for stem cells as there is a lack of convincing evidence to support the "stemness" of these cells [5] even though they have the propensity to differentiate into mesodermal lineage cells such as osteoblasts, chondrocytes, adipocytes, and muscle cells [135]. For example, cellular heterogeneity exists in the population of these isolated cells, and only a few of them contain colony-forming unit-fibroblastic (CFU-F) activity

H. Zhang • S. Huang • J. Hu • B. Wang
Center for Neuroscience Discovery, Institute for Healthy Aging, University of North Texas Health Science Center, Fort Worth, TX 76107, USA

K. Jin (✉)
Department of Pharmacology Neuroscience, University of North Texas Health Science Center, Fort Worth, TX, USA
e-mail: kunlin.jin@unthsc.edu

K. Jin et al. (eds.), *Bone Marrow Stem Cell Therapy for Stroke*,
DOI 10.1007/978-981-10-2929-5_4

[5]. Therefore, we use the term "bone marrow-derived multipotent mesenchymal stromal cells (BM-MSCs)" recommended by the International Society for Cellular Therapy as an alternative definition in this chapter, while the term "mesenchymal stem cells" will strictly be reserved for cells matching the stated criteria [5].

BM-MSCs are capable of promoting the regeneration of tissues such as bone, cartilage, fat, and muscle [4, 6, 7] and can also differentiate into neural cells such as neuron and astrocytes under both in vitro and in vivo conditions [8–12]. In recent years, the important roles of BM-MSCs have been demonstrated in cytological treatment for nervous system diseases through the expression of different neurotrophic and growth factors after transplantation [13]; these include the glial cell line-derived neurotrophic factor (GDNF), brain-derived neurotrophic factor (BDNF), nerve growth factor (NGF), and vascular endothelial growth factor (VEGF) [15–17]. Furthermore, BM-MSCs protect the neurovascular integrity between the basement membrane and astrocyte end feet, which efficiently ameliorates brain damage and significantly enhances functional recovery in animal models of various neurological disorders, since the neurovascular units (NVUs) that are composed of endothelial cells, astrocytes, and neurons are important for the maintenance of homeostasis in the central nervous system (CNS) [18]. Notably, the ability of BM-MSCs to interact with human leukocyte antigen (HLA)-unrelated immune cells and regulate their effects has important implications in transplantation biology. BM-MSCs possess remarkable immunosuppressive properties. The proliferation and function of the major immune cell populations including T cells, B cells, and natural killer (NK) cells can be inhibited by BM-MSCs. Also, BM-MSCs are able to modulate the activities of dendritic cells (DCs) and promote the generation of regulatory T cells (Tregs) both in vitro and in vivo. However, only a few studies have investigated the interaction of BM-MSCs with specific immune cell subtypes [19–21]. Moreover, transplanted BM-MSCs can benefit CNS injury by regulating the fate of NSCs; multipotent NSCs were originally viewed not only as an endogenous source for cellular replacement following CNS injury but also beneficial therapies for neurological immunomodulation of the damaged environment, which set into motion protective mechanisms that limit the degree of damage and enhance endogenous repair mechanisms following neurological insults [22, 23].

In the following sections, we will discuss the direct and indirect interactions of transplanted BM-MSCs with HSCs, immune cells, NSCs, neurons, astrocytes, and endothelial cells to better understand the mechanisms underlying BM-MSC-mediated neurological function reconstruction and preservation of the poststroke brain.

4.2 Bone Marrow Niche

4.2.1 Definition of Niche

The term "niche" refers to a specific and complex multidimensional position in a particular organ that provides both anatomical and functional depth to the concept of stem cells [24]. The niche for a certain type of stem cell is defined as a tissue-restricted local microenvironment that directly promotes stem cell maintenance and allows integration of signals from the periphery into appropriate stem cell behaviors [24–27]. The significance of the surrounding microenvironment in regulating and determining stem cell fate including stem cell renewal, expansion, and differentiation was first postulated in 1978 by Schofield based on the observation that the colony-forming unit-fibroblastic cells in the spleen (CFU-S) were less robust than the cells within the bone marrow at reconstituting hematopoiesis [26, 28]. Schofield proposed the concept of the niche as the specialized microenvironment housing HSCs in which the association between HSCs and BM-MSCs was seen to form a regulatory unit that was responsible for effective hematopoiesis [28]. Generally, adult or somatic stem cells will have limited functions when being placed in a dysfunctional or ex situ niche. The specific cues from specific sites guarantee the modulation in stem cell conditions when facing physiological challenges. The balance of the dynamic system based on the interplay between stem cells and their corresponding niche contributes to preventing stem cells from depletion and simultaneously protecting the host from overexuberant stem cell proliferation [24]. The imbalance of the dynamic system may accordingly lead to the occurrence of diseases.

4.2.2 HSC Niche in the Bone Marrow

The niche for HSCs in the bone marrow is widely accepted and experimentally proven. Knowledge of the identity and function of the cellular components (e.g., BM-MSCs, endothelial cells, functionally mature stromal cells) comprising the HSC niche has also been dramatically expanded despite certain controversies. The HSC niche is a complex functional unit. The illustration of the adult bone marrow HSC niche can be found in recently published reviews [26, 29]. Throughout the marrow, HSCs are preferentially located in perivascular regions such as sinusoids and arterioles, where CXC chemokine ligand (CXCL) 12 and stem cell factor (SCF) have been enriched to promote HSC maintenance by distinctly defined perivascular stromal cells and endothelial cells [30–36]. Other cells including macrophages, non-myelinating Schwann cells, sympathetic neuronal cells, and osteolineage cells also likely contribute to the niche. For example, macrophages, as functionally mature descendants of HSCs, can feed back to the niche to influence HSCs proliferation or migration, while primitive osteolineage cells might affect the differentiation of certain lymphoid progenitors. Mature/maturing osteolineage cells could also promote the formation of the HSC niche (e.g., control the size of the niche) even though they do not promote the maintenance of HSCs in a direct manner [26, 29]. Notably, BM-MSCs are known as the key components of the HSC niche.

4.3 BM-MSCs Contribute to the HSCs Niche

Recently, technical advances on high-precision histological imaging, targeted cell tracking, and genetic engineering have provided us more remarkable insights into how BM-MSCs and their progeny affect HSC maintenance and how they participate in regulating hematopoiesis. An explosion of interest in uncovering the nature of the HSC niche and its potential constituent is emerging.

For a better view of the HSC niche, attention should first be drawn to identifying HSCs and BM-MSCs. Exploring specific markers expressed by these two types of cells not only helps reliably identifying and distinguishing these cells from whole cell populations but also contributes to a deeper understanding of where they are located and what functions they are performing correspondingly. In the following sections, we will discuss the different roles of multiple BM-MSCs in the HSC niche on the basis of their respective molecular markers.

4.3.1 Identification of HSCs

HSCs were specifically isolated as Thy-1^{lo}Sca-1^{+}Lineage^{-}c-kit^{+} cells or CD34^{-}Sca-1^{+}Lineage^{-}c-kit^{+} cells in early studies [30, 37–39]. Such complex combinations of 10–12 surface markers limit their use in tissue sections. Identifying HSCs with only several markers holds more practical value. Simple combination of cell surface receptors of the SLAM family such as CD150 (positive in HSCs) and CD48 (negative in HSCs) allowed for high purification of HSCs using a two-color stain. About 21 % of injected CD150^{+}CD48^{-} cells yielded long-term multilineage reconstitution, which was similar to the results obtained with Thy-1^{lo}Sca-1^{+}Lineage^{-}c-kit^{+} cells. More so, an enhanced reliability to identify HSCs could be achieved when further combined with the lack of CD41 expression. CD41 distinguished CD150^{+}CD48^{-} HSCs from CD150^{+}CD48^{-} megakaryocytes; about 45 % of injected bone marrow CD150^{+}CD48^{-}CD41^{-} cells were detected to give long-term multilineage reconstitution. Using this method, CD150^{+}CD48^{-}CD41^{-} HSCs were shown to be located adjacent to the sinusoid vessels in the trabecular region of the bone marrow, thus providing evidence for an existence of a perivascular niche for bone marrow HSCs [30]. In a following study, which was deemed to be more precise, 58 % of the CD150^{+}CD48^{-}CD41^{-} cells were further detected adjacent to sinusoids, and up to 92–95 % of them were within five-cell diameters of a sinusoid [31].

4.3.2 Multiple BM-MSCs in the Perivascular HSC Niche

In consideration of the sinusoid dwelling of major HSCs, there appears to exist some key factors that are secreted by the perivascular cells for influencing HSCs.

By showing a severe reduction of HSCs in the adult bone marrow of CXCR4 (the primary physiological receptor for CXCL12) conditionally deficient mice, Sugiyama et al. highlighted the crucial role of CXCL12-CXCR4 signaling in maintaining the HSC pool [33]. CXCL12, also known as stromal cell-derived factor (SDF)-1 or pre-B-cell growth-stimulating factor (PBSF), is mainly expressed by a small population of CXCL12-abundant reticular (CAR) cells, which are predominantly surrounded by sinusoidal endothelial cells, with some also close to the endosteum [33, 40, 41]. Notably, CAR cells were among the population of multipotent stromal progenitors, with a potential to differentiate into adipocytic and osteoblastic cells both in vitro and in vivo [34]. CAR cells were further suggested to colocalize with the majority of HSCs throughout the bone marrow, playing an essential function in promoting HSC cycling and self-renewal [33, 34, 42]. As highly expressing CXCL12, CAR cells could be selectively eliminated or severely reduced by diphtheria toxin (DT) in CXCL12-DTR-GFP mice, in which a transgene encoding DT receptor-green fluorescent protein (DTR-GFP) fusion protein was knocked into the *Cxcl12* locus. In DT-treated CXCL12-DTR-GFP mice (or CAR cell-depleted mice), the number of HSCs was reduced about 50 % in the bone marrow; the expression of cell-cycle-promoting genes, e.g., those encoding cyclin D1, D2, A2, Cdc2a, and Cdc6, was significantly reduced, while mRNA expression of transcriptional repressor Mad1, known to inhibit HSCs proliferation, was markedly elevated in comparison with untreated wild-type mice or untreated CXCL12-DTR-GFP mice. However, the rapid reduction of HSCs in CAR cell-depleted mice cannot be simply explained by the findings that CAR cells maintain HSCs in a proliferative state. An enhanced differentiating capacity of HSCs caused by the short-term ablation of CAR cells should also be taken into consideration. Moreover, CAR cells were the major producers of SCF in the bone marrow. SCF rescued the expression of cyclin D1, D2, and Cdc6 in HSCs. In summary, CAR cells were suggested as marrow stromal adipo-osteogenic progenitors, which generated large amounts of CXCL12 and SCF to contribute to the maintenance of HSCs in the perivascular niche [34].

In addition to CAR cells, nestin$^+$ MSCs that are exclusively distributed in perivascular regions were also proposed to regulate HSCs maintenance. A landmark study by Méndez–Ferrer and colleagues found that the vast majority of CD150$^+$CD48$^-$Lin$^-$ cells (HSCs) were localized within five-cell diameters from and found to be directly adjacent to nestin$^+$ MSCs. During long-term bone marrow culture, nestin$^+$ MSCs were frequently present in areas where hematopoietic progenitors were enriched. Thus, these spatial and physical evidence confirmed the close relationship between nestin$^+$ MSCs and putative HSCs [35]. Strikingly, expressions of core genes regarding HSC maintenance and attraction such as *Cxcl12*, *Kitl* (encoding SCF), *Angpt1* (encoding angiopoietin-1), *IL7* (interleukin-7), *Vcam1* (encoding vascular cell adhesion molecule-1), and *Spp1* (encoding osteopontin) were extremely high in nestin$^+$ MSCs. These genes, except *Spp1*, could be significantly downregulated during the enforced HSCs mobilization or β3-adrenergic receptor (β3-AR) activation [35, 43]. Furthermore, selective depletion of nestin$^+$ MSCs by inducible expression of DTR in adult *Nes-cre^{ERT2}/iDTR* mice, treated with tamoxifen and DT, led to a severe reduction (50 %) of HSCs in the bone marrow but a proportional and selective increase in the spleen without any detectable differences in cell-cycle profile or apoptotic rate [35]. All the findings above supported the hypothesis that nestin$^+$ MSCs are functional components of the HSC niche, or could at least regulate the mobilization of HSCs toward extramedullary sites.

What should be noted is that the bone marrow resident nestin$^+$ MSCs could be divided into two distinct types on the basis of their GFP expression levels and cellular morphology i.e., Nes-GFP^{bright} and Nes-GFP^{dim} cells. Interestingly, the more abundant Nes-GFP^{dim} MSCs were reticular in shape and mostly associated with sinusoids, which showed a high similarity to CAR cells [33, 44]. Considering the high expression of CXCL12 in nestin$^+$ MSCs [35], it was likely that Nes-GFP^{dim} MSCs were largely overlapped with CAR cells and may even be a subpopulation of CAR cells. However, nestin$^+$ MSCs contained CFU-F activity, possessed the capacity of multilineage differentiation, and harbored high self-renewal activity both in vitro and in vivo [35]. Therefore, it appears that Nes-GFP^{dim} cells were more primitive than CAR cells.

To gain detailed insights into the source of factors for HSC maintenance, Ding et al. made selective deletions of SCF from different subpopulations of bone marrow cells in *Scf^{gfp}* knock in mice [32]. The resulting data showed that SCF was expressed largely by the stromal cells surrounding the sinusoids throughout the bone marrow and leptin receptor (LepR) expression was highly restricted within these SCF-GFP$^+$ perisinusoidal stromal cells. Conditional deletion of SCF in nestin$^+$ stromal cells, hematopoietic cells, and osteoblasts would not affect the frequency and functions of HSCs, while the selective ablation of SCF in LepR-expressing perivascular cells and endothelial cells would cause a marked reduction of HSCs, which indicated a selective requirement of SCF from perivascular cells and endothelial cells in promoting HSC maintenance [32]. As nestin$^+$ MSCs also produced large amounts of SCF [35], the overlapping condition between nestin$^+$ MSCs and LepR$^+$ cells became the subject of intensive investigations. In actuality, Nes-GFP^{dim} MSCs (about 80 %) largely overlapped with perisinusoidal LepR$^+$ cells, whereas no overlap was found between Nes-GFP^{bright} and LepR$^+$ cells [44]. Therefore, multiple perisinusoidal stromal cells including both LepR$^+$ cells and Nes-GFP^{dim} MSCs as well as partial CAR cells may contribute to maintaining the perisinusoidal HSC niche.

Moreover, Nes-GFP^{bright} cells were found exclusively along arterioles and showed significantly higher expression of genes regarding HSC niches in comparison with Nes-GFP^{dim} cells when subjected to RNA sequencing analysis [44]. In contrast, the classical pericyte marker NG2 appeared to largely label arteriolar Nes-GFP^{bright} stromal cells, but not LepR$^+$ Nes-GFP^{dim} and other stromal cells. The dormant HSCs were preferentially associated with arterioles; depletion of NG2$^+$ Nes-GFP^{bright} cells in tamoxifen-DT-treated *NG2-cre^{ERTM}/iDTR* mice not only obviously reduced the HSC pool but also caused HSCs to move away from arterioles and, likewise, switched them into a non-quiescent state. Thus, a bone marrow arteriolar niche for quiescent HSC maintenance that was potentially related to the rare NG2$^+$ Nes-GFP^{bright} stromal cells was strongly proposed [44].

4.4 Interaction Between Immune Cells and BMSCs

4.4.1 T Cells

BMSC transplantation can mediate immunomodulatory effects and reduce neuroinflammation [45, 46]. Several studies have now shown that MSCs in culture can suppress T-cell proliferation and alter the outcome of the immune cells' responses. MSCs suppress proliferation of allogeneic T cells in a MHC-independent manner [14, 47, 48]. Aggarwal et al. first showed that human MSCs reduced TH1 cell-derived interferon-γ (IFN-γ) and increased the secretion of TH2 cell-derived IL-4 and regulatory T cells (Tregs) [49]. Studies have demonstrated that MSCs suppress allogeneic T-cell proliferation in vitro and in vivo [14, 47, 50]. Inhibition of T-cell proliferation by MSCs appears to be mediated by both cell–cell interaction [14, 51, 52] and MSC-derived molecules that are believed to have immunomodulatory actions on T-cell responses e.g., transforming growth factor (TGF)-β1, IL-10 [53], hepatocyte growth factor (HGF) [47], indoleamine 2,3-dioxygenase (IDO) [54], prostaglandin E2 (PGE2) [49], nitric oxide (NO) [55], matrix metalloproteinases (in particular MMP-2 and MMP-9) [56], and human leukocyte antigen (HLA) class I molecule-G5 (HLA-G5) [57]. Moreover, MSCs inhibit stimulated T-cell proliferation by preventing their entry into the S phase of the cell cycle and by mediating irreversible G0/G1 phase arrest [58]. In a recent study, Ghannam et al. found that under inflammatory conditions, MSCs prevented the differentiation of naive $CD4^+$ T cells into Th17 cells and inhibited the function of Th17 cells in vitro by secreting PGE2 [59]. In addition, BMSCs also suppress T-cell activation and prevent expression and activation of a number of inflammatory factors [14]. It was reported that MSCs induced apoptosis of T cells in vitro by inducing NO production [60, 61]. In contrast to the strong inhibitory effects of MSCs on T-cell proliferation, activity, and apoptosis, there are only relatively minor and reversible effects on T-cell function and toxicity [62]. In addition, MSCs differentiated into various mesenchymal lineages do not appear to alter their interaction with T cells [63]. Moreover, differentiated MSCs had immunologic properties similar to the undifferentiated MSCs [64].

Tregs are a subpopulation of T cells that modulate the immune system, maintain tolerance to self-antigens, and control autoimmune disorders [65, 66]. Studies have shown that high levels of functional $CD4^+Foxp3^+$ Treg cells exist in the bone marrow and play important roles in regulating bone function [67–69]. Depletion of Tregs had no effect on MSCs-mediated inhibition of T-cell proliferation [70]. However, a recent study reported that MSCs could induce kidney allograft tolerance by inducing the generation of Tregs in vivo [71]. Additionally, MSCs have been reported to increase the formation of Tregs that are responsible for the inhibition of allogeneic lymphocyte proliferation [59, 72, 73]. It was demonstrated that Tregs inhibit excessive production of pro-inflammatory cytokines and modulate the invasion and/or activation of lymphocytes and microglia in the ischemic brain to prevent secondary infarct growth [74]. Recent studies indicated that BMSCs increased the production of Treg cells via a mechanism involving not only direct cell contact but also the secretion of soluble factors such as anti-inflammatory TGF-β, IL-10 [53], PGE2 [75], IDO, inducible NO synthase (iNOS), cyclooxygenase 2 (COX-2) [76], HLA-G5 [57], and the upregulation of developmental endothelial locus-1 (Del-1), an endogenous leukocyte-endothelial adhesion inhibitor [77]. BMSCs can also reprogram conventional T cells into Tregs [78].

Gamma delta T (γδT) cells have pivotal roles in the evolution of brain infarction and accompanying neurological deficits after stroke [79]. Activated γδT cells infiltrated the brain after ischemia, which, in turn, produced the pro-inflammatory cytokine IL-17 within hours. MSCs inhibited γδT-cell expansion from peripheral blood mononuclear cells in both cell-to-cell contact and Transwell systems, and MSCs were lysed by activated γδT cells through a T-cell receptor-dependent mechanism [80].

In in vivo studies, allogeneic bone marrow mesenchymal stem cell (allo-BMSC)-based tissue-engineered bone (TEB) has great potential for bone defect repair. The immunogenicities and biological roles of allo-BMSCs are still controversial. A recent finding indicated that allo-BMSCs can induce a transient immunoreaction, which may temporarily delay the osteogenesis of allo-BMSC/scaffold complex in the early stage of in vivo implantation, whereas the long-term engineered bone formation was not affected [81]. However, BMSCs significantly decrease the numbers of T cells and increase the numbers of Tregs as well as productions of pro- and anti-inflammatory cytokines, which play important roles in improving the neural function and decreasing infarct volume after ischemic stroke [82].

To summarize, through the interactions of MSCs with various immune cells, it appears that MSCs inhibit or limit inflammatory responses and promote anti-inflammatory pathways [49]. Although the cellular mechanisms underlying the immunosuppressive effects of MSCs on T cells have been elucidated previously, the underlying molecular mechanisms remain controversial. It is believed that the mechanisms underlying the suppressive effect of MSCs may differ by species [83]. In addition, the degree of the suppressive effect depends on the concentration of the MSCs; a low MSC/lymphocyte ratio is often accompanied by enhanced MSC proliferation [84].

4.4.2 B Cells

Deng et al. first reported that BMSCs had inhibitory effects on the proliferation, activation, and IgG secretion of B lymphocytes in BXSB mouse models of human systemic lupus erythematosus (SLE), with a slight enhancing effect on CD40 expression and an inhibitory effect on CD40L ectopic expression [85]. Similarly, MSCs have been shown to inhibit the proliferation of B cells activated with anti-CD40L, IL-4, IL-2, pokeweed mitogen, and anti-IgG antibodies [58, 86, 87]. In addition, MSCs also reduce the expression of chemokine receptors and immunoglobulin production by stimulated B cells. For example, MSCs can impair B-cell functions such as antibody production and secretion of the chemokine receptors (CXCR4, CXCR5, and CCR7), which are responsible for chemotaxis to CXCL12 and CXCL13 [87]. MMP-2 is required for bone marrow stromal cell support of chemotaxis [88]. However, exposure to one chemotherapeutic agent, etoposide (VP-16), can result in the reduced ability of BMSCs to support pro-B-cell chemotaxis [88]. Interestingly, MSCs do not appear to alter surface molecules involved in stimulatory cell cooperation such as HLA-DR, CD40, and the B7 family or inhibit the expressions of TNF-α, IFN-γ, IL-4, and IL-10 [49].

The nature of the mechanism involved in this inhibitory effect of MSCs has not yet been elucidated completely [83]. B-cell proliferation is inhibited by MSCs in a dose-dependent manner in vitro [87]. The inhibitory effect is partly attributable to physical contact and soluble factors, which lead to the blockade of the G0/G1 phases of the cell cycle, similar to what occurs with T cells [49].

4.4.3 NK Cells

MSCs have been shown to suppress NK cytotoxicity [89]. A study indicated that MSCs suppressed NK cell cytotoxicity against HLA class I-positive cells more effectively than HLA class I-negative cells [51]. MSCs impair NK cells to secrete the perforin-/granzyme-containing granules and lyse target cells. Conversely, NK cells are able to lyse MSCs even if IL-2-activated NK cell proliferation is inhibited [90]. However, MSCs did not inhibit the lysis of freshly isolated NK cells, and MSCs were not lysed by allogeneic NK cells [91]. Furthermore, numerous studies have shown that MSCs suppress NK cell proliferation and IFN-γ production driven by IL-2 or IL-15, but only partially inhibit the proliferation of activated NK cells [51, 91–93]. The mechanisms underlying the MSC-mediated immunosuppressive effects are still unclear. Cell–cell contact and soluble factors such as TGF-b1 and PGE2 are believed to play a role in the MSC-mediated suppression of NK cell proliferation [51].

4.4.4 Dendritic Cells

The effects of MSCs on the differentiation, maturation, and function of monocyte-derived DCs have also been reported. MSCs inhibit the differentiation of monocytes into immature DCs by reducing the expression of co-stimulatory molecules and hampering the ability of the former to stimulate naive T-cell proliferation and IL-12 secretion [52, 94]. MSCs strongly inhibited the maturation and functioning of DCs by interfering selectively with the generation of immature DCs via inhibitory mediator of MSCs-derived TNF-alpha, PGE2, but not IL-6 [21, 95]. However, the mechanism underlying the upregulation of PGE2 in monocyte-MSC co-cultures remains unclear. A recent study illustrates that MSCs alter the outcome of the immune response by inhibiting inflammatory DC1 signaling and promoting anti-inflammatory DC2 signaling by altering the cytokine secretion profile of DC subsets [49]. In addition, it was reported that the cell cycle in DCs was arrested in the G0/G1 phase upon interaction with MSCs [96].

4.5 Interaction of Bone Marrow Stem Cells with Neural Stem Cells

4.5.1 In Vitro

A number of studies have supported the notion that MSCs co-cultured with NSCs can influence the fate of NSCs. MSCs can indeed influence the proliferation and differentiation of NSCs indirectly, which may be mediated in part by notch signaling [97]. Using direct connection of NSCs onto the surface of MSCs, BMSCs could not only induce NSCs to differentiate into neurons but also enhance the survival of neurons as well as decrease the number of astrocytes [98]. In addition, other studies reported that MSCs promoted differentiation of NSCs into oligodendrocytes [99] and inhibited astrocyte differentiation [100, 101]. BMSCs mainly effected NSC differentiation, but not proliferation, which was why the oligodendrocyte proportion increased three times and astrocyte proportion decreased onefold, whereas the neuronal proportion remained unchanged [102].

Notch signaling is important in many aspects of CNS development and is most prominently involved in regulating the proliferation and differentiation of NSCs [103–105]. Wang et al. showed that the expressions of Notch1 and Hes1 increased in NSCs co-cultured with MSCs and decreased after adding the Notch signaling inhibitor, DAPT [97]. Similarly, culturing human MSCs with rat NSCs was found to stimulate differentiation into astrocytes and oligodendrocytes, along with increased expressions of Notch and TGFβ signaling in both cell types, suggesting that the differentiation of rat NSCs in the co-culture was driven through the Notch pathway and by increased secretion of soluble factors such as TGFβ1 by the human BMSCs [106].

MSCs are known to secrete many kinds of trophic factors including brain-derived neurotropic factor (BDNF), VEGF, and nerve growth factor (NGF) [107–111]. Earlier studies initially reported that the exposure of MSCs to certain chemicals caused neuron-like morphological changes in vitro [10, 112, 113], although the morphological change to neuron-like cells is due to cytoskeletal rearrangement rather than genuine acquisition of neuronal fate [114, 115], the neuronal differentiation of NSCs induced by BMSCs, and the neurogenic effect of BM-MSCs could be attributed to several factors such as cell surface molecules and extracellular matrix molecules in the microenvironment provided by BMSCs and soluble molecules secreted by BMSCs [116].

Bai et al. also suggested that cell surface interactions are is the primary mechanism by which MSCs influence the fate of NSCs, indicating that the major effect of MSCs on NSCs is mediated through soluble factors and this effect of differentiation of NSCs differs depending on how long the media is conditioned [99]. Ciliary neurotropic factor (CNTF) has a pro-oligodendrogenic effect on NSCs without affecting cell proliferation or survival [100]. Sygnecka and colleagues concluded that a higher BDNF secretion plays a role in primary MSC-mediated promotion of neuronal regeneration and axonal regrowth [117].

BMSCs can also promote the migration of NSC-differentiated cells. One striking effect of BMSC on neurosphere-derived stem cells is the rapid promotion of neural cell migration [99]. Previous research has shown that chemokine and their receptors are not only expressed in the developing brain but also have a role in orienting cell migration, trophic support, proliferation, and/or differentiation [118]. The chemokine (C-C motif) ligand 2 (CCL2) is a key paracrine factor in the neurogenic effects of BM-MSCs on NSCs in the brain [119].

4.5.2 In Vivo

Similar results have been obtained from in vivo studies. Yoo et al. first verified that transplantation of MSCs improved functional recovery along with increased proliferation of NSCs in the SVZ and survival of newly born neuroblasts [120]. Studies also showed that BMSC transplantation stimulates the differentiation and proliferation of NSCs after stroke and dramatically improves neurological function [112, 121]. In addition, the MSCs may transdifferentiate into neural lineages such as neurons and astrocytes in vivo [122]. Transplantation of BM-MSCs into the SVZ stimulated proliferation and maturation of endogenous progenitors toward the neuronal phenotype, indicating that newborn neurons were derived from endogenous neuronal progenitors, but not from the transplanted cells [119]. These findings are in agreement with previous investigations on the stimulatory effect of BM-MSCs on neurogenesis [109, 116, 120].

The enhanced functional recovery after MSC transplantation may reflect effective mobilization of NSCs to the area of insult. An intriguing recent observation was reported that following implantation into the hippocampi of immunodeficient mice, human MSCs stimulated the proliferation and dorsal migration of endogenous BrdU-labeled NSCs, suggesting that the transplanted BMSCs have an effect on the NSC fate in the SVZ and also the hippocampus [109, 120, 123]. Interestingly, cell surface molecules and extracellular matrix molecules on BMSCs were not involved in the neuronal differentiation of NSCs but soluble factors did instead [98]. However, BMSCs are capable of supporting nerve cells and increasing neurogenesis after cerebral ischemia by secreting neurotropic factors such as BDNF and basic fibroblast growth factor (bFGF) [124]. In addition, the presence of MSCs in areas of injury may enhance the number of NSCs recruited to injury sites and promote repair.

The in vitro studies suggest that the signals from MSCs are soluble, and several factors such as PDGF have been shown to promote NSC migration [125, 126].

Recent data have indicated the important role of chemokines such as MCP-1, SDF1, and its receptor CXCR4-CCL2 in migration and differentiation of neural progenitor cells after stroke [127, 128]. For instance, the chemotactic molecule SDF-1 may partially mediate the homing of transplanted BMSCs to the injury sites in the brain and then exert the beneficial effects evoked by BMSC treatment, which may be related to the induction of neurogenesis and reduction of glial scar formation [129].

MSCs may exert their effects on NSCs by attenuating inflammation. A recent study showed that human MSCs have a neuroprotective effect on neurons through anti-inflammatory actions mediated by the modulation of microglial activation, which could have major therapeutic implications in the treatment of neural diseases [130]. In addition, BMSC treatment reduced scar thickness and increased the number of proliferating cells and oligodendrocyte precursor cells of the ipsilateral SVZ [129]. Interestingly, it was revealed that the combination cell therapy of BMSCs and NSCs is more efficient in promoting functional recovery after cerebral stroke, suggesting that these stem cell types could work in synergy to produce additional benefits that would otherwise be absent if they acted alone [131].

4.6 Interrelation of Neuronal Cells and Transplanted BMSCs

BMSCs constitute a heterogeneous collection of mesenchymal stem and progenitor cells. Human BMSCs can transdifferentiate into neural and mesodermal cell lines [132–134]. BMSCs have been found to significantly promote neurite extension of neurons in an organotypic brain slice [135, 136]. Regeneration of the facial nerve was improved by both uBMSC and dBMSC in rats [137]. BMSCs also promoted neurite outgrowth in spinal neurons by secreting soluble factors such as BDNF and GDNF [138].

BMSCs significantly enhance neurogenesis in the SVZ. In fact, some transplanted BMSCs also express neuronal phenotypes in the neocortex, SVZ, corpus callosum, and peri-lesion area [139]. Other studies demonstrated that the transplanted BMSCs can express GABAA receptors and MAP2 in the peri-infarct neocortex, suggesting that the transplanted BMSCs may contribute to migration toward the peri-infarct area and acquire neuron-specific receptor function [140] along with improvement in sensorimotor function [113, 139, 141–144], synaptogenesis and nerve regeneration [145].

Other beneficial effects observed upon BMSC transplantation include enhanced structural neuroplasticity and increased axonal outgrowth from healthy brain tissue

[146]. At 28 days after intracarotid BMSC transplantation, axons, myelin, and white matter bundles were significantly increased in the striatum and corpus callosum. The remodeling of white matter in the cortical IBZ and corpus callosum is increased by axonal sprouting and remyelination [147]. It was revealed that axonal sprouting and remyelination in the cortical penumbra were significantly increased after transplantation of BMSCs directly into the striatum of MCAO mice [148]. Through ^{123}I-iomazenil single photon emission computed tomography, the neuronal integrity in the peri-infarct area can be visualized after the transplantation of BMSCs [149].

BMSC transplantation has the potential to repair the ischemia-damaged neural networks, restore lost neuronal connections [150], and significantly ameliorate the breakdown of neurovascular integrity [18]. Administration of BMSCs significantly increases the axonal restructuring on the deafferented red nucleus and the denervated spinal motor neurons, which increases axonal sprouting and rewiring of the corticospinal tract emanating from the uninjured motor cortex onto denervated spinal cord and the axonal connections from the intact motor cortex to the denervated spinal cord at both the cervical and lumbar levels [151]. Recently, many have reported that BMSCs markedly enhanced interhemispheric, intracortical, thalamocortical circuit connections [150]. More so, cortical neurons surviving in the peri-infarct motor cortex underwent axonal sprouting to restore connections between different cerebral areas [152]. Further, in a rat model of intracerebral hemorrhage, administration of BMSCs can significantly increase neuronal plasticity of the denervated corticospinal tract at bilateral forelimb areas of the cortex [153].

4.7 Interrelation Between Endothelial Cells and Transplanted BMSCs

Under indirect co-culture conditions, endothelial progenitor cells (EPCs) could enhance the proliferation of BMSCs but could not regulate cell apoptosis in vitro [154]. Transplanted BMSCs can selectively migrate to the site of the infarct area, stimulate angiogenesis and arteriogenesis as well as induce a neovascular response that results in a significant increase in local cerebral blood flow to the ischemic area, aid for the repair of the injured brain [155], and reduction of infarct size and blood–brain barrier disruption [156]. Interestingly, collection of transplanted BMSCs from the stroke rat saw significant increases in phosphorylated-Tie2 activity in brain-derived endothelial cells and capillary tube formation compared with BMSCs collected from normal rats [157]. Further, a clinical study showed that in ischemic stroke patients, the number of new vessels surrounding injured tissue correlated with longer survival [158]. Studies have shown the therapeutic potential of angiogenesis in the restoration of local blood flow and functional recovery in ischemic diseases [159].

Recent studies have claimed that appropriate application of pharmaceuticals (such as Z-VAD, SDF-1α, and statins) decreases the apoptosis of grafted BMSCs, induces migration into the ischemic area, modulates the expressions of protein/cell and trophic factors, promotes angiogenesis in the ischemic area, and enhances the therapeutic effect of BMSC transplantation. The intracerebral coadministration of SDF-1α and BMSCs promotes BMSC migration to the ischemic lesion, increases the density of blood vessels in the ischemic cortex, and enhances neuronal plasticity [160]. A further investigation revealed that DETA-NONOate increases the expressions of CXCR4 and MMP in BMSCs and promotes BMSC adhesion and migration to mouse brain endothelial cells and astrocytes [161]. In addition, combined intravenous injection of BMSCs with DETA upregulates angiopoietin-1 and its receptor Tie2 and restores neovascularization, which is regarded as the main therapeutic goal in ischemic stroke. Along with the angiogenesis in the infarct area and the improvement in the microenvironment, the proliferation of neurons and neuroglial cells and the growth of neurofibrils was also observed. These elements are reciprocal factors for the repair of the infarct area [162].

4.8 Interrelation of Glial Cells and Transplanted BMSC

BMSC treatment produced dramatic changes in the number and activation of brain astroglia and microglia, particularly in the region of the infarct [163]. An in vitro study found that BMSC-conditioned medium (BMSC-CM) significantly inhibited proliferation and secretion of pro-inflammatory factors by activated microglia and significantly induced apoptosis of microglia [164]. The intravenous injection of BMSCs noticeably increased the number of microglia/macrophages in the injured brain. Intracerebral transplantation of BMSCs significantly enhanced the number of astrocytes and in a lesser degree caused changes in the number of microglia/macrophages [165]. Furthermore, transplanted BMSCs reduce the thickness of the glial scar wall and the number of Nogo-A-positive cells along the scar border. BMSCs can also reduce neuron-specific enolase expression, increase reactive astrocyte density, and increase the number of proliferating cells in the SVZ following cerebral ischemia [166]. Hofstetter and colleagues transplanted BMSCs into the injured cord and found that the engrafted BMSCs were tightly associated with longitudinally arranged immature astrocytes and formed bundles bridging the epicenter of the injury [167]. Additionally, BMSC treatment can increase the number of oligodendrocyte precursor cells along the SVZ in the ischemic hemisphere [129]. BMSCs in the penumbra activate local astrocytes as well as increase GDNF levels in the ischemic hemisphere, which is released by reactive astrocytes to facilitate the self-repair efforts of the brain against ischemic attack [168]. As well, BMSCs can increase BMP2/BMP4 expression in ischemic astrocytes. These changes enhance SVZ progenitor cell gliogenesis by activating relevant signaling pathways.

BMSC-stimulated signaling of endogenous astrocytes may alter the ischemic environment, promoting remodeling of the brain after stroke [169]. BMSCs produce growth factors, activate the Akt pathway, and increase the survival of oligodendrocytes (OLG). BMSCs reduce p75 and caspase 3 expressions in OLGs, which lead to decreased OLG apoptosis. BMSCs participate in OLG protection that may occur with promoting growth factors/PI3K/Akt and inhibiting the p75/caspase pathways [170]. In addition, BMSCs have the ability to maintain the resting phenotype of microglia or to control microglial activation through their production of several factors [164]. Transplanted BMSCs could survive and improve neuronal behavior in rats with TBI and mediate neuroprotection and regeneration, which could be associated with the GDNF regulating apoptosis signals through BAX and BAD [171].

4.9 Conclusion

BMSC transplantation is a promising therapeutic strategy for nervous system diseases with unique advantages. Many obstacles associated with medical ethics and immunological rejection can be overcome, while the capability of self-renewal and multidirectional differentiation of BMSCs can be retained after transplantation. Through the interaction with other cells, BMSCs benefit the neurological dysfunction in either direct or indirect methods. Importantly, BMSCs-mediated immunomodulation may be an active component in inflammation modulation, tolerance induction, and reduction of transplantation complications such as rejection and GVHD, even though the complete mechanisms of immune modulation by BMSCs require further investigations. Furthermore, modification of BMSCs with neuroprotective factor- or neurotrophic factor-encoding genes enhances their therapeutic efficacies. Nevertheless, the clinical application of BMSC transplantation is limited, and safety issues must be considered. The fundamental properties of BMSCs and their potential for short- and long-term toxicity need to be determined before they can be widely used in clinical practice.

References

1. Friedenstein AJ, Chailakhjan RK, Lalykina KS. The development of fibroblast colonies in monolayer cultures of guinea-pig bone marrow and spleen cells. Cell Tissue Kinet. 1970;3(4):393–403.
2. Friedenstein AJ, Gorskaja JF, Kulagina NN. Fibroblast precursors in normal and irradiated mouse hematopoietic organs. Exp Hematol. 1976;4(5):267–74.
3. Friedenstein AJ, Deriglasova UF, Kulagina NN, Panasuk AF, Rudakowa SF, Luria EA, Ruadkow IA. Precursors for fibroblasts in different populations of hematopoietic cells as detected by the in vitro colony assay method. Exp Hematol. 1974;2(2):83–92.
4. Prockop DJ. Marrow stromal cells as stem cells for nonhematopoietic tissues. Science. 1997;276(5309):71–4.

5. Horwitz EM, Le Blanc K, Dominici M, Mueller I, Slaper-Cortenbach I, Marini FC, Deans RJ, Krause DS, Keating A, International Society for Cellular T. Clarification of the nomenclature for MSC: the International Society for Cellular Therapy position statement. Cytotherapy. 2005;7(5):393–5. doi:10.1080/14653240500319234.
6. Ferrari G, Cusella-De Angelis G, Coletta M, Paolucci E, Stornaiuolo A, Cossu G, Mavilio F. Muscle regeneration by bone marrow-derived myogenic progenitors. Science. 1998;279(5356):1528–30.
7. Pittenger MF, Mackay AM, Beck SC, Jaiswal RK, Douglas R, Mosca JD, Moorman MA, Simonetti DW, Craig S, Marshak DR. Multilineage potential of adult human mesenchymal stem cells. Science. 1999;284(5411):143–7.
8. Kopen GC, Prockop DJ, Phinney DG. Marrow stromal cells migrate throughout forebrain and cerebellum, and they differentiate into astrocytes after injection into neonatal mouse brains. Proc Natl Acad Sci U S A. 1999;96(19):10711–6.
9. Sanchez-Ramos J, Song S, Cardozo-Pelaez F, Hazzi C, Stedeford T, Willing A, Freeman TB, Saporta S, Janssen W, Patel N, Cooper DR, Sanberg PR. Adult bone marrow stromal cells differentiate into neural cells in vitro. Exp Neurol. 2000;164(2):247–56. doi:10.1006/exnr.2000.7389.
10. Woodbury D, Schwarz EJ, Prockop DJ, Black IB. Adult rat and human bone marrow stromal cells differentiate into neurons. J Neurosci Res. 2000;61(4):364–70.
11. Kohyama J, Abe H, Shimazaki T, Koizumi A, Nakashima K, Gojo S, Taga T, Okano H, Hata J, Umezawa A. Brain from bone: efficient "meta-differentiation" of marrow stroma-derived mature osteoblasts to neurons with Noggin or a demethylating agent. Differentiation. 2001;68(4–5):235–44.
12. Kwak DH, Yu K, Kim SM, Lee DH, Kim SM, Jung JU, Seo JW, Kim N, Lee S, Jung KY, You HK, Kim HA, Choo YK. Dynamic changes of gangliosides expression during the differentiation of embryonic and mesenchymal stem cells into neural cells. Exp Mol Med. 2006;38(6):668–76. doi:10.1038/emm.2006.79.
13. Malgieri A, Kantzari E, Patrizi MP, Gambardella S. Bone marrow and umbilical cord blood human mesenchymal stem cells: state of the art. Int J Clin Exp Med. 2010;3(4):248–69.
14. Tse WT, Pendleton JD, Beyer WM, Egalka MC, Guinan EC. Suppression of allogeneic T-cell proliferation by human marrow stromal cells: implications in transplantation. Transplantation. 2003;75(3):389–97. doi:10.1097/01.TP.0000045055.63901.A9.
15. Chen X, Li Y, Wang L, Katakowski M, Zhang L, Chen J, Xu Y, Gautam SC, Chopp M. Ischemic rat brain extracts induce human marrow stromal cell growth factor production. Neuropathology. 2002;22(4):275–9.
16. Dormady SP, Bashayan O, Dougherty R, Zhang XM, Basch RS. Immortalized multipotential mesenchymal cells and the hematopoietic microenvironment. J Hematother Stem Cell Res. 2001;10(1):125–40. doi:10.1089/152581601750098372.
17. Kurozumi K, Nakamura K, Tamiya T, Kawano Y, Ishii K, Kobune M, Hirai S, Uchida H, Sasaki K, Ito Y, Kato K, Honmou O, Houkin K, Date I, Hamada H. Mesenchymal stem cells that produce neurotrophic factors reduce ischemic damage in the rat middle cerebral artery occlusion model. Mol Ther. 2005;11(1):96–104. doi:10.1016/j.ymthe.2004.09.020.
18. Ito M, Kuroda S, Sugiyama T, Maruichi K, Kawabori M, Nakayama N, Houkin K, Iwasaki Y. Transplanted bone marrow stromal cells protect neurovascular units and ameliorate brain damage in stroke-prone spontaneously hypertensive rats. Neuropathology. 2012;32(5):522–33. doi:10.1111/j.1440-1789.2011.01291.x.
19. Le Blanc K, Rasmusson I, Gotherstrom C, Seidel C, Sundberg B, Sundin M, Rosendahl K, Tammik C, Ringden O. Mesenchymal stem cells inhibit the expression of CD25 (interleukin-2 receptor) and CD38 on phytohaemagglutinin-activated lymphocytes. Scand J Immunol. 2004;60(3):307–15. doi:10.1111/j.0300-9475.2004.01483.x.
20. Gotherstrom C, Ringden O, Westgren M, Tammik C, Le Blanc K. Immunomodulatory effects of human foetal liver-derived mesenchymal stem cells. Bone Marrow Transplant. 2003;32(3):265–72. doi:10.1038/sj.bmt.1704111.

21. Zhang W, Ge W, Li C, You S, Liao L, Han Q, Deng W, Zhao RC. Effects of mesenchymal stem cells on differentiation, maturation, and function of human monocyte-derived dendritic cells. Stem Cells Dev. 2004;13(3):263–71. doi:10.1089/154732804323099190.
22. Arvidsson A, Collin T, Kirik D, Kokaia Z, Lindvall O. Neuronal replacement from endogenous precursors in the adult brain after stroke. Nat Med. 2002;8(9):963–70. doi:10.1038/nm747.
23. Einstein O, Ben-Hur T. The changing face of neural stem cell therapy in neurologic diseases. Arch Neurol. 2008;65(4):452–6. doi:10.1001/archneur.65.4.452.
24. Scadden DT. The stem-cell niche as an entity of action. Nature. 2006;441(7097):1075–9. doi:10.1038/nature04957.
25. Lo Celso C, Scadden DT. The haematopoietic stem cell niche at a glance. J Cell Sci. 2011;124(Pt 21):3529–35. doi:10.1242/jcs.074112.
26. Morrison SJ, Scadden DT. The bone marrow niche for haematopoietic stem cells. Nature. 2014;505(7483):327–34. doi:10.1038/nature12984.
27. Morrison SJ, Spradling AC. Stem cells and niches: mechanisms that promote stem cell maintenance throughout life. Cell. 2008;132(4):598–611. doi:10.1016/j.cell.2008.01.038.
28. Schofield R. The relationship between the spleen colony-forming cell and the haemopoietic stem cell. Blood Cells. 1978;4(1–2):7–25.
29. Boulais PE, Frenette PS. Making sense of hematopoietic stem cell niches. Blood. 2015;125(17):2621–9. doi:10.1182/blood-2014-09-570192.
30. Kiel MJ, Yilmaz OH, Iwashita T, Yilmaz OH, Terhorst C, Morrison SJ. SLAM family receptors distinguish hematopoietic stem and progenitor cells and reveal endothelial niches for stem cells. Cell. 2005;121(7):1109–21. doi:10.1016/j.cell.2005.05.026.
31. Kiel MJ, Radice GL, Morrison SJ. Lack of evidence that hematopoietic stem cells depend on N-cadherin-mediated adhesion to osteoblasts for their maintenance. Cell Stem Cell. 2007;1(2):204–17. doi:10.1016/j.stem.2007.06.001.
32. Ding L, Saunders TL, Enikolopov G, Morrison SJ. Endothelial and perivascular cells maintain haematopoietic stem cells. Nature. 2012;481(7382):457–62. doi:10.1038/nature10783.
33. Sugiyama T, Kohara H, Noda M, Nagasawa T. Maintenance of the hematopoietic stem cell pool by CXCL12-CXCR4 chemokine signaling in bone marrow stromal cell niches. Immunity. 2006;25(6):977–88. doi:10.1016/j.immuni.2006.10.016.
34. Omatsu Y, Sugiyama T, Kohara H, Kondoh G, Fujii N, Kohno K, Nagasawa T. The essential functions of adipo-osteogenic progenitors as the hematopoietic stem and progenitor cell niche. Immunity. 2010;33(3):387–99. doi:10.1016/j.immuni.2010.08.017.
35. Mendez-Ferrer S, Michurina TV, Ferraro F, Mazloom AR, Macarthur BD, Lira SA, Scadden DT, Ma'ayan A, Enikolopov GN, Frenette PS. Mesenchymal and haematopoietic stem cells form a unique bone marrow niche. Nature. 2010;466(7308):829–34. doi:10.1038/nature09262.
36. Greenbaum A, Hsu YM, Day RB, Schuettpelz LG, Christopher MJ, Borgerding JN, Nagasawa T, Link DC. CXCL12 in early mesenchymal progenitors is required for haematopoietic stem-cell maintenance. Nature. 2013;495(7440):227–30. doi:10.1038/nature11926.
37. Spangrude GJ, Heimfeld S, Weissman IL. Purification and characterization of mouse hematopoietic stem cells. Science. 1988;241(4861):58–62.
38. Morrison SJ, Weissman IL. The long-term repopulating subset of hematopoietic stem cells is deterministic and isolatable by phenotype. Immunity. 1994;1(8):661–73.
39. Osawa M, Hanada K, Hamada H, Nakauchi H. Long-term lymphohematopoietic reconstitution by a single CD34-low/negative hematopoietic stem cell. Science. 1996;273(5272):242–5.
40. Nagasawa T, Kikutani H, Kishimoto T. Molecular cloning and structure of a pre-B-cell growth-stimulating factor. Proc Natl Acad Sci U S A. 1994;91(6):2305–9.
41. Tokoyoda K, Egawa T, Sugiyama T, Choi BI, Nagasawa T. Cellular niches controlling B lymphocyte behavior within bone marrow during development. Immunity. 2004;20(6):707–18. doi:10.1016/j.immuni.2004.05.001.

42. Ehninger A, Trumpp A. The bone marrow stem cell niche grows up: mesenchymal stem cells and macrophages move in. J Exp Med. 2011;208(3):421–8. doi:10.1084/jem.20110132.
43. Mendez-Ferrer S, Frenette PS. Hematopoietic stem cell trafficking: regulated adhesion and attraction to bone marrow microenvironment. Ann N Y Acad Sci. 2007;1116:392–413. doi:10.1196/annals.1402.086.
44. Kunisaki Y, Bruns I, Scheiermann C, Ahmed J, Pinho S, Zhang D, Mizoguchi T, Wei Q, Lucas D, Ito K, Mar JC, Bergman A, Frenette PS. Arteriolar niches maintain haematopoietic stem cell quiescence. Nature. 2013;502(7473):637–43. doi:10.1038/nature12612.
45. Yoo KH, Jang IK, Lee MW, Kim HE, Yang MS, Eom Y, Lee JE, Kim YJ, Yang SK, Jung HL, Sung KW, Kim CW, Koo HH. Comparison of immunomodulatory properties of mesenchymal stem cells derived from adult human tissues. Cell Immunol. 2009;259(2):150–6. doi:10.1016/j.cellimm.2009.06.010.
46. Sun Z, Satomoto M, Makita K. Therapeutic effects of intravenous administration of bone marrow stromal cells on sevoflurane-induced neuronal apoptosis and neuroinflammation in neonatal rats. Kor J Anesthesiol. 2015;68(4):397–401. doi:10.4097/kjae.2015.68.4.397.
47. Di Nicola M, Carlo-Stella C, Magni M, Milanesi M, Longoni PD, Matteucci P, Grisanti S, Gianni AM. Human bone marrow stromal cells suppress T-lymphocyte proliferation induced by cellular or nonspecific mitogenic stimuli. Blood. 2002;99(10):3838–43.
48. Le Blanc K, Tammik L, Sundberg B, Haynesworth SE, Ringden O. Mesenchymal stem cells inhibit and stimulate mixed lymphocyte cultures and mitogenic responses independently of the major histocompatibility complex. Scand J Immunol. 2003;57(1):11–20.
49. Aggarwal S, Pittenger MF. Human mesenchymal stem cells modulate allogeneic immune cell responses. Blood. 2005;105(4):1815–22. doi:10.1182/blood-2004-04-1559.
50. Bartholomew A, Sturgeon C, Siatskas M, Ferrer K, McIntosh K, Patil S, Hardy W, Devine S, Ucker D, Deans R, Moseley A, Hoffman R. Mesenchymal stem cells suppress lymphocyte proliferation in vitro and prolong skin graft survival in vivo. Exp Hematol. 2002;30(1):42–8.
51. Sotiropoulou PA, Perez SA, Gritzapis AD, Baxevanis CN, Papamichail M. Interactions between human mesenchymal stem cells and natural killer cells. Stem Cells. 2006;24(1):74–85. doi:10.1634/stemcells.2004-0359.
52. Jiang XX, Zhang Y, Liu B, Zhang SX, Wu Y, Yu XD, Mao N. Human mesenchymal stem cells inhibit differentiation and function of monocyte-derived dendritic cells. Blood. 2005;105(10):4120–6. doi:10.1182/blood-2004-02-0586.
53. Ye Z, Wang Y, Xie HY, Zheng SS. Immunosuppressive effects of rat mesenchymal stem cells: involvement of CD4+CD25+ regulatory T cells. Hepatobiliary Pancreat Dis Int. 2008;7(6):608–14.
54. Meisel R, Zibert A, Laryea M, Gobel U, Daubener W, Dilloo D. Human bone marrow stromal cells inhibit allogeneic T-cell responses by indoleamine 2,3-dioxygenase-mediated tryptophan degradation. Blood. 2004;103(12):4619–21. doi:10.1182/blood-2003-11-3909.
55. Sato K, Ozaki K, Oh I, Meguro A, Hatanaka K, Nagai T, Muroi K, Ozawa K. Nitric oxide plays a critical role in suppression of T-cell proliferation by mesenchymal stem cells. Blood. 2007;109(1):228–34. doi:10.1182/blood-2006-02-002246.
56. Ding Y, Xu D, Feng G, Bushell A, Muschel RJ, Wood KJ. Mesenchymal stem cells prevent the rejection of fully allogenic islet grafts by the immunosuppressive activity of matrix metalloproteinase-2 and -9. Diabetes. 2009;58(8):1797–806. doi:10.2337/db09-0317.
57. Selmani Z, Naji A, Zidi I, Favier B, Gaiffe E, Obert L, Borg C, Saas P, Tiberghien P, Rouas-Freiss N, Carosella ED, Deschaseaux F. Human leukocyte antigen-G5 secretion by human mesenchymal stem cells is required to suppress T lymphocyte and natural killer function and to induce CD4+CD25highFOXP3+ regulatory T cells. Stem Cells. 2008;26(1):212–22. doi:10.1634/stemcells.2007-0554.
58. Glennie S, Soeiro I, Dyson PJ, Lam EW, Dazzi F. Bone marrow mesenchymal stem cells induce division arrest anergy of activated T cells. Blood. 2005;105(7):2821–7. doi:10.1182/blood-2004-09-3696.

59. Ghannam S, Pene J, Moquet-Torcy G, Jorgensen C, Yssel H. Mesenchymal stem cells inhibit human Th17 cell differentiation and function and induce a T regulatory cell phenotype. J Immunol. 2010;185(1):302–12. doi:10.4049/jimmunol.0902007.
60. Jung KH, Song SU, Yi T, Jeon MS, Hong SW, Zheng HM, Lee HS, Choi MJ, Lee DH, Hong SS. Human bone marrow-derived clonal mesenchymal stem cells inhibit inflammation and reduce acute pancreatitis in rats. Gastroenterology. 2011;140(3):998–1008. doi:10.1053/j.gastro.2010.11.047.
61. Lim JH, Kim JS, Yoon IH, Shin JS, Nam HY, Yang SH, Kim SJ, Park CG. Immunomodulation of delayed-type hypersensitivity responses by mesenchymal stem cells is associated with bystander T cell apoptosis in the draining lymph node. J Immunol. 2010;185(7):4022–9. doi:10.4049/jimmunol.0902723.
62. Yi T, Song SU. Immunomodulatory properties of mesenchymal stem cells and their therapeutic applications. Arch Pharm Res. 2012;35(2):213–21. doi:10.1007/s12272-012-0202-z.
63. Deans RJ, Moseley AB. Mesenchymal stem cells: biology and potential clinical uses. Exp Hematol. 2000;28(8):875–84.
64. Le Blanc K, Tammik C, Rosendahl K, Zetterberg E, Ringden O. HLA expression and immunologic properties of differentiated and undifferentiated mesenchymal stem cells. Exp Hematol. 2003;31(10):890–6.
65. Gu L, Xiong X, Zhang H, Xu B, Steinberg GK, Zhao H. Distinctive effects of T cell subsets in neuronal injury induced by cocultured splenocytes in vitro and by in vivo stroke in mice. Stroke. 2012;43(7):1941–6. doi:10.1161/STROKEAHA.112.656611.
66. Baruch K, Ron-Harel N, Gal H, Deczkowska A, Shifrut E, Ndifon W, Mirlas-Neisberg N, Cardon M, Vaknin I, Cahalon L, Berkutzki T, Mattson MP, Gomez-Pinilla F, Friedman N, Schwartz M. CNS-specific immunity at the choroid plexus shifts toward destructive Th2 inflammation in brain aging. Proc Natl Acad Sci U S A. 2013;110(6):2264–9. doi:10.1073/pnas.1211270110.
67. Wan YY, Flavell RA. Identifying Foxp3-expressing suppressor T cells with a bicistronic reporter. Proc Natl Acad Sci U S A. 2005;102(14):5126–31. doi:10.1073/pnas.0501701102.
68. Zhao E, Wang L, Dai J, Kryczek I, Wei S, Vatan L, Altuwaijri S, Sparwasser T, Wang G, Keller ET, Zou W. Regulatory T cells in the bone marrow microenvironment in patients with prostate cancer. Oncoimmunology. 2012;1(2):152–61. doi:10.4161/onci.1.2.18480.
69. Zou L, Barnett B, Safah H, Larussa VF, Evdemon-Hogan M, Mottram P, Wei S, David O, Curiel TJ, Zou W. Bone marrow is a reservoir for CD4+CD25+ regulatory T cells that traffic through CXCL12/CXCR4 signals. Cancer Res. 2004;64(22):8451–5. doi:10.1158/0008-5472.CAN-04-1987.
70. Krampera M, Glennie S, Dyson J, Scott D, Laylor R, Simpson E, Dazzi F. Bone marrow mesenchymal stem cells inhibit the response of naive and memory antigen-specific T cells to their cognate peptide. Blood. 2003;101(9):3722–9. doi:10.1182/blood-2002-07-2104.
71. Ge W, Jiang J, Arp J, Liu W, Garcia B, Wang H. Regulatory T-cell generation and kidney allograft tolerance induced by mesenchymal stem cells associated with indoleamine 2,3-dioxygenase expression. Transplantation. 2010;90(12):1312–20. doi:10.1097/TP.0b013e3181fed001.
72. Tipnis S, Viswanathan C, Majumdar AS. Immunosuppressive properties of human umbilical cord-derived mesenchymal stem cells: role of B7-H1 and IDO. Immunol Cell Biol. 2010;88(8):795–806. doi:10.1038/icb.2010.47.
73. Djouad F, Plence P, Bony C, Tropel P, Apparailly F, Sany J, Noel D, Jorgensen C. Immunosuppressive effect of mesenchymal stem cells favors tumor growth in allogeneic animals. Blood. 2003;102(10):3837–44. doi:10.1182/blood-2003-04-1193.
74. Liesz A, Suri-Payer E, Veltkamp C, Doerr H, Sommer C, Rivest S, Giese T, Veltkamp R. Regulatory T cells are key cerebroprotective immunomodulators in acute experimental stroke. Nat Med. 2009;15(2):192–9. doi:10.1038/nm.1927.

75. English K, Ryan JM, Tobin L, Murphy MJ, Barry FP, Mahon BP. Cell contact, prostaglandin E(2) and transforming growth factor beta 1 play non-redundant roles in human mesenchymal stem cell induction of CD4+CD25(High) forkhead box P3+ regulatory T cells. Clin Exp Immunol. 2009;156(1):149–60. doi:10.1111/j.1365-2249.2009.03874.x.
76. Zhang Q, Shi S, Liu Y, Uyanne J, Shi Y, Shi S, Le AD. Mesenchymal stem cells derived from human gingiva are capable of immunomodulatory functions and ameliorate inflammation-related tissue destruction in experimental colitis. J Immunol. 2009;183(12):7787–98. doi:10.4049/jimmunol.0902318.
77. Yang N, Baban B, Isales CM, Shi XM. Crosstalk between bone marrow-derived mesenchymal stem cells and regulatory T cells through a glucocorticoid-induced leucine zipper/developmental endothelial locus-1-dependent mechanism. FASEB J. 2015;29(9):3954–63. doi:10.1096/fj.15-273664.
78. Le Blanc K, Mougiakakos D. Multipotent mesenchymal stromal cells and the innate immune system. Nat Rev Immunol. 2012;12(5):383–96. doi:10.1038/nri3209.
79. Shichita T, Sugiyama Y, Ooboshi H, Sugimori H, Nakagawa R, Takada I, Iwaki T, Okada Y, Iida M, Cua DJ, Iwakura Y, Yoshimura A. Pivotal role of cerebral interleukin-17-producing gammadelta T cells in the delayed phase of ischemic brain injury. Nat Med. 2009;15(8):946–50. doi:10.1038/nm.1999.
80. Prigione I, Benvenuto F, Bocca P, Battistini L, Uccelli A, Pistoia V. Reciprocal interactions between human mesenchymal stem cells and gammadelta T cells or invariant natural killer T cells. Stem Cells. 2009;27(3):693–702. doi:10.1634/stemcells.2008-0687.
81. Wu J, Wang Q, Fu X, Wu X, Gu C, Bi J, Xie F, Kang N, Liu X, Yan L, Cao Y, Xiao R. Influence of immunogenicity of allogeneic bone marrow mesenchymal stem cells on bone tissue engineering. Cell Transplant. 2016;25(2):229–42. doi:10.3727/096368915X687967.
82. Wang LQ, Lin ZZ, Zhang HX, Shao B, Xiao L, Jiang HG, Zhuge QC, Xie LK, Wang B, Su DM, Jin KL. Timing and dose regimens of marrow mesenchymal stem cell transplantation affect the outcomes and neuroinflammatory response after ischemic stroke. CNS Neurosci Ther. 2014;20(4):317–26. doi:10.1111/cns.12216.
83. Shi M, Liu ZW, Wang FS. Immunomodulatory properties and therapeutic application of mesenchymal stem cells. Clin Exp Immunol. 2011;164(1):1–8. doi:10.1111/j.1365-2249.2011.04327.x.
84. Le Blanc K, Ringden O. Immunomodulation by mesenchymal stem cells and clinical experience. J Intern Med. 2007;262(5):509–25. doi:10.1111/j.1365-2796.2007.01844.x.
85. Deng W, Han Q, Liao L, You S, Deng H, Zhao RC. Effects of allogeneic bone marrow-derived mesenchymal stem cells on T and B lymphocytes from BXSB mice. DNA Cell Biol. 2005;24(7):458–63. doi:10.1089/dna.2005.24.458.
86. Augello A, Tasso R, Negrini SM, Amateis A, Indiveri F, Cancedda R, Pennesi G. Bone marrow mesenchymal progenitor cells inhibit lymphocyte proliferation by activation of the programmed death 1 pathway. Eur J Immunol. 2005;35(5):1482–90. doi:10.1002/eji.200425405.
87. Corcione A, Benvenuto F, Ferretti E, Giunti D, Cappiello V, Cazzanti F, Risso M, Gualandi F, Mancardi GL, Pistoia V, Uccelli A. Human mesenchymal stem cells modulate B-cell functions. Blood. 2006;107(1):367–72. doi:10.1182/blood-2005-07-2657.
88. Clutter SD, Fortney J, Gibson LF. MMP-2 is required for bone marrow stromal cell support of pro-B-cell chemotaxis. Exp Hematol. 2005;33(10):1192–200. doi:10.1016/j.exphem.2005.06.022.
89. Powell RJ, Jenkins JS. Lymphocyte subpopulations. Postgrad Med J. 1987;63(745):931–5.
90. Spaggiari GM, Capobianco A, Becchetti S, Mingari MC, Moretta L. Mesenchymal stem cell-natural killer cell interactions: evidence that activated NK cells are capable of killing MSCs, whereas MSCs can inhibit IL-2-induced NK-cell proliferation. Blood. 2006;107(4):1484–90. doi:10.1182/blood-2005-07-2775.
91. Rasmusson I, Ringden O, Sundberg B, Le Blanc K. Mesenchymal stem cells inhibit the formation of cytotoxic T lymphocytes, but not activated cytotoxic T lymphocytes or natural killer cells. Transplantation. 2003;76(8):1208–13. doi:10.1097/01.TP.0000082540.43730.80.

92. Ryan JM, Barry F, Murphy JM, Mahon BP. Interferon-gamma does not break, but promotes the immunosuppressive capacity of adult human mesenchymal stem cells. Clin Exp Immunol. 2007;149(2):353–63. doi:10.1111/j.1365-2249.2007.03422.x.
93. Maccario R, Podesta M, Moretta A, Cometa A, Comoli P, Montagna D, Daudt L, Ibatici A, Piaggio G, Pozzi S, Frassoni F, Locatelli F. Interaction of human mesenchymal stem cells with cells involved in alloantigen-specific immune response favors the differentiation of CD4+ T-cell subsets expressing a regulatory/suppressive phenotype. Haematologica. 2005;90(4):516–25.
94. Nauta AJ, Kruisselbrink AB, Lurvink E, Willemze R, Fibbe WE. Mesenchymal stem cells inhibit generation and function of both CD34+-derived and monocyte-derived dendritic cells. J Immunol. 2006;177(4):2080–7.
95. Spaggiari GM, Abdelrazik H, Becchetti F, Moretta L. MSCs inhibit monocyte-derived DC maturation and function by selectively interfering with the generation of immature DCs: central role of MSC-derived prostaglandin E2. Blood. 2009;113(26):6576–83. doi:10.1182/blood-2009-02-203943.
96. Ramasamy R, Fazekasova H, Lam EW, Soeiro I, Lombardi G, Dazzi F. Mesenchymal stem cells inhibit dendritic cell differentiation and function by preventing entry into the cell cycle. Transplantation. 2007;83(1):71–6. doi:10.1097/01.tp.0000244572.24780.54.
97. Wang Y, Tu W, Lou Y, Xie A, Lai X, Guo F, Deng Z. Mesenchymal stem cells regulate the proliferation and differentiation of neural stem cells through Notch signaling. Cell Biol Int. 2009;33(11):1173–9. doi:10.1016/j.cellbi.2009.08.004.
98. Lou S, Gu P, Chen F, He C, Wang M, Lu C. The effect of bone marrow stromal cells on neuronal differentiation of mesencephalic neural stem cells in Sprague-Dawley rats. Brain Res. 2003;968(1):114–21.
99. Bai L, Caplan A, Lennon D, Miller RH. Human mesenchymal stem cells signals regulate neural stem cell fate. Neurochem Res. 2007;32(2):353–62. doi:10.1007/s11064-006-9212-x.
100. Rivera FJ, Kandasamy M, Couillard-Despres S, Caioni M, Sanchez R, Huber C, Weidner N, Bogdahn U, Aigner L. Oligodendrogenesis of adult neural progenitors: differential effects of ciliary neurotrophic factor and mesenchymal stem cell derived factors. J Neurochem. 2008;107(3):832–43. doi:10.1111/j.1471-4159.2008.05674.x.
101. Rivera FJ, Couillard-Despres S, Pedre X, Ploetz S, Caioni M, Lois C, Bogdahn U, Aigner L. Mesenchymal stem cells instruct oligodendrogenic fate decision on adult neural stem cells. Stem Cells. 2006;24(10):2209–19. doi:10.1634/stemcells.2005-0614.
102. Li X, Liu T, Song K, Yao L, Ge D, Bao C, Ma X, Cui Z. Culture of neural stem cells in calcium alginate beads. Biotechnol Prog. 2006;22(6):1683–9. doi:10.1021/bp060185z.
103. Androutsellis-Theotokis A, Leker RR, Soldner F, Hoeppner DJ, Ravin R, Poser SW, Rueger MA, Bae SK, Kittappa R, McKay RD. Notch signalling regulates stem cell numbers in vitro and in vivo. Nature. 2006;442(7104):823–6. doi:10.1038/nature04940.
104. Grandbarbe L, Bouissac J, Rand M, Hrabe de Angelis M, Artavanis-Tsakonas S, Mohier E. Delta-Notch signaling controls the generation of neurons/glia from neural stem cells in a stepwise process. Development. 2003;130(7):1391–402.
105. Chojnacki A, Shimazaki T, Gregg C, Weinmaster G, Weiss S. Glycoprotein 130 signaling regulates Notch1 expression and activation in the self-renewal of mammalian forebrain neural stem cells. J Neurosci. 2003;23(5):1730–41.
106. Robinson AP, Foraker JE, Ylostalo J, Prockop DJ. Human stem/progenitor cells from bone marrow enhance glial differentiation of rat neural stem cells: a role for transforming growth factor beta and Notch signaling. Stem Cells Dev. 2011;20(2):289–300. doi:10.1089/scd.2009.0444.
107. Kinnaird T, Stabile E, Burnett MS, Shou M, Lee CW, Barr S, Fuchs S, Epstein SE. Local delivery of marrow-derived stromal cells augments collateral perfusion through paracrine mechanisms. Circulation. 2004;109(12):1543–9. doi:10.1161/01.CIR.0000124062.31102.57.

108. Mahmood A, Lu D, Chopp M. Intravenous administration of marrow stromal cells (MSCs) increases the expression of growth factors in rat brain after traumatic brain injury. J Neurotrauma. 2004;21(1):33–9. doi:10.1089/089771504772695922.
109. Munoz JR, Stoutenger BR, Robinson AP, Spees JL, Prockop DJ. Human stem/progenitor cells from bone marrow promote neurogenesis of endogenous neural stem cells in the hippocampus of mice. Proc Natl Acad Sci U S A. 2005;102(50):18171–6. doi:10.1073/pnas.0508945102.
110. Haynesworth SE, Baber MA, Caplan AI. Cytokine expression by human marrow-derived mesenchymal progenitor cells in vitro: effects of dexamethasone and IL-1 alpha. J Cell Physiol. 1996;166(3):585–92.doi:10.1002/(SICI)1097-4652(199603)166:3<585::AID-JCP13>3.0.CO;2-6.
111. Labouyrie E, Dubus P, Groppi A, Mahon FX, Ferrer J, Parrens M, Reiffers J, de Mascarel A, Merlio JP. Expression of neurotrophins and their receptors in human bone marrow. Am J Pathol. 1999;154(2):405–15. doi:10.1016/S0002-9440(10)65287-X.
112. Chen J, Li Y, Katakowski M, Chen X, Wang L, Lu D, Lu M, Gautam SC, Chopp M. Intravenous bone marrow stromal cell therapy reduces apoptosis and promotes endogenous cell proliferation after stroke in female rat. J Neurosci Res. 2003;73(6):778–86. doi:10.1002/jnr.10691.
113. Zhao LR, Duan WM, Reyes M, Keene CD, Verfaillie CM, Low WC. Human bone marrow stem cells exhibit neural phenotypes and ameliorate neurological deficits after grafting into the ischemic brain of rats. Exp Neurol. 2002;174(1):11–20. doi:10.1006/exnr.2001.7853.
114. Neuhuber B, Gallo G, Howard L, Kostura L, Mackay A, Fischer I. Reevaluation of in vitro differentiation protocols for bone marrow stromal cells: disruption of actin cytoskeleton induces rapid morphological changes and mimics neuronal phenotype. J Neurosci Res. 2004;77(2):192–204. doi:10.1002/jnr.20147.
115. Lu P, Blesch A, Tuszynski MH. Induction of bone marrow stromal cells to neurons: differentiation, transdifferentiation, or artifact? J Neurosci Res. 2004;77(2):174–91. doi:10.1002/jnr.20148.
116. Croft AP, Przyborski SA. Mesenchymal stem cells expressing neural antigens instruct a neurogenic cell fate on neural stem cells. Exp Neurol. 2009;216(2):329–41. doi:10.1016/j.expneurol.2008.12.010.
117. Sygnecka K, Heider A, Scherf N, Alt R, Franke H, Heine C. Mesenchymal stem cells support neuronal fiber growth in an organotypic brain slice co-culture model. Stem Cells Dev. 2015;24(7):824–35. doi:10.1089/scd.2014.0262.
118. Bajetto A, Bonavia R, Barbero S, Florio T, Schettini G. Chemokines and their receptors in the central nervous system. Front Neuroendocrinol. 2001;22(3):147–84. doi:10.1006/frne.2001.0214.
119. Lee H, Kang JE, Lee JK, Bae JS, Jin HK. Bone-marrow-derived mesenchymal stem cells promote proliferation and neuronal differentiation of Niemann-Pick type C mouse neural stem cells by upregulation and secretion of CCL2. Hum Gene Ther. 2013;24(7):655–69. doi:10.1089/hum.2013.001.
120. Yoo SW, Kim SS, Lee SY, Lee HS, Kim HS, Lee YD, Suh-Kim H. Mesenchymal stem cells promote proliferation of endogenous neural stem cells and survival of newborn cells in a rat stroke model. Exp Mol Med. 2008;40(4):387–97. doi:10.3858/emm.2008.40.4.387.
121. Li Y, Chen J, Chen XG, Wang L, Gautam SC, Xu YX, Katakowski M, Zhang LJ, Lu M, Janakiraman N, Chopp M. Human marrow stromal cell therapy for stroke in rat: neurotrophins and functional recovery. Neurology. 2002;59(4):514–23.
122. Celil AB, Campbell PG. BMP-2 and insulin-like growth factor-I mediate Osterix (Osx) expression in human mesenchymal stem cells via the MAPK and protein kinase D signaling pathways. J Biol Chem. 2005;280(36):31353–9. doi:10.1074/jbc.M503845200.
123. Esneault E, Pacary E, Eddi D, Freret T, Tixier E, Toutain J, Touzani O, Schumann-Bard P, Petit E, Roussel S, Bernaudin M. Combined therapeutic strategy using erythropoietin and mesenchymal stem cells potentiates neurogenesis after transient focal cerebral ischemia in rats. J Cereb Blood Flow Metab. 2008;28(9):1552–63. doi:10.1038/jcbfm.2008.40.
124. Hosseini SM, Farahmandnia M, Razi Z, Delavarifar S, Shakibajahromi B. 12 hours after cerebral ischemia is the optimal time for bone marrow mesenchymal stem cell transplantation. Neural Regen Res. 2015;10(6):904–8. doi:10.4103/1673-5374.158354.

125. Andrae J, Afink G, Zhang XQ, Wurst W, Nister M. Forced expression of platelet-derived growth factor B in the mouse cerebellar primordium changes cell migration during midline fusion and causes cerebellar ectopia. Mol Cell Neurosci. 2004;26(2):308–21. doi:10.1016/j.mcn.2004.02.004.
126. Vasyutina E, Stebler J, Brand-Saberi B, Schulz S, Raz E, Birchmeier C. CXCR4 and Gab1 cooperate to control the development of migrating muscle progenitor cells. Genes Dev. 2005;19(18):2187–98. doi:10.1101/gad.346205.
127. Liu XS, Zhang ZG, Zhang RL, Gregg SR, Wang L, Yier T, Chopp M. Chemokine ligand 2 (CCL2) induces migration and differentiation of subventricular zone cells after stroke. J Neurosci Res. 2007;85(10):2120–5. doi:10.1002/jnr.21359.
128. Thored P, Arvidsson A, Cacci E, Ahlenius H, Kallur T, Darsalia V, Ekdahl CT, Kokaia Z, Lindvall O. Persistent production of neurons from adult brain stem cells during recovery after stroke. Stem Cells. 2006;24(3):739–47. doi:10.1634/stemcells.2005-0281.
129. Shen LH, Li Y, Chen J, Zacharek A, Gao Q, Kapke A, Lu M, Raginski K, Vanguri P, Smith A, Chopp M. Therapeutic benefit of bone marrow stromal cells administered 1 month after stroke. J Cereb Blood Flow Metab. 2007;27(1):6–13. doi:10.1038/sj.jcbfm.9600311.
130. Kim YJ, Park HJ, Lee G, Bang OY, Ahn YH, Joe E, Kim HO, Lee PH. Neuroprotective effects of human mesenchymal stem cells on dopaminergic neurons through anti-inflammatory action. Glia. 2009;57(1):13–23. doi:10.1002/glia.20731.
131. Hosseini SM, Farahmandnia M, Razi Z, Delavari S, Shakibajahromi B, Sarvestani FS, Kazemi S, Semsar M. Combination cell therapy with mesenchymal stem cells and neural stem cells for brain stroke in rats. Int J Stem Cells. 2015;8(1):99–105. doi:10.15283/ijsc.2015.8.1.99.
132. Friedenstein AJ, Chailakhyan RK, Gerasimov UV. Bone marrow osteogenic stem cells: in vitro cultivation and transplantation in diffusion chambers. Cell Tissue Kinet. 1987;20(3):263–72.
133. Bianco P, Gehron Robey P. Marrow stromal stem cells. J Clin Invest. 2000;105(12):1663–8. doi:10.1172/JCI10413.
134. Friedenstein AJ, Petrakova KV, Kurolesova AI, Frolova GP. Heterotopic of bone marrow. Analysis of precursor cells for osteogenic and hematopoietic tissues. Transplantation. 1968;6(2):230–47.
135. Kamei N, Tanaka N, Oishi Y, Ishikawa M, Hamasaki T, Nishida K, Nakanishi K, Sakai N, Ochi M. Bone marrow stromal cells promoting corticospinal axon growth through the release of humoral factors in organotypic cocultures in neonatal rats. J Neurosurg Spine. 2007;6(5):412–9. doi:10.3171/spi.2007.6.5.412.
136. Shichinohe H, Kuroda S, Tsuji S, Yamaguchi S, Yano S, Lee JB, Kobayashi H, Kikuchi S, Hida K, Iwasaki Y. Bone marrow stromal cells promote neurite extension in organotypic spinal cord slice: significance for cell transplantation therapy. Neurorehabil Neural Repair. 2008;22(5):447–57. doi:10.1177/1545968308315596.
137. Salomone R, Bento RF, Costa HJ, Azzi-Nogueira D, Ovando PC, Da-Silva CF, Zanatta DB, Strauss BE, Haddad LA. Bone marrow stem cells in facial nerve regeneration from isolated stumps. Muscle Nerve. 2013;48(3):423–9. doi:10.1002/mus.23768.
138. Gu W, Zhang F, Xue Q, Ma Z, Lu P, Yu B. Bone mesenchymal stromal cells stimulate neurite outgrowth of spinal neurons by secreting neurotrophic factors. Neurol Res. 2012;34(2):172–80. doi:10.1179/1743132811Y.0000000068.
139. Shichinohe H, Yamauchi T, Saito H, Houkin K, Kuroda S. Bone marrow stromal cell transplantation enhances recovery of motor function after lacunar stroke in rats. Acta Neurobiol Exp (Wars). 2013;73(3):354–63.
140. Shichinohe H, Kuroda S, Yano S, Ohnishi T, Tamagami H, Hida K, Iwasaki Y. Improved expression of gamma-aminobutyric acid receptor in mice with cerebral infarct and transplanted bone marrow stromal cells: an autoradiographic and histologic analysis. J Nucl Med. 2006;47(3):486–91.
141. Chen J, Zacharek A, Li Y, Li A, Wang L, Katakowski M, Roberts C, Lu M, Chopp M. N-cadherin mediates nitric oxide-induced neurogenesis in young and retired breeder neurospheres. Neuroscience. 2006;140(2):377–88. doi:10.1016/j.neuroscience.2006.02.064.

142. Weng JS, Liu N, Du HW, Chen RH, Zhang YX, Wang JH, Huang HP. Effects of bone marrow-derived mesenchymal stem cells transplantation on recovery of neurological functions and expression of synaptophysin in focal cerebral infarction in rats. Xi Bao Yu Fen Zi Mian Yi Xue Za Zhi. 2008;24(1):34–7.
143. Huang W, Mo X, Qin C, Zheng J, Liang Z, Zhang C. Transplantation of differentiated bone marrow stromal cells promotes motor functional recovery in rats with stroke. Neurol Res. 2013;35(3):320–8. doi:10.1179/1743132812Y.0000000151.
144. Chen J, Li Y, Wang L, Zhang Z, Lu D, Lu M, Chopp M. Therapeutic benefit of intravenous administration of bone marrow stromal cells after cerebral ischemia in rats. Stroke. 2001;32(4):1005–11.
145. Tohill M, Mantovani C, Wiberg M, Terenghi G. Rat bone marrow mesenchymal stem cells express glial markers and stimulate nerve regeneration. Neurosci Lett. 2004;362(3):200–3. doi:10.1016/j.neulet.2004.03.077.
146. Andrews EM, Tsai SY, Johnson SC, Farrer JR, Wagner JP, Kopen GC, Kartje GL. Human adult bone marrow-derived somatic cell therapy results in functional recovery and axonal plasticity following stroke in the rat. Exp Neurol. 2008;211(2):588–92. doi:10.1016/j.expneurol.2008.02.027.
147. Shen LH, Li Y, Chen J, Zhang J, Vanguri P, Borneman J, Chopp M. Intracarotid transplantation of bone marrow stromal cells increases axon-myelin remodeling after stroke. Neuroscience. 2006;137(2):393–9. doi:10.1016/j.neuroscience.2005.08.092.
148. Li Y, Chopp M, Chen J, Wang L, Gautam SC, Xu YX, Zhang Z. Intrastriatal transplantation of bone marrow nonhematopoietic cells improves functional recovery after stroke in adult mice. J Cereb Blood Flow Metab. 2000;20(9):1311–9. doi:10.1097/00004647-200009000-00006.
149. Saito H, Magota K, Zhao S, Kubo N, Kuge Y, Shichinohe H, Houkin K, Tamaki N, Kuroda S. 123I-iomazenil single photon emission computed tomography visualizes recovery of neuronal integrity by bone marrow stromal cell therapy in rat infarct brain. Stroke. 2013;44(10):2869–74. doi:10.1161/STROKEAHA.113.001612.
150. Song M, Mohamad O, Gu X, Wei L, Yu SP. Restoration of intracortical and thalamocortical circuits after transplantation of bone marrow mesenchymal stem cells into the ischemic brain of mice. Cell Transplant. 2013;22(11):2001–15. doi:10.3727/096368912X657909.
151. Liu Z, Li Y, Qu R, Shen L, Gao Q, Zhang X, Lu M, Savant-Bhonsale S, Borneman J, Chopp M. Axonal sprouting into the denervated spinal cord and synaptic and postsynaptic protein expression in the spinal cord after transplantation of bone marrow stromal cell in stroke rats. Brain Res. 2007;1149:172–80. doi:10.1016/j.brainres.2007.02.047.
152. Liu Z, Li Y, Zhang ZG, Cui X, Cui Y, Lu M, Savant-Bhonsale S, Chopp M. Bone marrow stromal cells enhance inter- and intracortical axonal connections after ischemic stroke in adult rats. J Cereb Blood Flow Metab. 2010;30(7):1288–95. doi:10.1038/jcbfm.2010.8.
153. Liang H, Yin Y, Lin T, Guan D, Ma B, Li C, Wang Y, Zhang X. Transplantation of bone marrow stromal cells enhances nerve regeneration of the corticospinal tract and improves recovery of neurological functions in a collagenase-induced rat model of intracerebral hemorrhage. Mol Cells. 2013;36(1):17–24. doi:10.1007/s10059-013-2306-9.
154. Wen L, Wen N, Wang Y, Ding Y. Effect of rat endothelial progenitor cell on proliferation and apoptosis of bone marrow mesenchymal stem cell. Zhonghua Kou Qiang Yi Xue Za Zhi. 2014;49(2):106–10.
155. Cui X, Chopp M, Zacharek A, Roberts C, Lu M, Savant-Bhonsale S, Chen J. Chemokine, vascular and therapeutic effects of combination Simvastatin and BMSC treatment of stroke. Neurobiol Dis. 2009;36(1):35–41. doi:10.1016/j.nbd.2009.06.012.
156. Wei ZZ, Gu X, Ferdinand A, Lee JH, Ji X, Ji XM, Yu SP, Wei L. Intranasal delivery of bone marrow mesenchymal stem cells improved neurovascular regeneration and rescued neuropsychiatric deficits after neonatal stroke in rats. Cell Transplant. 2015;24(3):391–402. doi:10.3727/096368915X686887.
157. Zacharek A, Shehadah A, Chen J, Cui X, Roberts C, Lu M, Chopp M. Comparison of bone marrow stromal cells derived from stroke and normal rats for stroke treatment. Stroke. 2010;41(3):524–30. doi:10.1161/STROKEAHA.109.568881.

158. Krupinski J, Kaluza J, Kumar P, Kumar S, Wang JM. Role of angiogenesis in patients with cerebral ischemic stroke. Stroke. 1994;25(9):1794–8.
159. Hamano K, Li TS, Kobayashi T, Kobayashi S, Matsuzaki M, Esato K. Angiogenesis induced by the implantation of self-bone marrow cells: a new material for therapeutic angiogenesis. Cell Transplant. 2000;9(3):439–43.
160. Shyu WC, Lin SZ, Yen PS, Su CY, Chen DC, Wang HJ, Li H. Stromal cell-derived factor-1 alpha promotes neuroprotection, angiogenesis, and mobilization/homing of bone marrow-derived cells in stroke rats. J Pharmacol Exp Ther. 2008;324(2):834–49. doi:10.1124/jpet.107.127746.
161. Cui X, Chen J, Zacharek A, Li Y, Roberts C, Kapke A, Savant-Bhonsale S, Chopp M. Nitric oxide donor upregulation of stromal cell-derived factor-1/chemokine (CXC motif) receptor 4 enhances bone marrow stromal cell migration into ischemic brain after stroke. Stem Cells. 2007;25(11):2777–85. doi:10.1634/stemcells.2007-0169.
162. Lennmyr F, Ata KA, Funa K, Olsson Y, Terent A. Expression of vascular endothelial growth factor (VEGF) and its receptors (Flt-1 and Flk-1) following permanent and transient occlusion of the middle cerebral artery in the rat. J Neuropathol Exp Neurol. 1998;57(9):874–82.
163. Goldmacher GV, Nasser R, Lee DY, Yigit S, Rosenwasser R, Iacovitti L. Tracking transplanted bone marrow stem cells and their effects in the rat MCAO stroke model. PLoS One. 2013;8(3):e60049. doi:10.1371/journal.pone.0060049.
164. Yan K, Zhang R, Sun C, Chen L, Li P, Liu Y, Peng L, Sun H, Qin K, Chen F, Huang W, Chen Y, Lv B, Du M, Zou Y, Cai Y, Qin L, Tang Y, Jiang X. Bone marrow-derived mesenchymal stem cells maintain the resting phenotype of microglia and inhibit microglial activation. PLoS One. 2013;8(12):e84116. doi:10.1371/journal.pone.0084116.
165. Opydo-Chanek M, Dabrowski Z. Response of astrocytes and microglia/macrophages to brain injury after bone marrow stromal cell transplantation: a quantitative study. Neurosci Lett. 2011;487(2):163–8. doi:10.1016/j.neulet.2010.10.014.
166. Pirzad Jahromi G, Shabanzadeh Pirsaraei A, Sadr SS, Kaka G, Jafari M, Seidi S, Charish J. Multipotent bone marrow stromal cell therapy promotes endogenous cell proliferation following ischemic stroke. Clin Exp Pharmacol Physiol. 2015;42(11):1158–67. doi:10.1111/1440-1681.12466.
167. el Gornicka-Pawlak B, Janowski M, Habich A, Jablonska A, Drela K, Kozlowska H, Lukomska B, Sypecka J, Domanska-Janik K. Systemic treatment of focal brain injury in the rat by human umbilical cord blood cells being at different level of neural commitment. Acta Neurobiol Exp (Wars). 2011;71(1):46–64.
168. Shen LH, Li Y, Chopp M. Astrocytic endogenous glial cell derived neurotrophic factor production is enhanced by bone marrow stromal cell transplantation in the ischemic boundary zone after stroke in adult rats. Glia. 2010;58(9):1074–81. doi:10.1002/glia.20988.
169. Xin H, Li Y, Chen X, Chopp M. Bone marrow stromal cells induce BMP2/4 production in oxygen-glucose-deprived astrocytes, which promotes an astrocytic phenotype in adult subventricular progenitor cells. J Neurosci Res. 2006;83(8):1485–93. doi:10.1002/jnr.20834.
170. Zhang J, Li Y, Zheng X, Gao Q, Liu Z, Qu R, Borneman J, Elias SB, Chopp M. Bone marrow stromal cells protect oligodendrocytes from oxygen-glucose deprivation injury. J Neurosci Res. 2008;86(7):1501–10. doi:10.1002/jnr.21617.
171. Shen Q, Yin Y, Xia QJ, Lin N, Wang YC, Liu J, Wang HP, Lim A, Wang TH. Bone marrow stromal cells promote neuronal restoration in rats with traumatic brain injury: involvement of GDNF regulating BAD and BAX signaling. Cell Physiol Biochem. 2016;38(2):748–62. doi:10.1159/000443031.

Chapter 5
Mesenchymal Stem Cells for Stroke Therapy

Yunqian Guan, Xunming Ji, Jieli Chen, Y. Alex Zhang, and Zhiguo Chen

Abstract Mesenchymal stem/stromal cells (MSCs) from different sources have been tested in experimental ischemic stroke and been proved effective in many studies. More understanding of the underlying mechanisms for MSC therapy is essential for improving the therapeutic efficacy and ameliorating or avoiding side effects. In this chapter, we reviewed progress in the field that includes the following aspects: (1) the neurotrophic and immunomodulatory functions of MSCs, the two main mechanisms through which MSCs exert effects in stroke treatment; (2) the aging and senescence of MSCs and their influence on outcome; (3) how autologous MSCs derived from diseased patients may affect the therapeutic effects; (4) different sources of MSCs and comparison of their effects in therapy; and (5) time of transplantation and the delivery methods.

Keywords Mesenchymal stem/stromal cells • Stroke • Transplantation • Neurotrophic • Immunomodulatory

Although a large number of preclinical and clinical tests/trials have been performed over the past 50 years, there are few drugs that have been proved effective for protection or repair of the damaged central nervous system (CNS) following ischemic stroke [1]. However, recent decades have seen a fast development in cell therapy which may potentially promote functional recovery from various CNS disorders, including ischemic stroke.

Y. Guan • Y.A. Zhang • Z. Chen (✉)
Cell Therapy Center, Xuanwu Hospital, Capital Medical University, Beijing, China
e-mail: chenzhiguo@gmail.com

X. Ji
Department of Neurosurgery, Xuanwu Hospital, Capital Medical University, Beijing, China

Institute of Hypoxia Medicine, Xuanwu Hospital, Capital Medical University, Beijing, China

China-America Joint Institute for Neuroscience, Xuanwu Hospital, Capital Medical University, Beijing, China
e-mail: jixm@ccmu.edu.cn

J. Chen
Department of Neurology, Henry Ford Hospital, Detroit, MI, USA

K. Jin et al. (eds.), *Bone Marrow Stem Cell Therapy for Stroke*,
DOI 10.1007/978-981-10-2929-5_5

A variety of cell types have been tested as cell sources of transplantation in animal models of CNS disorders, including embryonic stem (ES) cells, neural stem cells (NSCs), induced pluripotent stem (iPS) cells, and mesenchymal stem/stromal cells (MSCs). Accumulating evidence shows that MSCs may be a promising choice in tissue engineering and regenerative medicine due to their immunosuppressive and anti-inflammatory properties. Of these, bone marrow mesenchymal stem/stromal cells (BM-MSCs) have unique advantages in that they can be harvested from patients without posing ethical or immunological problems [2].

Basic and preclinical data support the translation of MSC therapy to clinical trials. The procedures of ex vivo expansion and transplantation of autologous MSCs are safe and well tolerated [3]. Stroke patients treated with autologous MSCs through an intravenous delivery route show improved functional recovery [4, 5]. MSC administration starting 24 h after stroke via different routes, such as intracerebral, intravenous, or intra-arterial, all promote functional outcome after stroke [6, 7]. In addition, delayed treatment of stroke with MSCs at 7 days or at 1 month after stroke onset also increases brain plasticity and improves long-term functional outcome [6, 8, 9].

Although MSCs are being tested in phase II clinical trials, many problems remain to be resolved, such as their precise mechanisms of action, the effects of long-term culture and aging on the functions of MSCs, the optimal delivery routes and time, and so on.

5.1 Mechanisms of the Neurorestorative Effect of MSCs in Stroke

MSCs consist of heterogeneous cell populations and may protect and repair the damaged CNS through multiple mechanisms [10]. Some early studies proposed a cell replacement mechanism for MSCs in the CNS; other studies suggested that MSCs may rescue neurons and promote the proliferation and maturation of local neural precursors through release of trophic factors. MSCs also have anti-inflammatory and immunomodulatory effects in periphery and the brain, providing a neuroprotective microenvironment.

5.1.1 Replacement May Not Be the Main Mechanism for Recovery

MSCs have a tendency to home to damaged tissue sites. When MSCs are systemically administered to humans and animals, they migrate to damaged tissue sites which are normally associated with inflammation [11]. In the context of stroke, MSCs are trapped in the lungs, spleen, and other organs outside the brain when administered systemically [12, 13]; only a small number may reach the ischemic sites.

BM-MSCs can be administrated into rats after stroke via intra-arterial, intravenous, and intraperitoneal routes. The dynamic distribution of infused MSCs was monitored by real-time imaging. After both intra-arterial and intravenous infusion, MSCs were detected in the lungs, spleen, liver, and other organs [14]. Part of infused MSCs were attracted by injury-induced chemokines and migrated toward the ischemic border [15–17]. Only a small portion (<5 %) were detected in brain parenchyma 14 days after transplantation. More than 80 % of MSCs that had reached the brain were localized to the ipsilateral hemisphere, the majority of which congregated in the ischemic border zones [18].

Controversy still exists regarding whether engrafted MSCs can replace damaged neurons. Some studies indicate that transplanted BM-MSCs can express proteins specific to neurons, astrocytes, and endothelial cells in the peri-infarct areas [19–22]. However, very few transplanted cells are found in the brain, and only a small percentage of these cells express neural markers [23–25]. In addition, when MSCs are administered 24 h after stroke, functional outcome is significantly improved from 7 days after treatment [6]. This benefit is not likely attributed to the very few MSCs differentiating into brain cells. Up to now, it is generally accepted that even though MSCs may trans-differentiate and replace some damaged neurons, it is not the main mechanism of therapeutic action.

In addition, some studies suggest that migration of transplanted MSCs into the brain is of some importance for the therapeutic effects. Compared with intravenous transplantation of MSCs, intra-arterial transplantation following ischemic stroke in rats results in smaller infarct volumes, better behavioral recovery, and a larger number of surviving cells in the brain. It is possible that by intravenous delivery, more cells are trapped in the peripheral organs, especially in the lungs, spleen, and lymphoid. The relatively higher number of cells that migrate to the injury sites through intra-arterial delivery route may presumably account for the observed therapeutic effects [26]. Furthermore, it looks like that the brain areas where MSCs migrate to also matter. A study reports that MSCs that have migrated to ischemic cortical regions, which are in charge of the motor functions of animals, act better than those that have migrated to the striatum, in facilitating the recovery of neurological functions [27, 28].

Other than the slim possibility of trans-differentiation into neurons or glia, the mechanisms underlying the beneficial effects from cells that have infiltrated into the brain may be closely related to the neurotrophic and immunomodulatory effects.

5.1.2 Neurotrophic Effects from MSCs

5.1.2.1 MSCs Secrete Trophic Factors

That functional recovery is detected often with very few transplanted MSCs surviving in the brain suggests that the cells may exert a persistent effect. MSCs may support the survival of the endogenous brain cells through paracrine production of soluble factors, including vascular endothelial growth factor (VEGF), Basic fibroblast growth factor (bFGF), glial-derived neurotrophic factor (GDNF), epidermal growth factor (EGF), brain-derived neurotrophic factor (BDNF), etc. [29, 30].

These growth factors have shown effects in promoting cell proliferation/migration and enhancing angiogenesis and vascular stabilization in the ischemic boundary [29–32]. In addition to cytokines, MSCs also produce factors responsible for extracellular matrix remodeling, such as collagen, matrix metalloproteinases (MMPs), and the tissue-derived inhibitors. With the soluble trophic factors, MSCs can prevent neurons from apoptosis, support angiogenesis/neurogenesis, and enhance synaptic plasticity, therefore promoting functional recovery after stroke [33, 34].

MSCs are also capable of producing certain cytokines to ameliorate inflammation-related injury. It has been documented that MSCs can interact with immune cells and produce more than ten soluble cytokines, such as TNF-stimulated gene-6 (TSG-6), hepatocyte growth factor (HGF), transforming growth factor (TGF)-β, prostaglandin (PG) E2, interleukin-(IL-)6, IL-10, IL-1 receptor antagonist, inducible nitric oxide synthase (iNOS), indoleamine 2,3-dioxygenase (IDO), galectin-1, and human leukocyte antigen (HLA)-G [35]. MSCs protect host cells from oxygen free radicals through secreting antioxidants and antiapoptotic molecules.

5.1.2.2 MSCs Induce Host to Secrete "Healing" Factors

As only a small portion (<5 %) of injected BM-MSCs can be detected in brain parenchyma 14 days after transplantation [18], it is unlikely that bioactive factors released by local MSCs per se are the only and direct contributors for the improvement of neurological functions. Intravenously administered cells may not even need to enter the brain to elicit an effect but rather can act in the periphery to increase trophic factor expression in the brain [8]. MSCs can secrete potent combinations of trophic factors to evoke responses from resident cells in the host, thus amplifying the endogenous levels of trophic factors in the brain.

Nowadays only limited knowledge is available about how MSCs modulate the molecular composition of the periphery system and CNS and how to induce the host cells to secrete neurotrophic factors.

In the CNS, astrocyte is one of the cell types that could be influenced by transplanted MSCs. Whether local reactive astrocytes act as the mediators between the limited number of MSCs infiltrating into the brain and the functional recovery has been studied. Being the most abundant cells in the brain, astrocytes provide many supportive activities essential for neuronal functions under physiological conditions [36, 37]. After various insults, including ischemia, local astrocytes are activated and undergo cellular hypertrophy. Previous studies show that MSCs can partially rescue damaged astrocytes, enhance astrocyte activation at the ischemic sites [38, 39], and induce astrocytic production of trophic factors like BDNF, bFGF, NGF, GDNF, and bone morphogenetic protein (BMP)2/4 [40–45]. The increased level of neurotrophins and growth factors produced directly from transplanted MSCs as well as those from evoked host cells could enhance the repair of injured cells around the ischemic tissue by increasing their viability and proliferation or by decreasing apoptosis of the injured neural cells.

In addition to astrocytes, brain endothelial cells also play an important role in MSC treatment of stroke. MSCs cocultured with mouse brain endothelial cells (MBECs) significantly increased MBEC expression of Angiopoietin-1/Tie2. MSCs treatment of stroke rats increases expression of Angiopoietin-1/Tie2 in brain endothelial cells as well as promotes angiogenesis and vascular stabilization [46].

Whether other cell types, such as neurons and microglia/macrophages, are involved requires further study. If so, the detailed mechanisms underlying the interaction of these cells with MSCs remain to be addressed.

Interestingly, the neurotrophic factors produced by host cells (triggered by infused MSCs) might be of more importance than those directly from the MSCs. When human MSCs (hMSCs) were transplanted into the peripheral circulation of rats, the neurotrophic factors of rat origin, not human, were positively correlated with the therapeutic outcome of rats [47]. It thus may be an efficacious approach to enhance the stimulating effect of infused MSCs to achieve a better outcome.

5.1.3 Immunomodulatory Functions

5.1.3.1 Anti-inflammatory Effects

Following CNS damage, inflammatory responses, which is characterized by pro-inflammatory cytokine release [48], lymphocyte infiltration [49, 50], and microglia activation [51, 52], are rapidly induced in the damaged CNS parenchyma. Clinical and preclinical studies suggest that inflammation has an important impact on stroke outcome and long-term prognosis. More severe neurological deficits were reported in stroke patients with preceding infections. A unique feature of MSCs is the immunomodulatory function, which may account for the beneficial effect in suppressing the pathological processes following stroke.

MSCs can suppress lymphocyte activation in response to allogeneic antigens by releasing immunosuppressive cytokines and factors [53, 54]. MSCs can also shift T-cell response from a T helper (Th)1 to a Th2 phenotype. In addition, MSCs are able to inhibit Th17 cell differentiation and function through cell–cell contact and by PGE2.

MSC-derived PGE2 and TGF-β 1 play a central role in the induction of CD4+CD25+FoxP3+Treg cells. Purified Treg cells in coculture with MSCs are able to suppress the proliferation of lymphocytes in response to alloantigen. MSCs do not only regulate Treg cells but are also capable of affecting B cells, natural killer (NK) cells , and dendritic cells (DCs). MSCs have been shown to alter B-cell proliferation/activation, differentiation, and antibody production. MSCs can inhibit the proliferation of NK cells [55, 56], prohibit the differentiation monocytes and $CD34^+$ progenitors into antigen-presenting DCs [57], as well as reduce expression of major histocompatibility complex class II (MHCII), CD40, and CD86 on DCs following maturation induction [34].

Furthermore, MSCs alter the cytokine secretion profile of immune-related cells, such as DCs, naive and effector T cells, and NK cells, to induce an anti-inflammatory or tolerant phenotype. For example, hMSCs can induce mature type 1 DCs to decrease TNFα secretion and mature type 2 DCs to increase IL-10 secretion. MSCs can also regulate TH1, TH2, and NK cells to decrease interferon-γ (IFN-γ) secretion and upregulate IL-4, respectively [58].

Accumulating evidence has confirmed the immunomodulatory effects of MSCs. Yet, the fully depicted molecular details still warrant further effort.

Among the mechanisms by which MSCs suppress the overzealous immune actions in diseases, the suppression of lymphocyte activation in response to allogeneic antigens by releasing immunosuppressive cytokines and factors has been widely documented. Furthermore, the peripheral immune organs, especially the spleen, take part in and play an important role in systemic inflammation after stroke. In this process, the modulatory functions of MSCs on the excessive inflammation and immune reactions after stroke may act through the spleen (will be discussed later).

5.1.3.2 How Do MSCs Suppress T Cells?

It has been a long time ongoing debate with regard to how MSCs suppress T cells. In previous work, several groups reported that MSCs inhibit the proliferation of T cells that are induced by alloantigens and nonspecific mitogens [59]. A number of studies established the concept that both soluble factors and cell–cell contact are required for the suppression [60]. In term of soluble factors, PGE2, IFN-γ, and NO are involved and of great importance.

The seemingly divergent concepts have been summarized and updated in Ren's work [61]. Ren and colleagues used MSCs derived from mice deficient in either the IFN-γ receptor or iNOS and showed that immunosuppression is mediated by INF-γ combined with any one of the pro-inflammatory cytokines, TNFα, IL-1α, or IL-1β. Such pro-inflammatory cytokine combinations lead to the release of chemokines from MSCs. The chemokines attract CXCR3-expressing T cells, and then the MSCs produce NO to suppress the proliferation of T cells. Both IFN-γ and NO are indispensable in exerting the suppressive effect. In addition to the in vitro work as mentioned above, the in vivo work using a mouse graft-versus-host disease (GvHD) model showed that the wild-type MSCs significantly reduce GvHD, but MSCs derived from IFNgammaR1(−/−) or iNOS(−/−) mice failed to prevent GvHD [61].

Further studies confirmed that the key players involved in MSC-mediated immunosuppression include indoleamine 2,3-dioxygenase (IDO) and iNOS. Human and other primate MSCs produce extremely high levels of IDO and very low levels of iNOS, whereas mouse MSCs express abundant iNOS and a low level of IDO. In vivo and in vitro studies showed that IDO plays an indispensable role in MSC-mediated immunosuppression in humans [62].

5.1.3.3 MSCs May Not Always Be Immunosuppressive

Many studies have shown that the immunosuppressive feature of MSCs is useful in treating immune disorders. As an encouraging example, in vitro expanded allogeneic MSCs successfully resolved severe grade IV acute GvHD [63]. However, several other reports showed that the effects of MSCs in GvHD are not always consistent. In some cases, MSC treatment could not prolong graft survival or suppress GvHD in vivo, although the MSCs are able to inhibit lymphocyte proliferation in vitro [64, 65].

These results suggest that the immunomodulatory effects of MSCs are conditional and may be dependent on the disease contexts [63]. As stated above, exposure to IFN-γ combined with any of the pro-inflammatory cytokines, TNFα, IL-1α, or IL-1β, is required to turn MSCs to acquire an immunosuppressive feature. When MSCs are not sufficiently or properly primed, they may become immunopromotive rather to be immunosuppressive. Crop et al. cultured human adipose tissue-derived autologous or allogeneic MSCs with nonactivated peripheral blood mononuclear cells (PBMCs) and found that 7 days of coculture with MSCs significantly increased the proliferation of PBMCs as high as threefold [66]. This immune-promoting effect suggests that, under some conditions, MSCs can enhance immune responses [67]. Exposure to IFN-γ combined with TNFα, IL-1α, or IL-1β is required to induce the immunosuppressive activity. However, when pro-inflammatory cytokines are inadequate to elicit sufficient production of NO or IDO or when iNOS activity is inhibited or genetically ablated, MSCs show an opposite effect and strongly enhance T-cell proliferation in vitro and promote a delayed-type hypersensitivity response in vivo. It seems that NO or IDO acts as a switch in MSC-mediated immunomodulation. Importantly, the dual effects on immune functions with IDO acting as a switch have also been observed in human MSCs [67].

Taken together, activation by pro-inflammatory cytokines is essential to bias MSCs into an immunosuppressive phenotype. If the environment cannot provide sufficient pro-inflammatory cytokines, MSCs may exert a different role and enhance immune responses. To take advantage of the immunosuppressive functions of MSCs, it may be useful to prime MSCs in vitro with IFN-γ and any of the pro-inflammatory cytokines, TNFα, IL-1α, or IL-1β, prior to transplantation.

Based on this theory, it is not difficult to understand that the critical determinants of anti-inflammation therapy for stroke patients are the time window and inflammatory cytokine levels. The levels of inflammatory cytokines in recipients change temporally, and at some time points may be insufficient to elicit the immunosuppressive phenotype of MSCs. Therefore, pretreatment of MSCs with inflammatory cytokines may enhance the therapeutic efficacy. It has been reported that IFN-γ-pretreated MSCs can enhance the therapeutic effect in animal models of acute myocardial ischemia [68] and protect 100 % of mice from GvHD-induced death [69]. Nevertheless, the therapeutic effect of pretreated MSCs in cerebral ischemia has not been investigated and warrants further research.

5.1.3.4 PGE2 Is a Key Factor in MSC-Mediated Immunosuppression

It was originally demonstrated by Nemeth et al. that the anti-inflammatory effects of lipopolysaccharides (LPS)- or TNFα-activated MSCs could be attributed to MSCs synthesizing and secreting PGE2 [70]. Others have shown that murine MSCs inhibit local inflammation in experimental arthritis through IL-6-dependent production of PGE2.

PGE2 is a lipid molecule derived from arachidonic acid through cyclooxygenase 1 (COX1) and cyclooxygenase 2 (COX2). These enzymes, and PGE2, are constitutively expressed by MSCs, and their expression increases in an inflammatory environment. For example, in vitro studies showed that MSCs produce more PGE2 when cocultured with ConA-induced T-cell blasts or pro-inflammatory cytokines [71]. PGE2 can turn activated macrophages into a regulatory-like phenotype and is able to inhibit the maturation and functions of monocyte-derived DCs [70]. MSCs can suppress LPS-induced glial activation in organotypic hippocampal slice cultures through secretion of PGE2 [72].

5.1.3.5 TSG-6 Is Another Mediator by Which MSCs Modulate Inflammation

Another important mediator in MSC-mediated immunosuppression, particularly in the spleen, is TNFα-stimulated gene/protein 6 (TSG-6). After settling down in the peripheral organs, mainly in the lungs and spleen, MSCs are stimulated to secret TSG-6 upon exposure to TNFα and other pro-inflammatory cytokines produced by resident immune cells. TSG-6 can reduce nuclear factor-κB (NF-κB) signaling in the resident macrophages as a negative feedback. Human umbilical cord cell treatment of stroke in the acute stage significantly reduces the inflammatory response through TSG-6, resulting in decreased necrotic and apoptotic cell death in the brain and increased motor and cognitive functions [73].

5.1.3.6 Immunomodulation Through the Spleen

Ischemia lesion in the brain, especially a severe ischemia, leads to inflammation not only locally in the brain but also in the peripheral immune organs. Studies have shown that the inflammatory status in the brain affects the periphery and vice versa.

The spleen is one of the organs that play an important role in the periphery after stroke. As the largest immune organ in the body that stores lots of T cells, B cells, NK cells, and monocytes/macrophages, the spleen is able to mobilize the immune cells and cytokines into the blood circulation immediately upon injury [74].

After the initial ischemic attack, the blood–brain barrier (BBB) is compromised. That the vascular endothelial cells express adhesion molecules permits an influx of peripheral immune cells including macrophages, neutrophils, leukocytes, T cells, and B cells into the brain [75]. This influx of peripheral immune cells exacerbates the local brain inflammatory responses, leading to enhanced neurodegeneration and cell loss. It was reported that the infiltrating immune cells and the increased pro-inflammatory cytokines negatively affect stroke outcome [76].

To confirm that the spleen is the main source of cells that infiltrate into the brain in an MCAO model, carboxyfluorescein diacetate succinimidyl ester (CFSE)-labeled splenocytes were traced. The number of labeled splenocytes decreased in the spleen and increased within the circulation at 48 h after MCAO. The labeled cells were found only in the injured hemisphere of MCAO animals and were mainly located in the vasculature. This phenomenon suggests that splenocytes may release factors at the BBB and facilitate immune cell infiltration into the brain [77].

The brain and the peripheral immune organs may communicate at least by two pathways, which are the hypothalamic–pituitary–adrenal axis and the sympathetic nervous system [78].

In the case of severe ischemia, the spleen can be affected through these pathways and release plenty of immune cells and pro-inflammatory cytokines. A significant raise of catecholamine concentration in blood circulation occurs following brain ischemia and activates the α1 adrenergic receptors expressed on the splenic smooth muscle cells which results in spleen contraction. In addition, the spleen size was significantly decreased following MCAO in rats and mice [77, 79] as well as in human stroke patients [80, 81]. Correspondingly, this reduction in spleen size is associated with increased release of immune cells and pro-inflammatory cytokines into the blood, a greater extent of infiltration of leukocytes and monocytes, and a higher level of microglia activation in the brain [77, 82].

Some follow-up studies showed that blockade of the α and β adrenergic receptors on the spleen significantly inhibits the reduction in spleen size and reduces the brain infarct volume [83]. Splenectomy prior to MCAO results in 50–80 % decrease in brain infarct volume, which is accompanied by a significant reduction in the number of infiltrating neutrophils and activated microglia/macrophages in the injured brain [81, 84].

The spleen is a highly vascularized immune organ. A large portion of systemically infused MSCs are trapped in the spleen, which makes the cell–cell contact possible between MSCs and immune cells. In this regard, the delivery of MSCs through the peripheral route and suppression of spleen-involved inflammation may influence stroke outcome and prognosis.

In the acute phase after stroke, transfusion of human umbilical cord blood cells (HUCBCs) not only reverses the reduction in spleen size but also alters the splenic cytokine expression profile from a pro-inflammatory (e.g., TNFα and IL-1β) to an anti-inflammatory (e.g., IL-10) one [79]. Not only in the acute ischemia, recently, Acosta et al. found that by intravenous transplantation of hMSCs at 60 days post stroke, labeled hMSCs preferentially migrated to the spleen and reduced striatal infarct and peri-infarct areas compared with those of sham animals. Treatment with MSCs significantly decreased by 75 % and 60 % major histocompatibility complex II-positive inflammatory cells in gray and white matter and reduced TNFα-positive cell density in the spleen of transplanted stroke animals. About 0.03 % hBM-MSCs survived in the spleen and only 0.0007 % hBM-MSCs survived in the brain. MSC migration to the spleen, but not to the brain, inversely correlated with inflammation and the infarct and peri-infarct volumes [85].

Another notable issue is the immunosuppression after stroke. The spleen usually decreases in size transiently at 1–4 days following MCAO in rats and recovers to its original size within 7 days [77]. Severe cerebral ischemia leads to prolonged spleen contraction, and if this situation persists, eventually, the storage and even the capacity to produce immune cells and pro-inflammatory cytokines will be exhausted. Furthermore, due to the feedback regulation of spleen functions after the surge release of splenocytes and pro-inflammatory cytokines, anti-inflammatory cytokines, such as IL-10 expression, is enhanced. Immunosuppression, which is often the cause of infection and death of stroke patients, may explain why in some cases, anti-inflammation therapy after stroke actually exacerbated the outcome [86].

It seems that the intravenous infusion of MSCs may suppress the "over-activated" inflammation and immune reaction in the spleen, reduce the influx of immune cells and pro-inflammatory cytokine into the brain, prevent the exhaustion of immune capacities of the spleen, and eventually avoid the immunosuppression in stroke patients. From this point of view, one of the major functions exerted by systemically transplanted MSCs may be acting as a buffer in the process of inflammatory events following stroke attack and preserving the normal immune functions of the spleen.

Another interesting notion is that the immunosuppressive effect of cell therapy is not restricted to MSCs. Other cell types, such as neural stem cells [87], hematopoietic stem cells [88], bone marrow mononuclear cells [89], and umbilical cord blood cells [90], all have shown some immunomodulatory functions in stroke models. Among these cell types, at least neural stem cells have been reported to have an immunosuppressive ability through the spleen [87].

5.1.3.7 Modulation of Microglia Activation in the Brain

The brain is an immune privileged organ separated from the periphery by BBB, which seals the brain off from toxins, pathogens, and the circulating immune cells. However, upon stroke attack, cerebral ischemia induces a robust neuroinflammatory response that includes marked changes in a variety of endogenous CNS cell types, as well as an influx of immune cells from the peripheral blood circulation. Among these cells, microglia are the main resident immune cells in the brain, which originate from the peripheral monocytes/macrophages during embryonic development [91].

Microglia maintain a constant survey of the immune status in the brain. Microglia can engulf pathogens, misfolded proteins, and scavenge dead cells in the event of infection, ischemia, and neurodegeneration [92, 93]. Microglia also clear away synapses that have been damaged by injuries and, under some conditions, even trim off weak connections/synapses between neurons [94]. Upon activation in acute ischemia, microglia contribute to the injury by the release of pro-inflammatory cytotoxins and other inflammatory mediators, such as TNFα, IL-1β, NO, and reactive oxygen species, which are the direct effectors of cerebral injury, and take part in the chronic ischemic injury by continuously maintaining the upregulation of these pro-inflammatory cytotoxins and other inflammatory mediators. Meanwhile, microglia may modulate inflammatory responses by releasing anti-inflammatory cytokines, protect neurons by releasing growth factors, and improve endogenous regeneration by clearing debris, and so on [95, 96].

According to the above two opposing status, microglia can be categorized into M1, which is a pro-inflammatory, and M2, an alternative anti-inflammatory phenotype. M1 microglia produce pro-inflammatory mediators such as TNFα , IL-1β , and IFN-γ . M1 microglia express CD80, CD86, and MHC class II on the cell membrane and can present antigens to T cells [97]. M2 microglia secrete anti-inflammatory mediators such as IL-10, TGF-β , IL-4, and IL-13, as well as various neurotrophic factors, such as insulin-like growth factor-1 (IGF-1). M2 microglia is regarded as "healing cells" that contribute to recovery after brain damage [98]. In contrast, M1 microglia tends to induce neuronal cell death more readily than M2 microglia. The M2 phenotype microglia are dominant in both resident microglia and newly recruited macrophages at early stages in ischemic stroke. At 24 h after ischemic attack, Ym-1 and CD206, markers for the M2 phenotype, are found in the ischemic core [99]. The expression of M2 markers is highest at 5 days after insult and start to taper off after 14 days [98], suggesting that microglia/macrophages participate in tissue repair in the ischemic core [100]. However, the M2 phenotype gradually transforms to M1 phenotype in the peri-infarct areas, which shows an exacerbating effect on neuronal death. Therefore, stroke therapies may need to be shifted from a strategy of simply suppressing microglia toward balancing the beneficial and detrimental effects of microglia [96].

MSCs are capable of increasing the mRNA and protein expression of M2 markers on microglia in the coculture system [101], and MSCs have been proven capable of inhibiting the shift of microglia from M2 to M1 or prompting the transition from M1 to M2 in different disease models, such as wound, lung disease, heart disease, renal disease, spinal cord injury, and brain trauma [102]. The pathological processes of some of these diseases are to some degree similar to stroke; therefore the available published data support the hypothesis that MSCs may be competent in modulating the functions of microglia in the ischemic brain.

5.2 Senescence of MSCs

Adult MSCs are a valuable resource for autologous and allogeneic cell therapies. However, patients suffering from stroke are often of advanced age. In light of this, the effects of the aged milieu on MSCs and the intrinsic aging of MSC in vivo are important questions to address. Furthermore, MSCs may require in vitro expansion before use, and their aging in vitro is thus also an important issue.

Compared to younger people, the extracellular microenvironment in older people may present with a more pro-inflammatory characteristics, rendering the aged milieu a more hostile environment for adult stem cells [103]. It has been shown that the functions of adult stem cells are suppressed in aged mice, suggesting a causal relationship between stem cell aging and organismic aging [104, 105].

To evaluate the correlation of age and the functional properties of MSCs, children and adult MSCs were compared. A shorter population doubling (PD) time and a higher colony-forming unit–fibroblast (CFU-F) count were found in children MSCs as compared with those of adult MSCs. The telomere length was also significantly greater in children MSCs vs. adult MSCs. These data suggest that children's BM-derived MSCs could be a more advantageous source for cell therapy [106].

Similar results were reported in an independent study [107], in which 17 BM samples were collected from young adult donors and 8 from pediatric donors. MSCs isolated from the two groups showed no morphological differences, while their cell growth was strictly related to the donor's age. The MSCs isolated from pediatric donors reached a cumulative population doubling times almost twice as high as MSCs isolated from young adult donors after 112 days [107]. The authors analyzed the surface marker expression from the two groups and no obvious difference was observed. Neither chromosomal alteration nor evidence of cellular senescence was observed in all the analyzed samples.

Meanwhile, accumulating evidence shows that MSCs undergo senescence itself in vitro and in vivo.

Kim et al. reported the replicative senescence of MSCs in vitro. The human MSCs (hMSCs) were cultured for more than 60 population doublings. In the long-term cultured hMSCs over PD 30, the adipogenic and osteogenic differentiation potentials were significantly reduced. Telomere length was shortened although the telomerase activity was unchanged. Oncogenic gene expression decreased in proportion as the PD number increased, and no tumor formation was observed when the long-term cultured hMSCs were injected into nude mice [108].

In another study using human tonsil-derived MSCs, Yu and colleagues investigated the MSCs for up to passage number 15. They found no alterations in the expression of MSC-specific surface markers, CD14, CD34, CD45, CD73, and CD90. However, the expression of CD146, recently identified as another MSC marker, dramatically decreased from ~23 % at passage 3 to ~1 % at passage 15. The average doubling time increased significantly from ~38 h at passage 10 to ~46 h at passage 15. From passage 10, the cell size increased slightly and SA-β-gal staining was evident. Also, the osteogenic differentiation ability increased up to passage 10 and decreased thereafter. However, the adipogenic and chondrogenic differentiation potentials decreased since the start of passaging in vitro [109].

The molecular mechanisms underlying MSC senescence have been explored. MSCs obtained from systemic lupus erythematosus (SLE) patients exhibit accelerated functional decline as part of the SLE phenotype. In these cells, expression of the senescence-related gene p16 is increased [110]. In other studies, senescence-related genes including p16, p21, and p53 were increased in long-term cultured MSCs [111–113].

Another mechanism that may be involved in MSC senescence is epigenetic regulation. Schellenberg analyzed the functional, genetic, and epigenetic status of long-term cultured MSCs [114]. Although chromosomal aberrations were not detected by karyotyping and single-nucleotide polymorphism (SNP) microarrays, highly consistent senescence-associated modifications at specific CpG sites were noted, such as trimethylation of H3K9, H3K27, and EZH2 targets. It seems that culture expansion of MSCs has profound functional implications, which may be hardly reflected by genomic instability but is associated with highly reproducible DNA methylation changes and repressive histone marks.

While the use of MSCs as cellular therapies for different clinical problems is still under investigation, it does appear that aged MSCs generally perform less effectively than their younger counterparts in various disease models [115, 116]. Many age-related changes in MSCs seem to be in agreement with this hypothesis. Collectively, the general age-associated changes may include MSC adhesion, migration, resistance to oxidative stress, and cytokine secretion. These changes could have important implications in tissue healing mediated by MSCs, through their homing, migration, and paracrine stimulation of other cells [115, 117, 118]. For example, MSCs isolated from aged rats showed a significantly reduced antioxidant and migration capacity [115, 119], as well as a decreased in vitro wound healing function [120]. Furthermore, MSCs obtained from 6-day-old to 6-week-old mice are significantly more adhesive than those from 1-year-old mice [121], while MSCs from 3-month-old mice secrete an extracellular matrix that is more proliferation- and osteogenesis-promoting than that from older mice [117].

However, not all studies support the notion that MSCs are affected by aging. For example, it was found that the ability of hMSCs to suppress proliferation of activated allogeneic T cells in vitro is not affected by biological age [122].

Despite the results discussed above, the information about the influence of aging is limited with regard to the roles of MSCs in stroke therapy. Li et al. found that ischemia-induced neurogenesis is enhanced to a higher extent by intravenous administration of hMSCs of early passage than those of later passage. The effects may be related to passage-dependent neurotrophic capacities of MSCs [123].

It is noteworthy that aging of MSCs in vitro vs. in vivo may be different. MSCs cultured in a dish may experience certain level of stress and lack the interaction with the microenvironment in a tissue. In addition, although less immunogenic, the interaction with immune cells in vivo may exert some pressure and screening effect on MSCs and influence the aging process and physiological roles of MSCs in vivo. Another thing that needs to be kept in mind is that MSCs are not a homogeneous population; it is possible that the effects observed by aging are only due to changes in a subpopulation of the MSCs. Further studies are required to address these issues.

The above findings suggest that genome test, at least karyotyping assay before cellular therapy, should be performed during the ex vivo culture of hMSCs, particularly when an extended propagation period is required.

5.3 Features of MSCs from Individuals with Diseases

Compared with allogeneic cell therapy, autologous MSC transplantation has certain merits, such as less risk of contamination, fewer ethical issues, and ease of acceptance by patients. However, since autologous MSCs are mostly derived from diseased people, it warrants a close examination whether the disease state affects the properties of MSCs.

Diabetes is a common disease. In type two diabetic and prediabetic patients, advanced glycation end products (AGEs) are normally present at an elevated level, which may suppress the proliferation, induce apoptosis, and increase the intracellular reactive oxygen species production in the MSCs [124].

Another disease condition/indication that has been investigated is heart failure. Dmitrieva and colleagues found that BM-MSCs from the heart failure patients demonstrate an early reduction in the proliferative activity and change of gene expression profiles that favor a pro-fibrotic phenotype of MSCs, which makes these cells less effective for therapeutic applications [125]. However, other study also found that treatment of stroke with BM-MSCs derived from brain ischemic stroke rats (isch-BMSCs) significantly improves functional outcome after stroke and shows better therapeutic effects than BM-MSCs derived from normal rats (Nor-BMSCs) [126]. Isch-BMSCs are superior to Nor-BMSCs for the neurorestorative treatment of stroke, which may be mediated by the enhanced trophic factor and angiogenic characteristics of Isch-BMSCs [126].

Cancer also influences BM-MSCs function. BM-MSCs from multiple myeloma patients have an increased expression of senescence-associated β-galactosidase, increased cell size, reduced proliferation capacity, and an aberrant secreting profile. In addition, multiple myeloma patient BM-MSCs more efficiently support the growth of $CD34^+$ cells, suggesting that not only BM-MSCs are altered by the disease state but that these cells may actively contribute to the disease progression [127].

Irradiation is an often chosen treatment for cancer patients, and it would be useful to learn whether irradiation affects the functions of BM-MSCs. Cmielova et al. studied the characteristics of BM-MSCs derived from patients who had received up to 20 Gy ironizing irradiation. These BM-MSCs showed significantly reduced proliferation but no obvious change in cell survival. Activation of p53 was detected from the first day of irradiation and remained elevated to day 13. The cell cycle was arrested in G2 phase, in agreement with the upregulated expression of cyclin-dependent kinases inhibitor 1A (p21Cip1/Waf1). Instead of apoptosis, the authors detected hallmarks of stress-induced premature senescence: increase in cyclin-dependent kinases inhibitor 2A (p16INK4a) and increased activity of senescence-associated β-galactosidase [128].

In another study, Li et al. examined the properties of BM-MSCs isolated from patients suffering from systemic lupus erythematosus (SLE), an autoimmune disease characterized by a mistaken immune attack on the patient's own internal organs/tissues. The MSCs exhibited impaired proliferation, differentiation, cytokine secretion profiles, and aberrant immunomodulatory functions. An increased frequency of apoptosis and aging in SLE BM-MSCs was also observed compared with those of normal controls. Moreover, intracellular reactive oxygen species levels of SLE BM-MSCs were higher than those of normal controls [110].

From the above studies, it seems that, despite the advantages associated with autologous transplantations, application of syngeneic MSCs from individuals suffering from certain diseases may not be the best choice. In such cases, allogeneic MSCs from healthy donors may be considered.

5.4 Different Sources of MSCs

Finding an appropriate and convenient cell source is a key step for cell therapy and tissue engineering.

Bone marrow has been one of the first reported cell sources for MSCs [129] and BM-MSCs remain a good choice for autologous transplantation [130]. However, several caveats exist regarding the clinical application of BM-MSCs. Firstly, there are only one MSC in around 10,000 cells in the bone marrow, giving rise to a relative low yield of MSCs; secondly, MSCs decrease in the proliferation and differentiation capacities with age, which may be a particular issue for old patients; thirdly, the isolation procedure is painful, and with allogeneic transplantation of BM-MSCs, the risk of viral exposure cannot be completely avoided.

Given the above aspects associated with BM-MSCs, researchers and clinicians have searched for alternative sources of MSCs. MSCs can also be extracted from adipose tissues (AT) and embryo-related tissues, such as umbilical cord (UC), chorionic plate (CP), and placenta.

Among the different sources, adipose tissue and bone marrow remain the most accessible sources to derive MSCs, and both can be used for autologous transplantation purposes. Compared with BM-MSCs, AT-MSCs are less invasive and less expensive to obtain and show a higher proliferative capacity with shorter population doubling times. But both AT-MSCs and BM-MSCs undergo senescence after certain number of passages in vitro. Compared to AT-MSCs and BM-MSCs, embryo-related MSCs (UC-MSCs, CP-MSCs, and placenta MSCs (P-MSCs)) are usually used for allogeneic purposes. Originated from tissues of early developmental stage, embryo-related MSCs retain a primitive stemness feature and possess a good proliferative capacity. Many studies compared the proliferative capacity between BM-MSCs and UC-MSCs and found that UC-MSCs show a much higher proliferative activity and no sign of senescence even after an extensive culture [131, 132]. Conconi et al. cultured UC-MSC over 16 serial passages and found no aberration in cell morphology or senescence. Mitchell et al. cultured porcine UC-MSC for more than 80 doubling times without detecting decrease in the proliferative capacity [133].

In terms of the immunosuppressive ability, Najar et al. observed a potent and dose-dependent inhibition of lymphocyte proliferation by these MSCs, regardless of the stimuli used to activate T cells. UC- and AT-MSCs display an inhibitory effect higher than that with BM-MSCs [134]. Ribeiro et al. compared the immunomodulatory abilities between human UC-, AT-, and BM-MSCs. They found that MSCs derived from all three tissues are able to inhibit the activation of $CD4^+$, $CD8^+$ T cells, and $CD56^{dim}$ NK cells, wherein AT-MSCs show a stronger inhibitory effect. Moreover, AT-MSCs block the T-cell activation process at an earlier phase than BM- or UCM-MSCs do, yielding a greater proportion of T cells in the nonactivated state [135].

Castro-Manrreza and colleagues compared BM-, UC-, and P-MSCs in their immunosuppressive properties. Except for P-MSCs, BM-MSCs and UCB-MSCs significantly inhibit the proliferation of both $CD4^+$- and $CD8^+$-activated T cells as well as increase the generation of $CD4^+CD25^+CTLA4^+$ Treg populations [136]. Therefore, in addition to BM-MSCs, at least UC-MSCs and AT-MSCs may be potent and reliable candidates for future therapeutic applications. However, which one is the best is not fully known and more work is needed to address this issue.

5.5 Multi- or Single Infusion of MSCs?

Many studies have demonstrated that systemic injection of hMSCs prepared from bone marrow or other sources have therapeutic benefits in rat ischemia models [47]. Yet, there is still a shortage of data comparing the therapeutic effects between a single and multiple injections of MSCs in stroke models.

Multiple infusions at various sites and times may lead to a cumulative response/effect. In the mouse GvHD model, multiple administrations of MSCs are often utilized to sustain and prolong their inhibitory effect [54]. Yet it is not clear whether repeated intravenous MSC administration would give a better outcome for stroke treatment.

To test this, Omori et al. used the MCAO model of rats, and infused hMSCs intravenously at a single time point 6 h post-ischemia (low and high cell doses) or at various multiple time points after MCAO. Ischemia lesion volume was reduced in all hMSC cell injection groups as compared to serum alone injection group. However, the greatest therapeutic benefit was achieved following a single high cell dose injection at 6 h post-ischemia rather than multiple lower cell infusions over multiple time points [137].

In another study, baboons were used to test whether allogeneic MSCs would elicit immune responses. The baboons were injected intravenously with MSCs (5×10^6/kg) followed by intramuscular injections of MSCs (5×10^6/kg) 6 weeks later from the same or a different donor. Almost 30 % of the animals produced alloantibodies that reacted with MSCs; but the host T-cell responses to donor alloantigens did not suppress the overall T-cell response in the host. This is the evidence showing that multiple administrations of high doses of allogeneic MSCs can induce alloreactive immune responses without compromising the overall immune system in nonhuman primates [138].

In a separate study, Isakova et al. found that even one-time intracranial administration of allogeneic MSCs can evoke immune reactions in the host rhesus macaques. In animals administered with allogeneic but not autologous MSCs, allo-recognition and allo-specific antibodies were detected. However, secondary antigen challenge failed to elicit a measurable response in allogeneic recipients. The authors indicated that although the behavior testing did not reveal significant changes related to transplanted MSCs at any doses, allogeneic MSCs were weakly immunogenic when allotransplanted in rhesus macaques, and this negatively influenced the MSC engraftment level and duration [139].

The two nonhuman primate experiments were performed using naive animals rather than using cerebral ischemia models. There is limited information on the immune safety and therapeutic efficacy of repeated intravenous injections of allogeneic MSCs in stroke models or patients. The influence of a single or multiple injections of MSCs, especially MSCs from different allogeneic donors, on the immune responses of the host should be taken into consideration in the clinical trials.

5.6 Time of Transplantation and Delivery Routes

Theoretically, the optimal timing and routes of cell delivery depend on the mechanisms of action of MSCs and the brain environment. As discussed above, the therapeutic mechanisms of MSCs are mainly related to the migration of transplanted cells, the modulation of inflammation/immune functions in the CNS and peripheral immune organs, and the neurotrophic effects from MSCs.

In literature, a wide range of intervals post-ischemia have been reported. Many of them demonstrated that functional recovery can be achieved by cell therapy within the first 2 days after ischemia. There are also studies showing a cell-enhanced recovery where cells are delivered up to 1 month post-ischemia. The optimal time window in the clinical setting has yet to be determined [85, 137].

The brain environment changes dynamically during the acute and subacute phases after stroke onset. In the acute phase (0–2 days after stroke in rodents), there is an increase in inflammation/immune reactions, represented by release of excitatory amino acids, pro-inflammatory cytokines, reactive oxygen species, etc. This is followed by a subacute phase (several days) during which the inflammation/immune reactions decrease continuously. Then the animals go into the chronic phase during which attempted endogenous brain repair may last several weeks to months.

If the cells are intended to be deployed into brain parenchyma, for example, by the intracerebral stereotaxic transplantation, survival of the engrafted cells would be critical; in this case, delivery at the subacute or later stages may be a good choice; at which time point, the inflammation in the brain has been ameliorated. As far as the inflammatory milieu is concerned, direct transplantation into the penumbra, where the microenvironment is not as hostile as the ischemia core, may enhance the survival of MSCs that in turn could modulate the ongoing inflammation and provide neurotrophic protection to neurons.

If the immunosuppressive function of MSCs is desired, exposure to inflammatory cytokines is required to induce the immunosuppressive phenotype of MSCs and should be taken into account. At the acute stage after stoke, inflammatory cytokines are produced at peak levels, at which time infusion of MSCs may lead to a good outcome for immunosuppression.

Other treatment strategies, no matter aiming at neuroprotection or at enhancing the endogenous neuroregeneration, are related to the neurotrophic mechanisms. The knowledge is limited about what factors regulate the secretion of neurotrophic factors from MSCs and how MSCs stimulate the host cells to participate in neuroprotection and neurorestoration. Delivery at acute or later stages may be both beneficial in these conditions; but an earlier intervention may be desirable since it may be good for the neurotrophic protection to start as early as possible.

Nevertheless, thorough data concerning the optimal timing of intervention are still lacking and the abovementioned hypotheses need to be further tested.

In terms of the delivery routes, MSCs can be transplanted intracerebrally, intravenously, or intra-arterially.

The delivery routes most likely affect the trafficking of MSCs to the target organs. Intracerebral injection is the most efficient method for delivering MSCs to the damaged brain tissue, but other methods also have their merits.

Intravenous and intra-arterial injections are the two widely used approaches of systemic administration. As stated above, by intravenous injection of MSCs, MSCs tend to migrate to the peripheral immune organs, including the lung, spleen, liver, and thymus, whereas the intra-arterial method is favorable for the aggregation of MSCs in the injured brain areas.

There are a limited number of studies that directly compared the therapeutic effects of different delivery routes under the same conditions.

Both intravenous and intra-arterial transplantation of bone marrow mononuclear cells (BM-MNC) and BMSCs improved functional recovery and reduced lesion size in stroke rats as compared with saline treatment [6, 7, 28]. However, no significant differences between intravenous and intra-arterial groups were observed.

Savetz et al. found that intra-arterial delivery led to more BM-MNCs in the peri-infarct area at 1 and 6 h after ischemic attack; but unexpectedly, the presence of a larger number of cells in the brain did not correlate with functional improvement or tissue repair as previously suggested by others. The authors suggested that, once reaching the quantity threshold, more cells present in the brain may not lead to a greater recovery [28].

Compared to BM-MNCs, delivery of MSCs through different routes may be a different story. MSCs are much larger than MNCs in size and will be trapped to a higher extent in spleen and lymphoid organs after being delivered intravenously. Accordingly, intra-arterial infusion of MSCs will give rise to more cells in the brain. This raises an interesting question about whether more MSCs in the spleen or in the brain would be more beneficial for stroke recovery. This comparison involves different mechanisms of action exerted by MSCs through the spleen or brain, and such effects may be time and cell dose dependent. Each delivery route may have multiple therapeutic effects at both the CNS and the peripheral organs. The collective outcome of cell therapy is thus the sum of these therapeutic effects.

Nevertheless, which route of administration is the best for stroke treatment, especially for clinical use, is yet not clear. The therapeutic effects of MSCs at different sites and time points require further studies.

5.7 Conclusion and Perspectives

MSCs can be isolated from different tissues and consist of a heterogeneous cell population. Normally, MSCs need to be passaged a few times in vitro before use. After transplantation through certain routes, MSCs migrate and distribute at different areas, and part of them may reach the injured sites. The microenvironments at different sites vary and change dynamically, and MSCs interact with this complex microenvironment through different mechanisms, presenting with an immunosuppressive, immunopromotive, and/or neurotrophic phenotype, dependent on the specific context.

Given the complexity involved in cell therapy for stroke, an optimal treatment strategy has yet to be worked out. Nevertheless, an increasing body of evidence suggests that using MSCs to treat stroke is safe and to some degree might be efficacious. Compared to conventional drug treatment, cell therapy using MSCs certainly has its advantages. However, to maximize the efficacy with MSCs, the treatment strategy may need to be individualized, taking into consideration at least the disease stages of the patients. In perspective, the stroke patients can be roughly categorized into three groups according to the disease severity, and the treatment strategies may be different accordingly. Still, many questions remain to be addressed in the field.

For example, in severe stroke in which pro-inflammatory cytokines are produced at high levels, infused MSCs may be primed by the cytokines and act in an immunosuppressive way. In this context, will MSC treatment give a better outcome than with anti-inflammatory drugs, in attenuating the acute inflammation and preventing stroke-induced immune suppression?

With regard to stoke of medium severity, it is not yet clear if MSC infusion would dampen or promote the inflammatory damage in the brain, given the condition that the pro-inflammatory cytokine levels may not be high enough to prime MSCs to an immunosuppressive phenotype. If this is the case, will pretreatment with pro-inflammatory cytokines before MSC transplantation improve the outcome?

In the case of mild stroke, in which the inflammation level is not high, how to balance the immunosuppressive, immunopromotive, and the neurotrophic features of transplanted MSCs?

Another closely related question is how to categorize the patients into the three groups of severities. In addition to the clinical symptoms, can we use levels of pro-inflammatory cytokines as a criterion to group the patients? If so, shall we use cytokines in blood or in cerebrospinal fluid as a reference? And which cytokines should be used?

Answering these questions will definitely accelerate the translation of MSC therapy from bench to bedside.

References

1. Savitz SI, Fisher M. Future of neuroprotection for acute stroke: in the aftermath of the SAINT trials. Ann Neurol. 2007;61(5):396–402.
2. Borlongan CV, et al. Potential of stem/progenitor cells in treating stroke: the missing steps in translating cell therapy from laboratory to clinic. Regen Med. 2008;3(3):249–50.
3. Mazzini L, et al. Stem cell therapy in amyotrophic lateral sclerosis: a methodological approach in humans. Amyotroph Lateral Scler Other Motor Neuron Disord. 2003;4(3):158–61.
4. Bang OY, et al. Autologous mesenchymal stem cell transplantation in stroke patients. Ann Neurol. 2005;57(6):874–82.
5. Lee JS, et al. A long-term follow-up study of intravenous autologous mesenchymal stem cell transplantation in patients with ischemic stroke. Stem Cells. 2010;28(6):1099–106.
6. Chen J, et al. Therapeutic benefit of intravenous administration of bone marrow stromal cells after cerebral ischemia in rats. Stroke. 2001;32(4):1005–11.
7. Li Y, et al. Treatment of stroke in rat with intracarotid administration of marrow stromal cells. Neurology. 2001;56(12):1666–72.
8. Li CD, et al. Mesenchymal stem cells derived from human placenta suppress allogeneic umbilical cord blood lymphocyte proliferation. Cell Res. 2005;15(7):539–47.
9. Li Y, et al. Gliosis and brain remodeling after treatment of stroke in rats with marrow stromal cells. Glia. 2005;49(3):407–17.
10. Hokari M, et al. Bone marrow stromal cells protect and repair damaged neurons through multiple mechanisms. J Neurosci Res. 2008;86(5):1024–35.
11. Mahmood A, et al. Treatment of traumatic brain injury in adult rats with intravenous administration of human bone marrow stromal cells. Neurosurgery. 2003;53(3):697–702. discussion 702-3.
12. Lee RH, et al. Intravenous hMSCs improve myocardial infarction in mice because cells embolized in lung are activated to secrete the anti-inflammatory protein TSG-6. Cell Stem Cell. 2009;5(1):54–63.
13. Garbuzova-Davis S, et al. Intravenous administration of human umbilical cord blood cells in a mouse model of amyotrophic lateral sclerosis: distribution, migration, and differentiation. J Hematother Stem Cell Res. 2003;12(3):255–70.
14. Gao J, et al. The dynamic in vivo distribution of bone marrow-derived mesenchymal stem cells after infusion. Cells Tissues Organs. 2001;169(1):12–20.
15. Hill WD, et al. SDF-1 (CXCL12) is upregulated in the ischemic penumbra following stroke: association with bone marrow cell homing to injury. J Neuropathol Exp Neurol. 2004;63(1):84–96.
16. Newman MB, et al. Stroke-induced migration of human umbilical cord blood cells: time course and cytokines. Stem Cells Dev. 2005;14(5):576–86.
17. Shen LH, et al. Therapeutic benefit of bone marrow stromal cells administered 1 month after stroke. J Cereb Blood Flow Metab. 2007;27(1):6–13.
18. Tyndall A, Gratwohl A. Adult stem cell transplantation in autoimmune disease. Curr Opin Hematol. 2009;16(4):285–91.
19. Sugiyama T, et al. Near-infrared fluorescence labeling allows noninvasive tracking of bone marrow stromal cells transplanted into rat infarct brain. Neurosurgery. 2011;68(4):1036–47. discussion 1047.

20. Osanai T, et al. Therapeutic effects of intra-arterial delivery of bone marrow stromal cells in traumatic brain injury of rats – in vivo cell tracking study by near-infrared fluorescence imaging. Neurosurgery. 2012;70(2):435–44. discussion 444.
21. Lee J, et al. Migration and differentiation of nuclear fluorescence-labeled bone marrow stromal cells after transplantation into cerebral infarct and spinal cord injury in mice. Neuropathology. 2003;23(3):169–80.
22. Shen LH, et al. One-year follow-up after bone marrow stromal cell treatment in middle-aged female rats with stroke. Stroke. 2007;38(7):2150–6.
23. Roybon L, et al. Failure of transdifferentiation of adult hematopoietic stem cells into neurons. Stem Cells. 2006;24(6):1594–604.
24. Wagers AJ, et al. Little evidence for developmental plasticity of adult hematopoietic stem cells. Science. 2002;297(5590):2256–9.
25. Castro RF, et al. Failure of bone marrow cells to transdifferentiate into neural cells in vivo. Science. 2002;297(5585):1299.
26. Toyoshima A, et al. Intra-arterial transplantation of allogeneic mesenchymal stem cells mounts neuroprotective effects in a transient ischemic stroke model in rats: analyses of therapeutic time window and its mechanisms. PLoS One. 2015;10(6):e0127302.
27. Lee SH, et al. Differential migration of mesenchymal stem cells to ischemic regions after middle cerebral artery occlusion in rats. PLoS One. 2015;10(8):e0134920.
28. Yang B, et al. Intra-arterial delivery is not superior to intravenous delivery of autologous bone marrow mononuclear cells in acute ischemic stroke. Stroke. 2013;44(12):3463–72.
29. Li Y, et al. Human marrow stromal cell therapy for stroke in rat: neurotrophins and functional recovery. Neurology. 2002;59(4):514–23.
30. Borlongan CV, et al. Central nervous system entry of peripherally injected umbilical cord blood cells is not required for neuroprotection in stroke. Stroke. 2004;35(10):2385–9.
31. Kurozumi K, et al. Mesenchymal stem cells that produce neurotrophic factors reduce ischemic damage in the rat middle cerebral artery occlusion model. Mol Ther. 2005;11(1):96–104.
32. Modo M, et al. Effects of implantation site of stem cell grafts on behavioral recovery from stroke damage. Stroke. 2002;33(9):2270–8.
33. Erdo F, et al. Host-dependent tumorigenesis of embryonic stem cell transplantation in experimental stroke. J Cereb Blood Flow Metab. 2003;23(7):780–5.
34. Toda H, et al. Grafting neural stem cells improved the impaired spatial recognition in ischemic rats. Neurosci Lett. 2001;316(1):9–12.
35. Pittenger M. Sleuthing the source of regeneration by MSCs. Cell Stem Cell. 2009;5(1):8–10.
36. Privat A. Astrocytes as support for axonal regeneration in the central nervous system of mammals. Glia. 2003;43(1):91–3.
37. Pekny M, Nilsson M. Astrocyte activation and reactive gliosis. Glia. 2005;50(4):427–34.
38. Li L, et al. Protective role of reactive astrocytes in brain ischemia. J Cereb Blood Flow Metab. 2008;28(3):468–81.
39. Li Y, et al. The role of astrocytes in mediating exogenous cell-based restorative therapy for stroke. Glia. 2014;62(1):1–16.
40. Spaggiari GM, et al. MSCs inhibit monocyte-derived DC maturation and function by selectively interfering with the generation of immature DCs: central role of MSC-derived prostaglandin E2. Blood. 2009;113(26):6576–83.
41. Raffaghello L, et al. Human mesenchymal stem cells inhibit neutrophil apoptosis: a model for neutrophil preservation in the bone marrow niche. Stem Cells. 2008;26(1):151–62.
42. Friedenstein AJ, et al. Stromal cells responsible for transferring the microenvironment of the hemopoietic tissues. Cloning in vitro and retransplantation in vivo. Transplantation. 1974;17(4):331–40.
43. Gennery AR, Cant AJ. Cord blood stem cell transplantation in primary immune deficiencies. Curr Opin Allergy Clin Immunol. 2007;7(6):528–34.

44. Tyndall A, Uccelli A. Multipotent mesenchymal stromal cells for autoimmune diseases: teaching new dogs old tricks. Bone Marrow Transplant. 2009;43(11):821–8.
45. Ramasamy R, et al. Mesenchymal stem cells inhibit proliferation and apoptosis of tumor cells: impact on in vivo tumor growth. Leukemia. 2007;21(2):304–10.
46. Zacharek A, et al. Angiopoietin1/Tie2 and VEGF/Flk1 induced by MSC treatment amplifies angiogenesis and vascular stabilization after stroke. J Cereb Blood Flow Metab. 2007;27(10):1684–91.
47. Chen J, et al. Intravenous administration of human bone marrow stromal cells induces angiogenesis in the ischemic boundary zone after stroke in rats. Circ Res. 2003;92(6):692–9.
48. Tuttolomondo A, et al. Inflammatory cytokines in acute ischemic stroke. Curr Pharm Des. 2008;14(33):3574–89.
49. Gomes-Leal W, Corkill DJ, Picanco-Diniz CW. Systematic analysis of axonal damage and inflammatory response in different white matter tracts of acutely injured rat spinal cord. Brain Res. 2005;1066(1–2):57–70.
50. Souza-Rodrigues RD, et al. Inflammatory response and white matter damage after microinjections of endothelin-1 into the rat striatum. Brain Res. 2008;1200:78–88.
51. Thored P, et al. Long-term accumulation of microglia with proneurogenic phenotype concomitant with persistent neurogenesis in adult subventricular zone after stroke. Glia. 2009;57(8):835–49.
52. Gomes-Leal W, et al. Astrocytosis, microglia activation, oligodendrocyte degeneration, and pyknosis following acute spinal cord injury. Exp Neurol. 2004;190(2):456–67.
53. Nasef A, et al. Identification of IL-10 and TGF-beta transcripts involved in the inhibition of T-lymphocyte proliferation during cell contact with human mesenchymal stem cells. Gene Expr. 2007;13(4–5):217–26.
54. Yanez R, et al. Adipose tissue-derived mesenchymal stem cells have in vivo immunosuppressive properties applicable for the control of the graft-versus-host disease. Stem Cells. 2006;24(11):2582–91.
55. Sotiropoulou PA, et al. Interactions between human mesenchymal stem cells and natural killer cells. Stem Cells. 2006;24(1):74–85.
56. Spaggiari GM, et al. Mesenchymal stem cell-natural killer cell interactions: evidence that activated NK cells are capable of killing MSCs, whereas MSCs can inhibit IL-2-induced NK-cell proliferation. Blood. 2006;107(4):1484–90.
57. Djouad F, et al. Mesenchymal stem cells inhibit the differentiation of dendritic cells through an interleukin-6-dependent mechanism. Stem Cells. 2007;25(8):2025–32.
58. Aggarwal S, Pittenger MF. Human mesenchymal stem cells modulate allogeneic immune cell responses. Blood. 2005;105(4):1815–22.
59. Keating A. Mesenchymal stromal cells. Curr Opin Hematol. 2006;13(6):419–25.
60. Krampera M, et al. Bone marrow mesenchymal stem cells inhibit the response of naive and memory antigen-specific T cells to their cognate peptide. Blood. 2003;101(9):3722–9.
61. Ren G, et al. Mesenchymal stem cell-mediated immunosuppression occurs via concerted action of chemokines and nitric oxide. Cell Stem Cell. 2008;2(2):141–50.
62. Ren G, et al. Species variation in the mechanisms of mesenchymal stem cell-mediated immunosuppression. Stem Cells. 2009;27(8):1954–62.
63. Zheng ZH, et al. Allogeneic mesenchymal stem cell and mesenchymal stem cell-differentiated chondrocyte suppress the responses of type II collagen-reactive T cells in rheumatoid arthritis. Rheumatology (Oxford). 2008;47(1):22–30.
64. Sudres M, et al. Bone marrow mesenchymal stem cells suppress lymphocyte proliferation in vitro but fail to prevent graft-versus-host disease in mice. J Immunol. 2006;176(12):7761–7.
65. Inoue S, et al. Immunomodulatory effects of mesenchymal stem cells in a rat organ transplant model. Transplantation. 2006;81(11):1589–95.
66. Crop MJ, et al. Human adipose tissue-derived mesenchymal stem cells induce explosive T-cell proliferation. Stem Cells Dev. 2010;19(12):1843–53.

67. Li W, et al. Mesenchymal stem cells: a double-edged sword in regulating immune responses. Cell Death Differ. 2012;19(9):1505–13.
68. Herrmann JL, et al. Preconditioning mesenchymal stem cells with transforming growth factor-alpha improves mesenchymal stem cell-mediated cardioprotection. Shock. 2010;33(1):24–30.
69. Polchert D, et al. IFN-gamma activation of mesenchymal stem cells for treatment and prevention of graft versus host disease. Eur J Immunol. 2008;38(6):1745–55.
70. Nemeth K, et al. Bone marrow stromal cells attenuate sepsis via prostaglandin E(2)-dependent reprogramming of host macrophages to increase their interleukin-10 production. Nat Med. 2009;15(1):42–9.
71. Hegyi B, et al. Activated T-cells and pro-inflammatory cytokines differentially regulate prostaglandin E2 secretion by mesenchymal stem cells. Biochem Biophys Res Commun. 2012;419(2):215–20.
72. Ylostalo JH, et al. Human mesenchymal stem/stromal cells cultured as spheroids are self-activated to produce prostaglandin E2 that directs stimulated macrophages into an anti-inflammatory phenotype. Stem Cells. 2012;30(10):2283–96.
73. Sun T, Ma QH. Repairing neural injuries using human umbilical cord blood. Mol Neurobiol. 2013;47(3):938–45.
74. Swirski FK, et al. Identification of splenic reservoir monocytes and their deployment to inflammatory sites. Science. 2009;325(5940):612–6.
75. Emsley HC, et al. An early and sustained peripheral inflammatory response in acute ischaemic stroke: relationships with infection and atherosclerosis. J Neuroimmunol. 2003;139(1–2):93–101.
76. Lucas SM, Rothwell NJ, Gibson RM. The role of inflammation in CNS injury and disease. Br J Pharmacol. 2006;147 Suppl 1:S232–40.
77. Seifert HA, et al. A transient decrease in spleen size following stroke corresponds to splenocyte release into systemic circulation. J Neuroimmune Pharmacol. 2012;7(4):1017–24.
78. Woiciechowsky C, et al. Mechanisms of brain-mediated systemic anti-inflammatory syndrome causing immunodepression. J Mol Med (Berl). 1999;77(11):769–80.
79. Vendrame M, et al. Cord blood rescues stroke-induced changes in splenocyte phenotype and function. Exp Neurol. 2006;199(1):191–200.
80. Sahota P, et al. Changes in spleen size in patients with acute ischemic stroke: a pilot observational study. Int J Stroke. 2013;8(2):60–7.
81. Zhang BJ, et al. Splenectomy protects experimental rats from cerebral damage after stroke due to anti-inflammatory effects. Chin Med J (Engl). 2013;126(12):2354–60.
82. Offner H, et al. Experimental stroke induces massive, rapid activation of the peripheral immune system. J Cereb Blood Flow Metab. 2006;26(5):654–65.
83. Ajmo Jr CT, et al. Blockade of adrenoreceptors inhibits the splenic response to stroke. Exp Neurol. 2009;218(1):47–55.
84. Ajmo Jr CT, et al. The spleen contributes to stroke-induced neurodegeneration. J Neurosci Res. 2008;86(10):2227–34.
85. Acosta SA, et al. Intravenous bone marrow stem cell grafts preferentially migrate to spleen and abrogate chronic inflammation in stroke. Stroke. 2015;46(9):2616–27.
86. Chamorro A, Urra X, Planas AM. Infection after acute ischemic stroke: a manifestation of brain-induced immunodepression. Stroke. 2007;38(3):1097–103.
87. Lee ST, et al. Anti-inflammatory mechanism of intravascular neural stem cell transplantation in haemorrhagic stroke. Brain. 2008;131(Pt 3):616–29.
88. Schwarting S, et al. Hematopoietic stem cells reduce postischemic inflammation and ameliorate ischemic brain injury. Stroke. 2008;39(10):2867–75.
89. Yang B, et al. Therapeutic time window and dose response of autologous bone marrow mononuclear cells for ischemic stroke. J Neurosci Res. 2011;89(6):833–9.

90. Karlupia N, et al. Intraarterial transplantation of human umbilical cord blood mononuclear cells is more efficacious and safer compared with umbilical cord mesenchymal stromal cells in a rodent stroke model. Stem Cell Res Ther. 2014;5(2):45.
91. Kreutzberg GW. Microglia: a sensor for pathological events in the CNS. Trends Neurosci. 1996;19(8):312–8.
92. Thomas WE. Brain macrophages: evaluation of microglia and their functions. Brain Res Brain Res Rev. 1992;17(1):61–74.
93. Lalancette-Hebert M, et al. Selective ablation of proliferating microglial cells exacerbates ischemic injury in the brain. J Neurosci. 2007;27(10):2596–605.
94. Hughes V. Microglia: the constant gardeners. Nature. 2012;485(7400):570–2.
95. Weinstein JR, Koerner IP, Moller T. Microglia in ischemic brain injury. Future Neurol. 2010;5(2):227–46.
96. Taylor RA, Sansing LH. Microglial responses after ischemic stroke and intracerebral hemorrhage. Clin Dev Immunol. 2013;2013:746068.
97. Starossom SC, et al. Galectin-1 deactivates classically activated microglia and protects from inflammation-induced neurodegeneration. Immunity. 2012;37(2):249–63.
98. Hu X, et al. Microglia/macrophage polarization dynamics reveal novel mechanism of injury expansion after focal cerebral ischemia. Stroke. 2012;43(11):3063–70.
99. Perego C, Fumagalli S, De Simoni MG. Temporal pattern of expression and colocalization of microglia/macrophage phenotype markers following brain ischemic injury in mice. J Neuroinflammation. 2011;8:174.
100. Vorwald-Denholtz PP, De Robertis EM. Temporal pattern of the posterior expression of Wingless in Drosophila blastoderm. Gene Expr Patterns. 2011;11(7):456–63.
101. Zanier ER, et al. Bone marrow mesenchymal stromal cells drive protective M2 microglia polarization after brain trauma. Neurotherapeutics. 2014;11(3):679–95.
102. Zheng G, et al. Mesenchymal stromal cells affect disease outcomes via macrophage polarization. Stem Cells Int. 2015;2015:989473.
103. Baker N, Boyette LB, Tuan RS. Characterization of bone marrow-derived mesenchymal stem cells in aging. Bone. 2015;70:37–47.
104. Carlson ME, Conboy IM. Loss of stem cell regenerative capacity within aged niches. Aging Cell. 2007;6(3):371–82.
105. Lavasani M, et al. Muscle-derived stem/progenitor cell dysfunction limits healthspan and lifespan in a murine progeria model. Nat Commun. 2012;3:608.
106. Choumerianou DM, et al. Comparative study of stemness characteristics of mesenchymal cells from bone marrow of children and adults. Cytotherapy. 2010;12(7):881–7.
107. Mareschi K, et al. Expansion of mesenchymal stem cells isolated from pediatric and adult donor bone marrow. J Cell Biochem. 2006;97(4):744–54.
108. Kim J, et al. Biological characterization of long-term cultured human mesenchymal stem cells. Arch Pharm Res. 2009;32(1):117–26.
109. Yu Y, et al. Characterization of long-term in vitro culture-related alterations of human tonsil-derived mesenchymal stem cells: role for CCN1 in replicative senescence-associated increase in osteogenic differentiation. J Anat. 2014;225(5):510–8.
110. Li X, et al. Enhanced apoptosis and senescence of bone-marrow-derived mesenchymal stem cells in patients with systemic lupus erythematosus. Stem Cells Dev. 2012;21(13):2387–94.
111. Stolzing A, et al. Age-related changes in human bone marrow-derived mesenchymal stem cells: consequences for cell therapies. Mech Ageing Dev. 2008;129(3):163–73.
112. Zhang DY, Wang HJ, Tan YZ. Wnt/beta-catenin signaling induces the aging of mesenchymal stem cells through the DNA damage response and the p53/p21 pathway. PLoS One. 2011;6(6):e21397.
113. Hao H, et al. Culturing on Wharton's jelly extract delays mesenchymal stem cell senescence through p53 and p16INK4a/pRb pathways. PLoS One. 2013;8(3):e58314.

114. Schellenberg A, et al. Replicative senescence of mesenchymal stem cells causes DNA-methylation changes which correlate with repressive histone marks. Aging (Albany NY). 2011;3(9):873–88.
115. Geissler S, et al. Functional comparison of chronological and in vitro aging: differential role of the cytoskeleton and mitochondria in mesenchymal stromal cells. PLoS One. 2012;7(12):e52700.
116. Bustos ML, et al. Aging mesenchymal stem cells fail to protect because of impaired migration and antiinflammatory response. Am J Respir Crit Care Med. 2014;189(7):787–98.
117. Sun Y, et al. Rescuing replication and osteogenesis of aged mesenchymal stem cells by exposure to a young extracellular matrix. FASEB J. 2011;25(5):1474–85.
118. Yao J, et al. Tissue inhibitor of matrix metalloproteinase-3 or vascular endothelial growth factor transfection of aged human mesenchymal stem cells enhances cell therapy after myocardial infarction. Rejuvenation Res. 2012;15(5):495–506.
119. Yu JM, et al. Age-related changes in mesenchymal stem cells derived from rhesus macaque bone marrow. Aging Cell. 2011;10(1):66–79.
120. Choudhery MS, et al. Bone marrow derived mesenchymal stem cells from aged mice have reduced wound healing, angiogenesis, proliferation and anti-apoptosis capabilities. Cell Biol Int. 2012;36(8):747–53.
121. Kretlow JD, et al. Donor age and cell passage affects differentiation potential of murine bone marrow-derived stem cells. BMC Cell Biol. 2008;9:60.
122. Siegel G, et al. Phenotype, donor age and gender affect function of human bone marrow-derived mesenchymal stromal cells. BMC Med. 2013;11:146.
123. Li WY, et al. Mesenchymal stem cells for ischemic stroke: changes in effects after ex vivo culturing. Cell Transplant. 2008;17(9):1045–59.
124. Leslie WD, et al. Type 2 diabetes and bone. J Bone Miner Res. 2012;27(11):2231–7.
125. Dmitrieva RI, et al. Functional properties of bone marrow derived multipotent mesenchymal stromal cells are altered in heart failure patients, and could be corrected by adjustment of expansion strategies. Aging (Albany NY). 2015;7(1):14–25.
126. Zacharek A, et al. Comparison of bone marrow stromal cells derived from stroke and normal rats for stroke treatment. Stroke. 2010;41(3):524–30.
127. Andre T, et al. Evidences of early senescence in multiple myeloma bone marrow mesenchymal stromal cells. PLoS One. 2013;8(3):e59756.
128. Cmielova J, et al. Gamma radiation induces senescence in human adult mesenchymal stem cells from bone marrow and periodontal ligaments. Int J Radiat Biol. 2012;88(5):393–404.
129. Kern S, et al. Comparative analysis of mesenchymal stem cells from bone marrow, umbilical cord blood, or adipose tissue. Stem Cells. 2006;24(5):1294–301.
130. Fernandez Vallone VB, et al. Mesenchymal stem cells and their use in therapy: what has been achieved? Differentiation. 2013;85(1–2):1–10.
131. Bartholomew A, et al. Mesenchymal stem cells suppress lymphocyte proliferation in vitro and prolong skin graft survival in vivo. Exp Hematol. 2002;30(1):42–8.
132. Djouad F, et al. Immunosuppressive effect of mesenchymal stem cells favors tumor growth in allogeneic animals. Blood. 2003;102(10):3837–44.
133. Glennie S, et al. Bone marrow mesenchymal stem cells induce division arrest anergy of activated T cells. Blood. 2005;105(7):2821–7.
134. Najar M, et al. Mesenchymal stromal cells use PGE2 to modulate activation and proliferation of lymphocyte subsets: combined comparison of adipose tissue, Wharton's jelly and bone marrow sources. Cell Immunol. 2010;264(2):171–9.
135. Ribeiro A, et al. Mesenchymal stem cells from umbilical cord matrix, adipose tissue and bone marrow exhibit different capability to suppress peripheral blood B, natural killer and T cells. Stem Cell Res Ther. 2013;4(5):125.
136. Castro-Manrreza ME, et al. Human mesenchymal stromal cells from adult and neonatal sources: a comparative in vitro analysis of their immunosuppressive properties against T cells. Stem Cells Dev. 2014;23(11):1217–32.

137. Omori Y, et al. Optimization of a therapeutic protocol for intravenous injection of human mesenchymal stem cells after cerebral ischemia in adult rats. Brain Res. 2008;1236:30–8.
138. Beggs KJ, et al. Immunologic consequences of multiple, high-dose administration of allogeneic mesenchymal stem cells to baboons. Cell Transplant. 2006;15(8–9):711–21.
139. Isakova IA, et al. Allo-reactivity of mesenchymal stem cells in rhesus macaques is dose and haplotype dependent and limits durable cell engraftment in vivo. PLoS One. 2014;9(1):e87238.

Chapter 6
Bone Marrow Hematopoietic Stem Cell Therapy in Stroke

Michael Jiang and Ling Wei

Abstract Mammalian bone marrow is host to an active stem cell population with the responsibility of maintaining and replenishing a variety of high turnover cells that are vital for survival, regeneration, and support of the rest of the body. The major outputs of this hematopoietic system are the erythrocytes and platelets of the blood and the granulocytes and macrophages of the immune system. These stem cells also give rise to antibody-producing B-lymphocytes and antibody-producing T-lymphocytes that consume foreign or inflammatory substrates in the body. As all of these substituent cell types have a very limited life span, they must be constantly and consistently replenished. This rapid turnover of up to 100 billion new cells from progenitors forms the hematopoietic system (Mohammadi et al. Int J Stem Cell Res Transplant, 2014 2(02), 59–62). In recent years, bone marrow hematopoietic stem cells have drawn increasing attention for their therapeutic potential used in cell transplantation therapy for neurological disorders including stroke and traumatic brain injury. This review article is to summarize recent progress in basic and preclinical investigations on these cells including hematopoietic and mesenchymal stem cells.

Keywords Bone marrow • Hematopoietic stem cell • Therapy • Stroke

6.1 History of Hematopoietic Research

Hematopoietic stem cells have been studied for more than a half century as a therapeutic approach for patients suffering from compromised immune systems [34]. The hematopoietic system was particularly sensitive to acute radiation with death resulting from either pathogenic invasion or loss of blood due to low levels of platelets or other reasons. In 1952, Lorenz et al. demonstrated that radiation-damaged hematopoietic systems could be rescued by injections of bone marrow cell

M. Jiang • L. Wei (✉)
Department of Anesthesiology, Emory University School of Medicine,
101 Woodruff Circle, WMR 617, Atlanta, GA 30322, USA
e-mail: lwei7@emory.edu

K. Jin et al. (eds.), *Bone Marrow Stem Cell Therapy for Stroke*,
DOI 10.1007/978-981-10-2929-5_6

suspensions [36, 64]. This discovery formed the foundation of modern-day treatment for hematopoietic system failure by transplantation of bone marrow hematopoietic stem cells [13].

Loss of cellular regulatory substitutes can occur permanently or temporarily in neurological diseases including ischemic stroke.

The selective vulnerability of the hematopoietic system to low doses of radiation also gained relevance as it was compromised in patients with the advent of chemical or radioactive chemotherapeutic agents that target rapidly dividing cells [54, 57]. The hematopoietic system is comprised of cells that divide rapidly as is the case of many cancer cells. Therefore, drugs and therapies used to treat cancers can cause patients suffering a weakened immune system that requires secondary hematopoietic therapies to restore the immune activity [57]. Much work in the hematopoietic system originated and continues in murine models. Hematopoietic stem cells (HSCs) can be dissected from the yolk sac or the intrabody aorta-gonad-mesonephros (AMG) region of mouse embryos [37, 43]. Surprisingly, adult hematopoietic stem cells are not present during the earliest stages of development but arise after the formation of the hematopoietic system [45].

6.2 Hematopoietic Stem Cells and Stroke

Stroke remains a leading cause of death and disability in the United States and around the world [41]. The current limited availability of effective therapeutics for human stroke patients has inspired multifaceted and increasingly complex approaches to limit ischemic damage and/or enhance post-stroke regeneration. Treatments that can do both have become attractive candidates in stroke research [35]. Among these are various cell-based transplantation therapies which have had demonstrable success in animal models of stroke [3, 5, 67]. The investigation has looked at the effects of transplanting different stem cells and neural progenitor cells, timing of the transplantation, and the delivery method of the desired cells. Cellular therapies have multifaceted effects during both acute and chronic phases of neurological disease. Major cell types currently under investigation including but not limited to induced pluripotent stem (iPS) cells [44], embryonic stem cells [18], bone marrow-derived mesenchymal stem cells [33], and bone marrow hematopoietic stem cells (HSC) [47] offer unique advantages with its own set of drawbacks. Among the various potential cell types used for cell-based stroke therapeutics, HSCs maintain the advantage of being available from autologous donation without risk of graft rejection, low immunoreaction, and self-renewal [1, 15].

Ischemic stroke is characterized by rapid deregulation of normal cellular processes resulting from lack of oxygen. The resulting cell loss due to excitotoxicity and various apoptotic and/or necrotic mechanisms generates a highly toxic environment with compromised access to the hematopoietic system [9]. As such, a temporary ischemic insult can result in a chronic condition within a brain loci that mimics many of the pathological hallmarks characteristic of systemic hematopoietic failure.

Interestingly, ischemic stroke and myocardial infarction can mobilize CD34+ hematopoietic stem cells into blood circulation [47]. Augmenting the body's hematopoietic system by stimulating bone marrow stem cells into the blood and homing to the ischemic region is an attractive means of assuaging stroke pathogenesis during acute and chronic phases of injury.

Brain ischemia is known to trigger immune responses including activation of resident microglia within 24 h [17]. Bone marrow-derived monocytes can enter the brain through the blood-brain barrier near the ischemic region where they also exhibit a microglia phenotype [26]. Both local immune cells and ischemic tissues are known to release cytokines, complement factors, and free radicals that escalate the immune response [30]. While some of these factors have been found to be anti-inflammatory such as IL-10 [20], the general environmental tone is pro-inflammatory. Consequently, both the ischemic region and the peri-infarct regions following stroke harbor a pro-apoptotic and/or pro-necrotic tone. In the days to weeks following ischemia, this toxic environment leads to the expansion of neuronal and glial loss [21].

6.3 Hematopoietic Stem Cell Transplantation

Ischemia results in massive loss of not only neuronal cells but also a host of supportive cell types necessary for maintenance and survival of brain tissue [14]. Together, the loss of these cells represents loss of entire neurovascular units which includes cell types such as endothelial cells, pericytes, astrocytes, and other glial cells [14, 23]. One of the major goals of transplantation of hematopoietic stem cells for stroke is to augment the body's natural ability to supply supportive substituent cell types to the ischemic region. Rather than delivery of a single differentiated cell type to the ischemic region, HSCs have been demonstrated to differentiate into a variety of supportive cell types including astrocytes, oligodendrocytes, neural precursors, macrophages, and microphages [31, 56]. The multipotent characteristic of HSCs makes them a preferable cell therapy without need for combination with other cell types.

Systemic injection of HSCs can be traced using pre-labeling of injected cells to differentiate them from host tissues. Injected GFP-tagged HSCs can be observed in the spleen within the first 24 h of delivery [32]. HSCs typically home to the bone marrow and spleen regardless of delivery methods [24, 50]. A limited number of injected HSCs can be observed in the brain where their most common phenotype is that of microglia-like cells reminiscent of host microglia found in a healthy state [24, 51, 55]. Modification of HSCs by exogenous overexpression of Sca-1, Thy-1, and c-kit in a spinal cord injury model demonstrated differentiation of HSCs into astrocytes, oligodendrocytes, and neuronal precursors [65].

As a cardiovascular disease, ischemic stroke damages both the brain and the circulatory arteries and veins in and around the infarct. Regenerative therapy for ischemic stroke favors approaches that enhance both neurogenesis and angiogenesis.

Toward the latter goal, HSCs include a subset of CD34+ endothelial stem and progenitor cells. HSC therapy via systemic injection or local intracerebral injection after stroke are both known to increase angiogenesis in the peri-infarct region. Along with angiogenesis, these studies also indicate that HSC therapy promotes enhanced functional recovery and reduced infarct size. Specifically, CD34+ cells from HSCs which provide the strongest angiogenic effects can be enriched using immunoselection for acute ischemic stroke [59].

Post-ischemia neurogenesis in the brain stems from two well-studied regions: the subventricular zone and the subgranular zone of the dentate gyrus [22]. While neurogenesis has been well documented to play a role following ischemic stroke in animal models [42, 48], its effect has been minimal with reports of only 0.2 % of neurons being replaced by endogenous neurogenesis [2]. These results can be verified by delivery of CD34 cells from HSC preparations to the post-ischemic brain. Furthermore, consistent with these findings are reports that pro-angiogenic agents such as erythropoietin (EPO) produce similar results as CD34+ HSC delivery [16]. Suppression of endothelial proliferation by endostatin can also directly mask the beneficial effects observed from CD34+ cell delivery [59]. These findings indicate a role for HSC therapy in augmenting endogenous post-stroke regeneration by enhancing neovascularization to support new neurons which have otherwise shown to migrate but quickly die in the toxic environment surrounding stroke [2].

The relationship between vascular support and neuro-regeneration can be linked to effective angiogenic therapies studied in other organs including the liver and thyroid. Supporting regenerating tissues with neovascularization can significantly improve endogenous cell replacement both *in vivo* and *in vitro* [53, 62]. It is hypothesized that improved vascularization may help brain regeneration by providing additional routes for the removal of dead tissue which also doubles as additional migration routes into the ischemic zone. Examination of this hypothesis and various factors that may play a role in promoting the survival of both endogenous and transplanted neuro-progenitors has been extensively investigated including but not limited to fibroblast growth factor 2 [29], platelet-derived growth factor [58], brain-derived neurotrophic factor [10], interleukin-8 [64], vascular endothelial growth factor [63], insulin growth factor-1 [68], and focal adhesion kinase [27]. Generally, multifactorial therapies aim to improve both the migration and trophic support provided to the ischemic tissue following stroke. In animal models, augmenting just one of these by aforementioned and other factors has consistently demonstrated enhanced functional behavior and reduction in stroke severity [66].

6.4 The Advantage and Disadvantage of Using BMSCs

Stem cells are characterized not only by their multipotent differentiation ability but also by their faculty for self-replication [12]. Hematopoietic stem cells are able to reproduce themselves while also differentiating into hematopoietic progenitor cells which are still multipotent but without the ability for self-replication [52]. HSCs removed from bone marrow of rodents are capable of surviving and perpetuating in vitro cultures beyond the life span of the original host. However, self-replication in HSCs is not indefinite, and this phenomenon has been the focus of studies examining the effect of telomere length and telomerase activity [39]. Once removed from murine hosts, *in vitro* HSC preparations can survive up to three or four lifetimes of the original host but cannot be maintained much past this length. The inability for indefinite self-renewal may be related to telomerase activity [40].

Unlike induced pluripotent stem cells or bone marrow mesenchymal stem cells, HSCs exhibit limited capacity for expansion *in vitro*. Consequently reproducing large quantities of HSCs is challenging as they must be harvested from bone marrow itself [49], umbilical cord blood [6], or mobilized blood [7, 38]. HSCs are penchant to differentiation in cell culture rather than self-perpetuity [60]. Investigation into HSC self-renewal mechanisms *in vitro* that are absent *in vivo* is currently ongoing.

While systemic delivery of HSC via intravenous injection has proven effective in the treatment of both myocardial infarction and ischemic stroke, the mechanisms governing separate protective and restorative or regenerative effects are difficult to isolate. A key role of HSC infiltration to the ischemic brain is the anti-inflammatory effects conveyed by HSC-differentiated glia and immune cells. Tracking of transplanted BM-derived cells was traditionally conducted by Till and McCulloch by irradiating HSC from donors to induce chromosomal repairs and breaks making them distinguishable from host HSCs. In this way, fully differentiated cells with various fates could be mapped back to a single transplanted donor colony bearing the same radiation-induced chromosomal marker [4]. In contrast, recent studies examining HSC infiltration after stroke use HSCs isolated from green fluorescent protein (GFP) expressing transgenic mice that can be tracked once injected into C57/BL6 recipient mice.

Unfortunately, CD34+ cells make up approximately just 1 % of BMHSC preparations making it difficult to discern the role played by these cells. In support of the effect of CD34+ cell fraction from HSC preparations, Taguchi et al. demonstrated that cell delivery in the CD34+ group but not CD34-negative group after stroke had enhanced neurogenesis via angiogenesis. Using human CD34+ cells in an immunocompromised cell line, Taguchi et al. reported enhanced neovascularization in the peri-infarcted tissue along with enhanced neurogenesis [59].

In addition to HSCs, two other bone marrow-derived cell types have also been studied for the treatment of ischemic stroke: hematopoietic progenitor stem cells and bone marrow-derived mesenchymal stem cells. Interestingly, injection of bone marrow stem cell-conditioned media alone has been demonstrated to convey functional recovery benefits in a rodent model of stroke [11]. Thus it has been proposed that HSC transplantation can act as an augmented source for paracrine trophic support to the ischemic brain [8, 25]. This hypothesis supports the efficacy observed using HSC transplantation studies despite relatively poor homing of the transplanted cells to the brain and ischemic region itself. Despite this, the study of HSC therapy continues to seek methods for improving survival, homing, and efficacy of transplanted HSCs.

An effective approach to improving HSC transplantation therapy is the preconditioning of transplanted cells by hypoxia immediately prior to administration in an ischemic stroke model [1]. Hypoxia preconditioning (HP) has been leveraged in stem cell therapies for its multifactorial effects on improving cell survival following transplantation [28]. Both embryonic stem cells and bone marrow stem cells which undergo this hypoxia preconditioning protocol are significantly more resistant to necrotic and apoptotic insults making them better able to survive *in vitro* and *in vivo* ischemic conditions [19, 46, 61]. Preconditioning bone marrow-derived stem cells is also an effective means of improving the homing of intravenously injected stem cell preparations for the treatment of stroke and myocardial infarction [28]. Hu et al. describe a potassium channel-kinase interaction (Kv2.1-FAK)-mediated mechanism whereby hypoxia enhances the expression of Kv2.1, thereby phosphorylating FAK and boosting BMSC migration ability. Selective knockdown of the Kv2.1 channel or applying a K+ channel inhibitor significantly dampened the ability of BMSCs to migrate to the infarcted tissue. Altering gene expression in HSCs related to survival, migration, and differentiation of transplanted cells remains a popular avenue for investigation.

References

1. Aggarwal R, Pompili VJ, Das H. Genetic modification of ex-vivo expanded stem cells for clinical application. Front Biosci (Landmark Edition). 2009;15:854–71.
2. Arvidsson A, Collin T, Kirik D, Kokaia Z, Lindvall O. Neuronal replacement from endogenous precursors in the adult brain after stroke. Nat Med. 2002;8(9):963–70.
3. Bang OY, Lee JS, Lee PH, Lee G. Autologous mesenchymal stem cell transplantation in stroke patients. Ann Neurol. 2005;57(6):874–82.
4. Becker AJ, McCulloch EA, Till JE. Cytological demonstration of the clonal nature of spleen colonies derived from transplanted mouse marrow cells. 1963.
5. Bliss T, Guzman R, Daadi M, Steinberg GK. Cell transplantation therapy for stroke. Stroke. 2007;38(2):817–26.
6. Broxmeyer HE, Douglas GW, Hangoc G, Cooper S, Bard J, English D, … Boyse EA. Human umbilical cord blood as a potential source of transplantable hematopoietic stem/progenitor cells. Proc Natl Acad Sci. 1989;86(10):3828–32.
7. Broxmeyer HE, Orschell CM, Clapp DW, Hangoc G, Cooper S, Plett PA, … Campbell TB. Rapid mobilization of murine and human hematopoietic stem and progenitor cells with AMD3100, a CXCR4 antagonist. J Exp Med. 2005;201(8):1307–18.
8. Caplan AI, Dennis JE. Mesenchymal stem cells as trophic mediators. J Cell Biochem. 2006;98(5):1076–84.
9. Castillo J, Dávalos A, Noya M. Progression of ischaemic stroke and excitotoxic amino acids. Lancet. 1997;349(9045):79–82.
10. Chen J, Zhang C, Jiang H, Li Y, Zhang L, Robin A, … Chopp M. Atorvastatin induction of VEGF and BDNF promotes brain plasticity after stroke in mice. J Cereb Blood Flow Metab. 2005;25(2):281–90.
11. Chen X, Li Y, Wang L, Katakowski M, Zhang L, Chen J, … Chopp M. Ischemic rat brain extracts induce human marrow stromal cell growth factor production. Neuropathology, 2002;22(4):275–79.

12. Clarke ML, Frampton J. Hematopoietic stem cells. Regenerative medicine. Springer; 2013. p. 251–77
13. Deeg HJ, Gooley TA, Flowers ME, Sale GE, Slattery JT, Anasetti C, … Kiem H-P. Allogeneic hematopoietic stem cell transplantation for myelofibrosis. Blood. 2003;102(12):3912–18.
14. Del Zoppo G. The neurovascular unit in the setting of stroke. J Intern Med. 2010;267(2):156–71.
15. Domen J, Weissman IL. Self-renewal, differentiation or death: regulation and manipulation of hematopoietic stem cell fate. Mol Med Today. 1999;5(5):201–8.
16. Ehrenreich H, Weissenborn K, Prange H, Schneider D, Weimar C, Wartenberg K, … Wegrzyn M. Recombinant human erythropoietin in the treatment of acute ischemic stroke. Stroke. 2009;40(12):e647–56.
17. Ekdahl C, Kokaia Z, Lindvall O. Brain inflammation and adult neurogenesis: the dual role of microglia. Neuroscience. 2009;158(3):1021–9.
18. Erdö F, Bührle C, Blunk J, Hoehn M, Xia Y, Fleischmann B, … Hescheler J. Host-dependent tumorigenesis of embryonic stem cell transplantation in experimental stroke. J Cereb Blood Flow Metab. 2003;23(7):780–85.
19. Francis K, Wei L. Human embryonic stem cell neural differentiation and enhanced cell survival promoted by hypoxic preconditioning. Cell Death Dis. 2010;1(2), e22.
20. Frenkel D, Huang Z, Maron R, Koldzic DN, Hancock WW, Moskowitz MA, Weiner HL. Nasal vaccination with myelin oligodendrocyte glycoprotein reduces stroke size by inducing IL-10-producing CD4+ T cells. J Immunol. 2003;171(12):6549–55.
21. Furlan M, Marchal G, Derlon JM, Baron JC, Viader F. Spontaneous neurological recovery after stroke and the fate of the ischemic penumbra. Ann Neurol. 1996;40(2):216–26.
22. Gage FH. Mammalian neural stem cells. Science. 2000;287(5457):1433–8.
23. Hawkins BT, Davis TP. The blood-brain barrier/neurovascular unit in health and disease. Pharmacol Rev. 2005;57(2):173–85.
24. Henschler R, Fehervizyova Z, Bistrian R, Seifried E. A mouse model to study organ homing behaviour of haemopoietic progenitor cells reveals high selectivity but low efficiency of multipotent progenitors to home into haemopoietic organs. Br J Haematol. 2004;126(1):111–9.
25. Hess DC, Borlongan C. Stem cells and neurological diseases. Cell Prolif. 2008;41(s1):94–114.
26. Hill WD, Hess DC, Martin-Studdard A, Carothers JJ, Zheng J, Hale D, … Conway SJ. SDF-1 (CXCL12) Is Upregulated in the ischemic penumbra following stroke: association with bone marrow cell homing to injury. J Neuropathol Exp Neurol. 2004;63(1):84–96.
27. Hu X, Wei L, Taylor TM, Wei J, Zhou X, Wang J-A, Yu SP. Hypoxic preconditioning enhances bone marrow mesenchymal stem cell migration via Kv2. 1 channel and FAK activation. Am J Phys Cell Phys. 2011;301(2):C362–72.
28. Hu X, Yu SP, Fraser JL, Lu Z, Ogle ME, Wang J-A, Wei L. Transplantation of hypoxia-preconditioned mesenchymal stem cells improves infarcted heart function via enhanced survival of implanted cells and angiogenesis. J Thorac Cardiovasc Surg. 2008;135(4):799–808.
29. Jin K, Sun Y, Xie L, Batteur S, Mao XO, Smelick C, … Greenberg DA. Neurogenesis and aging: FGF-2 and HB-EGF restore neurogenesis in hippocampus and subventricular zone of aged mice. Aging Cell. 2003;2(3):175–83.
30. Kamel H, Iadecola C. Brain-immune interactions and ischemic stroke: clinical implications. Arch Neurol. 2012;69(5):576–81.
31. Kopen GC, Prockop DJ, Phinney DG. Marrow stromal cells migrate throughout forebrain and cerebellum, and they differentiate into astrocytes after injection into neonatal mouse brains. Proc Natl Acad Sci. 1999;96(19):10711–6.
32. Li Y, Chen J, Chen X, Wang L, Gautam S, Xu Y, … Janakiraman N. Human marrow stromal cell therapy for stroke in rat neurotrophins and functional recovery. Neurology. 2002;59(4):514–23.

33. Li Y, Chopp M, Chen J, Wang L, Gautam SC, Xu Y-X, Zhang Z. Intrastriatal transplantation of bone marrow nonhematopoietic cells improves functional recovery after stroke in adult mice. J Cereb Blood Flow Metab. 2000;20(9):1311–9.
34. Lidman B, Cohn C. Effect of radar emmanations on the hematopoietic system. Air Surg Bull US Army Air Forces Office Air Surg. 1945;2:448.
35. Lo EH. A new penumbra: transitioning from injury into repair after stroke. Nat Med. 2008;14(5):497–500.
36. Lorenz E, Congdon C, Uphoff D. Modification of acute irradiation injury in mice and Guinea-Pigs by bone marrow injections 1. Radiology. 1952;58(6):863–77.
37. Matsuoka S, Tsuji K, Hisakawa H, Xu M-j, Ebihara Y, Ishii T, … Ikeda Y. Generation of definitive hematopoietic stem cells from murine early yolk sac and paraaortic splanchnopleures by aorta-gonad-mesonephros region–derived stromal cells. Blood. 2001;98(1):6–12.
38. Mohammadi H, Mohammadnejad J, Yavari K. Human peripheral blood derived hematopoietic stem cell: history, the isolation methods and investigation of different parameters effects on their differentiation to the body cells. Int J Stem Cell Res Transplant. 2014;2(02):59–62.
39. Morrison SJ, Prowse KR, Ho P, Weissman IL. Telomerase activity in hematopoietic cells is associated with self-renewal potential. Immunity. 1996;5(3):207–16.
40. Morrison SJ, Wandycz AM, Akashi K, Globerson A, Weissman IL. The aging of hematopoietic stem cells. Nat Med. 1996;2(9):1011–6.
41. Mozaffarian D, Benjamin EJ, Go AS, Arnett DK, Blaha MJ, Cushman M, … Fullerton HJ. Heart disease and stroke statistics—2016 update a report from the American Heart Association. Circulation. 2015; CIR. 0000000000000350.
42. Ohab JJ, Fleming S, Blesch A, Carmichael ST. A neurovascular niche for neurogenesis after stroke. J Neurosci. 2006;26(50):13007–16.
43. Ohneda O, Fennie C, Zheng Z, Donahue C, La H, Villacorta R, … Lasky LA. Hematopoietic stem cell maintenance and differentiation are supported by embryonic aorta-gonad-mesonephros region–derived endothelium. Blood. 1998;92(3):908–19.
44. Oki K, Tatarishvili J, Wood J, Koch P, Wattananit S, Mine Y, … Ladewig J. Human-induced pluripotent stem cells form functional neurons and improve recovery after grafting in stroke-damaged brain. Stem Cells. 2012;30(6):1120–133.
45. Orkin SH. Development of the hematopoietic system. Curr Opin Genet Dev. 1996;6(5):597–602.
46. Pacary E, Legros H, Valable S, Duchatelle P, Lecocq M, Petit E, … Bernaudin M. Synergistic effects of CoCl2 and ROCK inhibition on mesenchymal stem cell differentiation into neuron-like cells. J Cell Sci. 2006;119(13):2667–78.
47. Paczkowska E, Larysz B, Rzeuski R, Karbicka A, Jałowiński R, Kornacewicz-Jach Z, … Machaliński B. Human hematopoietic stem/progenitor-enriched CD34+ cells are mobilized into peripheral blood during stress related to ischemic stroke or acute myocardial infarction. Eur J Haematol. 2005;75(6):461–67.
48. Parent JM, Vexler ZS, Gong C, Derugin N, Ferriero DM. Rat forebrain neurogenesis and striatal neuron replacement after focal stroke. Ann Neurol. 2002;52(6):802–13.
49. Parmar K, Mauch P, Vergilio J-A, Sackstein R, Down JD. Distribution of hematopoietic stem cells in the bone marrow according to regional hypoxia. Proc Natl Acad Sci. 2007;104(13):5431–6.
50. Plett PA, Frankovitz SM, Orschell CM. Distribution of marrow repopulating cells between bone marrow and spleen early after transplantation. Blood. 2003;102(6):2285–91.
51. Recio JS, Alvarez-Dolado M, Díaz D, Baltanás FC, Piquer-Gil M, Alonso JR, Weruaga E. Bone marrow contributes simultaneously to different neural types in the central nervous system through different mechanisms of plasticity. Cell Transplant. 2011;20(8):1179–92.
52. Reya T, Morrison SJ, Clarke MF, Weissman IL. Stem cells, cancer, and cancer stem cells. Nature. 2001;414(6859):105–11.

53. Ross MA, Sander CM, Kleeb TB, Watkins SC, Stolz DB. Spatiotemporal expression of angiogenesis growth factor receptors during the revascularization of regenerating rat liver. Hepatology. 2001;34(6):1135–48.
54. Siena S, Bregni M, Brando B, Belli N, Ravagnani F, Gandola L, … Gianni AM. Flow cytometry for clinical estimation of circulating hematopoietic progenitors for autologous transplantation in cancer patients. Blood. 1991;77(2):400–09.
55. Simard AR, Rivest S. Bone marrow stem cells have the ability to populate the entire central nervous system into fully differentiated parenchymal microglia. FASEB J. 2004;18(9):998–1000.
56. Spangrude GJ. Hematopoietic stem-cell differentiation. Curr Opin Immunol. 1991;3(2):171–8.
57. Stadtmauer EA, O'Neill A, Goldstein LJ, Crilley PA, Mangan KF, Ingle JN, … Erban JK. Conventional-dose chemotherapy compared with high-dose chemotherapy plus autologous hematopoietic stem-cell transplantation for metastatic breast cancer. N Engl J Med. 2000;342(15):1069–76.
58. Su EJ, Fredriksson L, Geyer M, Folestad E, Cale J, Andrae J, … Yepes M. Activation of PDGF-CC by tissue plasminogen activator impairs blood-brain barrier integrity during ischemic stroke. Nat Med. 2008;14(7):731–37.
59. Taguchi A, Soma T, Tanaka H, Kanda T, Nishimura H, Yoshikawa H, … Stern DM. Administration of CD34+ cells after stroke enhances neurogenesis via angiogenesisin a mouse model. J Clin Investig. 2004;114(3):330–38.
60. Terada N, Hamazaki T, Oka M, Hoki M, Mastalerz DM, Nakano Y, … Scott EW. Bone marrow cells adopt the phenotype of other cells by spontaneous cell fusion. Nature. 2002;416(6880):542–45.
61. Theus MH, Wei L, Cui L, Francis K, Hu X, Keogh C, Yu SP. In vitro hypoxic preconditioning of embryonic stem cells as a strategy of promoting cell survival and functional benefits after transplantation into the ischemic rat brain. Exp Neurol. 2008;210(2):656–70.
62. Toda S, Nishimura T, Yamada S, Koike N, Yonemitsu N, Watanabe K, … Sugihara H. Immunohistochemical expression of growth factors in subacute thyroiditis and their effects on thyroid folliculogenesis and angiogenesis in collagen gel matrix culture. J Pathol. 1999;188(4):415–22.
63. van Bruggen N, Thibodeaux H, Palmer JT, Lee WP, Fu L, Cairns B, … van Lookeren Campagne M. VEGF antagonism reduces edema formation and tissue damage after ischemia/reperfusion injury in the mouse brain. J Clin Investig. 1999;104(11):1613–20.
64. Wang L, Li Y, Chen X, Chen J, Gautam SC, Xu Y, Chopp M. MCP-1, MIP-1, IL-8 and ischemic cerebral tissue enhance human bone marrow stromal cell migration in interface culture. Hematology. 2002;7(2):113–7.
65. Weissman IL, Anderson DJ, Gage F. Stem and progenitor cells: origins, phenotypes, lineage commitments, and transdifferentiations. Annu Rev Cell Dev Biol. 2001;17(1):387–403.
66. Zhang ZG, Chopp M. Neurorestorative therapies for stroke: underlying mechanisms and translation to the clinic. Lancet Neurol. 2009;8(5):491–500.
67. Zhao L-R, Duan W-M, Reyes M, Keene CD, Verfaillie CM, Low WC. Human bone marrow stem cells exhibit neural phenotypes and ameliorate neurological deficits after grafting into the ischemic brain of rats. Exp Neurol. 2002;174(1):11–20.
68. Zhu W, Fan Y, Frenzel T, Gasmi M, Bartus RT, Young WL, … Chen Y. Insulin growth factor-1 gene transfer enhances neurovascular remodeling and improves long-term stroke outcome in mice. Stroke. 2008;39(4):1254–61.

Chapter 7
Endothelial Progenitor Cell Therapy in Stroke

Yaying Song, Zhijun Zhang, and Guo-Yuan Yang

Abstract Endothelial progenitor cells (EPCs) are unique stem cells in circulating blood. Studies showed strong beneficial evidence using EPC therapy in experimental animal models and clinical trials. In this review, we discussed the characteristics of EPCs in the stroke therapy. We summarized the effect of EPCs on the treatment of cerebrovascular diseases including ischemic and hemorrhagic stroke, moyamoya disease, and vascular tumor, etc. Although the molecular mechanisms of EPC therapy are not fully understood, the function of EPCs included releasing growth factors, regulating microenvironment in the injury territory, and maintaining blood-brain barrier integrity. Clinical application of stem cell-based therapy is still in its infancy. The next decade of EPC research in the stroke field needs to focus on the studying the molecular mechanism or combining other type stem cells to enhance the potential of this therapeutic avenue, and translate to clinical application.

Keywords Blood-brain barrier • EPC • Ischemia • Stroke • Therapy

Abbreviations

BAVMs	Brain arteriovenous malformations
BBB	Blood-brain barrier
BM	Bone marrow
EOCs	Endothelial outgrowth cells
EPCs	Endothelial progenitor cells

Y. Song • G.-Y. Yang (✉)
Department of Neurology, Shanghai Ruijin Hospital, School of Medicine, Shanghai Jiao Tong University, Shanghai 200030, China

Neuroscience and Neuroengineering Center, Med-X Research Institute and School, Biomedical Engineering, Shanghai Jiao Tong University,
1954 Hua Shan Rd, Shanghai 200030, China
e-mail: gyyang0626@163.com

Z. Zhang
Department of Neurology, Shanghai Ruijin Hospital, School of Medicine, Shanghai Jiao Tong University, Shanghai 200030, China

K. Jin et al. (eds.), *Bone Marrow Stem Cell Therapy for Stroke*,
DOI 10.1007/978-981-10-2929-5_7

ESCs	Embryonic stem cells
GCSFs	Granulocyte colony-stimulating factors
HCC	Hepatocellular carcinoma
HGF	Hepatocyte growth factor
*h*UCB	Human umbilical cord blood
MIF	Macrophage migration inhibitory factor
MSCs	Mesenchymal stem cells
NIHSS	NIH Stroke Scale
NSCs	Neural stem cells
OGD	Oxygen glucose depletion
SMPCs	Smooth muscle progenitor cells
tPA	Tissue plasminogen activator

7.1 Stroke and Neurorepair After Stroke

Ischemic stroke is a leading cause of morbidity and mortality among cerebral disease. Irreversible injury happened both at the molecular and cellular level. These damages furthermore affect microvascular endothelium, glial cells, pericytes, and extracellular matrix, which constitute the neurovascular unit. Although collateral arteries could transiently open and supply oxygen and glucose for cell viability in the penumbra, it also trigger worsen secondary damage without an appropriate reperfusion for the timely restoration.

Neurorepair plays the most important role in the functional recovery after ischemic stroke; its target is to attenuate cerebral parenchyma injury in the acute ischemic phase and reestablish neuronal function [1]. The neurorepair processes include angiogenesis, neurogenesis, oligodendrogenesis, remodeling of neurovascular unit, and neuronal network formation. Angiogenesis involves endothelial cell (EC) proliferation, migration, microvessel sprout, capillary connection formation, and the increases of the vascular bed resistance in afflicted artery [2]. Arteriogenesis contributed to the collateral flow in the early phase of ischemic stroke, while angiogenesis improved cerebral blood perfusion in the late phase of ischemic stroke [3]. Cerebral ischemia can also induce neurogenesis in both hippocampal subgranular zone (SGZ) and sub-ventricular zone (SVZ) within the adult mammalian [4, 5]. New myelination of oligodendrocytes induced by stroke is also involved in cerebral ischemia repair [3]. Neurovascular unit, consisting of astrocytes, neurons, and vascular structures and their interactions, regulates the blood-brain barrier (BBB) and plays a critical role in physiological and pathological response; its remodeling is the basis for functional recovery after injury [6]. Thus, the longer-time recovery of neurological function regulated by all of the elements and enhancing neurorepair of the damaged tissue could be the promising therapeutic approach.

Currently, tissue plasminogen activator (tPA) is the only US FDA approved drug for patients with ischemic stroke onset within 4.5 h. Patients who have the time to receive this treatment are only 2–5 % [7]. Although a clinical study showed that thrombolytic treatment remains effective at the first 24 h [8], its use still has limitation for the therapy of patient suffering acute ischemic stroke.

Cell-based therapy, especially stem cell based, seem attractive to neurorestoration due to their wide therapeutic window and high rehabilitation potential. Various cell types such as mesenchymal stem cells (MSCs), neural stem cells (NSCs), embryonic stem cells (ESCs), and EPCs show their efficiency in both experiment and clinical trials [8]. Although it still has practical and clinical issues like transplantation time windows, routes of cell administration, and potential detrimental effects, current studies show that stem or progenitor cells play beneficial role by paracrine and systemic effects, mostly via releasing chemokine, growth factors, and nanoparticle [9].

7.2 Current Clinical Trials of EPC Therapy in Stroke

A series of clinical trials focus on EPC in the past decade. The results showed EPC may be a characteristic biomarker in vascular diseases [10–13]. Currently, two clinical trials are ongoing using EPCs as treatment for clinical stroke patients, while no clinical trials have been published yet. Chen et al. conducted a multicenter, single-blind, randomized, parallel-controlled clinical trial (NCT01468064) on stroke patients aged 18–80 years, within 7 days of the onset of symptoms. They intravenously infused 2.5 million autologous EPCs per kg suspended in 100 ml normal saline with 5 % autologous serum approximately 4 weeks after bone marrow (BM) aspiration. Second study used the same amount of cells transplanted approximately 1 week after initial boosting. NCT02605707 is a single-blind trial studying the effects of autologous EPCs plus conventional treatment including rehabilitation on patient aged 18–80 with stroke history between 6 and 60 months and NIH Stroke Scale (NIHSS) score of 7 or more points. The outcome showed the change of a serial neurological examination after transplantation (ClinicalTrials.gov identifiers: NCT01468064 and NCT02605707).

7.3 Endothelial Progenitor Cells

EPCs were initially isolated from adult peripheral blood and capable of differentiating into an endothelial-like phenotype cells in 1997. Traditional view that ECs differentiation was occurred only in embryonic development was challenged [14]. $CD34^{+}$ cells derived from bone marrow can mobilize to the peripheral circulation and involve in vascular recovery [15]. Data from Takahashi Laboratory had shown that circulating EPCs mobilize endogenously as natural response foci to promote neovascular formation [16]. The level of circulating EPC is considered as a biomarker in cardiovascular and cerebral vascular disease. The reduction of EPCs indicates the risk factor or more susceptibility to development of disease [17, 18]. The beneficial role of EPCs has been focused the valuable capability to improve the treatment of ischemic disease as a cell-based therapy.

EPCs as one of the adult stem cells originate from hematopoietic cells. It can also be found in the peripheral blood in adults. Evidence suggest that hematological stem cells (HSCs) and EPCs are derived from a common precursor [19, 20]; thus it has been speculated that circulating EPCs originate from hematopoietic stem cell [14]. Human adipose-derived stem cells can also be differentiated into EPCs via microenvironment, and the protective effect of EPC was further demonstrated in the hypoxic-ischemic injury model [21]. MSCs are capable to differentiate into cells with phenotypic and function features of endothelial cells *in vitro* [22]. Other alternative sources of EPCs include ESCs, fetal umbilical cord blood, and induced pluripotent stem (iPS) cell [23].

In in vitro cell culture, two main protocols have been used to isolate EPCs: one is primary cell culture and the other is colony assays based on surface markers. EPCs can be isolated from bone marrow, peripheral circulating blood, human umbilical cord blood (*h*UCB) [24], fetal liver, etc. Currently, the specific marker for EPC is controversial, and early or late endothelial progenitor cells have their putative markers. Pathological conditions such as cerebral ischemia, limb ischemia, and acute myocardial infarction can increase the number of EPCs rapidly in the circulation [16, 25, 26]. These EPCs were supposed to migrate from the bone marrow.

Considering the uncertain specific markers, studies use groups of markers to identify the proximate population. Asahara et al. first described EPCs as $CD34^{+}$/vascular endothelial growth factor receptor $(VEGFR)^{+}$ cells [14]. CD34 is widely recognized as the principal marker to identify EPCs, but it can also be expressed in ECs. Some researchers believed $CD133^{+}$ is the marker for more immature or putative EPCs [27, 28]. Human KDR and mouse Flk-1 are the receptors for VEGFR, widely expressed on mesoderm-derived lineages [29]. Mihail supposed that the loss of $CD133^{+}$ and a parallel expressing von Willebrand factor (vWF) may indicate EPC maturation [30]. EPC mobilization from bone marrow is regulated by a variety of factors, and it may be a part of the inflammatory response [31, 32]. EPCs can be activated by angiogenic growth factors from the periphery, especially associate with the level of endogenous VEGF [33]. EPCs then migrate to the systemic circulation, resident to the injury endothelium via the induction of cytokines and their receptors. Furthermore, EPCs participate in the cardiovascular disease and the cerebral vascular disease via promotion of angiogenesis, neovascularization, and regeneration after ischemic injury [34–36].

EPC tracking is a property approach for monitoring in vivo. MRI is an effective tool in tracking transplanted stem cells following cell labeling with superparamagnetic iron oxide (SPIO) nanoparticles [37]. Intravenously injected magnetically labeled EPCs with ferumoxide-protamine sulfate (FePro) in mice can detect tumor development by MRI [38]. *Lentivirus* packaging fluorescein is also a candidate for the track location lenti-PGK-TdTomato or GFP transduction to EPCs. Thébaud developed a qualification of fluoresce label, which has less chromosomal aberration and suitable for experiments in vivo [39]. EPCs stained with Dil-Ac-LDL also have been used for tracking after injecting into animals [40]. Hence, these tracking technologies provide useful approaches for the EPC-based therapy.

7.4 The Function of EPCs

Generally, EPCs are generated from the BM and stay quiescently in the stem cell niche [41]. When endothelial injury such as heart attack or ischemic stroke occurs, EPCs are capable to mobilize into circulation and participate in endothelial repair and neovascularization through the differentiation into ECs and secretion of various growth factors and cytokines [42, 43]. EPCs can secrete protective cytokines and growth factors to promote the function of repair after ischemic injury [44–46]. Neurovascular niche refers to factors release after brain injury to create a supported niche and thus lead to promote post-stroke neurogenesis [5]. Stromal-derived factor 1 (SDF-1) and angiopoietin-1 (Ang-1) are upregulated after ischemic stroke that induce angiogenesis and furthermore link to neurogenesis in the unique "neurovascular niche." It is noted that neuroblast migration was abolished when the angiogenesis was inhibited. ECs also are critical components to secrete soluble factors that maintain CNS stem cell self-renewal and neurogenic potential [47].

EPCs in ischemic injury regeneration have the ability of mobilization, migration, differentiation. Several signal transduction pathways are involved in these processes and regulate the repair of injury. SDF-1/CXCR4 axis is one of signal for stem cell homing [48]. Our group demonstrated that SDF-1α promoted angiogenesis and neurogenesis during the post-acute phase of ischemia [49]. SDF-1/CXCR4 can also cooperate with VEGF/VEGFR, GCSF/stem cell factor (SCF), etc. [50, 51]. EPCs can express endothelial nitric oxide synthase (eNOS) [52], and eNOS-dependent signal influences EPC migration and correlation to insulin-like growth factor-binding protein (IGFBP-3), and estrogen receptors, modulating the function of bone marrow-derived EPCs in both autocrine and paracrine, contributing to vascular repair [53, 54]. Many factors such as soluble intercellular adhesion molecule, granulocyte-monocyte colony-stimulating factor, hepatocyte growth factor, interleukin-6 (IL-6), IL-10, estrogen, and endothelial nitric oxide synthase (eNOS) participate in these activating processes. Both tissue factors and EPC surface receptors are involved in the EPC homing process. In addition, many factors such as CXCR2 and its ligands CXCL1 and CXCL7, CCL5/CCR5 promote EPCs migration [55–57]. In a hypoxia precondition of *h*UCB, HIF-1α-to-Epac1-to-matrix metalloproteinase (MMP) signaling pathway is required for the improved therapeutic efficacy [55–58]. MMPs play a key role in the formation of vascular networks by EPCs; MMP-9 deficiency disturbed the function of EPCs and further impaired neovascularization [59]. MMP-9 is found as a novel functional cell surface marker to identify proangiogenic cells from early EPCs [60]. Hypoxia-controlled MMP-9 overexpression shows its promising approach of gene therapy for ischemic stroke during the delayed phase of ischemia [61].

To maintain normal brain activity and reaction of physiological and pathological conditions, several types of brain resident cells and long-distance cells to work together are required. Studies suggested EPC regeneration depended on their differentiation ability or a paracrine approach [14, 16, 62]. In the interaction of EPCs and ECs, an in vitro study showed that RAGE plays a positive role [63]. Considering astrocytes as an important component of BBB, astrocytes can cross talk to EPCs and increase EPC proliferation both in vitro and in vivo [64]. A combined transplantation of EPCs and MSCs suggests the cross talk happened between progenitor and stem cells [65, 66]. An in vitro study uses a coculture model that showed a higher proliferation and vascular network formation [67]. Smooth muscle cells (SMCs) also communicated to EPCs. Cocultured EPCs and SMCs stimulated abundant cytokines to release in vitro and increase capillary density and improve blood perfusion in a model of cerebral ischemia in rat [68]. It is noted that EPCs could also give a rise to smooth muscle-like progeny.

BBB disruption occurs in the acute phase of ischemic stroke. Neovascularization stimulating agent like VEGF could decrease infarct volume in the ischemic penumbra and improve outcome; however it also increased BBB permeability and resulted in brain edema. EPCs might reverse the hypoxic state by neovascularization and repair ischemia-induced endothelial injury. EPCs transplantation could improve vascular integrity, which indicating their ability in the BBB repair after injury [40]. Although EPC-based therapy may damage BBB, there are many studies showing that EPCs improved the reconstitution of BBB via angiogenesis and vasculogenesis [69].

7.5 EPC Therapy in Hypoxia and Hypoglycemia In Vitro

Generally, ECs sense hypoxia and subsequently result in redox-mediated cell activation and trigger a series of inflammation by releasing cytokines, chemokines, and growth factors. EPCs can respond to hypoxia, hypoxia-inducible factor-1α, VEGF, or erythropoietin-induced EPC activation and mobilization from the BM [69, 70].

EPC function could be regulated by various factors, which acts in a paracrine fashion leading to endogenous EC proliferation. The increase of circulating EPCs is endogenously mobilized in the BM in response to tissue ischemia or exogenously by cytokine therapy; this increase will enhance neovascularization of ischemic tissues [16]. Angiogenesis occurred in response to cerebral ischemia. Hypoxia triggers inflammatory cell infiltration and furthermore promotes local neovascularization [16, 71]. Insulin-like growth factor (IGF)-binding protein-2 (IGFBP-2)-treated human umbilical vein ECs (HUVECs) significantly enhance the incorporation of EPC tubule formation and facilitate neovascularization under oxygen glucose depletion (OGD) condition [72]. Two types of EPCs were used most in in vitro study. "Early EPCs" obtained from a short-time culture for 4–7 days are thought to derive from myeloid lineage and "late EPC" also called "outgrowth endothelial cells (OECs)" from a long-term culture for 2–4 weeks. Late EPCs showed more capable to promote the formation of capillary-like structure when placed on Matrigel-coated dish [73].

EPC therapy in OGD model limits the progression of secondary axonal injury [74]. After a 90 min OGD treatment in primary neuron cells, the number of dead cells significantly decreases when EPCs were cocultured, while an increase in axonal degeneration appeared when subsequently cocultured with EPCs after subjected to OGD. The development of neurons is guided by molecules, which are provided by blood vessels. Some molecules can be the inductor to guide both neovasculature and neurogenesis. The interdependence between vasculatures and neurons suggested that EPCs are an important mediator of axon survival and vascular protection. The therapeutic effects on axons showed that EPCs played its role by releasing paracrine trophic factor. $CD34^+$ cells provided molecules such as VEGF, FGF-2, and IGF-1 and showed to accelerate endogenous neurogenesis [75–77]. EPCs had both vaso- and neuroprotective potential after OGD on hippocampal organotypic slices (OHC), which suggested paracrine mechanism of EPC-dependent protection [78]. The release of TGF-ß1 and IL-6 is slightly enhanced accompanied by the increase of TLR-3/4 ligands.

In vitro studies of EPCs combing growth factors or cytokines may enhance EPCs function [79]. For example, as chemoattractants, both SDF-1 and VEGF increased EPC migration; a combination of VEGF and SDF-1 caused an additive increase in EPC migration. SDF-1, VEGF, and FGF-2 could promote the formation of tubelike structures in Matrigel by testing tubelike formation of EPCs. Hypoxia upregulated chemokine receptors CXCR2 and CXCR4 in EPCs and triggered SDF-1, VEGF, CXCL1, and macrophage migration inhibitory factor (MIF) release. Consequently stimulating EPCs enhanced tube formation [80]. Gene transduction indicated that hepatocyte growth factor (HGF) and microRNAs promoted a much greater tubelike structure in ischemia, which suggested microRNA and related proteins are involved in EPC-based treatment [81].

7.6 EPC Therapy in Ischemic Brain In Vivo

Cell-based therapy has been used as a strategy to promote neovascularization and regeneration after ischemia injury. Delivery of culture-expanded EPCs for transplantation required property approaches; three different strategies have been principally used. The first strategy is systemic administration or intravenous delivery of EPCs, which is injected usually from tail vein, femoral vein, or jugular vein. This strategy improved neo-vascularization in the cerebral ischemic zone and functional recovery after middle cerebral artery occlusion (MCAO) [82]. The second strategy is intra-arterial delivery; studies showed that mice got a better outcome after the injection of ex vivo-expanded autologous BM-derived EPCs by internal carotid artery (ICA) [54]. Since intra-arterial infusion could cause embolism, intravenous infusion is believed to be the optimal route, which is proper for the clinical application [83]. The third strategy is striatal delivery, which means cells are directly injected to the striatal area, or the ischemic core. Administration directly to the injury core is assumed to be more effective. Therefore, both routes of cell administration were assessed for producing the greater recovery. The result showed that intravenous delivery may be more effective than striatal delivery in producing long-term functional benefits to the stroked animal [84, 85]. Intracerebral CD34+ cells transplantation could enhance endogenous stem cell mobilization, homing and engraftment into the brain [86]. Furthermore, a fourth approach, nasal administration, was also mentioned in cell-based brain ischemia therapy. Intranasally administered cells could penetrate the BBB into the brain parenchyma and cerebrospinal fluid (CSF) [87]. Intranasal delivery of MSCs improved functional outcome in a model of ischemic brain injury, indicating the efficiency of the nasal route [88]. As for the dose, the range of $0.2–3.0 \times 10^4$ EPCs per gram body weight shows satisfactory efficacy in various ischemic animal models [82, 89–91]. For human application in the future, the dose of EPC and the administration time window shall be carefully considered. The clinical trial NCT01468064 is designed to intravenously administrate 2.5×10^6 EPCs per kilogram body weight. They transplanted EPC at 4 weeks after BM aspiration, and second transplantation was implemented at 1 week after initial boosting. Another clinical trial NCT02605707 is designed to use autologous EPC transplantation for the chronic ischemic stroke, but it did not show the detailed information.

EPCs have the stem/progenitor cell and hematopoietic characteristics and could be stimulated by ischemia circumstance. Once EPCs are recruited to the injured tissue, it undergoes endothelial regeneration and eventually has a better outcome. Circulating BM-derived EPCs contribute to neovascularization and microvascular structure after MCAO in the adult brain [34]. Circulating human EPCs showed homing to the ischemic brain in 24 h after delivery, therefore EPCs play beneficial role both at delayed and acute phase [82]. Autologous peripheral blood-derived EPCs have a minimal risk to the host during EPC isolation and required minimal manipulation before transplantation [92]. Autologous transfusion of peripheral blood-derived EPCs via the internal carotid artery had been shown to reduce the brain infarct zone (BIZ) and neurological deficit in a rat model of MCAO [93]. Transplantation of *h*UCB-derived EPC promotes functional recovery through the magnitude of endogenous proliferation, angiogenesis, and neurogenesis; considering the controversial marker to definite EPCs, some study use *h*UCB $CD34^+$ and $AC133^+$ cells [94, 95]. These kinds of EPCs may also serve as a property agent for homologous or autologous transplantation in ischemic tissue.

Considering SDF-1/CXCR4 signaling played an important role in the activation of EPCs, overexpression of SDF-1/CXCR4 genes showed better outcomes compared to the controls. Hiasa demonstrated that SDF-1α gene transfer enhanced recruitment of EPCs and promoted vasculogenesis in vivo [96]. The effectiveness of SDF-1α gene transfer into a mouse MCAO model suggested that SDF-1α facilitated to recruit progenitor cells in the brain and consequently benefit stroke recovery. AAV-SDF-1α expression represents a promising avenue for ischemic stroke therapy [49]. Furthermore, insulin or the combination of SDF-1α and VEGF increases EPC-mediated neovascularization [97, 98]. microRNAs are small, noncoding, single-stranded RNAs that play a key role in regulating EPC function [99]. For example, miR-126, miR-130α, and miR-221/miR-222 regulate the proliferation of EPC; miR-21, miR-34a, and miR-22 regulated senescence of EPC; miR-126 and miR-150 regulate cell migration of EPC; and miRNA-107, miRNA-16, miRNA-34a, and miRNA-126 are related to EPC differentiation [1, 100–107].

Several studies have shown that the administration of early EPCs is associated with a better outcome after ischemic injury. Endothelial outgrowth cells (EOCs) also known as endothelial colony-forming cell populations of human EPCs are particularly promising for vascular tissue engineering applications [108]. OECs are also known as late outgrowth or endothelial colony-forming cells (ECFCs). Although there is no single specific marker for this cell population, OECs could be obtained under cell culture and are capable to form vessel-like structures [109, 110]. OECs isolated from *h*UCB could improve functional recovery by intravenous injection to MCAO rat after 1 h reperfusion [89].

Moyamoya disease (MMD) is a cerebral vascular disease in which certain arteries are constricted and thus creates a chronic ischemic circumstances. Abnormal neovasculature in moyamoya undergo vasculogenesis, and the differentiated mature ECs go through a pathological way [111]. It was reported children with MMD had abnormal cluster formation [112]. The number of EPCs was significantly reduced with less tube formation and increased senescent-like phenotype. MMD patients had a significant increase in circulating EPC mobilization. Circulating EPC number reflected mixed conditions of abnormal vasculogenesis, while OEC number was frequently detected in moyamoya vessels [113]. As circulating EPCs were closely involved in the initiation and development of moyamoya disease, it was possible to manipulate EPC behaviors to lead to a better outcome [114].

7.7 EPC Therapy in Other Cerebrovascular Diseases

In order to prevent secondary neuronal injury in intracerebral hemorrhage, vascular repair could be a potential target for the nerve regeneration [115]. Neurological deficit was diminished when EPCs were intravenously administrated on the first day after intracerebral hemorrhage in rats [116]. One of limitations of EPC therapy was the number of EPC injection. Therefore, searching for molecules to regulate EPC function may be essential targets in the ICH therapy. Furthermore, SDF-1α was always considered to play a key role in the recruitment of EPCs and supported revascularization in ischemic tissue. In a rodent model of ICH, delivering SDF-1α together with EPC resulted in significantly increased blood vessel formation [117]. In a hemorrhagic stroke model, EPO pretreatment promoted EPC mobilization and therefore significantly attenuated tissue injury and dysfunction [118]. The level of late EPCs could serve as a biomarker in patients with hemorrhagic stroke [119]. It is noted that EPCs had a therapy potential in aneurysm, which promoted endothelialization of the coiled aneurysm neck via induction of EPCs [120].

Brain arteriovenous malformations (BAVMs) were congenital vascular lesion that resulted in deficiency to form the capillary bed and involved the vasculogenesis of blood vessels and could cause intracranial hemorrhage [121]. Since there are a higher number of EPCs and SDF-1 expression in BAVM, SDF-1 in EPC-treated BAVM may promote vasculogenesis and inflammatory cell migration [122]. The number of EPCs and the expression of SDF-1 were increased in BAVM patients. EPCs are activated by SDF-1/CXCR4 signaling, which are recruited to form abnormal vasculature cluster in BAVM patients. Furthermore, EPCs may mediate pathological vascular remodeling and affect the clinical development of AVMs due to its presence in the nidus of the brain and spinal cord AVMs [123]. The higher-staged BAVMs increased attractive factors and then recruit EPCs [124]. The number of proliferating ECs changed in stage III and recruited EPCs and finally resulted in significantly increased abnormal vasculature cluster. Further evidence is warranted to determine whether EPCs are a suitable potential agent for the BAVM to turn to normal structure.

EPCs could mobilize in response to hypoxia condition, then promote vasculogenesis, and result in vascular generation in the pathological tissue [125, 126]. During tumor development, SDF-1/CXCR4 and VEGF/VEGFR pathways are known as mediators of EPC mobilization. Therefore, potential target may focus on new anti-vasculogenic approach [127]. Characterization of tumor-associated EPCs in hematological tumor provides the possibility of specific anti-vasculogenic therapy [128]. EPCs mediated the "angiogenic switch" and induced cancer cell metastases, which suggested EPCs represent a therapeutic target both in early and late stage of cancer progression [129]. A study showed circulating EPCs are able to predict an increased risk of non-small cell lung (NSCL) carcinoma recurrence and death in an early post-surgery [130]. BM-EPCs were reported close to hepatocellular carcinoma (HCC). Injection of EPCs specifically homes to tumor tissue and plays a prominent role in HCC neovascularization. Blockade of BM-EPC-mediated vasculogenesis improved the therapeutic efficacy for the HCC [131]. EPCs could be used as a candidate prognostic and predictive biomarker for gastric cancer [132]. Systemic delivery EPCs transduce with an immune-activating gene could alter the tumor immune microenvironment and lead to a therapeutic effect of murine ovarian cancer [133]. Thus, therapy aimed at EPCs should be taken into consideration for cell-based therapy for tumor or as a potential vehicle for gene or molecule therapy. Studies also established to use labeled EPCs to investigate the migration and incorporation of EPCs into tumor neovasculature [134].

7.8 Combination Therapy to Improve EPC Efficacy

Although EPC-based therapy attracts attention on experiment and clinical trials, results were not satisfactory enough. Tumorigenicity and limited resources confined application of EPCs. Strategies to enhance the efficiency of EPCs are combination therapy including co-transplantation of different types of cells, adjunct treatment, or ex vivo modification to improve EPC function by molecule factors. Using genomic and proteomic combination approach showed that TNF-α could increase EPC migration and incorporation into vessel-like structures [135]. HMGB1 upregulation could promote exogenous *h*PB-EPC-mediated stroke recovery by modulating paracrine function of *h*PB-EPCs in a model of MCAO [136]. Osteopontin (OPN) played a role in the homing and incorporation of EPCs to the site of endothelial injury [137]. Endothelial nitric oxide synthase (eNOS)-mediated angioprotective function and the combination of eNOS transfection and OPN produced more focal adhesions and spreading area. Combination of autologous EPCs and OPN coatings could be a promising method of developing functional endothelialized surfaces [138, 139]. The combinatorial granulocyte colony-stimulating factors (GCSFs) and EPCs were also proven to produce synergistic effects in ischemic stroke [140]. In diabetic ischemic stroke model, EPCs could be accelerated to senescence and apoptosis mainly due to p38 mitogen-activated protein kinase (MAPK) activation. Bai showed the synergistic efficiency of combination of EPC transplantation and a p38 inhibitor RWJ 67657 administration. Suppression of p38 signaling increased the number of EPCs and improved neurological function and thus accelerated recovery from diabetic stroke via promotion of angiogenesis and neurogenesis [141]. Erythropoietin (EPO) promoted post-stroke angiogenesis by stimulating endogenous EPC proliferation and migration [142]. Furthermore, EPO administration in combination with EOCs showed synergistic effects on neurological recovery, angiogenesis, and neurogenesis in rat MCAO model [52]. Study also showed that EPCs possibly favor the proliferation of mononuclear cells (MNCs) and HSCs. The concentrations of IL-6 and VEGF were upregulated and might be related to hematopoietic reconstitution and homing ability [143].

As mentioned above, EPCs have a late population, called OECs. OECs have different functions compared to EPCs; the combination of EPCs and OECs showed synergism during neovascularization. These two cell types augmented the angiogenic capability in vitro and showed better outcome compared to any single-cell-type transplantation [144]. Neuroblast survival was needed to mature vascular network [145], as smooth muscle cells/pericytes were needed for the maintenance of BBB integrity during cerebral ischemia [146]. The intravenous coadministration of *h*UCB-derived smooth muscle progenitor cells (SMPCs) and EPCs in mice got better outcomes than administration of one single type alone. The results showed that more mature vascular remodeling, more cell proliferation, and neuroblast migration occurred in both peri-infarct and infarct area [145]. Transplantation of NSCs or neural precursor cells (NPCs) is also used to repair the nerve system because of the poor self-regeneration ability in CNS [147]. Adipose-derived stem cells differentiated toward EPCs and NPCs; and the combination of both cells produced significant reduction of hypoxic-ischemic brain injury [21]. It is possible that NPCs and EPCs have cell-cell interaction during cell migration; and hypoxic microenvironment could enhance cell-cell connection. A clinical trial in refractory ischemia used the combination infusion of MSCs and EPCs which showed significant improvement in quality of life according to the phase I [148]. The improved outcome suggested that this therapy is safe and feasible.

Besides induced by molecule and combination with various cells in EPC-based therapy, magnetic guide also used to enhance therapeutic effect. Labeled EPC with SiO_4@SPIONs showed the efficient guide by exterior magnetic field toward ischemic hemisphere [149].

7.9 Mechanism of EPC Therapy in Stroke

EPCs possessed the capability of differentiation and neovascularization. Studies provided evidence that neurological benefits coupled with angiogenesis [89]. After ischemic stroke, EPCs were stimulated and recruited to the ischemia brain and promoted recovery. When ischemic injury happen, SDF-1α was highly increased, which resulted in recruitment of CXCR4 that generally express in EPCs. Therefore, EPCs could perform their functions to promote the repair of injured brain. SDF-1/CXCR4 axis also cooperated with growth factors such as VEGF/VEGFR, KDR/CD34, and GCSF/stem cell factor (SCF) [13, 50, 51]. SDF-1α/CXCR4 axis possessed capability triggering cell proliferation and anti-apoptosis signals such as MAPKs and PI3K and the serine/threonine kinase Akt [150]. High CXCR4-expressed EPCs increased capillary density in ischemic hind limbs [151].

Endogenous nitric oxide (eNOS), increased following ischemic stroke, is one of the factors promoting EPC mobilization [54]. Chen demonstrated that eNOS was a downstream mediator for VEGF release and angiogenesis. Furthermore, eNOS also regulated BNDF expression and neuronal progenitor cell proliferation, migration, and neurite outgrowth, finally affecting function recovery after cerebral ischemia [152].

HIF-1α is a transcription factor that regulates the adaptive response to hypoxia in cells. HIF-1α expression upregulated VEGF and EPO levels in hypoxia brain [134]. VEGF/VEGFR signaling further leads to the growth of new vessels after ischemic stroke [153].

Early EPCs secreted abundant proangiogenic cytokines including VEGF, transforming growth factor-β (TGF-β), macrophage colony-stimulating factor (MCSF), placental growth factor (PGF), HGF, fibroblast growth factor (FGF), MIF, thrombopoietin (TPO), and interleukin-8 (IL-8). It could also secrete a few anti-angiogenic cytokines and neurotrophic and neuroregulatory cytokines including brain-derived neurotrophic factor (BNDF) [154, 155]. EPC mobilization and recruitment to ischemic area could be a very complicated process. Identification of these regulators and understanding of their function provided clear knowledge of EPC-based therapy.

EPCs are involved in growth factor secretion, cell proliferation, neovascularization, and neurovascular unit repair, which suggested that EPCs played an important role under certain conditions such as tumor progression, myocardial infarction, and stroke [14, 15, 34, 156, 157]. However, circulating EPCs promoted the vascular repair or angiogenesis mainly by releasing paracrine factors such as VEGF or HGF to activate resident ECs rather than self-proliferation from the myeloid-monocytic lineage. Resident ECs primarily contributed to endothelial regeneration [158]. EPC-induced angiogenesis may include a large quantity of ECs by proliferation and differentiation or an increased supply of growth factors to activate resident mature ECs [159]. Therefore, the secretion of angiogenic growth factors such as VEGF, HGF, and granulocyte-macrophage colony-stimulating factors (GM-CSFs) from EPCs contributed to the proangiogenic effect. That means circulating cells may involve the regulation of vascular repair via paracrine mechanisms. In a model of MCAO-based diabetic animals, monocyte function could be reduced and the neovascularization is attenuated [160].

Neurorepair required the replacement of dead neurons and reconnection of neuronal network. Ischemia triggered the activation of several endogenous progenitor cells such as NPCs, EPCs, or oligodendrocyte progenitor cells (OPCs). Neurogenesis in the adult brain occurred in two areas: the hippocampal subgranular zone (SGZ) and sub-ventricular zone (SVZ) [4, 5]. Hypoxia was a strong activator of neurogenesis in CNS, since hypoxia-inducible VEGF and its receptor VEGFR-2/Flk are expressed in NSCs in vitro. VEGF was supposed to acts as trophic factor for NSCs in vitro and for sustained neurogenesis in the adult CNS [161]. Human NSCs expressed CXCR4 and migrated toward the infarct area, where SDF-1α is upregulated [161, 162]. Thus secretive factors of EPCs such as VEGF and SDF-1 provide a suitable microenvironment for neuronal regeneration and survival. Evidence showed that progenitor cells like NPCs were capable to differentiate to matured neurons and glia; the angiogenic environment was required to promote effective neurogenesis and survival [75]. Neurons could be formed in the SVZ and stimulated to migrate toward the ischemic boundary region [163]. Neuronal migration is influenced by cell-secreted factors and by cell-bound molecules including gamma-aminobutyric acid (GABA), VEGF, BDNF, polysialylated neural cell adhesion molecule (PSA-NCAM), matrix metalloproteinases (MMPs), β1-integrins, angiopoietin (Ang), and extracellular matrix components [164]. Angiogenesis, neurogenesis, and synaptic plasticity were endogenous processes, which could be activated under pathological conditions in adult brain. Administration of $CD34^+$ cells enhanced angiogenesis and increased neuroblast migration from SVZ to the ischemic region, which suggested that a stable neovasculature was essential to support neuronal regeneration [91]. Palmer provided for the first time that adult neurogenesis occurred within an angiogenic niche [165]. Signals that recruit neurogenesis are known as angiogenic factors such as FGF-2, and some factors responded to angiogenic factors including FGF, TGF-α and TGF-β, and PDGF [166–168]. These may modulate both angiogenesis and neural regeneration within a "vascular niche."

Recently, novel modes of neurovascular regulation were being highlighted, such as the microRNA (miRNA) modulation and microparticle or microvesicle release. miRNAs are small, mainly 20–24-nucleotide, noncoding, single-stranded RNAs with gene regulatory activities [169]. miRNAs regulated EPC functions including EPC proliferation, mobilization, migration, differentiation, and tube formation [99]. miRNA-221 and miRNA-222 are transcribed from the same miRNA cluster and showed to regulate EPC proliferation. It has been reported that overexpression of miR-221 in EPCs could attenuate EPC proliferation [170]. It is demonstrated that the level of miR-221 had a correlation to the number of EPCs in vascular disease like cardiovascular disease and atherosclerosis [107].

MiR-126 as one of the most abundant miRNAs in ECs governs vascular integrity and angiogenesis [171]. miR-126 regulated VEGF-dependent PI3K and MAPK signal pathway, known as the negative regulators of VEGF. Inhibition of VEGF signal resulted in defects similar to the miR-126 knockdown in a zebrafish, which indicated that miR-126 could be a target for modulating vascular integrity and angiogenesis [172]. Further study demonstrated that miR-126 could have an ability to promote HUCBC-induced neuro-restorative effect after ischemic stroke in type 2 diabetes mellitus rodents. miR-126 expression was increased and coupled with M2 macrophage polarization after HUBCB treatment in a stroke model of type 2 diabetes mellitus rodents. These data supported that the functional recovery attributed to miR-126 mainly resulted from the neuro-restorative rather than neuroprotection [173].

Other miRNAs are also involved in EPC function. Knowing that CXCR4 is a target of miR-150, Rolland-Turner demonstrated that miR-150 is involved in EPC recruitment to the ischemic heart under ischemic condition [174, 175]. miR-31-miR-720 pathway is shown critical to EPC activation and regulated EPC-induced angiogenesis after brain ischemia [176]. miR-31-5p induces EPC migration/invasion by targeting FAT4 and promoted angiogenesis [177].

Microvesicles (MVs) have small particle size, 0.1–1 μm, carrying proteins and mRNAs, miRNAs, DNAs, and cytokines and serving as veritable vectors for intercellular exchange of information, which could be shed from various cell types [178]. MVs released from EPCs had the capability of triggering angiogenesis via a horizontal transfer of mRNAs or miRNAs [179]. MVs from serum-free EPCs medium showed beneficial effects on hypoxia/reoxygenation human brain, which may be due to carry their parent proteins and genetic materials [180]. EPC-derived MVs incorporated in ECs could promote EC survival and proliferation and form capillary-like structures by a horizontal transfer of mRNAs [179]. These findings indicated that EPC-MVs could not only be a potential mechanism of EPC-based therapy, but also provide a novel vehicle for cerebral ischemia therapy.

7.10 Conclusions

During recent years, numerous studies demonstrated that EPCs played an important part in cerebrovascular disease including ischemic stroke, moyamoya disease, hemorrhagic stroke, vascular tumor, etc. EPCs showed as a promising tool in cerebrovascular therapy. Stem cell-based therapy in ischemic stroke is still in its infancy. EPC-based therapy in ischemic stroke has been highlighted not only because of its capability to establish connection in angiogenesis, neurogenesis, oligodendrogenesis, and whole neurorepair niche but also of its involvement in BBB repair and functional recovery. Although the pivotal role was investigated in ischemic response and therapy potential both in vitro and in vivo, the in-depth mechanisms are still needed to be discovered. As rapid technological advances and novel modes of molecule regulator is being uncovered such as paracrine factors, miRNAs, and microvesicles, the strategies of combination therapy to optimize the function of EPC-based therapy could be the promising application. Further studies should be focused on the definitive markers of different types of EPCs, mechanisms pertaining to function of EPCs, and standardization of EPC administration in ischemic stroke. Challenges remain when facing the translation from in vitro or in vivo animal studies to clinical trials. Preclinical studies and clinical research were needed to improve the benefit and minimize risks as well.

Acknowledgments This study was supported by research grants from the National Natural Science Foundation of China (#81471178 and #U1232205 GYY), the Science and Technology Commission of Shanghai Municipality (#13ZR1422600 ZJZ), and KC Wong Foundation (GYY).

References

1. Moretti A, Ferrari F, Villa RF. Neuroprotection for ischaemic stroke: current status and challenges. Pharmacol Ther. 2015;146:23–34.
2. Troidl K, Schaper W. Arteriogenesis versus angiogenesis in peripheral artery disease. Diabetes Metab Res Rev. 2012;28 Suppl 1:27–9.
3. Liu J, et al. Vascular remodeling after ischemic stroke: mechanisms and therapeutic potentials. Prog Neurobiol. 2014;115:138–56.
4. Ohab JJ, Carmichael ST. Poststroke neurogenesis: emerging principles of migration and localization of immature neurons. Neuroscientist. 2008;14(4):369–80.
5. Ohab JJ, et al. A neurovascular niche for neurogenesis after stroke. J Neurosci. 2006;26(50):13007–16.
6. Vallon M, et al. Developmental and pathological angiogenesis in the central nervous system. Cell Mol Life Sci. 2014;71(18):3489–506.
7. Ma F, et al. Endothelial progenitor cells and revascularization following stroke. Brain Res. 2015;1623:150–9.
8. Darby DG, et al. Pathophysiological topography of acute ischemia by combined diffusion-weighted and perfusion MRI. Stroke. 1999;30(10):2043–52.
9. Janowski M, Wagner DC, Boltze J. Stem cell-based tissue replacement after stroke: factual necessity or notorious fiction? Stroke. 2015;46(8):2354–63.
10. Pelliccia F, et al. Endothelial progenitor cells predict long-term prognosis in patients with stable angina treated with percutaneous coronary intervention: five-year follow-up of the PROCREATION study. Circ J. 2013;77(7):1728–35.
11. Sobrino T, et al. Increased levels of circulating endothelial progenitor cells in patients with ischaemic stroke treated with statins during acute phase. Eur J Neurol. 2012;19(12):1539–46.
12. Ni G, et al. Increased levels of circulating SDF-1alpha and CD34+ CXCR4+ cells in patients with moyamoya disease. Eur J Neurol. 2011;18(11):1304–9.
13. Yip HK, et al. Effect of erythropoietin on level of circulating endothelial progenitor cells and outcome in patients after acute ischemic stroke. Crit Care. 2011;15(1):R40.
14. Asahara T, et al. Isolation of putative progenitor endothelial cells for angiogenesis. Science. 1997;275(5302):964–7.
15. Shi Q, et al. Evidence for circulating bone marrow-derived endothelial cells. Blood. 1998;92(2):362–7.

16. Takahashi T, et al. Ischemia- and cytokine-induced mobilization of bone marrow-derived endothelial progenitor cells for neovascularization. Nat Med. 1999;5(4):434–8.
17. Xiao Q, et al. Endothelial progenitor cells, cardiovascular risk factors, cytokine levels and atherosclerosis – results from a large population-based study. PLoS ONE. 2007;2(10):e975.
18. Fadini GP, et al. Endothelial progenitor cells in the natural history of atherosclerosis. Atherosclerosis. 2007;194(1):46–54.
19. Bailey AS, et al. Transplanted adult hematopoietic stems cells differentiate into functional endothelial cells. Blood. 2004;103(1):13–9.
20. Flamme I, Risau W. Induction of vasculogenesis and hematopoiesis in vitro. Development. 1992;116(2):435–9.
21. Hsueh YY, et al. Synergy of endothelial and neural progenitor cells from adipose-derived stem cells to preserve neurovascular structures in rat hypoxic-ischemic brain injury. Sci Rep. 2015;5:14985.
22. Oswald J, et al. Mesenchymal stem cells can be differentiated into endothelial cells in vitro. Stem Cells. 2004;22(3):377–84.
23. Margariti A, et al. Direct reprogramming of fibroblasts into endothelial cells capable of angiogenesis and reendothelialization in tissue-engineered vessels. Proc Natl Acad Sci U S A. 2012;109(34):13793–8.
24. Murohara T, et al. Transplanted cord blood-derived endothelial precursor cells augment postnatal neovascularization. J Clin Invest. 2000;105(11):1527–36.
25. Vasa M, et al. Number and migratory activity of circulating endothelial progenitor cells inversely correlate with risk factors for coronary artery disease. Circ Res. 2001;89(1):E1–7.
26. Shintani S, et al. Mobilization of endothelial progenitor cells in patients with acute myocardial infarction. Circulation. 2001;103(23):2776–9.
27. Bryder D, Rossi DJ, Weissman IL. Hematopoietic stem cells: the paradigmatic tissue-specific stem cell. Am J Pathol. 2006;169(2):338–46.
28. Gehling UM, et al. In vitro differentiation of endothelial cells from AC133-positive progenitor cells. Blood. 2000;95(10):3106–12.
29. Keller G. Embryonic stem cell differentiation: emergence of a new era in biology and medicine. Genes Dev. 2005;19(10):1129–55.
30. Hristov M, Erl W, Weber PC. Endothelial progenitor cells: mobilization, differentiation, and homing. Arterioscler Thromb Vasc Biol. 2003;23(7):1185–9.
31. Schaper W, Scholz D. Factors regulating arteriogenesis. Arterioscler Thromb Vasc Biol. 2003;23(7):1143–51.
32. Sunderkotter C, et al. Macrophages and angiogenesis. J Leukoc Biol. 1994;55(3):410–22.
33. Kalka C, et al. Vascular endothelial growth factor(165) gene transfer augments circulating endothelial progenitor cells in human subjects. Circ Res. 2000;86(12):1198–202.
34. Zhang ZG, et al. Bone marrow-derived endothelial progenitor cells participate in cerebral neovascularization after focal cerebral ischemia in the adult mouse. Circ Res. 2002;90(3):284–8.
35. Murohara T. Therapeutic vasculogenesis using human cord blood-derived endothelial progenitors. Trends Cardiovasc Med. 2001;11(8):303–7.
36. Kawamoto A, et al. Therapeutic potential of ex vivo expanded endothelial progenitor cells for myocardial ischemia. Circulation. 2001;103(5):634–7.
37. Wei MQ, et al. Experimental study of endothelial progenitor cells labeled with superparamagnetic iron oxide in vitro. Mol Med Rep. 2015;11(5):3814–9.
38. Arbab AS, et al. Magnetic resonance imaging and confocal microscopy studies of magnetically labeled endothelial progenitor cells trafficking to sites of tumor angiogenesis. Stem Cells. 2006;24(3):671–8.
39. Thebaud NB, et al. Labeling and qualification of endothelial progenitor cells for tracking in tissue engineering: an in vitro study. Int J Artif Organs. 2015;38(4):224–32.
40. Aicher A, et al. Assessment of the tissue distribution of transplanted human endothelial progenitor cells by radioactive labeling. Circulation. 2003;107(16):2134–9.

41. Lapidot T, Petit I. Current understanding of stem cell mobilization: the roles of chemokines, proteolytic enzymes, adhesion molecules, cytokines, and stromal cells. Exp Hematol. 2002;30(9):973–81.
42. Tilling L, Chowienczyk P, Clapp B. Progenitors in motion: mechanisms of mobilization of endothelial progenitor cells. Br J Clin Pharmacol. 2009;68(4):484–92.
43. Ingram DA, et al. Vessel wall-derived endothelial cells rapidly proliferate because they contain a complete hierarchy of endothelial progenitor cells. Blood. 2005;105(7):2783–6.
44. Zhao YH, et al. Endothelial progenitor cells: therapeutic perspective for ischemic stroke. CNS Neurosci Ther. 2013;19(2):67–75.
45. Yamashita T, Abe K. Mechanisms of endogenous endothelial repair in stroke. Curr Pharm Des. 2012;18(25):3649–52.
46. Navarro-Sobrino M, et al. Mobilization, endothelial differentiation and functional capacity of endothelial progenitor cells after ischemic stroke. Microvasc Res. 2010;80(3):317–23.
47. Shen Q, et al. Endothelial cells stimulate self-renewal and expand neurogenesis of neural stem cells. Science. 2004;304(5675):1338–40.
48. Shen L, et al. A novel mechanism for endothelial progenitor cells homing: the SDF-1/CXCR4-Rac pathway may regulate endothelial progenitor cells homing through cellular polarization. Med Hypotheses. 2011;76(2):256–8.
49. Li Y, et al. Postacute stromal cell-derived factor-1alpha expression promotes neurovascular recovery in ischemic mice. Stroke. 2014;45(6):1822–9.
50. Bogoslovsky T, et al. Stromal-derived factor-1[alpha] correlates with circulating endothelial progenitor cells and with acute lesion volume in stroke patients. Stroke. 2011;42(3):618–25.
51. Toth ZE, et al. The combination of granulocyte colony-stimulating factor and stem cell factor significantly increases the number of bone marrow-derived endothelial cells in brains of mice following cerebral ischemia. Blood. 2008;111(12):5544–52.
52. Pellegrini L, et al. Therapeutic benefit of a combined strategy using erythropoietin and endothelial progenitor cells after transient focal cerebral ischemia in rats. Neurol Res. 2013;35(9):937–47.
53. Kielczewski JL, et al. Insulin-like growth factor binding protein-3 mediates vascular repair by enhancing nitric oxide generation. Circ Res. 2009;105(9):897–905.
54. Ohta T, et al. Administration of ex vivo-expanded bone marrow-derived endothelial progenitor cells attenuates focal cerebral ischemia-reperfusion injury in rats. Neurosurgery. 2006;59(3):679–86. discussion 679–86.
55. Hristov M, et al. Importance of CXC chemokine receptor 2 in the homing of human peripheral blood endothelial progenitor cells to sites of arterial injury. Circ Res. 2007;100(4):590–7.
56. Walenta KL, et al. Differential chemokine receptor expression regulates functional specialization of endothelial progenitor cell subpopulations. Basic Res Cardiol. 2011;106(2):299–305.
57. Ishida Y, et al. Pivotal role of the CCL5/CCR5 interaction for recruitment of endothelial progenitor cells in mouse wound healing. J Clin Invest. 2012;122(2):711–21.
58. Lin CH, et al. Role of HIF-1alpha-activated Epac1 on HSC-mediated neuroplasticity in stroke model. Neurobiol Dis. 2013;58:76–91.
59. Huang PH, et al. Matrix metalloproteinase-9 is essential for ischemia-induced neovascularization by modulating bone marrow-derived endothelial progenitor cells. Arterioscler Thromb Vasc Biol. 2009;29(8):1179–84.
60. Kanayasu-Toyoda T, et al. Cell-surface MMP-9 protein is a novel functional marker to identify and separate pro-angiogenic cells from early endothelial progenitor cells derived from CD133 cells. Stem Cells. 2016;34(5):1251–62.
61. Cai H, et al. Hypoxia-controlled matrix metalloproteinase-9 hyperexpression promotes behavioral recovery after ischemia. Neurosci Bull. 2015;31(5):550–60.
62. Orlic D, et al. Transplanted adult bone marrow cells repair myocardial infarcts in mice. Ann N Y Acad Sci. 2001;938:221–9. discussion 229–30.

63. Hayakawa K, et al. Reactive astrocytes promote adhesive interactions between brain endothelium and endothelial progenitor cells via HMGB1 and beta-2 integrin signaling. Stem Cell Res. 2014;12(2):531–8.
64. Hayakawa K, et al. High-mobility group box 1 from reactive astrocytes enhances the accumulation of endothelial progenitor cells in damaged white matter. J Neurochem. 2013;125(2):273–80.
65. Ball SG, Shuttleworth AC, Kielty CM. Direct cell contact influences bone marrow mesenchymal stem cell fate. Int J Biochem Cell Biol. 2004;36(4):714–27.
66. Lemischka IR, Moore KA. Stem cells: interactive niches. Nature. 2003;425(6960):778–9.
67. Aguirre A, Planell JA, Engel E. Dynamics of bone marrow-derived endothelial progenitor cell/mesenchymal stem cell interaction in co-culture and its implications in angiogenesis. Biochem Biophys Res Commun. 2010;400(2):284–91.
68. Shudo Y, et al. Spatially oriented, temporally sequential smooth muscle cell-endothelial progenitor cell bi-level cell sheet neovascularizes ischemic myocardium. Circulation. 2013;128(11 Suppl 1):S59–68.
69. Kaneko Y, et al. Cell therapy for stroke: emphasis on optimizing safety and efficacy profile of endothelial progenitor cells. Curr Pharm Des. 2012;18(25):3731–4.
70. Hoenig MR, Bianchi C, Sellke FW. Hypoxia inducible factor-1 alpha, endothelial progenitor cells, monocytes, cardiovascular risk, wound healing, cobalt and hydralazine: a unifying hypothesis. Curr Drug Targets. 2008;9(5):422–35.
71. Kwon HM, et al. Enhanced coronary vasa vasorum neovascularization in experimental hypercholesterolemia. J Clin Invest. 1998;101(8):1551–6.
72. Feng N, et al. Insulin-like growth factor binding protein-2 promotes adhesion of endothelial progenitor cells to endothelial cells via integrin alpha5beta1. J Mol Neurosci. 2015;57(3):426–34.
73. Minami Y, et al. Angiogenic potential of early and late outgrowth endothelial progenitor cells is dependent on the time of emergence. Int J Cardiol. 2015;186:305–14.
74. Park KJ, et al. Bone marrow-derived endothelial progenitor cells protect postischemic axons after traumatic brain injury. J Cereb Blood Flow Metab. 2014;34(2):357–66.
75. Nakatomi H, et al. Regeneration of hippocampal pyramidal neurons after ischemic brain injury by recruitment of endogenous neural progenitors. Cell. 2002;110(4):429–41.
76. Jin K, et al. Vascular endothelial growth factor (VEGF) stimulates neurogenesis in vitro and in vivo. Proc Natl Acad Sci U S A. 2002;99(18):11946–50.
77. Drago J, et al. Fibroblast growth factor-mediated proliferation of central nervous system precursors depends on endogenous production of insulin-like growth factor I. Proc Natl Acad Sci U S A. 1991;88(6):2199–203.
78. Obtulowicz P, et al. Induction of endothelial phenotype from Warton Jelly-derived MSC and comparison of their vaso- and neuro-protective potential with primary WJ-MSC in CA1 hippocampal region ex vivo. Cell Transplant. 2016;25(4):715–27.
79. Peplow PV. Influence of growth factors and cytokines on angiogenic function of endothelial progenitor cells: a review of in vitro human studies. Growth Factors. 2014;32(3–4):83–116.
80. Kanzler I, et al. Differential roles of angiogenic chemokines in endothelial progenitor cell-induced angiogenesis. Basic Res Cardiol. 2013;108(1):310.
81. Song MB, et al. Transfection of HGF gene enhances endothelial progenitor cell (EPC) function and improves EPC transplant efficiency for balloon-induced arterial injury in hypercholesterolemic rats. Vascul Pharmacol. 2009;51(2–3):205–13.
82. Fan Y, et al. Endothelial progenitor cell transplantation improves long-term stroke outcome in mice. Ann Neurol. 2010;67(4):488–97.
83. Honmou O, et al. Intravenous administration of auto serum-expanded autologous mesenchymal stem cells in stroke. Brain. 2011;134(Pt 6):1790–807.
84. Willing AE, et al. Intravenous versus intrastriatal cord blood administration in a rodent model of stroke. J Neurosci Res. 2003;73(3):296–307.

85. Chen J, et al. Intravenous administration of human umbilical cord blood reduces behavioral deficits after stroke in rats. Stroke. 2001;32(11):2682–8.
86. Naruse K, et al. Therapeutic neovascularization using cord blood-derived endothelial progenitor cells for diabetic neuropathy. Diabetes. 2005;54(6):1823–8.
87. Danielyan L, et al. Intranasal delivery of cells to the brain. Eur J Cell Biol. 2009;88(6):315–24.
88. van Velthoven CT, et al. Nasal administration of stem cells: a promising novel route to treat neonatal ischemic brain damage. Pediatr Res. 2010;68(5):419–22.
89. Moubarik C, et al. Transplanted late outgrowth endothelial progenitor cells as cell therapy product for stroke. Stem Cell Rev. 2011;7(1):208–20.
90. Asahara T, Kawamoto A, Masuda H. Concise review: circulating endothelial progenitor cells for vascular medicine. Stem Cells. 2011;29(11):1650–5.
91. Taguchi A, et al. Administration of CD34+ cells after stroke enhances neurogenesis via angiogenesis in a mouse model. J Clin Invest. 2004;114(3):330–8.
92. Perin EC, Willerson JT. CD34+ autologous human stem cells in treating refractory angina. Circ Res. 2011;109(4):351–2.
93. Chen YL, et al. Intra-carotid arterial administration of autologous peripheral blood-derived endothelial progenitor cells improves acute ischemic stroke neurological outcomes in rats. Int J Cardiol. 2015;201:668–83.
94. Liang CC, et al. The protective effect of human umbilical cord blood CD34+ cells and estradiol against focal cerebral ischemia in female ovariectomized rat: cerebral MR imaging and immunohistochemical study. PLoS ONE. 2016;11(1):e0147133.
95. Iskander A, et al. Intravenous administration of human umbilical cord blood-derived AC133+ endothelial progenitor cells in rat stroke model reduces infarct volume: magnetic resonance imaging and histological findings. Stem Cells Transl Med. 2013;2(9):703–14.
96. Hiasa K, et al. Gene transfer of stromal cell-derived factor-1alpha enhances ischemic vasculogenesis and angiogenesis via vascular endothelial growth factor/endothelial nitric oxide synthase-related pathway: next-generation chemokine therapy for therapeutic neovascularization. Circulation. 2004;109(20):2454–61.
97. Dong L, et al. Insulin modulates ischemia-induced endothelial progenitor cell mobilization and neovascularization in diabetic mice. Microvasc Res. 2011;82(3):227–36.
98. Yu JX, et al. Combination of stromal-derived factor-1alpha and vascular endothelial growth factor gene-modified endothelial progenitor cells is more effective for ischemic neovascularization. J Vasc Surg. 2009;50(3):608–16.
99. Qu K, et al. MicroRNAs: key regulators of endothelial progenitor cell functions. Clin Chim Acta. 2015;448:65–73.
100. Zheng Y, Xu Z. MicroRNA-22 induces endothelial progenitor cell senescence by targeting AKT3. Cell Physiol Biochem. 2014;34(5):1547–55.
101. Cheng BB, et al. MicroRNA-34a targets Forkhead box j2 to modulate differentiation of endothelial progenitor cells in response to shear stress. J Mol Cell Cardiol. 2014;74:4–12.
102. Zhang J, et al. microRNA 126 inhibits the transition of endothelial progenitor cells to mesenchymal cells via the PIK3R2-PI3K/Akt signalling pathway. PLoS ONE. 2013;8(12):e83294.
103. Goretti E, et al. MicroRNA-16 affects key functions of human endothelial progenitor cells. J Leukoc Biol. 2013;93(5):645–55.
104. Tabuchi T, et al. MicroRNA-34a regulates the longevity-associated protein SIRT1 in coronary artery disease: effect of statins on SIRT1 and microRNA-34a expression. Clin Sci (Lond). 2012;123(3):161–71.
105. Meng S, et al. Downregulation of microRNA-126 in endothelial progenitor cells from diabetes patients, impairs their functional properties, via target gene Spred-1. J Mol Cell Cardiol. 2012;53(1):64–72.
106. Meng S, et al. MicroRNA 107 partly inhibits endothelial progenitor cells differentiation via HIF-1beta. PLoS ONE. 2012;7(7):e40323.

107. Minami Y, et al. Effect of atorvastatin on microRNA 221/222 expression in endothelial progenitor cells obtained from patients with coronary artery disease. Eur J Clin Invest. 2009;39(5):359–67.
108. George AL, et al. Endothelial progenitor cell biology in disease and tissue regeneration. J Hematol Oncol. 2011;4:24.
109. Hur J, et al. Characterization of two types of endothelial progenitor cells and their different contributions to neovasculogenesis. Arterioscler Thromb Vasc Biol. 2004;24(2):288–93.
110. Akita T, et al. Hypoxic preconditioning augments efficacy of human endothelial progenitor cells for therapeutic neovascularization. Lab Invest. 2003;83(1):65–73.
111. Sherma AK, Bendok BR. Endothelial progenitor cells in moyamoya disease. Neurosurgery. 2009;64(5):N11.
112. Kim JH, et al. Decreased level and defective function of circulating endothelial progenitor cells in children with moyamoya disease. J Neurosci Res. 2010;88(3):510–8.
113. Jung KH, et al. Circulating endothelial progenitor cells as a pathogenetic marker of moyamoya disease. J Cereb Blood Flow Metab. 2008;28(11):1795–803.
114. Sugiyama T, et al. Bone marrow-derived endothelial progenitor cells participate in the initiation of moyamoya disease. Neurol Med Chir (Tokyo). 2011;51(11):767–73.
115. Pereira Lopes FR, et al. Double gene therapy with granulocyte colony-stimulating factor and vascular endothelial growth factor acts synergistically to improve nerve regeneration and functional outcome after sciatic nerve injury in mice. Neuroscience. 2013;230:184–97.
116. Kioi M, et al. Inhibition of vasculogenesis, but not angiogenesis, prevents the recurrence of glioblastoma after irradiation in mice. J Clin Invest. 2010;120(3):694–705.
117. Li B, et al. The effect of CXCL12 on endothelial progenitor cells: potential target for angiogenesis in intracerebral hemorrhage. J Interferon Cytokine Res. 2015;35(1):23–31.
118. Nandra KK, et al. Pharmacological preconditioning with erythropoietin attenuates the organ injury and dysfunction induced in a rat model of hemorrhagic shock. Dis Model Mech. 2013;6(3):701–9.
119. Paczkowska E, et al. Increased circulating endothelial progenitor cells in patients with haemorrhagic and ischaemic stroke: the role of endothelin-1. J Neurol Sci. 2013;325(1–2):90–9.
120. Liu P, et al. Rosuvastatin for enhancement of aneurysm neck endothelialization after coil embolization: promotion of endothelial progenitor cells in a rodent model. J Neurosurg. 2016;124(5):1265–74.
121. Asif K, Leschke J, Lazzaro MA. Cerebral arteriovenous malformation diagnosis and management. Semin Neurol. 2013;33(5):468–75.
122. Wang L, et al. The role of SDF-1/CXCR4 in the vasculogenesis and remodeling of cerebral arteriovenous malformation. Ther Clin Risk Manag. 2015;11:1337–44.
123. Gao P, et al. Evidence of endothelial progenitor cells in the human brain and spinal cord arteriovenous malformations. Neurosurgery. 2010;67(4):1029–35.
124. Lu L, et al. Increased endothelial progenitor cells and vasculogenic factors in higher-staged arteriovenous malformations. Plast Reconstr Surg. 2011;128(4):260e–9.
125. Ribatti D. The involvement of endothelial progenitor cells in tumor angiogenesis. J Cell Mol Med. 2004;8(3):294–300.
126. Rafii S. Circulating endothelial precursors: mystery, reality, and promise. J Clin Invest. 2000;105(1):17–9.
127. Moschetta M, et al. Role of endothelial progenitor cells in cancer progression. Biochim Biophys Acta. 2014;1846(1):26–39.
128. Marcola M, Rodrigues CE. Endothelial progenitor cells in tumor angiogenesis: another brick in the wall. Stem Cells Int. 2015;2015:832649.
129. Ammendola M, et al. Targeting endothelial progenitor cells in cancer as a novel biomarker and anti-angiogenic therapy. Curr Stem Cell Res Ther. 2015;10(2):181–7.
130. Pirro M, et al. Baseline and post-surgery endothelial progenitor cell levels in patients with early-stage non-small-cell lung carcinoma: impact on cancer recurrence and survival. Eur J Cardiothorac Surg. 2013;44(4):e245–52.

131. Zhu H, et al. The mobilization, recruitment and contribution of bone marrow-derived endothelial progenitor cells to the tumor neovascularization occur at an early stage and throughout the entire process of hepatocellular carcinoma growth. Oncol Rep. 2012;28(4):1217–24.
132. Ahn JB, et al. Circulating endothelial progenitor cells (EPC) for tumor vasculogenesis in gastric cancer patients. Cancer Lett. 2010;288(1):124–32.
133. Hamanishi J, et al. Activated local immunity by CC chemokine ligand 19-transduced embryonic endothelial progenitor cells suppresses metastasis of murine ovarian cancer. Stem Cells. 2010;28(1):164–73.
134. Althaus J, et al. Expression of the gene encoding the pro-apoptotic BNIP3 protein and stimulation of hypoxia-inducible factor-1alpha (HIF-1alpha) protein following focal cerebral ischemia in rats. Neurochem Int. 2006;48(8):687–95.
135. Prisco AR et al. TNFalpha regulates endothelial progenitor cell migration via CADM1 and NF-kB. Stem Cells. 2016;34(7):1922–33.
136. Chen C, et al. Effect of HMGB1 on the paracrine action of EPC promotes post-ischemic neovascularization in mice. Stem Cells. 2014;32(10):2679–89.
137. Molin DG, van den Akker NM, Post MJ. Affirmative action of osteopontin on endothelial progenitors. Arterioscler Thromb Vasc Biol. 2008;28(12):2099–100.
138. Murohara T, et al. Role of endothelial nitric oxide synthase in endothelial cell migration. Arterioscler Thromb Vasc Biol. 1999;19(5):1156–61.
139. Yuan Y, et al. Derivation of human peripheral blood derived endothelial progenitor cells and the role of osteopontin surface modification and eNOS transfection. Biomaterials. 2013;34(30):7292–301.
140. Pena I, Borlongan CV. Translating G-CSF as an adjunct therapy to stem cell transplantation for stroke. Transl Stroke Res. 2015;6(6):421–9.
141. Bai YY, et al. Synergistic effects of transplanted endothelial progenitor cells and RWJ 67657 in diabetic ischemic stroke models. Stroke. 2015;46(7):1938–46.
142. Kim YJ, Jung YW. Systemic injection of recombinant human erythropoietin after focal cerebral ischemia enhances oligodendroglial and endothelial progenitor cells in rat brain. Anat Cell Biol. 2010;43(2):140–9.
143. Chi Y, et al. Detection of cytokines in supernatant from hematopoietic stem/progenitor cells co-cultured with mesenchymal stem cells and endothelial progenitor cells. Cell Tissue Bank. 2014;15(3):397–402.
144. Yoon CH, et al. Synergistic neovascularization by mixed transplantation of early endothelial progenitor cells and late outgrowth endothelial cells: the role of angiogenic cytokines and matrix metalloproteinases. Circulation. 2005;112(11):1618–27.
145. Nih LR, et al. Neuroblast survival depends on mature vascular network formation after mouse stroke: role of endothelial and smooth muscle progenitor cell co-administration. Eur J Neurosci. 2012;35(8):1208–17.
146. Daneman R, et al. Pericytes are required for blood-brain barrier integrity during embryogenesis. Nature. 2010;468(7323):562–6.
147. Schwarz SC, Schwarz J. Translation of stem cell therapy for neurological diseases. Transl Res. 2010;156(3):155–60.
148. Lasala GP, et al. Combination stem cell therapy for the treatment of medically refractory coronary ischemia: a Phase I study. Cardiovasc Revasc Med. 2011;12(1):29–34.
149. Li Q, et al. Silica-coated superparamagnetic iron oxide nanoparticles targeting of EPCs in ischemic brain injury. Biomaterials. 2013;34(21):4982–92.
150. Ganju RK, et al. The alpha-chemokine, stromal cell-derived factor-1alpha, binds to the transmembrane G-protein-coupled CXCR-4 receptor and activates multiple signal transduction pathways. J Biol Chem. 1998;273(36):23169–75.
151. Oh BJ, et al. Differences in donor CXCR4 expression levels are correlated with functional capacity and therapeutic outcome of angiogenic treatment with endothelial colony forming cells. Biochem Biophys Res Commun. 2010;398(4):627–33.

152. Chen J, et al. Endothelial nitric oxide synthase regulates brain-derived neurotrophic factor expression and neurogenesis after stroke in mice. J Neurosci. 2005;25(9):2366–75.
153. Marti HJ, et al. Hypoxia-induced vascular endothelial growth factor expression precedes neovascularization after cerebral ischemia. Am J Pathol. 2000;156(3):965–76.
154. Peplow PV. Growth factor- and cytokine-stimulated endothelial progenitor cells in post-ischemic cerebral neovascularization. Neural Regen Res. 2014;9(15):1425–9.
155. He T, et al. Transplantation of circulating endothelial progenitor cells restores endothelial function of denuded rabbit carotid arteries. Stroke. 2004;35(10):2378–84.
156. Asahara T, et al. Bone marrow origin of endothelial progenitor cells responsible for postnatal vasculogenesis in physiological and pathological neovascularization. Circ Res. 1999;85(3):221–8.
157. Kawamoto A, et al. Intramyocardial transplantation of autologous endothelial progenitor cells for therapeutic neovascularization of myocardial ischemia. Circulation. 2003;107(3):461–8.
158. Rehman J, et al. Peripheral blood "endothelial progenitor cells" are derived from monocyte/macrophages and secrete angiogenic growth factors. Circulation. 2003;107(8):1164–9.
159. Nolan DJ, et al. Molecular signatures of tissue-specific microvascular endothelial cell heterogeneity in organ maintenance and regeneration. Dev Cell. 2013;26(2):204–19.
160. Waltenberger J, Lange J, Kranz A. Vascular endothelial growth factor-A-induced chemotaxis of monocytes is attenuated in patients with diabetes mellitus: a potential predictor for the individual capacity to develop collaterals. Circulation. 2000;102(2):185–90.
161. Schanzer A, et al. Direct stimulation of adult neural stem cells in vitro and neurogenesis in vivo by vascular endothelial growth factor. Brain Pathol. 2004;14(3):237–48.
162. Imitola J, et al. Directed migration of neural stem cells to sites of CNS injury by the stromal cell-derived factor 1alpha/CXC chemokine receptor 4 pathway. Proc Natl Acad Sci U S A. 2004;101(52):18117–22.
163. Kokaia Z, Lindvall O. Neurogenesis after ischaemic brain insults. Curr Opin Neurobiol. 2003;13(1):127–32.
164. Kahle MP, Bix GJ. Neuronal restoration following ischemic stroke: influences, barriers, and therapeutic potential. Neurorehabil Neural Repair. 2013;27(5):469–78.
165. Palmer TD, Willhoite AR, Gage FH. Vascular niche for adult hippocampal neurogenesis. J Comp Neurol. 2000;425(4):479–94.
166. Wolswijk G, Riddle PN, Noble M. Platelet-derived growth factor is mitogenic for O-2Aadult progenitor cells. Glia. 1991;4(5):495–503.
167. Barres BA, et al. Multiple extracellular signals are required for long-term oligodendrocyte survival. Development. 1993;118(1):283–95.
168. Cameron HA, Hazel TG, McKay RD. Regulation of neurogenesis by growth factors and neurotransmitters. J Neurobiol. 1998;36(2):287–306.
169. Bartel DP. MicroRNAs: genomics, biogenesis, mechanism, and function. Cell. 2004;116(2):281–97.
170. Zhang X, et al. Increased expression of microRNA-221 inhibits PAK1 in endothelial progenitor cells and impairs its function via c-Raf/MEK/ERK pathway. Biochem Biophys Res Commun. 2013;431(3):404–8.
171. Wang S, et al. The endothelial-specific microRNA miR-126 governs vascular integrity and angiogenesis. Dev Cell. 2008;15(2):261–71.
172. Fish JE, et al. miR-126 regulates angiogenic signaling and vascular integrity. Dev Cell. 2008;15(2):272–84.
173. Chen J, et al. MiR-126 contributes to human umbilical cord blood cell-induced neurorestorative effects after stroke in type-2 diabetic mice. Stem Cells. 2016;34(1):102–13.
174. Tano N, Kim HW, Ashraf M. microRNA-150 regulates mobilization and migration of bone marrow-derived mononuclear cells by targeting Cxcr4. PLoS ONE. 2011;6(10):e23114.
175. Rolland-Turner M, et al. Adenosine stimulates the migration of human endothelial progenitor cells. Role of CXCR4 and microRNA-150. PLoS ONE. 2013;8(1):e54135.

176. Wang HW, et al. Deficiency of the microRNA-31-microRNA-720 pathway in the plasma and endothelial progenitor cells from patients with coronary artery disease. Arterioscler Thromb Vasc Biol. 2014;34(4):857–69.
177. Cheng CC, et al. Genetic module and miRNome trait analyses reflect the distinct biological features of endothelial progenitor cells from different anatomic locations. BMC Genomics. 2012;13:447.
178. Mause SF, Weber C. Microparticles: protagonists of a novel communication network for intercellular information exchange. Circ Res. 2010;107(9):1047–57.
179. Deregibus MC, et al. Endothelial progenitor cell derived microvesicles activate an angiogenic program in endothelial cells by a horizontal transfer of mRNA. Blood. 2007;110(7):2440–8.
180. Wang J, et al. Effects of endothelial progenitor cell-derived microvesicles on hypoxia/reoxygenation-induced endothelial dysfunction and apoptosis. Oxid Med Cell Longev. 2013;2013:572729.

Chapter 8
Co-transplantation Strategies and Combination Therapies for Stroke

Aurel Popa-Wagner, Raluca Elena Sandu, and Ovidiu Ciobanu

Abstract Worldwide cerebrovascular disease (CVD) is increasing in parallel with modernization, changes in lifestyle, and the growing elderly population. The incidence of stroke increases significantly with age both in men and women with incidence rates accelerating above 70 years. Since stroke afflicts mostly the elderly comorbid patients, it is highly desirable to test the efficacy of cell therapies in an appropriate animal stroke model. All monotherapeutic attempts to prevent or ameliorate brain damage following stroke have failed so far. In view of previous findings indicating that stroke impacts a wide range of mechanisms, ranging from central nervous system (CNS) physiology over CNS regeneration and plasticity to the adaptive immune system in an age-dependent manner, the failure of monotherapies is perhaps not unlikely. Bone marrow-derived mesenchymal stem cells (BM MSCs) and hematopoietic stem/progenitor cells (HSPC) are the most frequently cells used in preclinical and clinical neurorestorative studies in stroke therapy. Therefore co-transplantation of BM MSCs with other cells may be a better strategy to improve microenvironment, make the grafting more efficient, and improve functional recovery after stroke. Current knowledge includes: (1) the potential for neurogenesis is also preserved in aged, stroke-injured brains; (2) the environment of the aged brain is not hostile to transplantation of BM MSC; and (3) the extent of recovery is successful in some but not all behavioral tests. However, there remain significant developmental and translational issues to be resolved in

A. Popa-Wagner (✉)
Department of Psychiatry, University of Medicine Rostock,
Gehlsheimer Straße 20, 18147 Rostock, Germany
e-mail: aurel.popa-wagner@med.uni-rostock.de;
aurel.popa-wagner@geriatrics-healthyageing.com

R.E. Sandu
Department of Functional Sciences, Center of Clinical and Experimental Medicine,
University of Medicine and Pharmacy of Craiova, Craiova, Romania

O. Ciobanu
Vivantes Humboldt-Klinikum Psyhiatrie, Psychotherapie und Psychosomatik Zentrum für affektive Erkrankungen, Berlin, Germany

K. Jin et al. (eds.), *Bone Marrow Stem Cell Therapy for Stroke*,
DOI 10.1007/978-981-10-2929-5_8

future studies such as: (1) Understanding the differentiation into specific phenotypes. Upon transplantation, the differentiated cells often de-differentiate (Kalladka and Muir Stem Cells Cloning 7:31–44, 2014). (2) Tumorigenesis remains a significant concern (Riess et al. J Neurotrauma 24(1):216–225, 2007). (3) Anti-neuroinflammatory therapies are a potential target to promote regeneration and repair in diverse injury and neurodegenerative conditions by stem cell therapy. (iv) Efficacy of cell therapy can be enhanced by physical rehabilitation (Dunnett Clin Neurol 110:43–59, 2013). We recommend that in a real clinical practice involving older poststroke patients, successful regenerative therapies would have to be carried out for a much longer time. The BM MSC therapy in aged rodents warrants further investigation including repeated administrations of therapeutic cells at several time points after stroke and using various combinations with G-CSF or other relevant growth factors/cytokines.

Finally, a better understanding of potential risks of stem cell therapies in strokes shall makes the translation of cell therapies safer. Likewise, awareness of their biology may help improve their efficacy to achieve therapeutic success.

Keywords Stroke • Strategy • Combination therapies • Co-transplantation

8.1 Introduction

Cerebrovascular diseases (CVDs) represent a leading cause of death worldwide. Worldwide CVD is increasing in parallel with modernization, changes in lifestyle, and the growing elderly population. Stroke not only has a very high mortality rate but also results in debilitating neurological impairments or permanent disability in survivors associated with huge economic losses. Etiologically, ischemic strokes represent approximately 80 % of all cerebrovascular infarctions, and hemorrhagic strokes account for the remaining 20 % [51].

In demographically developed countries, the average age at which stroke occurs is around 73 years reflecting the older age structure of these countries. The probability of a first stroke or first transitory ischemic attack is around 1.6 per 1000 and 0.42 per 1000, respectively. In less developed regions, the average age of stroke will be younger due to the different population age structure resulting from higher mortality rates and competing causes of death.

Stroke patients are at highest risk of death in the first weeks after the event, and between 20 % and 50 % die within the first month depending on type, severity, age, comorbidity, and effectiveness of treatment of complications. Patients who survive may be left with no disability or with mild, moderate, or severe disability. Considerable spontaneous recovery occurs up to about 6 months [22]. However, patients with a history of stroke are at risk of a subsequent event of around 10 % in the first year and 5 % per year thereafter [31].

The negative consequences of stroke extend well beyond the victims themselves, ultimately including families, caregivers, social networks, and employers. The proportion of patients achieving independence in self-care by 1 year after a stroke ranges from around 60–83 %. This wide variation relates to whether the studies are community based or hospital based, in which activities are considered in estimating independence and the methods used to rate ability. In established marked economies (EMEs), depending on the organization of hospital services, between 10 and 15 % of survivors are resident in an institution at 1 year [5].

8.1.1 Age Is the Principal Risk Factor for Stroke

The incidence of stroke increases significantly with age both in men and women with incidence rates accelerating above 70 years [155]. However, there are age-related gender differences in the incidence of stroke. Men aged up to 75 years old are more likely to be hit by stroke than women. The risk to have a stroke then becomes higher in women than men aged 85 years or older [155]. This may be attributed to sex-related differences in the life expectancy of women and the development of age-related comorbidities. Effects of age and gender on stroke incidence, functional recovery, and mortality have not only been shown in humans but also in animal models [16, 64]. Indeed, the age-dependent increase in the evolution of ischemic tissue into infarction strongly suggests that age is a biological marker for the variability in tissue outcome in acute human stroke [7].

It should be noted that the age-associated decline in functional brain reserve is most pronounced after the age of 85 and implies an impaired response to stressors and illnesses [61]. Importantly, age-associated changes show great variability among individuals, which are modified by genetic and long-term lifestyle factors [175, 192].

8.1.2 Elderly, Comorbidities, and Stroke

Our knowledge about the molecular and cellular mechanisms underlying accelerated infarct progression in subjects with metabolic syndrome is still poor. Some studies report a strong connection between nutrition and body weight, on one hand, and increased oxidative stress or pro-inflammatory changes in the brain, which promotes neural imbalance and glucose level elevation, on the other hand [32]. Zhang and colleagues suggested that metabolic inflammatory changes in the brain are linked to the IKK/NF-kB signaling pathway [199].

Observational studies have shown a strong correlation between blood lipid levels and stroke [80]. In animal models it could be shown that VEGF-induced angiogenesis is compromised by hyperlipidemia and provided an explanation of poor efficacy of angiogenic therapies in patients suffering from hyperlipidemia [196].

Overnutrition and hypercholesterolemia may not only be responsible for metabolic inflammation of the brain but can also induce mitochondrial dysfunction and increased oxidative stress (NADPH oxidase dependent), which may promote metabolic syndrome and related diseases [32, 70]. Cellular stress and hyperglycemia are known to accelerate the aging process. In this light, a better understanding of molecular factors and signaling pathways underlying the metabolic syndrome as well as the contribution of comorbidities to stroke-induced sequelae may be translated into more successful treatments or prevention therapies against age-associated diseases which, in turn, would improve lifespan and quality of life.

8.1.3 Stroke Models Using Aged Animals Are Clinically More Relevant

Since stroke afflicts mostly the elderly comorbid patients, it is highly desirable to test the efficacy of cell therapies in an appropriate animal stroke model. Animal models of stroke often ignore age and comorbidities frequently associated with senescence, and this could be one of the explanations for unsuccessful bench-to-bedside translation of neuroprotective strategies. Worldwide stroke is increasing in parallel with modernization, changes in lifestyle, and the growing elderly population. In particular, rates in Eastern Europe have been increasing, such that currently the highest rates are found in countries such as Bulgaria, Romania, and Hungary. Among the women and men, individuals with a low-risk lifestyle (exercising daily, consuming a prudent diet including moderate alcohol and having a healthy weight during midlife) had a significantly lower risk of stroke than individuals without a low-risk lifestyle. Therefore the relatively high incidence of stroke may be due in part to the impact of numerous known risk factors in [51] arterial hypertension, diabetes, high cholesterol, smoking, alcoholism, obesity, stress, and a sedentary lifestyle.

Comorbidities like diabetes, arterial hypertension, or comorbidity factors such as hypercholesterolemia are common in elderly persons and are associated with higher risk of stroke [65]. Moreover, simultaneous presence of vascular diabetic complications and associated comorbidities like hypertension and chronic diabetes significantly increased the level of ischemic damage [152].

Currently, there are several different rodent models with comorbidities such as the spontaneously hypertensive rat (SHR) model, stroke-prone spontaneously hypertensive rat (SHRSP), the streptozotocin rat model for diabetes, and the high-fat diet or high-sugar diet Sprague Dawley rats. Diffusion (DWI)- and perfusion (PWI)-weighted magnetic resonance imaging of the ischemic penumbra in SHRSP showed increased infarct size early after stroke as compared to the normotensive WKY rats. Moreover, the infarct volume after 60 min MCAO was greater in SHRSP ($36 \pm 4\,\%$ of hemisphere volume) than in SHR ($19 \pm 5\,\%$) or WKY normotensive rats ($5 \pm 2\,\%$) [116].

High blood pressure is a major risk factor for stroke. Large clinical trials have shown that ACE inhibitors reduce the incidence of stroke by up to 43 % [195]. However, because normotensive patients also benefit from ACE inhibition, it has been suggested that these effects may also be independent of the blood pressure-lowering effects of ACE inhibition [171]. Indeed, neither short (7 days) nor long term (42 days) prior to stroke administration of ACE inhibitors to SHR reduced the infarct size despite lowering the blood pressure, while WKY normotensive rats showed, paradoxically, marked reductions in infarct volume [142].

Over the past 10 years, a variety of models of middle cerebral artery occlusion (MCAO) have been established in rodents [8]. MCAO in aged rodents has been produced with permanent or transient occlusion for 30–120 min using (1) MCA ligation after craniectomy [189] and (2) intraluminal thread occlusion [174], a hook attached to a micromanipulator [140], cauterization [50, 87], photothrombosis [200], and endothelin injection [172, 183], injection of a thrombus via ECA [50], or intraluminal thrombus formation by thrombin injection (using occlusion of distal branches of the middle cerebral artery (MCA)).

Since focal cerebral ischemia is technically difficult to perform in very old rats and since based on epidemiological studies human stroke occurs more often in late middle-aged (60–70 years old) subjects, it is advisable to use middle-aged instead of very old animals for stroke research [137].

8.1.4 Spontaneous Stroke Recovery in Aged Patients and Animals

Stroke patients regain some of their lost neurological functions during the first weeks or months after the stroke. In contrast, in animal models of stroke, complete spontaneous recovery may occur in young rats, depending on the size and location of the ischemic lesion.

In clinical practice, physical therapy is widely used for stimulating poststroke recovery [69, 75, 100] which is thought to occur via recruitment of adjacent cortical regions [67].

In animal models of stroke, young rats begin to show improvements of neurological function starting by day 2 poststroke, whereas in aged rats, neurological recovery is hardly detectable before days 4–5, with about 75 % of the functional improvement observed in young rats [28].

Housing experimental animals in an enriched environment enhances the recovery from brain damage both in young and aged animals [28]. When aged rats were allowed to recover in an enriched environment, the delay period was shortened, and behavioral performance was significantly improved. The improvement in task performance positively correlated with slower infarct development, fewer proliferating astrocytes, and smaller size of the glial scar [28]. Even more effective rehabilitation of the contralateral forelimb could be achieved by Corbett and colleagues by combining enriched environment with training [72].

8.1.5 Bone Marrow-Derived Cells and the Therapy of Subcortical Stroke

Despite improving knowledge about stroke pathology, therapeutic benefits for stroke patients are limited. However, spontaneous recovery is common, whenever the infarct is located in the striatum, a subcortical structure that exhibits activity-dependent plasticity related to movement and motor learning. This may explain why patients with subcortical lacunar stroke are more likely to have early functional recovery after stroke [15, 158]. Other studies suggest that the beneficial effect could be due to in situ secretion of neuroprotective factors by the transplanted cells. For example, human-derived inducible pluripotent cells (iPSC) implanted into the striatum of young animals at 1 week after MCAO protected substantia nigra from atrophy, probably through a trophic effect [134]. Neurological recovery in patients with subcortical stroke is associated with neuroplasticity in the contralesional striatum [148] and axonal plasticity in contralesional motor cortex [150], which may explain why patients with subcortical stroke are likely to exhibit functional neurological recovery [15, 158].

However, whether endogenous neurogenesis in the perilesional subventricular zone contributes to spontaneous recovery after stroke has not yet been established. In addition, age, comorbidities, physical condition of the patient, and severity of disease could substantially influence these steps and, therefore, the outcome of the healing process.

Several studies on aging have established that the neurogenic subventricular zone become disorganized with increasing age. During aging, ependymal cells accumulate dense bodies, become flattened and gradually lose their cilia. Both ependymal cells and astrocytes accumulate dense bodies and intermediate filaments in their cytoplasm. The number of newly born neurons within the SVZ decreases over time, while the generation of oligodendroglial cells seems to be preserved in the aged brain [34]. Similarly, the number of DCX-positive cells in the RMS and the OB from fetal to adult stages decreases with age, while the number of the newly generated non-neuronal cells, such as oligodendrocytes, seems to remain relatively constant [17, 160, 188].

8.1.6 Cell Therapy of Cortical Stroke Using Mesenchymal Stem Cells in Aged Animal Models

Due to the ethical concerns and limited availability of using pluripotent embryonic stem cells (ES) and induced pluripotent cells (iPS) in the clinic, great interest has developed in mesenchymal stem cells (MSCs), which are free of both ethical concerns and teratoma formation.

Recent studies suggest that modified mesenchymal stem cells (MSCs) are able to form a "biobridge" between neurogenic subventricular zone (SVZ) and the ischemic cortex area (penumbra). Using this road, endogenous stem cells can migrate from the neurogenic area to the site of lesion and may ameliorate outcome in experimental models of cerebral ischemia [54, 75, 90, 178]. Several studies also showed that grafting of BM MSCs in the peripheral circulation improved functional neurological outcome and reduced infarct volume [75].

Mesenchymal stem cells (MSCs) and hematopoietic stem/progenitor cells (HSPC) that are most frequently used in preclinical and clinical neurorestorative studies in stroke therapy augment this endogenous response. MSC can be obtained either from the bone marrow or adipose tissue [129]. HSPC can be isolated from bone marrow or from umbilical cord blood (UCB) or can be mobilized into the blood by the administration of G-CSF. Amniotic fluid has been investigated as a new cell source for mesenchymal stem cells in the development of cell-based transplantation. Earlier studies have demonstrated the ability of amniotic fluid-derived stem cells to differentiate along a neurogenic pathway [44].

8.1.7 Combination Therapy of Stroke in Young Animals

All monotherapeutic attempts to prevent or ameliorate brain damage following stroke have failed so far. In view of previous findings indicating that stroke impacts a wide range of mechanisms, ranging from central nervous system (CNS) physiology over CNS regeneration and plasticity to the adaptive immune system in an age-dependent manner, the failure of monotherapies is perhaps not unlikely.

Tissue recovery after brain ischemia requires both neural and vascular components in the damaged brain area. The poor survival and differentiation of both the transplanted cells and their progenies in the inhospitable environment of the infarcted cortex have prompted the search for alternatives and new concepts like the neurovascular unit to limit the severe death of transplanted cells. Therefore, co-transplantation of NSCs with other cells in the niche may be a better strategy to improve microenvironment and make the grafting more efficient. To this end co-transplantation of neural stem cells and endothelial cells (ECs) within a mouse model of stroke enhanced the survival, proliferation, and differentiation of transplanted cells and improved functional recovery [126]. Taking a step further, co-transplantation of ESC-VPCs with NPCs after ischemic stroke supplied not only neural cells but also ECs and pericytes, thus providing nearly all important components for recovery of the neurovascular unit at the infarcted area [97].

The combination of mesenchymal stem cells and neural stem cells could improve also functional recovery after stroke if given prior to stroke. To this end, a mix of MSCs isolated from the femurs and tibias of rats and NSCs isolated from rat embryo ganglion eminence was labeled with PKH26-GL and administered one day before stroke into the lateral ventricle and neurological recovery evaluated for 28 days after stroke. The results indicate that the combination cell therapy is more efficient in promoting brain recovery after stroke than each stem cell alone [77].

More recently, triple cell co-transplantation with rat NSCs, astrocytes, and brain microvascular endothelial cells (BMECs) has been proposed. After grafting these cells into the brain of MACO/R rat model with different combinations, we found that the learning and memory ability of these rats improved to some extent. The rats which transplantation of NSCs with astrocyte and BMECs together have got the most achievement and they are better than those who grafted only two cells. The rats grafted only one cell have got the worst improvement. From our experiment, we get the conclusion that co-transplantation of NSCs with astrocytes and BMECs can improve learning and memory in the water maze test, probably due to the microenvironment improvement by the transplanted astrocytes and BMECs [33].

In an interesting approach, the combination therapy was given in sequence. First, mesenchymal stem cells were transplanted during the acute phase after stroke (1 day) in an attempt to diminish the inflammation and provide an appropriate microenvironment for regeneration after ischemia. Then, the neural stem cells were transplanted at 7 days after stroke to help regeneration by differentiation into neurons, oligodendrocytes, or astrocytes [77].

8.1.8 Co-administration of G-CSF and BMMNC in the Poststroke Aged Rats

Multimodal approach is an alternative strategy to promote functional recovery after stroke in aging brain, using systemic manipulation strategies. Several studies showed that grafting of autologous BM mononuclear cells (BM MNCs), a heterogeneous population containing mesenchymal and hematopoietic stem/progenitor cell fractions, in the peripheral circulation beneficially influences functional recovery as well as infarct volume in various species and is safe for stroke patients [21, 26, 161]. We therefore reasoned that timely application of BM MNC after stroke could bridge the gap until G-CSF-driven mobilization of endogenous BM cells comes to full effect, leading to improved outcome of this experimental combination treatment for stroke in aged rats.

G-CSF exerts a wide range of potential effects [120] and can reduce the number of fatal hemorrhages after experimental thrombolysis [46]. It has been suggested that G-CSF exerts a therapeutic effect after stroke by anti-apoptotic properties and by reducing excitotoxicity-driven penumbral apoptosis [163]. The latter effect was considered strong enough to reduce lesion size in young animals [24]. However, the aging brain is in need for increased glutamate signaling, which is reflected by abundant expression of Na^+-dependent membrane glutamate transporters, particularly in white matter areas. This in turn renders the aging brain highly susceptible to ischemic excitotoxicity, swiftly exhausting mitochondrial capacities [11]. Despite some positive impact of G-CSF on the aged brain apart from anti-excitotoxicity, this situation may have limited G-CSF monotherapy benefits [141] although a remaining benefit was clearly shown. We tested the hypothesis that treating poststroke aged rats with the combination of bone marrow-derived mononuclear cells (BMMNCs)

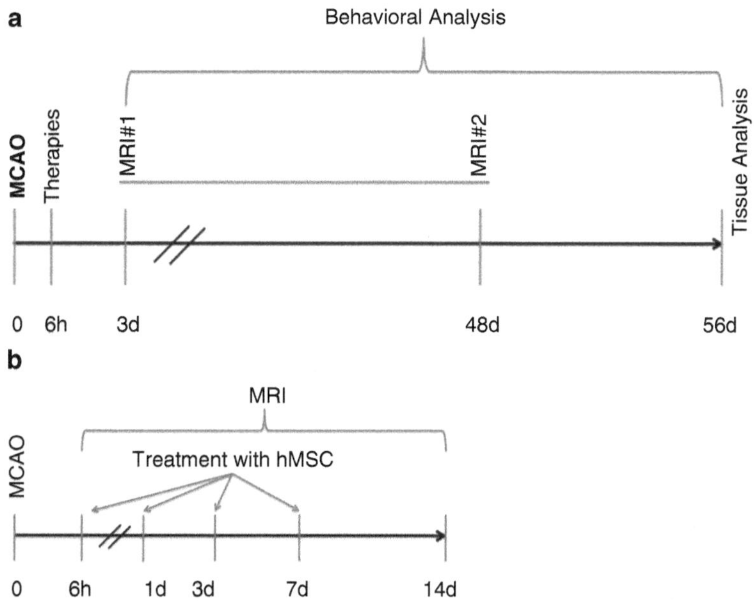

Fig. 8.1 Time window of BM MSC therapies. Most therapies start early after stroke with a single dose (**a**). The efficacy of several doses of cells given at different time points has also been tested (**b**). Current knowledge suggests that administered BMMNC provide indirect neuroprotection leading to infarct size reduction after ischemic damage in a time window of up to 1 month

and G-CSF improves the long-term (56 days) functional outcome by compensating the delay before G-CSF comes to full effect. To this end, 1×10^6 syngeneic BMMNC per kg bodyweight (BW) in combination with G-CSF (50 μg/kg, intraperitoneal application, continued for 28 days) was administered via the jugular vein to aged Sprague Dawley rats at 6 h poststroke (Fig. 8.1a). Infarct volume was measured by magnetic resonance imaging at 3 and 48 days poststroke and additionally by immunohistochemistry at day 56 (Fig. 8.1a). Functional recovery was tested during the entire poststroke survival period. Daily G-CSF treatment led to robust and consistent improvement of neurological function, but did not alter final infarct volumes. This result was unexpected since benefits of G-CSF and BMMNC treatment paradigms in stroke, independently from each other, have been repeatedly reported by independent experiments and groups and were hypothesized to work synergistically especially in the aged, stroke-lesioned brain. The combination of G-CSF and BMMNC did not further improve poststroke recovery. The lack of an additional benefit may be due to a hitherto not well-investigated interaction between both approaches and, to a minor extent, to the insensitivity of the aged brains to regenerative mechanisms. Also considering recent findings on other tandem approaches involving G-CSF in animal models featuring relevant comorbidities, we conclude that such combination therapies are not the optimal approach to treat the acutely injured aged brain.

Current knowledge suggests that administered BMMNCs provide indirect neuroprotection leading to infarct size reduction after ischemic damage in a time window of up to 1 month [93]. G-CSF in turn induces BMMNC mobilization, while the SDF-1/CXCR4 system causes BMMNC to invade the ischemic brain [164, 170], where they are believed to exert therapeutic effects. However, the initiation of this potentially beneficial effect may take simply too much time: although a granulocyte boost is seen after about 48 h, peaking of G-CSF-based mobilization can take up to 9 days [73], which is beyond the therapeutic time window for BMMNC. Since endogenous G-CSF is not available in sufficient concentrations directly after the ischemic event [24], a combination therapy providing (i) G-CSF in sufficient amounts to act neuroprotectively and (ii) exogenous BMMNC early enough to bridge the time gap until G-CSF-based endogenous BMMNC mobilization comes to full effect seemed promising – but failed to fulfill the expectations.

One may assume that either the lesioned and aged rat brain environment was insensitive to regenerative mechanisms by BMMNC or cell treatment has been mainly ineffective. Indeed, the aggravated impact of ischemic damage on the aged brain is well known, while potential detrimental effects of aging on BMMNC have been anticipated [136, 137, 185]. Moreover, technical complications may come in play as well: a limiting influence of long-term cryopreservation on the therapeutic efficacy of umbilical cord blood MNC, a population being very similar to BMMNC, has been discussed recently [191]. However, deriving syngeneic cells from young animals and limiting cryopreservation to no more than 4 weeks in our experiment might have limited such aging and cryopreservation effects on the donor side. Although a remaining impact cannot be excluded per se, a complete failure of the BMMNC treatment seems unlikely. An alternative explanation for the reduced efficacy of the combination treatment could be interference between both treatment regimes. A recent study in hypertensive animals demonstrated that intravenously administered BMMNCs occupy splenic granulocyte clearance capacities for apoptotic cells [143]. This clearance system usually removes apoptotic granulocytes from the circulation, which represents an important anti-inflammatory mechanism. Being already compromised by externally administered BMMNC, the swift and early granulocyte boost from the BM by G-CSF may have completely exhausted the clearance system in our treatment scenario. This detrimental interaction may have caused a sustained systemic and central pro-inflammatory bias, leading to subtle additional damage, not enhancing but partly reducing the neuroprotective G-CSF effect. It remains for further investigation whether this interaction is even more relevant in the aged brain.

False-negative results are a common phenomenon when selected sample sizes are too small to reveal small-scale treatment effects with G-CSF and G-CSF+BMMNC group outcomes just differing randomly from each other. To prevent such scenarios, we chose relatively large samples ($n=21$ animals per group). This is close to detect an effect size of 20 %, being recommended to assume when assessing experimental stroke therapies [111]. We therefore consider that insufficient study power is not very likely to have "masked" a positive effect of G-CSF+BMMNC.

8.1.9 Clinical Trials

Mobilization of hematopoietic stem cells by administration of granulocyte colony-stimulating factor has been assessed and not shown to have a therapeutic benefit in patients after stroke either [143, 154, 176]. Therefore the first goal of clinical trials was to assess the feasibility and safety of transplanting autologous bone marrow mononuclear cells into stroke patients [144, 145, 156, 161], and both positive [177] and negative [145] stroke outcomes have been reported. Recently, intravenous administration of autologous bone marrow mononuclear cells to patients with severe embolic stroke was shown to be feasible and safe with a tendency for improved neurologic recovery and improvement in cerebral blood flow and metabolism in poststroke [176].

8.1.10 Co-transplantation of G-CSF and BM MSC

Several studies showed that grafting of bone marrow-derived stem cells in the peripheral circulation improved functional neurological outcome and reduced infarct volume. Most of these studies used bone marrow mesenchymal cells (BM MSCs) [69, 124]. In addition, MSCs can serve as an excellent cellular delivery system. In a previous work, we have shown that application of G-CSF shortly after stroke in aged rats increases neurogenesis and improves some of the behavioral indices [141]. Even though hMSC S-TRAIL cells were seen within the injection site, a large number of cells migrated toward the tumor along the corpus callosum. In vitro, paracrine factors secreted by MSCs protect neurons from apoptotic cell death in the OGD model of cerebral ischemia [162].

Therefore we reasoned that the efficiency of the bone marrow-derived cell therapy may be increased by simultaneous application of G-CSF. In particular we tested the hypothesis that grafting of pre-differentiated bone marrow mesenchymal stem cells (BM MSC) in G-CSF-treated animals increases the odds of long-term functional outcome in aged rodents. To this end, 10^6 rat BM MSCs isolated from young Sprague Dawley rats were administered a single dose of BM MSCs (10^6/kg) given in combination with G-CSF (50 μg/kg for 28 days) via the jugular vein or intrathecally to Sprague Dawley rats at 6 h poststroke (Fig. 8.1a). The phenotypes of BMSCs used in this study were positive for CD105, CD166, CD29, and CD44. Cells tested negative for CD14, CD34, and CD45. Cells were tested for purity by flow cytometry and for their ability to differentiate into osteogenic, chondrogenic, and adipogenic lineages. The control groups received daily injections of either G-CSF 50 μg/kg or vehicle (5 % glucose) for 28 days. To investigate the localization of injected cells, a separate group of aged animals were injected with mesenchymal cells of human origin (hBM MSCs). Although hBM MSCs are poorly immunogenic [38] and the rats survived just 4 days after administration, the animals were given cyclosporine A (s.c., Sandimmune, Novartis, 10 mg/kg) diluted in Cremophor EL (Sigma) to prevent graft rejection.

Infarct volume was measured by MRI at 3 and 48 days poststroke and additionally by immunohistochemistry at day 56 (Fig. 8.1a). Functional recovery was tested during the entire poststroke survival period of 56 days. Daily treatment of poststroke aged rats with G-CSF led to a robust and consistent improvement of neurological function. The combination therapy of GG-CSF+BM MSC in aged rats showed, surprisingly, no additional improvement in recuperation of the sensory function (adhesive tape), although recuperation of more complex motor (rotating pole) and spatial reference memory tasks was improved both by G-CSF and the combination. Paradoxically, MCAO rats swam slightly faster than unoperated animals, probably due to a poststroke excitatory state, an observation confirmed in previous studies [18].

However, of the treated groups the best recovery rate was seen for the G-CSF group which showed significant improvement in the water maze spatial reference memory task between days 21 and 42, suggesting that the beneficial effect of the G-CSF treatment is restricted to the G-CSF treatment period [141]. We hypothesized that the improved functional recuperation of the G-CSF group may have been helped by the stimulation of endogenous neurogenesis by G-CSF as previously reported [141]. In the combined treatment study, we found increased cellularity in the formerly infarct core of the G-CSF+BM MSC group at day 56 poststroke and intact neurogenesis in the lateral ventricle region. However, there was a clear regional separation of the DCX^+ cells which emanated from the ventricular wall and BrdU-labeled nuclei which were localized mainly in the vascular network of the lateral ventricle. Since BrdU was administered for the first 14 days after stroke, we hypothesize that at 2 months poststroke, DCX^+ cells with BrdU nuclei did not survive. Instead, $BrdU^+$ nuclei survived most likely in endothelial cells of the vascular wall [138]. Further, the combination therapy significantly improved recuperation and microvessel density in the formerly infarct core and beyond. Finally, we found that the aged brain environment is permissive for the migration of human BM MSCs toward the lesion site and concluded that, in a real clinical situation involving older poststroke patients, successful regenerative therapies may have to be delivered throughout a prolonged period, perhaps for 6–12 months.

8.1.11 Clinical Studies Using Combination Therapies

Novastem, a leader in regenerative medicine, treated stroke patients with ischemia-tolerant mesenchymal stem cells (itMSCs) in combination with ischemia-tolerant neural stem cells (itNSCs), both of which are proprietary products of Stemedica, given intrathecally to patients with motor aphasia due to ischemic stroke. The study aimed to evaluate functional changes on subjects at the baseline, at 90 days, and at 180 days, and it is measured using the United States National Institute of Health Stroke Scale (NIHSS), Stroke and Aphasia Quality of Life Scale-39 (SAQCOL-39), and the Boston Diagnostic Aphasia Examination (BDAE) neuropsychological evaluation for diagnosis. In addition, an MRI taken with a gadolinium-based contrast agent (GBCA) is performed as a baseline analysis and at the study's conclusion. The end point is to evaluate the safety and tolerance of the two-cell treatment, as well as observing initial changes in functionality.

8.1.12 *Therapeutic Window and Route of Administration for Combination Therapies*

Several studies showed that grafting of bone marrow-derived stem cells in the peripheral circulation improved functional neurological outcome and reduced infarct volume [75]. Most of these studies used bone marrow mesenchymal cells (BM MSCs), but feasibility and safety in clinical trials were also shown for the bone marrow mononuclear cells (BMMNCs) [69, 124]. A conclusive result on optimal timing and dosing is, however, still missing.

The ideal therapeutic protocol in terms of number of cells and at what times to be delivered is uncertain. Different routes of MSC administration have been used to treat damaged ischemic brain tissue.

Numerous studies have shown that the intracerebral, intravenous, and intra-arterial routes of transplantation of stem cells are effective for ischemic stroke [12, 23, 26, 76, 93, 95, 121, 122, 188, 190]. The advantage of intracerebral transplantation is the targeted deposition of stem cells into the lesioned brain, compared to intra-arterial or intravenous transplantation resulting in cells being trapped in peripheral organs such as the spleen [1, 66, 184]. However, local stem cell delivery increases risks and side effects, such as bleeding and tissue injury. Therefore, in clinical practice, intravascular transplantation remains the preferred route of administration being much less invasive than intracerebral transplantation. Direct tissue injection via arterial administration may produce occlusion and embolization [49, 130, 168], while the intravenous approach would allow a broad biodistribution and easy access [66, 169]. Recent studies indicate that intra-arterial stem cell transplantation for ischemic stroke could induce functional recovery in ischemic stroke animals [26, 66, 79, 86, 89, 121, 157, 165, 166, 186], with a clinical phase I/II trial demonstrating its safety and feasibility in stroke patients [124].

The efficacy of mesenchymal stem cells (MSCs) to restore neurological function has been demonstrated repeatedly in animal models of stroke and might be a feasible and safe mode for treatment of stroke patients [39]. However, it is unclear how the timing of intra-arterial MSC transplantation to ischemic stroke affects the functional recovery. In a recent study intra-arterial allogeneic MSC transplantation in ischemic stroke model of rats at 24 h after MCAO led to robust functional recovery that was paralleled by reduced infarct volumes and a high number of integrated MSCs successfully migrating toward the ischemic cortex [182].

Omori and colleagues did a comparative study on the therapeutic benefits of multiple injections vs. single-dose application of therapeutic cells. Thus it was reported that a relatively large dose of 3×10^6 cells administered at an early time point is more effective than multiple injections of 1 × 106 cells at 6, 24, and 48 h and showed nearly identical efficacy as a single-cell dose at 6 h and suggested that relatively high dosing of hMSCs within the first 2 days is the most effective protocol (Fig. 8.1b) [127].

In another study the beneficial effect of 3 × 107 bone marrow-derived mononuclear cells (BMMNCs) and BM MSC injected into the jugular vein after ischemia has been evaluated at 1, 7, 14, or 30 days poststroke in a rat model of focal ischemia induced by thermocoagulation [53]. The results indicate that the treatment of sensorimotor cortical ischemia with intravenously administered BMMNCs is only effective in the acute/subacute phases of the disease with only a few showing an improved functional recovery when MSCs were administered 1 month after ischemia [165]. Some other studies have shown that, on the contrary, intrathecal delivery by lumbar puncture may be a more efficient approach for the BM MSC treatment of stroke [101].

Some other studies have shown that intrathecal delivery by lumbar puncture may be useful and feasible for MSCs treatment of stroke [101]. However, we did not find major differences in the efficacy of the two routes of administration. In our model, CD166-positive cells in the ipsilateral hemisphere migrated all the way from the wall of the lateral ventricle (Fig. 8.2a) and its vicinity (Fig. 8.2b) through the corpus callosum (Fig. 8.2c) to the peri-infarcted region (Fig. 8.2d) and to the infarcted area (Fig. 8.2e). Noteworthy is the presence of human nuclei that were intermingled with rat nuclei in the infarcted area (Fig. 8.2f).

8.1.13 The Mechanisms by Which MSCs May Ameliorate Infarcted Brain Tissue

The mechanism of action of stem cell transplantation remains not fully understood and has been related more to the capacity of MSCs to release neuroprotective factors (paracrine mechanism) than to their capacity to replace damaged neural cells through their transdifferentiation properties. Administration of MSCs in acute stroke animal models markedly decreased brain infarct size, improved neurological function by enhancing neurogenesis, and showed anti-inflammatory and anti-apoptotic effects. Additionally, initial clinical studies using intravenously delivered MSCs have been initiated in human subjects with stroke [96].

Cellular therapy using mesenchymal stem cells (MSCs) can, in addition, enhance the endogenous restorative mechanisms of the injured brain by supporting processes of neovascularization and neurogenesis [12, 40, 42, 68, 78, 101]. Thus, beyond the formerly infarct core, several groups noted vigorous sprouting angiogenesis as evidenced by RECA/BrdU double-positive immunostaining of the blood vessels during the resolution phase of angiogenesis (Fig. 8.3). By number of laminin/BrdU co-localizations, the density of the newly formed blood vessels was significantly higher in the brains of aged animals treated with the combination G-CSF + BM MSC as compared to controls and G-CSF alone [10].

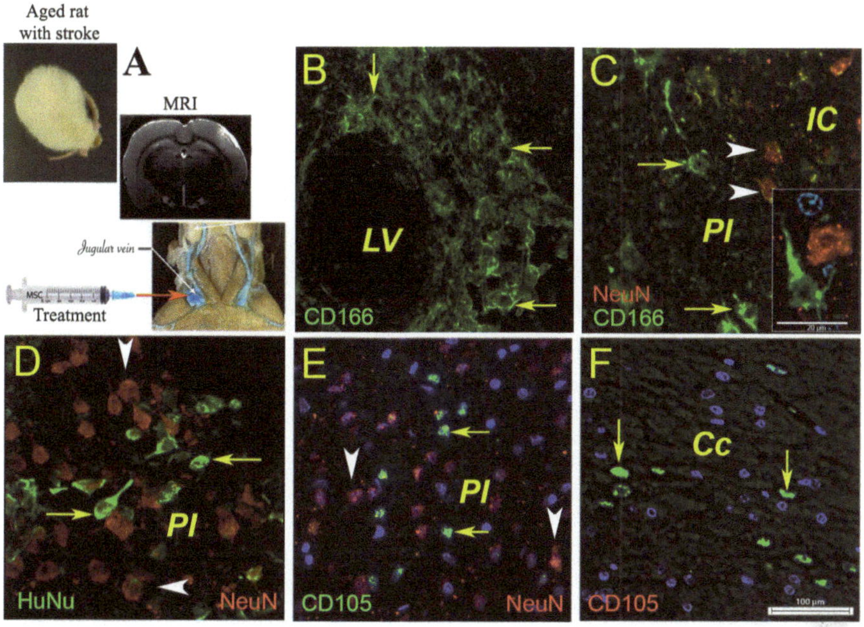

Fig. 8.2 Combination therapies of stroke. (**a**) Treatment with BM MSC starts early after stroke; (**b**) in our model the cells probably entered the injured brain via the lateral ventricle as shown by the CD166-positive cells; (**c**) about 1 % of the injected CD166-positive cells reached the infarcted area where they were intermingled with surviving or degenerating neuronal nuclei showing NeuN immunopositivity; (**d**, **e**) note the presence of immunopositivity for human nuclei (**d**; HuNu, *arrows*) and CD105 (**e**) that were dispersed between the rat nuclei in the infarcted area (NeuN; *arrowheads*); (**f**) vigorous sprouting angiogenesis has been reported as evidenced by RECA/BrdU double-positive immunostaining of the blood vessels (*violet*) as well as numerous $BrdU^+$ nuclei in the newly formed endothelium (*blue*) and reconstruction of the basal lamina (laminin, *green*) during the resolution phase of angiogenesis

These results strongly suggest that the BM MSC promoted angiogenesis rather than neurogenesis in the lesioned area of aged rats. Indeed, previous studies have shown that delayed intracerebral injection of hMSCs modified the cerebral microvasculature after transient ischemia [27, 122] and improved the cerebral blood flow (CBF) [82]. In our model, the MSC reached the peri-infarcted region after 4 days. However, 56 days after the administration of BM MSCs, the proliferation marker BrdU was incorporated preferentially in the "pinwheel" architecture of the ventricular epithelium [62, 99] and blood vessel in the formerly infarct core, while the DCX^+ cells occupied an adjacent, distinct position in SVZ and were not detectable at all in the lesioned area (Fig. 8.3). Since BrdU was administered for the first 14 days after stroke, we concluded that at 2 months poststroke, DCX^+ cells with BrdU nuclei did not survive, strongly suggesting that the early neuronal progenitors did not survive in the hostile environment of the poststroke aged brain. Instead, in animals treated with the combination of G-CSF and BM MSCs, we noted an increased number of newly formed blood vessels in the formerly infarct core and the

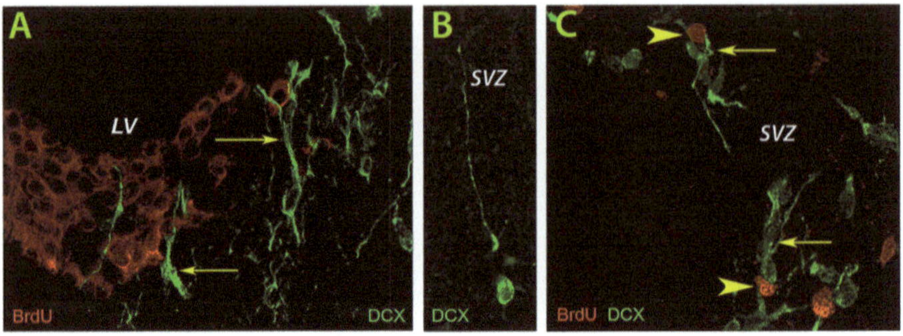

Fig. 8.3 Poststroke neurogenesis and angiogenesis following combined therapy, BM MSC with G-CSF. At 8 weeks poststroke, none of the DCX^+ cells in the SVZ of control animals co-localized with BrdU-labeled nuclei. Instead, the BrdU-positive nuclei were distributed mainly in the "pinwheel" architecture of the ventricular epithelium (**a**). The DCX^+ cells occupied an adjacent, distinct position (**a**, *arrows*). Some of the DCX^+ migrated away from the ventricular wall (**b**). We noted vigorous neurogenesis with many DCX^+ (*arrows*) co-localizing with BrdU nuclei in the G-CSF-treated animals (**c**; *arrowheads*)

region beyond it which we dubbed the "islet of regeneration" [30]. These results strongly suggest that BM MSCs promoted rather neurogenesis in the lesioned area [27]. Indeed, previous studies have shown that delayed intracerebral injection of hMSCs or intravenous administration of MSCs prepared from the bone marrow modified the cerebral microvasculature after transient ischemia [122] and improved the neovascularization along with enhanced functional recovery and cerebral blood flow (CBF) [82].

The current understanding of mechanisms underlying stroke treatment with bone marrow-derived cells is that homing of cells into the infarcted brain may cause trophic support and hereby enhanced poststroke recovery. Transplanted NSCs may promote recovery also without differentiating to neurons and even without long-term survival through several other mechanisms, e.g., neurogenesis [180], modulation of inflammation [76], neuroprotection [8], and stimulation of angiogenesis [27] and brain plasticity [105]. Delayed and persistent functional improvement induced by intracerebral NSC transplantation after stroke, without cell survival and tissue replacement at 6 months after the insult, was recently documented by electrophysiology, fMRI, and behavioral testing [149].

Therapeutic effects mediated by cell transplantation are likely related to the secretion of growth factors and cytokines. Mechanistically, studies done on young subjects using human umbilical cord blood-derived MSCs or bone marrow-derived cell treatment for stroke suggest that homing of cells into the infarcted brain may cause trophic support for host neurons and hereby enhanced poststroke recovery. Indeed, MSCs enriched from adult human umbilical cord and bone marrow have demonstrated therapeutic efficacy for treatment of stroke in a rat model, presumably by increasing the expression of cytokines CXCL2 and CXCL5 and growth factors BDNF, NT-3, FGF9, HBEGF, and VEGF in the ischemic brain that was accompanied by reduction of the infarct volume, increased neurogenesis, and improved neurological function [12, 39, 42, 68, 78]. However, our combination therapy did not further enhance neurogenesis in SVZ of the aged animals.

Reports on the differentiation of transplanted cells yielded conflicting results. Earlier studies have shown that MSCs injected into the lateral ventricle of neonatal mice persistently engraft and migrate throughout the brain and adopt an astrocyte-, oligodendrocyte- [38, 108, 203], or endothelial cell-like phenotypes (ECs) some of them being incorporated into newly formed brain vessels [60, 71, 197]. Occasionally a neuronal phenotype has been reported in vivo [25, 94, 146] and in vitro [47] and by supporting axonal growth and provide tissue protection at sites of injury [4, 109]. Committed neural progenitor cells, NS-MSCs, produced in vitro from rat and human MSCs, differentiated into neuronal cells after transplantation and became immunoreactive to various neurotransmitter-related markers within the host tissue.

Neural stem cells (NSCs) originate from the ectoderm, and mesenchymal stem cells (MSCs) originate from the mesoderm; therefore, it is highly unlikely that stem cells isolated from one niche could form terminally differentiated cells from the other [159]. Nevertheless, recent data showed that undifferentiated BMSCs express not only mesodermal but also endodermal and ectodermal genes. Therefore, BMSCs are capable, in principle, of producing mesodermal derivatives and perhaps also ectodermal derivatives through transformation [103]. Further, during human neural development, the initiation of neuroectoderm begins with *Pax6* expression, followed by *Sox1*, which subsequently regulates the expression of downstream genes, such as *Nestin* [131, 198]. This knowledge led to generation of functional NSCs from hAD-MSCs by activating transcriptional factor for early neural development [59].

To this end BMSCs were transfected with the BDNF gene. The transfected BMSCs displayed a nerve-like phenotype and expressed the neural cell markers nestin and GFAP [3, 59].

Reports on the differentiation of transplanted cells yielded conflicting results. Earlier studies have shown that MSCs injected into the lateral ventricle of neonatal mice persistently engraft and migrate throughout the brain and adopt an astrocyte-, oligodendrocyte- [20, 108, 203], or endothelial cell-like phenotypes (ECs), some of them being incorporated into newly formed brain vessels [40, 60, 71, 197]. Occasionally, a neuronal phenotype has been reported in vivo [25, 94, 146] at the site of injury [4, 40, 109] and in vitro [47]. Committed neural progenitor cells, NS-MSCs, produced in vitro from rat and human MSCs, differentiated into neuronal cells after transplantation and became immunoreactive to various neurotransmitter-related markers within the host tissue.

It has been shown that MSCs are able to release several angiogenic and neurotrophic factors as well as anti-inflammatory molecules [48]. At this regard, it has been shown that when stimulated by inflammatory cytokines, MSCs increased their anti-inflammatory capacity, suggesting that MSCs may even improve their efficacy when localized in an inflammatory microenvironment in vivo [19].

8.1.14 Can Stem Cell Therapy After Stroke Limit Neuroinflammation?

The efficacy of all cell therapies so far is discouragingly low mainly because the time course of interactions between host neuroinflammation, which is considered to be a major obstacle to exogenous-mediated neuronal precursor cells, and stroke-induced neurogenesis or exogenously administered stem cells is virtually unknown.

Cell therapy itself can be used during the first week poststroke to limit neuroinflammation in animal models [8, 14, 52, 147]. Thus, syngeneic systemic delivery of mouse NSCs into mice subjected to stroke gave rise to an anti-inflammatory effect lasting up to 30 days [8].

A recent study emphasized the crucial importance of timing and cell dose for successful poststroke treatment using BM MSC. The study found that transplantation of BMSCs at 3 and 24 h, but not 7 days after focal ischemia, significantly reduced the lesion volume and improved motor deficits. Similarly, transplanted cells in the range 1×10(6) to 10(7), but not at 1×10(4) to 10(5), significantly improved functional outcome after stroke. In addition to inhibiting macrophage/microglia activation in the ischemic brain, BMSC transplantation profoundly reduced infiltration of gamma delta T (γδT) cells, which are detrimental to the ischemic brain, and significantly increased regulatory T cells (Tregs), along with altered Treg-associated cytokines in the ischemic brain [188]. Furthermore, the survival of MSCs was found to be very low (between 0 and 30%) and was highly dependent on the injection route/site (intravenous, intracerebral, or intrathecal) and cell source (auto-, allo-, or xenogeneic) [45]. Immune-related factors that impair neuronal survival and induce neuronal death also inhibit regeneration; therefore, immunomodulation should benefit stem cell therapy. Both in vitro and after transplantation in vivo, NSPCs not only form neural cells for replacement but also exert immunomodulatory and trophic effects, the so-called therapeutic plasticity [27, 92, 115].

Studies done in humans have largely confirmed animal studies. Thus implantation of human fetal NSCs into cortex or striatum after stroke may have immunomodulatory effects on the host tissue. Transplanted cells may act neuroprotectively by suppressing microglia/macrophage activation in the stroke-injured cortex both at 1 and at 6 and 14 weeks [51, 76, 118].

8.1.15 The Role of the Aged Microenvironment for the Combination Therapy

One conclusion from the heterochronic parabiosis studies is that the regenerative capacity of old tissue can be enhanced by the young systemic milieu [41]. However, it has been observed that rejuvenation of old tissue may require the neutralization or removal of inhibitory factors in the microenvironment [35, 36, 81, 194].

The aged brain is particularly refractory to growth phenomena after injuries. We and others have shown that potential mechanisms for self-repair also operate in the post-ischemic aged brain. The major factors involving the loss of regenerative capacity in the aged brain are an age-related decrease in neurogenesis and a loss of environmental hostility disturbing regeneration and migration of neuronal precursors toward the ischemic lesion due to the inflammatory response to stroke that is most evident in aged brains [136]. The current understanding of the involved mechanisms of bone marrow-derived cell treatment for stroke is that homing of cells into the infarcted brain may cause trophic support and hereby enhances poststroke recovery. Structural reorganization is mainly thought to be mediated by trophic factors released by the homed cells as well as neurotrophins [88, 163] and angiogenesis [115, 181, 197]. In addition, G-CSF obviously exhibits autocrine functions with anti-apoptotic and pro-differentiative effects in the brain [163].

The major factors involving the loss of regenerative capacity of the aged brain are an age-related decrease in neurogenesis and the environmental hostility created by the inflammatory response to stroke [139].

The microenvironment within the neurogenic niche allows the production of new neurons and provides permissive cues for their migration along the rostral pathway [160, 201, 202]. Since the neurogenic niche is localized around blood vessels [37, 166], the possibility arises that diminished neurogenesis during aging may be modulated by both CNS-derived cues [98, 123, 151] and cues extrinsic to the CNS delivered by blood.

In animal models, the number of new neurons in the injured striatum, a brain region located in close proximity to the neurogenic SVZ, in aged rodents after stroke was similar to that in young animals [2, 43] despite 50 % decline in neurogenesis in the subventricular zone of elderly rodents compared to young adult animals [57]. However, the proportion of surviving neurons is discouragingly low [6, 102, 132]. Similar findings have been reported in humans [83, 91, 110, 114]. Earlier studies on postmortem human brains provided evidence of cell proliferation and neuroblast formation after stroke even in aged patients, too [83, 110, 119]. The finding that new neurons are continuously added in the adult human striatum [58] along with the presence of an increased number of putative neuroblasts in the human striatum after stroke lends support to this hypothesis [110]. However, whether endogenous neurogenesis contributes to spontaneous recovery after stroke has not yet been established. In addition, age, comorbidities, physical condition of the patient, and severity of disease could substantially influence these steps and, therefore, the outcome of the healing process [135].

The process of cellular senescence can be an important additional contributor to chronic poststroke injuries by creating a "primed" inflammatory environment in the brain [29, 63, 84]. Persistent neuronal death causes a prolonged neuroinflammatory response in the infarcted area of comorbid subjects, too [70]. Previously we have shown that aged brains develop a fulminant inflammatory response to stroke [9]. The early phase of the inflammatory response (the first 10 days after injury) comprises infiltration of PMNs and T cells, and each cell population reaches its maximal abundance at specific time points. The second phase of cellular inflammation initiates

2 weeks after injury, peaks at 2 months, and remains detectable at 6 months post-injury [13]. Unfortunately, this time frame also coincides with the initiation of regenerative processes [28, 136]. Therefore poststroke inflammation may be one of the major factors that limits the efficacy of cell therapy especially with NPCs.

Although mild neuroinflammation can be beneficial for regenerative events aimed at functional restoration after stroke [133], persistent poststroke neuroinflammation results in decreased proliferation of the newly born NSPCs and ineffective integration into the circuitry of the reorganized brain area [106].

Earlier studies have suggested that neuroinflammation alone inhibits neurogenesis and that inflammatory blockade with indomethacin, a common nonsteroidal anti-inflammatory drug, restores neurogenesis after endotoxin-induced inflammation and augments neurogenesis after cranial irradiation [56, 125]. Recent studies have reported that targeting the inflammatory response to ischemic injury limits the expansion of the lesion and increases the survival of neurons after stroke [92, 112, 193]. For example, ibuprofen was found to downregulate the TBI-induced inflammatory response. In addition, migrating neuroblasts from transplanted cells were observed near the contusion and in the ipsilateral hippocampus in ibuprofen-treated animals only, suggesting that the anti-inflammatory treatment had beneficial effects on graft survival and/or differentiation especially in aged subjects [187].

In a recent study conducted on aged animals, the grafted hiPSC suppressed microglia/macrophage activation in the stroke-injured cortex as evidenced by differential morphological changes of these cells in the cell-grafted and vehicle-injected animals [179]. Although it is not clear how microglia/macrophages were affected at earlier time points after stroke in aged animals, it seems possible that the observed immunomodulatory action of the grafts could contribute to both neuroprotective and beneficial plastic responses in the host brain. Consistent with our findings, previous studies have shown that implantation of human fetal NSCs into the cortex or striatum after stroke can suppress the number of microglia/macrophages in the peri-infarcted area both at 1 [76] and at 6 and 14 weeks [118]. Importantly, also syngeneic systemic delivery of mouse NSCs into mice subjected to stroke gave rise to an anti-inflammatory effect lasting up to 30 days [8].

Recent experiments using both bone marrow-derived mesenchymal cells and neural precursor cells for stroke therapy suggest that the aged rat brain may not be refractory to cell survival and remodeling and that it also supports plasticity and remodeling. An open question remains, however, if transplanted cells have any beneficial effect on behavioral recovery. MSCs exhibit intrinsic homing properties to sites of injury, inflammation, and hypoxia [74, 113, 128, 173] that can be used for targeted delivery of therapeutic factors. In a mouse model of glioblastoma, a large number of cells migrated toward the tumor along the corpus callosum [117]. Our study shows that the aged rat brain environment still supports this migratory pathway in the ischemic brain as suggested by the presence of several markers of human MSCs (CD166 and CD105) in the corpus callosum and peri-infarcted area.

The combination therapy between G-CSF and stem cells in the present study did not fulfill all expectations. A potential explanation could be the fact that marrow stromal cells which may be mainly responsible for the supportive/trophic effect cannot be mobilized by G-CSF treatment. Even more important could be the fact that G-CSF-induced stem cell mobilization from the bone marrow occurs typically with a delay of 6–9 days which was probably beyond the therapeutic time window for this approach [73, 107]. Nevertheless, the combination treatments did not show clear benefits for stroke in aged animals suggesting that either (1) G-CSF did not fulfill its chemoattractant role for the administered cells, or (2) one time cell administration is not sufficient to improve tissue recovery after stroke, or (3) the aged rat brain environment was hostile to regenerative events, or (4) the treatment was too short in duration.

8.2 Conclusions

To date, all monotherapeutic attempts to prevent or lessen brain damage following stroke have failed. In view of our findings that stroke impacts a wide range of systems in an age-dependent manner, from CNS physiology to CNS regeneration and plasticity, the failure of therapies aimed at only a single target system is perhaps inevitable.

To conclude, (1) the potential for neurogenesis is also preserved in aged, stroke-injured brains; (2) the environment of the aged brain is not hostile to transplantation of BM MSC; and (3) the extent of recovery is successful in some but not all behavioral tests [8, 28, 104, 179].

There remain significant developmental and translational issues to be resolved in future studies such as: (1) Understanding the differentiation into specific phenotypes. Upon transplantation, the differentiated cells often de-differentiate [85]. (2) Tumorigenesis remains a significant concern [153]. (3) Anti-neuroinflammatory therapies are a potential target to promote regeneration and repair in diverse injury and neurodegenerative conditions by stem cell therapy. (4) Efficacy of cell therapy can be enhanced by physical rehabilitation [55]. We recommend that in a real clinical practice involving older poststroke patients, successful regenerative therapies would have to be carried out for a much longer time. The BM MSC therapy in aged rodents warrants further investigation including repeated administrations of therapeutic cells at several time points after stroke and using various combinations with G-CSF or other relevant growth factors/cytokines.

Finally, a better understanding of potential risks of stem cell therapies in strokes hall makes the translation of cell therapies safer. Likewise, awareness of may help improve their efficacy to achieve therapeutic success [20].

References

1. Acosta SA, Tajiri N, Hoover J, Kaneko Y, Borlongan CV. Intravenous bone marrow stem cell grafts preferentially migrate to spleen and abrogate chronic inflammation in stroke. Stroke. 2015;46(9):2616–27. doi:10.1161/strokeaha.115.009854.

2. Ahlenius H, Visan V, Kokaia M, Lindvall O, Kokaia Z. Neural stem and progenitor cells retain their potential for proliferation and differentiation into functional neurons despite lower number in aged brain. J Neurosci. 2009;29(14):4408–19. doi:10.1523/jneurosci.6003-08.2009.
3. Akiyama H, Chaboissier MC, Martin JF, Schedl A, de Crombrugghe B. The transcription factor Sox9 has essential roles in successive steps of the chondrocyte differentiation pathway and is required for expression of Sox5 and Sox6. Genes Dev. 2002;16(21):2813–28. doi:10.1101/gad.1017802.
4. Ankeny DP, McTigue DM, Jakeman LB. Bone marrow transplants provide tissue protection and directional guidance for axons after contusive spinal cord injury in rats. Exp Neurol. 2004;190(1):17–31. doi:10.1016/j.expneurol.2004.05.045.
5. Appelros P, Nydevik I, Viitanen M. Poor outcome after first-ever stroke: predictors for death, dependency, and recurrent stroke within the first year. Stroke. 2003;34(1):122–6.
6. Arvidsson A, Collin T, Kirik D, Kokaia Z, Lindvall O. Neuronal replacement from endogenous precursors in the adult brain after stroke. Nat Med. 2002;8(9):963–70. doi:10.1038/nm747.
7. Ay H, Koroshetz WJ, Vangel M, Benner T, Melinosky C, Zhu M, Menezes N, Lopez CJ, Sorensen AG. Conversion of ischemic brain tissue into infarction increases with age. Stroke. 2005;36(12):2632–6. doi:10.1161/01.str.0000189991.23918.01.
8. Bacigaluppi M, Pluchino S, Peruzzotti-Jametti L, Kilic E, Kilic U, Salani G, Brambilla E, West MJ, Comi G, Martino G, Hermann DM. Delayed post-ischaemic neuroprotection following systemic neural stem cell transplantation involves multiple mechanisms. Brain. 2009;132(Pt 8):2239–51. doi:10.1093/brain/awp174.
9. Badan I, Buchhold B, Hamm A, Gratz M, Walker LC, Platt D, Kessler C, Popa-Wagner A. Accelerated glial reactivity to stroke in aged rats correlates with reduced functional recovery. J Cereb Blood Flow Metab. 2003;23(7):845–54. doi:10.1097/01.wcb.0000071883.63724.a7.
10. Balseanu AT, Buga AM, Catalin B, Wagner DC, Boltze J, Zagrean AM, Reymann K, Schaebitz W, Popa-Wagner A. Multimodal approaches for regenerative stroke therapies: combination of granulocyte colony-stimulating factor with bone marrow mesenchymal stem cells is not superior to G-CSF alone. Front Aging Neurosci. 2014;6:130. doi:10.3389/fnagi.2014.00130.
11. Baltan S. Excitotoxicity and mitochondrial dysfunction underlie age-dependent ischemic white matter injury. Adv Neurobiol. 2014;11:151–70. doi:10.1007/978-3-319-08894-5_8.
12. Bao X, Wei J, Feng M, Lu S, Li G, Dou W, Ma W, Ma S, An Y, Qin C, Zhao RC, Wang R. Transplantation of human bone marrow-derived mesenchymal stem cells promotes behavioral recovery and endogenous neurogenesis after cerebral ischemia in rats. Brain Res. 2011;1367:103–13. doi:10.1016/j.brainres.2010.10.063.
13. Beck H, Raab S, Copanaki E, Heil M, Scholz A, Shibuya M, Deller T, Machein M, Plate KH. VEGFR-1 signaling regulates the homing of bone marrow-derived cells in a mouse stroke model. J Neuropathol Exp Neurol. 2010;69(2):168–75. doi:10.1097/NEN.0b013e3181c9c05b.
14. Bedi SS, Walker PA, Shah SK, Jimenez F, Thomas CP, Smith P, Hetz RA, Xue H, Pati S, Dash PK, Cox Jr CS. Autologous bone marrow mononuclear cells therapy attenuates activated microglial/macrophage response and improves spatial learning after traumatic brain injury. J Trauma Acute Care Surg. 2013;75(3):410–6. doi:10.1097/TA.0b013e31829617c6.
15. Bejot Y, Catteau A, Caillier M, Rouaud O, Durier J, Marie C, Di Carlo A, Osseby GV, Moreau T, Giroud M. Trends in incidence, risk factors, and survival in symptomatic lacunar stroke in Dijon, France, from 1989 to 2006: a population-based study. Stroke. 2008;39(7):1945–51. doi:10.1161/strokeaha.107.510933.
16. Bergerat A, Decano J, Wu CJ, Choi H, Nesvizhskii AI, Moran AM, Ruiz-Opazo N, Steffen M, Herrera VL. Prestroke proteomic changes in cerebral microvessels in stroke-prone, transgenic[hCETP]-Hyperlipidemic, Dahl salt-sensitive hypertensive rats. Mol Med. 2011;17(7–8):588–98. doi:10.2119/molmed.2010.00228.

17. Bergmann O, Liebl J, Bernard S, Alkass K, Yeung MS, Steier P, Kutschera W, Johnson L, Landen M, Druid H, Spalding KL, Frisen J. The age of olfactory bulb neurons in humans. Neuron. 2012;74(4):634–9. doi:10.1016/j.neuron.2012.03.030.
18. Bingham D, Martin SJ, Macrae IM, Carswell HV. Watermaze performance after middle cerebral artery occlusion in the rat: the role of sensorimotor versus memory impairments. J Cereb Blood Flow Metab. 2012;32(6):989–99. doi:10.1038/jcbfm.2012.16.
19. Blasi A, Martino C, Balducci L, Saldarelli M, Soleti A, Navone SE, Canzi L, Cristini S, Invernici G, Parati EA, Alessandri G. Dermal fibroblasts display similar phenotypic and differentiation capacity to fat-derived mesenchymal stem cells, but differ in anti-inflammatory and angiogenic potential. Vasc Cell. 2011;3(1):5. doi:10.1186/2045-824x-3-5.
20. Boltze J, Arnold A, Walczak P, Jolkkonen J, Cui L, Wagner DC. The dark side of the force – constraints and complications of cell therapies for stroke. Front Neurol. 2015;6:155. doi:10.3389/fneur.2015.00155.
21. Boltze J, Nitzsche B, Geiger KD, Schoon HA. Histopathological investigation of different MCAO modalities and impact of autologous bone marrow mononuclear cell administration in an ovine stroke model. Transl Stroke Res. 2011;2(3):279–93. doi:10.1007/s12975-011-0101-5.
22. Bonita R, Beaglehole R. Recovery of motor function after stroke. Stroke. 1988;19(12):1497–500.
23. Borlongan CV, Glover LE, Sanberg PR, Hess DC. Permeating the blood brain barrier and abrogating the inflammation in stroke: implications for stroke therapy. Curr Pharm Des. 2012;18(25):3670–6.
24. Bratane BT, Bouley J, Schneider A, Bastan B, Henninger N, Fisher M. Granulocyte-colony stimulating factor delays PWI/DWI mismatch evolution and reduces final infarct volume in permanent-suture and embolic focal cerebral ischemia models in the rat. Stroke. 2009;40(9):3102–6. doi:10.1161/strokeaha.109.553958.
25. Brazelton TR, Rossi FM, Keshet GI, Blau HM. From marrow to brain: expression of neuronal phenotypes in adult mice. Science. 2000;290(5497):1775–9.
26. Brenneman M, Sharma S, Harting M, Strong R, Cox Jr CS, Aronowski J, Grotta JC, Savitz SI. Autologous bone marrow mononuclear cells enhance recovery after acute ischemic stroke in young and middle-aged rats. J Cereb Blood Flow Metab. 2010;30(1):140–9. doi:10.1038/jcbfm.2009.198.
27. Bronckaers A, Hilkens P, Martens W, Gervois P, Ratajczak J, Struys T, Lambrichts I. Mesenchymal stem/stromal cells as a pharmacological and therapeutic approach to accelerate angiogenesis. Pharmacol Ther. 2014;143(2):181–96. doi:10.1016/j.pharmthera.2014.02.013.
28. Buchhold B, Mogoanta L, Suofu Y, Hamm A, Walker L, Kessler C, Popa-Wagner A. Environmental enrichment improves functional and neuropathological indices following stroke in young and aged rats. Restor Neurol Neurosci. 2007;25(5–6):467–84.
29. Buga AM, Di Napoli M, Popa-Wagner A. Preclinical models of stroke in aged animals with or without comorbidities: role of neuroinflammation. Biogerontology. 2013;14(6):651–62. doi:10.1007/s10522-013-9465-0.
30. Buga AM, Margaritescu C, Scholz CJ, Radu E, Zelenak C, Popa-Wagner A. Transcriptomics of post-stroke angiogenesis in the aged brain. Front Aging Neurosci. 2014;6:44. doi:10.3389/fnagi.2014.00044.
31. Burn J, Dennis M, Bamford J, Sandercock P, Wade D, Warlow C. Long-term risk of recurrent stroke after a first-ever stroke. The Oxfordshire Community Stroke Project. Stroke. 1994;25(2):333–7.
32. Cai D, Liu T. Inflammatory cause of metabolic syndrome via brain stress and NF-kappaB. Aging (Albany NY). 2012;4(2):98–115.
33. Cai Q, Chen Z, Song P, Wu L, Wang L, Deng G, Liu B, Chen Q. Co-transplantation of hippocampal neural stem cells and astrocytes and microvascular endothelial cells improve the memory in ischemic stroke rat. Int J Clin Exp Med. 2015;8(8):13109–17.

34. Capilla-Gonzalez V, Herranz-Perez V, Garcia-Verdugo JM. The aged brain: genesis and fate of residual progenitor cells in the subventricular zone. Front Cell Neurosci. 2015;9:365. doi:10.3389/fncel.2015.00365.
35. Carlson ME, Conboy MJ, Hsu M, Barchas L, Jeong J, Agrawal A, Mikels AJ, Agrawal S, Schaffer DV, Conboy IM. Relative roles of TGF-beta1 and Wnt in the systemic regulation and aging of satellite cell responses. Aging Cell. 2009;8(6):676–89. doi:10.1111/j.1474-9726.2009.00517.x.
36. Carlson ME, Suetta C, Conboy MJ, Aagaard P, Mackey A, Kjaer M, Conboy I. Molecular aging and rejuvenation of human muscle stem cells. EMBO Mol Med. 2009;1(8–9):381–91. doi:10.1002/emmm.200900045.
37. Carpentier PA, Palmer TD. Immune influence on adult neural stem cell regulation and function. Neuron. 2009;64(1):79–92. doi:10.1016/j.neuron.2009.08.038.
38. Chamberlain G, Fox J, Ashton B, Middleton J. Concise review: mesenchymal stem cells: their phenotype, differentiation capacity, immunological features, and potential for homing. Stem Cells. 2007;25(11):2739–49. doi:10.1634/stemcells.2007-0197.
39. Chen J, Zhang ZG, Li Y, Wang L, Xu YX, Gautam SC, Lu M, Zhu Z, Chopp M. Intravenous administration of human bone marrow stromal cells induces angiogenesis in the ischemic boundary zone after stroke in rats. Circ Res. 2003;92(6):692–9. doi:10.1161/01.res.0000063425.51108.8d.
40. Chen Y, Ai Y, Slevin JR, Maley BE, Gash DM. Progenitor proliferation in the adult hippocampus and substantia nigra induced by glial cell line-derived neurotrophic factor. Exp Neurol. 2005;196(1):87–95. doi:10.1016/j.expneurol.2005.07.010.
41. Conboy IM, Conboy MJ, Rebo J. Systemic problems: a perspective on stem cell aging and rejuvenation. Aging (Albany NY). 2015;7(10):754–65.
42. Crigler L, Robey RC, Asawachaicharn A, Gaupp D, Phinney DG. Human mesenchymal stem cell subpopulations express a variety of neuro-regulatory molecules and promote neuronal cell survival and neuritogenesis. Exp Neurol. 2006;198(1):54–64. doi:10.1016/j.expneurol.2005.10.029.
43. Darsalia V, Heldmann U, Lindvall O, Kokaia Z. Stroke-induced neurogenesis in aged brain. Stroke. 2005;36(8):1790–5. doi:10.1161/01.STR.0000173151.36031.be.
44. De Coppi P, Bartsch Jr G, Siddiqui MM, Xu T, Santos CC, Perin L, Mostoslavsky G, Serre AC, Snyder EY, Yoo JJ, Furth ME, Soker S, Atala A. Isolation of amniotic stem cell lines with potential for therapy. Nat Biotechnol. 2007;25(1):100–6. doi:10.1038/nbt1274.
45. De Vocht N, Lin D, Praet J, Hoornaert C, Reekmans K, Le Blon D, Daans J, Pauwels P, Goossens H, Hens N, Berneman Z, Van der Linden A, Ponsaerts P. Quantitative and phenotypic analysis of mesenchymal stromal cell graft survival and recognition by microglia and astrocytes in mouse brain. Immunobiology. 2013;218(5):696–705. doi:10.1016/j.imbio.2012.08.266.
46. dela Pena IC, Yoo A, Tajiri N, Acosta SA, Ji X, Kaneko Y, Borlongan CV. Granulocyte colony-stimulating factor attenuates delayed tPA-induced hemorrhagic transformation in ischemic stroke rats by enhancing angiogenesis and vasculogenesis. J Cereb Blood Flow Metab. 2015;35(2):338–46. doi:10.1038/jcbfm.2014.208.
47. Deng J, Petersen BE, Steindler DA, Jorgensen ML, Laywell ED. Mesenchymal stem cells spontaneously express neural proteins in culture and are neurogenic after transplantation. Stem Cells. 2006;24(4):1054–64. doi:10.1634/stemcells.2005-0370.
48. Dharmasaroja P. Bone marrow-derived mesenchymal stem cells for the treatment of ischemic stroke. J Clin Neurosci. 2009;16(1):12–20. doi:10.1016/j.jocn.2008.05.006.
49. Dimmeler S, Burchfield J, Zeiher AM. Cell-based therapy of myocardial infarction. Arterioscler Thromb Vasc Biol. 2008;28(2):208–16. doi:10.1161/atvbaha.107.155317.
50. DiNapoli VA, Huber JD, Houser K, Li X, Rosen CL. Early disruptions of the blood-brain barrier may contribute to exacerbated neuronal damage and prolonged functional recovery following stroke in aged rats. Neurobiol Aging. 2008;29(5):753–64. doi:10.1016/j.neurobiolaging.2006.12.007.

51. Donnan GA, Davis SM. Breaking the 3 h barrier for treatment of acute ischaemic stroke. Lancet Neurol. 2008;7(11):981–2. doi:10.1016/s1474-4422(08)70230-8.
52. Dooley D, Vidal P, Hendrix S. Immunopharmacological intervention for successful neural stem cell therapy: new perspectives in CNS neurogenesis and repair. Pharmacol Ther. 2014;141(1):21–31. doi:10.1016/j.pharmthera.2013.08.001.
53. dos Santos AP, Milatovic D, Au C, Yin Z, Batoreu MC, Aschner M. Rat brain endothelial cells are a target of manganese toxicity. Brain Res. 2010;1326:152–61. doi:10.1016/j.brainres.2010.02.016.
54. Duncan K, Gonzales-Portillo GS, Acosta SA, Kaneko Y, Borlongan CV, Tajiri N. Stem cell-paved biobridges facilitate stem transplant and host brain cell interactions for stroke therapy. Brain Res. 2015;1623:160–5. doi:10.1016/j.brainres.2015.03.007.
55. Dunnett SB. Neural tissue transplantation, repair, and rehabilitation. Handb Clin Neurol. 2013;110:43–59. doi:10.1016/b978-0-444-52901-5.00004-6.
56. Ekdahl CT, Claasen JH, Bonde S, Kokaia Z, Lindvall O. Inflammation is detrimental for neurogenesis in adult brain. Proc Natl Acad Sci U S A. 2003;100(23):13632–7. doi:10.1073/pnas.2234031100.
57. Enwere E, Shingo T, Gregg C, Fujikawa H, Ohta S, Weiss S. Aging results in reduced epidermal growth factor receptor signaling, diminished olfactory neurogenesis, and deficits in fine olfactory discrimination. J Neurosci. 2004;24(38):8354–65. doi:10.1523/jneurosci.2751-04.2004.
58. Ernst A, Alkass K, Bernard S, Salehpour M, Perl S, Tisdale J, Possnert G, Druid H, Frisen J. Neurogenesis in the striatum of the adult human brain. Cell. 2014;156(5):1072–83. doi:10.1016/j.cell.2014.01.044.
59. Feng N, Han Q, Li J, Wang S, Li H, Yao X, Zhao RC. Generation of highly purified neural stem cells from human adipose-derived mesenchymal stem cells by Sox1 activation. Stem Cells Dev. 2014;23(5):515–29. doi:10.1089/scd.2013.0263.
60. Foteinos G, Hu Y, Xiao Q, Metzler B, Xu Q. Rapid endothelial turnover in atherosclerosis-prone areas coincides with stem cell repair in apolipoprotein E-deficient mice. Circulation. 2008;117(14):1856–63. doi:10.1161/circulationaha.107.746008.
61. Freret T, Gaudreau P, Schumann-Bard P, Billard JM, Popa-Wagner A. Mechanisms underlying the neuroprotective effect of brain reserve against late life depression. J Neural Transm (Vienna). 2015;122 Suppl 1:S55–61. doi:10.1007/s00702-013-1154-2.
62. Gajera CR, Emich H, Lioubinski O, Christ A, Beckervordersandforth-Bonk R, Yoshikawa K, Bachmann S, Christensen EI, Gotz M, Kempermann G, Peterson AS, Willnow TE, Hammes A. LRP2 in ependymal cells regulates BMP signaling in the adult neurogenic niche. J Cell Sci. 2010;123(Pt 11):1922–30. doi:10.1242/jcs.065912.
63. Gauberti M, Montagne A, Marcos-Contreras OA, Le Behot A, Maubert E, Vivien D. Ultra-sensitive molecular MRI of vascular cell adhesion molecule-1 reveals a dynamic inflammatory penumbra after strokes. Stroke. 2013;44(7):1988–96. doi:10.1161/strokeaha.111.000544.
64. Gokcay F, Arsava EM, Baykaner T, Vangel M, Garg P, Wu O, Singhal AB, Furie KL, Sorensen AG, Ay H. Age-dependent susceptibility to infarct growth in women. Stroke. 2011;42(4):947–51. doi:10.1161/strokeaha.110.603902.
65. Goldstein LB, Samsa GP, Matchar DB, Horner RD. Charlson Index comorbidity adjustment for ischemic stroke outcome studies. Stroke. 2004;35(8):1941–5. doi:10.1161/01.STR.0000135225.80898.1c.
66. Guzman R, Choi R, Gera A, De Los Angeles A, Andres RH, Steinberg GK. Intravascular cell replacement therapy for stroke. Neurosurg Focus. 2008;24(3–4):E15. doi:10.3171/foc/2008/24/3-4/e14.
67. Hallett M. Plasticity of the human motor cortex and recovery from stroke. Brain Res Brain Res Rev. 2001;36(2–3):169–74.
68. Hayase M, Kitada M, Wakao S, Itokazu Y, Nozaki K, Hashimoto N, Takagi Y, Dezawa M. Committed neural progenitor cells derived from genetically modified bone marrow stromal cells ameliorate deficits in a rat model of stroke. J Cereb Blood Flow Metab. 2009;29(8):1409–20. doi:10.1038/jcbfm.2009.62.

69. Hermann DM, Chopp M. Promoting brain remodelling and plasticity for stroke recovery: therapeutic promise and potential pitfalls of clinical translation. Lancet Neurol. 2012;11(4):369–80. doi:10.1016/s1474-4422(12)70039-x.
70. Herz J, Hagen SI, Bergmuller E, Sabellek P, Gothert JR, Buer J, Hansen W, Hermann DM, Doeppner TR. Exacerbation of ischemic brain injury in hypercholesterolemic mice is associated with pronounced changes in peripheral and cerebral immune responses. Neurobiol Dis. 2014;62:456–68. doi:10.1016/j.nbd.2013.10.022.
71. Hess DC, Hill WD, Martin-Studdard A, Carothers J, Brailer J, Carroll J. Blood into brain after stroke. Trends Mol Med. 2002;8(9):452–3.
72. Hicks AU, Hewlett K, Windle V, Chernenko G, Ploughman M, Jolkkonen J, Weiss S, Corbett D. Enriched environment enhances transplanted subventricular zone stem cell migration and functional recovery after stroke. Neuroscience. 2007;146(1):31–40. doi:10.1016/j.neuroscience.2007.01.020.
73. Hill QA, Buxton D, Pearce R, Gesinde MO, Smith GM, Cook G. An analysis of the optimal timing of peripheral blood stem cell harvesting following priming with cyclophosphamide and G-CSF. Bone Marrow Transplant. 2007;40(10):925–30. doi:10.1038/sj.bmt.1705847.
74. Hofstetter CP, Schwarz EJ, Hess D, Widenfalk J, El Manira A, Prockop DJ, Olson L. Marrow stromal cells form guiding strands in the injured spinal cord and promote recovery. Proc Natl Acad Sci U S A. 2002;99(4):2199–204. doi:10.1073/pnas.042678299.
75. Honmou O, Onodera R, Sasaki M, Waxman SG, Kocsis JD. Mesenchymal stem cells: therapeutic outlook for stroke. Trends Mol Med. 2012;18(5):292–7. doi:10.1016/j.molmed.2012.02.003.
76. Horie N, Pereira MP, Niizuma K, Sun G, Keren-Gill H, Encarnacion A, Shamloo M, Hamilton SA, Jiang K, Huhn S, Palmer TD, Bliss TM, Steinberg GK. Transplanted stem cell-secreted vascular endothelial growth factor effects poststroke recovery, inflammation, and vascular repair. Stem Cells. 2011;29(2):274–85. doi:10.1002/stem.584.
77. Hosseini SM, Farahmandnia M, Kazemi S, Shakibajahromi B, Sarvestani FS, Khodabande Z. A novel cell therapy method for recovering after brain stroke in rats. Int J Stem Cells. 2015;8(2):191–9. 10.15283/ijsc.2015.8.2.191.
78. Hsieh JY, Wang HW, Chang SJ, Liao KH, Lee IH, Lin WS, Wu CH, Lin WY, Cheng SM. Mesenchymal stem cells from human umbilical cord express preferentially secreted factors related to neuroprotection, neurogenesis, and angiogenesis. PLoS One. 2013;8(8):e72604. doi:10.1371/journal.pone.0072604.
79. Ishizaka S, Horie N, Satoh K, Fukuda Y, Nishida N, Nagata I. Intra-arterial cell transplantation provides timing-dependent cell distribution and functional recovery after stroke. Stroke. 2013;44(3):720–6. doi:10.1161/strokeaha.112.677328.
80. Iso H, Jacobs Jr DR, Wentworth D, Neaton JD, Cohen JD. Serum cholesterol levels and six-year mortality from stroke in 350,977 men screened for the multiple risk factor intervention trial. N Engl J Med. 1989;320(14):904–10. doi:10.1056/nejm198904063201405.
81. Jeong J, Conboy MJ, Conboy IM. Pharmacological inhibition of myostatin/TGF-beta receptor/pSmad3 signaling rescues muscle regenerative responses in mouse model of type 1 diabetes. Acta Pharmacol Sin. 2013;34(8):1052–60. doi:10.1038/aps.2013.67.
82. Jiang Q, Thiffault C, Kramer BC, Ding GL, Zhang L, Nejad-Davarani SP, Li L, Arbab AS, Lu M, Navia B, Victor SJ, Hong K, Li QJ, Wang SY, Li Y, Chopp M. MRI detects brain reorganization after human umbilical tissue-derived cells (hUTC) treatment of stroke in rat. PLoS One. 2012;7(8):e42845. doi:10.1371/journal.pone.0042845.
83. Jin K, Wang X, Xie L, Mao XO, Zhu W, Wang Y, Shen J, Mao Y, Banwait S, Greenberg DA. Evidence for stroke-induced neurogenesis in the human brain. Proc Natl Acad Sci U S A. 2006;103(35):13198–202. doi:10.1073/pnas.0603512103.
84. Joseph C, Buga AM, Vintilescu R, Balseanu AT, Moldovan M, Junker H, Walker L, Lotze M, Popa-Wagner A. Prolonged gaseous hypothermia prevents the upregulation of phagocytosis-specific protein annexin 1 and causes low-amplitude EEG activity in the aged rat brain after

cerebral ischemia. J Cereb Blood Flow Metab. 2012;32(8):1632–42. doi:10.1038/jcbfm.2012.65.
85. Kalladka D, Muir KW. Brain repair: cell therapy in stroke. Stem Cells Cloning. 2014;7:31–44. doi:10.2147/sccaa.s38003.
86. Kamiya N, Ueda M, Igarashi H, Nishiyama Y, Suda S, Inaba T, Katayama Y. Intra-arterial transplantation of bone marrow mononuclear cells immediately after reperfusion decreases brain injury after focal ischemia in rats. Life Sci. 2008;83(11–12):433–7. doi:10.1016/j.lfs.2008.07.018.
87. Katsman D, Zheng J, Spinelli K, Carmichael ST. Tissue microenvironments within functional cortical subdivisions adjacent to focal stroke. J Cereb Blood Flow Metab. 2003;23(9):997–1009. doi:10.1097/01.wcb.0000084252.20114.be.
88. Kawada H. Cell therapy using bone marrow-derived stem cells; the possibility of its application in regenerative medicine. Rinsho Ketsueki. 2006;47(6):431–8.
89. Keimpema E, Fokkens MR, Nagy Z, Agoston V, Luiten PG, Nyakas C, Boddeke HW, Copray JC. Early transient presence of implanted bone marrow stem cells reduces lesion size after cerebral ischaemia in adult rats. Neuropathol Appl Neurobiol. 2009;35(1):89–102. doi:10.1111/j.1365-2990.2008.00961.x.
90. Kocsis JD, Honmou O. Bone marrow stem cells in experimental stroke. Prog Brain Res. 2012;201:79–98. doi:10.1016/b978-0-444-59544-7.00005-6.
91. Kojima T, Hirota Y, Ema M, Takahashi S, Miyoshi I, Okano H, Sawamoto K. Subventricular zone-derived neural progenitor cells migrate along a blood vessel scaffold toward the post-stroke striatum. Stem Cells. 2010;28(3):545–54. doi:10.1002/stem.306.
92. Kokaia Z, Lindvall O. Stem cell repair of striatal ischemia. Prog Brain Res. 2012;201:35–53. doi:10.1016/b978-0-444-59544-7.00003-2.
93. Komatsu K, Honmou O, Suzuki J, Houkin K, Hamada H, Kocsis JD. Therapeutic time window of mesenchymal stem cells derived from bone marrow after cerebral ischemia. Brain Res. 2010;1334:84–92. doi:10.1016/j.brainres.2010.04.006.
94. Kopen GC, Prockop DJ, Phinney DG. Marrow stromal cells migrate throughout forebrain and cerebellum, and they differentiate into astrocytes after injection into neonatal mouse brains. Proc Natl Acad Sci U S A. 1999;96(19):10711–6.
95. Kurozumi K, Nakamura K, Tamiya T, Kawano Y, Kobune M, Hirai S, Uchida H, Sasaki K, Ito Y, Kato K, Honmou O, Houkin K, Date I, Hamada H. BDNF gene-modified mesenchymal stem cells promote functional recovery and reduce infarct size in the rat middle cerebral artery occlusion model. Mol Ther. 2004;9(2):189–97. doi:10.1016/j.ymthe.2003.10.012.
96. Lee JS, Hong JM, Moon GJ, Lee PH, Ahn YH, Bang OY. A long-term follow-up study of intravenous autologous mesenchymal stem cell transplantation in patients with ischemic stroke. Stem Cells. 2010;28(6):1099–106. doi:10.1002/stem.430.
97. Li J, Tang Y, Wang Y, Tang R, Jiang W, Yang GY, Gao WQ. Neurovascular recovery via co-transplanted neural and vascular progenitors leads to improved functional restoration after ischemic stroke in rats. Stem Cell Rep. 2014;3(1):101–14. doi:10.1016/j.stemcr.2014.05.012.
98. Lie DC, Colamarino SA, Song HJ, Desire L, Mira H, Consiglio A, Lein ES, Jessberger S, Lansford H, Dearie AR, Gage FH. Wnt signalling regulates adult hippocampal neurogenesis. Nature. 2005;437(7063):1370–5. doi:10.1038/nature04108.
99. Liebner S, Corada M, Bangsow T, Babbage J, Taddei A, Czupalla CJ, Reis M, Felici A, Wolburg H, Fruttiger M, Taketo MM, von Melchner H, Plate KH, Gerhardt H, Dejana E. Wnt/beta-catenin signaling controls development of the blood-brain barrier. J Cell Biol. 2008;183(3):409–17. doi:10.1083/jcb.200806024.
100. Liepert J, Hamzei F, Weiller C. Lesion-induced and training-induced brain reorganization. Restor Neurol Neurosci. 2004;22(3–5):269–77.
101. Lim JY, Jeong CH, Jun JA, Kim SM, Ryu CH, Hou Y, Oh W, Chang JW, Jeun SS. Therapeutic effects of human umbilical cord blood-derived mesenchymal stem cells after intrathecal administration by lumbar puncture in a rat model of cerebral ischemia. Stem Cell Res Ther. 2011;2(5):38. doi:10.1186/scrt79.

102. Lindvall O, Kokaia Z. Neurogenesis following stroke affecting the adult brain. Cold Spring Harb Perspect Biol 7 (11). 2015. doi:10.1101/cshperspect.a019034
103. Liu Q, Cheng G, Wang Z, Zhan S, Xiong B, Zhao X. Bone marrow-derived mesenchymal stem cells differentiate into nerve-like cells in vitro after transfection with brain-derived neurotrophic factor gene. In Vitro Cell Dev Biol Anim. 2015;51(3):319–27. doi:10.1007/s11626-015-9875-1.
104. Liu X, Ye R, Yan T, Yu SP, Wei L, Xu G, Fan X, Jiang Y, Stetler RA, Liu G, Chen J. Cell based therapies for ischemic stroke: from basic science to bedside. Prog Neurobiol. 2014;115:92–115. doi:10.1016/j.pneurobio.2013.11.007.
105. Liu Z, Li Y, Zhang X, Savant-Bhonsale S, Chopp M. Contralesional axonal remodeling of the corticospinal system in adult rats after stroke and bone marrow stromal cell treatment. Stroke. 2008;39(9):2571–7. doi:10.1161/strokeaha.107.511659.
106. London A, Cohen M, Schwartz M. Microglia and monocyte-derived macrophages: functionally distinct populations that act in concert in CNS plasticity and repair. Front Cell Neurosci. 2013;7:34. doi:10.3389/fncel.2013.00034.
107. Lowenthal RM, Ragg SJ, Anderson J, Nicholson L, Harrup RA, Tuck D. A randomized controlled clinical trial to determine the optimum duration of G-CSF priming prior to BM stem cell harvesting. Cytotherapy. 2007;9(2):158–64. doi:10.1080/14653240601182820.
108. Lu L, Chen X, Zhang CW, Yang WL, Wu YJ, Sun L, Bai LM, Gu XS, Ahmed S, Dawe GS, Xiao ZC. Morphological and functional characterization of predifferentiation of myelinating glia-like cells from human bone marrow stromal cells through activation of F3/Notch signaling in mouse retina. Stem Cells. 2008;26(2):580–90. doi:10.1634/stemcells.2007-0106.
109. Lu P, Jones LL, Tuszynski MH. BDNF-expressing marrow stromal cells support extensive axonal growth at sites of spinal cord injury. Exp Neurol. 2005;191(2):344–60. doi:10.1016/j.expneurol.2004.09.018.
110. Macas J, Nern C, Plate KH, Momma S. Increased generation of neuronal progenitors after ischemic injury in the aged adult human forebrain. J Neurosci. 2006;26(50):13114–9. doi:10.1523/jneurosci.4667-06.2006.
111. Macleod MR, van der Worp HB, Sena ES, Howells DW, Dirnagl U, Donnan GA. Evidence for the efficacy of NXY-059 in experimental focal cerebral ischaemia is confounded by study quality. Stroke. 2008;39(10):2824–9.
112. Macrez R, Ali C, Toutirais O, Le Mauff B, Defer G, Dirnagl U, Vivien D. Stroke and the immune system: from pathophysiology to new therapeutic strategies. Lancet Neurol. 2011;10(5):471–80. doi:10.1016/s1474-4422(11)70066-7.
113. Mahmood A, Lu D, Lu M, Chopp M. Treatment of traumatic brain injury in adult rats with intravenous administration of human bone marrow stromal cells. Neurosurgery. 2003;53(3):697–702. discussion 702–693.
114. Marti-Fabregas J, Romaguera-Ros M, Gomez-Pinedo U, Martinez-Ramirez S, Jimenez-Xarrie E, Marin R, Marti-Vilalta JL, Garcia-Verdugo JM. Proliferation in the human ipsilateral subventricular zone after ischemic stroke. Neurology. 2010;74(5):357–65. doi:10.1212/WNL.0b013e3181cbccec.
115. Martino G, Pluchino S. The therapeutic potential of neural stem cells. Nat Rev Neurosci. 2006;7(5):395–406. doi:10.1038/nrn1908.
116. McCabe C, Gallagher L, Gsell W, Graham D, Dominiczak AF, Macrae IM. Differences in the evolution of the ischemic penumbra in stroke-prone spontaneously hypertensive and Wistar-Kyoto rats. Stroke. 2009;40(12):3864–8. doi:10.1161/strokeaha.109.559021.
117. Menon LG, Kelly K, Yang HW, Kim SK, Black PM, Carroll RS. Human bone marrow-derived mesenchymal stromal cells expressing S-TRAIL as a cellular delivery vehicle for human glioma therapy. Stem Cells. 2009;27(9):2320–30. doi:10.1002/stem.136.
118. Mine Y, Tatarishvili J, Oki K, Monni E, Kokaia Z, Lindvall O. Grafted human neural stem cells enhance several steps of endogenous neurogenesis and improve behavioral recovery after middle cerebral artery occlusion in rats. Neurobiol Dis. 2013;52:191–203. doi:10.1016/j.nbd.2012.12.006.

119. Minger SL, Ekonomou A, Carta EM, Chinoy A, Perry RH, Ballard CG. Endogenous neurogenesis in the human brain following cerebral infarction. Regen Med. 2007;2(1):69–74. doi:10.2217/17460751.2.1.69.
120. Minnerup J, Heidrich J, Wellmann J, Rogalewski A, Schneider A, Schabitz WR. Meta-analysis of the efficacy of granulocyte-colony stimulating factor in animal models of focal cerebral ischemia. Stroke. 2008;39(6):1855–61. doi:10.1161/strokeaha.107.506816.
121. Mitkari B, Kerkela E, Nystedt J, Korhonen M, Mikkonen V, Huhtala T, Jolkkonen J. Intra-arterial infusion of human bone marrow-derived mesenchymal stem cells results in transient localization in the brain after cerebral ischemia in rats. Exp Neurol. 2013;239:158–62. doi:10.1016/j.expneurol.2012.09.018.
122. Moisan A, Pannetier N, Grillon E, Richard MJ, de Fraipont F, Remy C, Barbier EL, Detante O. Intracerebral injection of human mesenchymal stem cells impacts cerebral microvasculature after experimental stroke: MRI study. NMR Biomed. 2012;25(12):1340–8. doi:10.1002/nbm.2806.
123. Molofsky AV, Slutsky SG, Joseph NM, He S, Pardal R, Krishnamurthy J, Sharpless NE, Morrison SJ. Increasing p16INK4a expression decreases forebrain progenitors and neurogenesis during ageing. Nature. 2006;443(7110):448–52. doi:10.1038/nature05091.
124. Moniche F, Gonzalez A, Gonzalez-Marcos JR, Carmona M, Pinero P, Espigado I, Garcia-Solis D, Cayuela A, Montaner J, Boada C, Rosell A, Jimenez MD, Mayol A, Gil-Peralta A. Intra-arterial bone marrow mononuclear cells in ischemic stroke: a pilot clinical trial. Stroke. 2012;43(8):2242–4. doi:10.1161/strokeaha.112.659409.
125. Monje ML, Toda H, Palmer TD. Inflammatory blockade restores adult hippocampal neurogenesis. Science. 2003;302(5651):1760–5. doi:10.1126/science.1088417.
126. Nakagomi N, Nakagomi T, Kubo S, Nakano-Doi A, Saino O, Takata M, Yoshikawa H, Stern DM, Matsuyama T, Taguchi A. Endothelial cells support survival, proliferation, and neuronal differentiation of transplanted adult ischemia-induced neural stem/progenitor cells after cerebral infarction. Stem Cells. 2009;27(9):2185–95. doi:10.1002/stem.161.
127. Omori Y, Honmou O, Harada K, Suzuki J, Houkin K, Kocsis JD. Optimization of a therapeutic protocol for intravenous injection of human mesenchymal stem cells after cerebral ischemia in adult rats. Brain Res. 2008;1236:30–8. doi:10.1016/j.brainres.2008.07.116.
128. Orlic D, Kajstura J, Chimenti S, Limana F, Jakoniuk I, Quaini F, Nadal-Ginard B, Bodine DM, Leri A, Anversa P. Mobilized bone marrow cells repair the infarcted heart, improving function and survival. Proc Natl Acad Sci U S A. 2001;98(18):10344–9. doi:10.1073/pnas.181177898.
129. Otero-Ortega L, Gutierrez-Fernandez M, Ramos-Cejudo J, Rodriguez-Frutos B, Fuentes B, Sobrino T, Hernanz TN, Campos F, Lopez JA, Cerdan S, Vazquez J, Diez-Tejedor E. White matter injury restoration after stem cell administration in subcortical ischemic stroke. Stem Cell Res Ther. 2015;6:121. doi:10.1186/s13287-015-0111-4.
130. Ott HC, McCue J, Taylor DA. Cell-based cardiovascular repair – the hurdles and the opportunities. Basic Res Cardiol. 2005;100(6):504–17. doi:10.1007/s00395-004-0558-z.
131. Pankratz MT, Li XJ, Lavaute TM, Lyons EA, Chen X, Zhang SC. Directed neural differentiation of human embryonic stem cells via an obligated primitive anterior stage. Stem Cells. 2007;25(6):1511–20. doi:10.1634/stemcells.2006-0707.
132. Parent JM, Vexler ZS, Gong C, Derugin N, Ferriero DM. Rat forebrain neurogenesis and striatal neuron replacement after focal stroke. Ann Neurol. 2002;52(6):802–13. doi:10.1002/ana.10393.
133. Perry VH. Contribution of systemic inflammation to chronic neurodegeneration. Acta Neuropathol. 2010;120(3):277–86. doi:10.1007/s00401-010-0722-x.
134. Polentes J, Jendelova P, Cailleret M, Braun H, Romanyuk N, Tropel P, Brenot M, Itier V, Seminatore C, Baldauf K, Turnovcova K, Jirak D, Teletin M, Come J, Tournois J, Reymann K, Sykova E, Viville S, Onteniente B. Human induced pluripotent stem cells improve stroke outcome and reduce secondary degeneration in the recipient brain. Cell Transplant. 2012;21(12):2587–602. doi:10.3727/096368912x653228.

135. Popa-Wagner A, Buga AM, Doeppner TR, Hermann DM. Stem cell therapies in preclinical models of stroke associated with aging. Front Cell Neurosci. 2014;8:347. doi:10.3389/fncel.2014.00347.
136. Popa-Wagner A, Buga AM, Kokaia Z. Perturbed cellular response to brain injury during aging. Ageing Res Rev. 2011;10(1):71–9. doi:10.1016/j.arr.2009.10.008.
137. Popa-Wagner A, Carmichael ST, Kokaia Z, Kessler C, Walker LC. The response of the aged brain to stroke: too much, too soon? Curr Neurovasc Res. 2007;4(3):216–27.
138. Popa-Wagner A, Dinca I, Yalikun S, Walker L, Kroemer H, Kessler C. Accelerated delimitation of the infarct zone by capillary-derived nestin-positive cells in aged rats. Curr Neurovasc Res. 2006;3(1):3–13.
139. Popa-Wagner A, Mitran S, Sivanesan S, Chang E, Buga AM. ROS and brain diseases: the good, the bad, and the ugly. Oxidative Med Cell Longev. 2013;2013:963520. doi:10.1155/2013/963520.
140. Popa-Wagner A, Schroder E, Walker LC, Kessler C. beta-Amyloid precursor protein and ss-amyloid peptide immunoreactivity in the rat brain after middle cerebral artery occlusion: effect of age. Stroke. 1998;29(10):2196–202.
141. Popa-Wagner A, Stocker K, Balseanu AT, Rogalewski A, Diederich K, Minnerup J, Margaritescu C, Schabitz WR. Effects of granulocyte-colony stimulating factor after stroke in aged rats. Stroke. 2010;41(5):1027–31. doi:10.1161/strokeaha.109.575621.
142. Porritt MJ, Chen M, Rewell SS, Dean RG, Burrell LM, Howells DW. ACE inhibition reduces infarction in normotensive but not hypertensive rats: correlation with cortical ACE activity. J Cereb Blood Flow Metab. 2010;30(8):1520–6. doi:10.1038/jcbfm.2010.57.
143. Posel C, Scheibe J, Kranz A, Bothe V, Quente E, Frohlich W, Lange F, Schabitz WR, Minnerup J, Boltze J, Wagner DC. Bone marrow cell transplantation time-dependently abolishes efficacy of granulocyte colony-stimulating factor after stroke in hypertensive rats. Stroke. 2014;45(8):2431–7. doi:10.1161/strokeaha.113.004460.
144. Prasad K, Mohanty S, Bhatia R, Srivastava MV, Garg A, Srivastava A, Goyal V, Tripathi M, Kumar A, Bal C, Vij A, Mishra NK. Autologous intravenous bone marrow mononuclear cell therapy for patients with subacute ischaemic stroke: a pilot study. Indian J Med Res. 2012;136(2):221–8.
145. Prasad K, Sharma A, Garg A, Mohanty S, Bhatnagar S, Johri S, Singh KK, Nair V, Sarkar RS, Gorthi SP, Hassan KM, Prabhakar S, Marwaha N, Khandelwal N, Misra UK, Kalita J, Nityanand S. Intravenous autologous bone marrow mononuclear stem cell therapy for ischemic stroke: a multicentric, randomized trial. Stroke. 2014;45(12):3618–24. doi:10.1161/strokeaha.114.007028.
146. Priller J, Persons DA, Klett FF, Kempermann G, Kreutzberg GW, Dirnagl U. Neogenesis of cerebellar Purkinje neurons from gene-marked bone marrow cells in vivo. J Cell Biol. 2001;155(5):733–8. doi:10.1083/jcb.200105103.
147. Prockop DJ, Oh JY. Medical therapies with adult stem/progenitor cells (MSCs): a backward journey from dramatic results in vivo to the cellular and molecular explanations. J Cell Biochem. 2012;113(5):1460–9. doi:10.1002/jcb.24046.
148. Qin L, Jing D, Parauda S, Carmel J, Ratan RR, Lee FS, Cho S. An adaptive role for BDNF Val66Met polymorphism in motor recovery in chronic stroke. J Neurosci. 2014;34(7):2493–502. doi:10.1523/jneurosci.4140-13.2014.
149. Ramos-Cabrer P, Justicia C, Wiedermann D, Hoehn M. Stem cell mediation of functional recovery after stroke in the rat. PLoS One. 2010;5(9):e12779. doi:10.1371/journal.pone.0012779.
150. Reitmeir R, Kilic E, Reinboth BS, Guo Z, ElAli A, Zechariah A, Kilic U, Hermann DM. Vascular endothelial growth factor induces contralesional corticobulbar plasticity and functional neurological recovery in the ischemic brain. Acta Neuropathol. 2012;123(2):273–84. doi:10.1007/s00401-011-0914-z.
151. Renault VM, Rafalski VA, Morgan AA, Salih DA, Brett JO, Webb AE, Villeda SA, Thekkat PU, Guillerey C, Denko NC, Palmer TD, Butte AJ, Brunet A. FoxO3 regulates neural stem cell homeostasis. Cell Stem Cell. 2009;5(5):527–39. doi:10.1016/j.stem.2009.09.014.

152. Rewell SS, Fernandez JA, Cox SF, Spratt NJ, Hogan L, Aleksoska E, van Raay L, Liberatore GT, Batchelor PE, Howells DW. Inducing stroke in aged, hypertensive, diabetic rats. J Cereb Blood Flow Metab. 2010;30(4):729–33. doi:10.1038/jcbfm.2009.273.
153. Riess P, Molcanyi M, Bentz K, Maegele M, Simanski C, Carlitscheck C, Schneider A, Hescheler J, Bouillon B, Schafer U, Neugebauer E. Embryonic stem cell transplantation after experimental traumatic brain injury dramatically improves neurological outcome, but may cause tumors. J Neurotrauma. 2007;24(1):216–25. doi:10.1089/neu.2006.0141.
154. Ringelstein EB, Thijs V, Norrving B, Chamorro A, Aichner F, Grond M, Saver J, Laage R, Schneider A, Rathgeb F, Vogt G, Charisse G, Fiebach JB, Schwab S, Schabitz WR, Kollmar R, Fisher M, Brozman M, Skoloudik D, Gruber F, Serena Leal J, Veltkamp R, Kohrmann M, Berrouschot J. Granulocyte colony-stimulating factor in patients with acute ischemic stroke: results of the AX200 for Ischemic Stroke trial. Stroke. 2013;44(10):2681–7. doi:10.1161/strokeaha.113.001531.
155. Roger VL, Go AS, Lloyd-Jones DM, Benjamin EJ, Berry JD, Borden WB, Bravata DM, Dai S, Ford ES, Fox CS, Fullerton HJ, Gillespie C, Hailpern SM, Heit JA, Howard VJ, Kissela BM, Kittner SJ, Lackland DT, Lichtman JH, Lisabeth LD, Makuc DM, Marcus GM, Marelli A, Matchar DB, Moy CS, Mozaffarian D, Mussolino ME, Nichol G, Paynter NP, Soliman EZ, Sorlie PD, Sotoodehnia N, Turan TN, Virani SS, Wong ND, Woo D, Turner MB. Heart disease and stroke statistics – 2012 update: a report from the American Heart Association. Circulation. 2012;125(1):e2–220. doi:10.1161/CIR.0b013e31823ac046.
156. Rosado-de-Castro PH, Schmidt Fda R, Battistella V, Lopes de Souza SA, Gutfilen B, Goldenberg RC, Kasai-Brunswick TH, Vairo L, Silva RM, Wajnberg E, Alvarenga Americano do Brasil PE, Gasparetto EL, Maiolino A, Alves-Leon SV, Andre C, Mendez-Otero R, Rodriguezde Freitas G, Barbosa da Fonseca LM. Biodistribution of bone marrow mononuclear cells after intra-arterial or intravenous transplantation in subacute stroke patients. Regen Med. 2013;8(2):145–55. doi:10.2217/rme.13.2.
157. Rosenblum S, Wang N, Smith TN, Pendharkar AV, Chua JY, Birk H, Guzman R. Timing of intra-arterial neural stem cell transplantation after hypoxia-ischemia influences cell engraftment, survival, and differentiation. Stroke. 2012;43(6):1624–31. doi:10.1161/strokeaha.111.637884.
158. Rothrock JF, Clark WM, Lyden PD. Spontaneous early improvement following ischemic stroke. Stroke. 1995;26(8):1358–60.
159. Salgado AJ, Sousa JC, Costa BM, Pires AO, Mateus-Pinheiro A, Teixeira FG, Pinto L, Sousa N. Mesenchymal stem cells secretome as a modulator of the neurogenic niche: basic insights and therapeutic opportunities. Front Cell Neurosci. 2015;9:249. doi:10.3389/fncel.2015.00249.
160. Sanai N, Tramontin AD, Quinones-Hinojosa A, Barbaro NM, Gupta N, Kunwar S, Lawton MT, McDermott MW, Parsa AT, Manuel-Garcia Verdugo J, Berger MS, Alvarez-Buylla A. Unique astrocyte ribbon in adult human brain contains neural stem cells but lacks chain migration. Nature. 2004;427(6976):740–4. doi:10.1038/nature02301.
161. Savitz SI, Misra V, Kasam M, Juneja H, Cox Jr CS, Alderman S, Aisiku I, Kar S, Gee A, Grotta JC. Intravenous autologous bone marrow mononuclear cells for ischemic stroke. Ann Neurol. 2011;70(1):59–69. doi:10.1002/ana.22458.
162. Scheibe F, Klein O, Klose J, Priller J. Mesenchymal stromal cells rescue cortical neurons from apoptotic cell death in an in vitro model of cerebral ischemia. Cell Mol Neurobiol. 2012;32(4):567–76. doi:10.1007/s10571-012-9798-2.
163. Schneider A, Kruger C, Steigleder T, Weber D, Pitzer C, Laage R, Aronowski J, Maurer MH, Gassler N, Mier W, Hasselblatt M, Kollmar R, Schwab S, Sommer C, Bach A, Kuhn HG, Schabitz WR. The hematopoietic factor G-CSF is a neuronal ligand that counteracts programmed cell death and drives neurogenesis. J Clin Invest. 2005;115(8):2083–98. doi:10.1172/jci23559.
164. Sharma S, Yang B, Strong R, Xi X, Brenneman M, Grotta JC, Aronowski J, Savitz SI. Bone marrow mononuclear cells protect neurons and modulate microglia in cell culture models of ischemic stroke. J Neurosci Res. 2010;88(13):2869–76. doi:10.1002/jnr.22452.

165. Shen LH, Li Y, Chen J, Cui Y, Zhang C, Kapke A, Lu M, Savant-Bhonsale S, Chopp M. One-year follow-up after bone marrow stromal cell treatment in middle-aged female rats with stroke. Stroke. 2007;38(7):2150–6. doi:10.1161/strokeaha.106.481218.
166. Shen LH, Li Y, Chen J, Zhang J, Vanguri P, Borneman J, Chopp M. Intracarotid transplantation of bone marrow stromal cells increases axon-myelin remodeling after stroke. Neuroscience. 2006;137(2):393–9. doi:10.1016/j.neuroscience.2005.08.092.
167. Shen Q, Goderie SK, Jin L, Karanth N, Sun Y, Abramova N, Vincent P, Pumiglia K, Temple S. Endothelial cells stimulate self-renewal and expand neurogenesis of neural stem cells. Science. 2004;304(5675):1338–40. doi:10.1126/science.1095505.
168. Sherman W, Martens TP, Viles-Gonzalez JF, Siminiak T. Catheter-based delivery of cells to the heart. Nat Clin Pract Cardiovasc Med. 2006;3 Suppl 1:S57–64. doi:10.1038/ncpcardio0446.
169. Shinozuka K, Dailey T, Tajiri N, Ishikawa H, Kaneko Y, Borlongan CV. Stem cell transplantation for neuroprotection in stroke. Brain Sci. 2013;3(1):239–61. doi:10.3390/brainsci3010239.
170. Shyu WC, Lin SZ, Yang HI, Tzeng YS, Pang CY, Yen PS, Li H. Functional recovery of stroke rats induced by granulocyte colony-stimulating factor-stimulated stem cells. Circulation. 2004;110(13):1847–54. doi:10.1161/01.cir.0000142616.07367.66.
171. Sleight P. The role of angiotensin-converting enzyme inhibitors in the treatment of hypertension. Curr Cardiol Rep. 2001;3(6):511–8.
172. Soleman S, Yip P, Leasure JL, Moon L. Sustained sensorimotor impairments after endothelin-1 induced focal cerebral ischemia (stroke) in aged rats. Exp Neurol. 2010;222(1):13–24. doi:10.1016/j.expneurol.2009.11.007.
173. Spaeth E, Klopp A, Dembinski J, Andreeff M, Marini F. Inflammation and tumor microenvironments: defining the migratory itinerary of mesenchymal stem cells. Gene Ther. 2008;15(10):730–8. doi:10.1038/gt.2008.39.
174. Sutherland GR, Dix GA, Auer RN. Effect of age in rodent models of focal and forebrain ischemia. Stroke. 1996;27:1663–7. doi:10.1161/01.STR.27.9.1663.
175. Tacutu R, Budovsky A, Fraifeld VE. The NetAge database: a compendium of networks for longevity, age-related diseases and associated processes. Biogerontology. 2010;11(4):513–22. doi:10.1007/s10522-010-9265-8.
176. Taguchi A, Kasahara Y, Matsuyama T. Letter by Taguchi et al regarding article, "Granulocyte colony-stimulating factor in patients with acute ischemic stroke: results of the AX200 for Ischemic Stroke Trial". Stroke. 2014;45(1):e8. doi:10.1161/strokeaha.113.003683.
177. Taguchi A, Ohtani M, Soma T, Watanabe M, Kinosita N. Therapeutic angiogenesis by autologous bone-marrow transplantation in a general hospital setting. Eur J Vasc Endovasc Surg. 2003;25(3):276–8. doi:10.1053/ejvs.2002.1831.
178. Tajiri N, Quach DM, Kaneko Y, Wu S, Lee D, Lam T, Hayama KL, Hazel TG, Johe K, Wu MC, Borlongan CV. Behavioral and histopathological assessment of adult ischemic rat brains after intracerebral transplantation of NSI-566RSC cell lines. PLoS One. 2014;9(3):e91408. doi:10.1371/journal.pone.0091408.
179. Tatarishvili J, Oki K, Monni E, Koch P, Memanishvili T, Buga AM, Verma V, Popa-Wagner A, Brustle O, Lindvall O, Kokaia Z. Human induced pluripotent stem cells improve recovery in stroke-injured aged rats. Restor Neurol Neurosci. 2014;32(4):547–58. doi:10.3233/rnn-140404.
180. Tfilin M, Sudai E, Merenlender A, Gispan I, Yadid G, Turgeman G. Mesenchymal stem cells increase hippocampal neurogenesis and counteract depressive-like behavior. Mol Psychiatry. 2010;15(12):1164–75. doi:10.1038/mp.2009.110.
181. Toth ZE, Leker RR, Shahar T, Pastorino S, Szalayova I, Asemenew B, Key S, Parmelee A, Mayer B, Nemeth K, Bratincsak A, Mezey E. The combination of granulocyte colony-stimulating factor and stem cell factor significantly increases the number of bone marrow-derived endothelial cells in brains of mice following cerebral ischemia. Blood. 2008;111(12):5544–52. doi:10.1182/blood-2007-10-119073.

182. Toyoshima A, Yasuhara T, Kameda M, Morimoto J, Takeuchi H, Wang F, Sasaki T, Sasada S, Shinko A, Wakamori T, Okazaki M, Kondo A, Agari T, Borlongan CV, Date I. Intra-arterial transplantation of allogeneic mesenchymal stem cells mounts neuroprotective effects in a transient ischemic stroke model in rats: analyses of therapeutic time window and its mechanisms. PLoS One. 2015;10(6):e0127302. doi:10.1371/journal.pone.0127302.
183. Trueman RC, Harrison DJ, Dwyer DM, Dunnett SB, Hoehn M, Farr TD. A critical re-examination of the intraluminal filament MCAO model: impact of external carotid artery transection. Transl Stroke Res. 2011;2(4):651–61. doi:10.1007/s12975-011-0102-4.
184. Vendrame M, Gemma C, Pennypacker KR, Bickford PC, Davis Sanberg C, Sanberg PR, Willing AE. Cord blood rescues stroke-induced changes in splenocyte phenotype and function. Exp Neurol. 2006;199(1):191–200. doi:10.1016/j.expneurol.2006.03.017.
185. Wagner DC, Bojko M, Peters M, Lorenz M, Voigt C, Kaminski A, Hasenclever D, Scholz M, Kranz A, Weise G, Boltze J. Impact of age on the efficacy of bone marrow mononuclear cell transplantation in experimental stroke. Exp Transl Stroke Med. 2012;4(1):17. doi:10.1186/2040-7378-4-17.
186. Walczak P, Zhang J, Gilad AA, Kedziorek DA, Ruiz-Cabello J, Young RG, Pittenger MF, van Zijl PC, Huang J, Bulte JW. Dual-modality monitoring of targeted intraarterial delivery of mesenchymal stem cells after transient ischemia. Stroke. 2008;39(5):1569–74. doi:10.1161/strokeaha.107.502047.
187. Wallenquist U, Holmqvist K, Hanell A, Marklund N, Hillered L, Forsberg-Nilsson K. Ibuprofen attenuates the inflammatory response and allows formation of migratory neuroblasts from grafted stem cells after traumatic brain injury. Restor Neurol Neurosci. 2012;30(1):9–19. doi:10.3233/rnn-2011-0606.
188. Wang F, Yasuhara T, Shingo T, Kameda M, Tajiri N, Yuan WJ, Kondo A, Kadota T, Baba T, Tayra JT, Kikuchi Y, Miyoshi Y, Date I. Intravenous administration of mesenchymal stem cells exerts therapeutic effects on parkinsonian model of rats: focusing on neuroprotective effects of stromal cell-derived factor-1alpha. BMC Neurosci. 2010;11:52. doi:10.1186/1471-2202-11-52.
189. Wang LC, Futrell N, Wang DZ, Chen FJ, Zhai QH, Schultz LR. A reproducible model of middle cerebral infarcts, compatible with long-term survival, in aged rats. Stroke. 1995;26(11):2087–90.
190. Wang LQ, Lin ZZ, Zhang HX, Shao B, Xiao L, Jiang HG, Zhuge QC, Xie LK, Wang B, Su DM, Jin KL. Timing and dose regimens of marrow mesenchymal stem cell transplantation affect the outcomes and neuroinflammatory response after ischemic stroke. CNS Neurosci Ther. 2014;20(4):317–26. doi:10.1111/cns.12216.
191. Weise G, Lorenz M, Posel C, Maria Riegelsberger U, Storbeck V, Kamprad M, Kranz A, Wagner DC, Boltze J. Transplantation of cryopreserved human umbilical cord blood mononuclear cells does not induce sustained recovery after experimental stroke in spontaneously hypertensive rats. J Cereb Blood Flow Metab. 2014;34(1):e1–9. doi:10.1038/jcbfm.2013.185.
192. Wolfson M, Budovsky A, Tacutu R, Fraifeld V. The signaling hubs at the crossroad of longevity and age-related disease networks. Int J Biochem Cell Biol. 2009;41(3):516–20. doi:10.1016/j.biocel.2008.08.026.
193. Yilmaz G, Granger DN. Cell adhesion molecules and ischemic stroke. Neurol Res. 2008;30(8):783–93. doi:10.1179/174313208x341085.
194. Yousef H, Conboy MJ, Morgenthaler A, Schlesinger C, Bugaj L, Paliwal P, Greer C, Conboy IM, Schaffer D. Systemic attenuation of the TGF-beta pathway by a single drug simultaneously rejuvenates hippocampal neurogenesis and myogenesis in the same old mammal. Oncotarget. 2015;6(14):11959–78. 10.18632/oncotarget.3851.
195. Yusuf S. After the HOPE Study. ACE inhibitor now for every diabetic patient? Interview by Dr. Dirk Einecke. MMW Fortschr Med. 2000;142(44):10.
196. Zechariah A, ElAli A, Doeppner TR, Jin F, Hasan MR, Helfrich I, Mies G, Hermann DM. Vascular endothelial growth factor promotes pericyte coverage of brain capillaries,

improves cerebral blood flow during subsequent focal cerebral ischemia, and preserves the metabolic penumbra. Stroke. 2013;44(6):1690–7. doi:10.1161/strokeaha.111.000240.
197. Zhang C, Harder DR. Cerebral capillary endothelial cell mitogenesis and morphogenesis induced by astrocytic epoxyeicosatrienoic acid. Stroke. 2002;33(12):2957–64.
198. Zhang SC. Neural subtype specification from embryonic stem cells. Brain Pathol. 2006;16(2):132–42. doi:10.1111/j.1750-3639.2006.00008.x.
199. Zhang X, Zhang G, Zhang H, Karin M, Bai H, Cai D. Hypothalamic IKKbeta/NF-kappaB and ER stress link overnutrition to energy imbalance and obesity. Cell. 2008;135(1):61–73. doi:10.1016/j.cell.2008.07.043.
200. Zhang ZG, Zhang L, Ding G, Jiang Q, Zhang RL, Zhang X, Gan WB, Chopp M. A model of mini-embolic stroke offers measurements of the neurovascular unit response in the living mouse. Stroke. 2005;36(12):2701–4. doi:10.1161/01.STR.0000190007.18897.e3.
201. Zhao C, Deng W, Gage FH. Mechanisms and functional implications of adult neurogenesis. Cell. 2008;132(4):645–60. doi:10.1016/j.cell.2008.01.033.
202. Zhao CS, Puurunen K, Schallert T, Sivenius J, Jolkkonen J. Behavioral effects of photothrombotic ischemic cortical injury in aged rats treated with the sedative-hypnotic GABAergic drug zopiclone. Behav Brain Res. 2005;160(2):260–6. doi:10.1016/j.bbr.2004.12.007.
203. Zhao LR, Duan WM, Reyes M, Keene CD, Verfaillie CM, Low WC. Human bone marrow stem cells exhibit neural phenotypes and ameliorate neurological deficits after grafting into the ischemic brain of rats. Exp Neurol. 2002;174(1):11–20. doi:10.1006/exnr.2001.7853.

Chapter 9
Modification of Bone Marrow Stem Cells for Homing and Survival During Cerebral Ischemia

Yaohui Tang

Abstract Over the last decade, major advances have been made in stem cell-based therapy for ischemic stroke, which is one of the leading causes of death and disability worldwide. Various stem cells from bone marrow, such as mesenchymal stem cells (MSCs), hematopoietic stem cells (HSCs), and endothelial progenitor cells (EPCs), have shown therapeutic potential for stroke. Concomitant with these exciting findings are some fundamental bottlenecks that must be overcome in order to accelerate their clinical translation, including the low survival and engraftment caused by the harsh microenvironment after transplantation. In this chapter, strategies such as gene modification, hypoxia/growth factor preconditioning, and biomaterial-based methods to improve cell survival and homing are summarized, and the potential strategies for their future application are also discussed.

Keywords Bone marrow stem cells • Modification • Homing • Survival • Ischemia

9.1 Introduction

Stroke is the third leading cause of mortality and the leading cause of long-term disability in the United States. Approximately 8,000,000 people suffer a stroke, and more than 140,000 people die each year. Ischemic stroke accounts for over 80 % of total stroke patients [126]. Though extensive neuroprotection and regenerative studies have been performed, only tissue plasminogen activator (tPA) has been proven to be effective. However, due to its narrow therapeutic time window (less than 4.5 h) and hemorrhagic complication, fewer than 5 % of stroke patients are able to benefit from tPA, and even among those, only 10 % return to independent living [82].

Y. Tang (✉)
Department of Orthopaedic Surgery and Bioengineering, School of Medicine, Stanford University, 300 Pasteur Drive, Stanford, CA 94305, USA
e-mail: yaohui@stanford.edu

K. Jin et al. (eds.), *Bone Marrow Stem Cell Therapy for Stroke*,
DOI 10.1007/978-981-10-2929-5_9

Recently, growing evidence suggests that stem cells, including MSCs, neural stem cells (NSCs), EPCs, and induced pluripotent stem cells (iPS), are beneficial for cerebral ischemia [77, 112]. Among these, bone marrow-derived stem cells (BMSCs) have the most promising therapeutic potential, because a large quantity can be harvested autologously without ethical or immunological issues [120]. However, a number of problems remain unresolved and need specific attention prior to general clinical translation. For example, it is still challenging that stem cell survival, homing, and engraftment rates are low after transplantation in the pathological environment subjected to multiple insults, including ischemia/hypoxia, inflammatory response, and so on, which hamper the benefits and applications of cell-based therapy.

This chapter first summarized recent progress in basic and translational research in the field of BMSC transplantation for ischemic stroke. It then critically discussed how to enhance BMSC-based therapy by improving grafted cell survival and homing to further establish BMSC transplantation therapy as a scientifically proven method in clinical applications.

9.1.1 Basic Concept of Bone Marrow Stem Cells (BMSCs) in Stroke

Bone marrow (BM) consists of heterogeneous stem cell populations, including MSCs, HSCs, EPCs, and very small embryonic-like cells (VSELs). Their neuronal differentiation potential as well as neurotrophic factor secretion capacity has prompted interest in using BM as stem cell donor source for cell-based therapy in stroke.

9.1.1.1 Bone Marrow-Derived Mesenchymal Stem Cells (MSCs)in Stroke

BM-derived MSCs are a population of plastic-adherent fibroblastic cells, with CD29, CD105, and CD73 positive, but lack of hematopoietic surface markers such as CD34 and CD45. MSCs have the potential to differentiate into mesodermal cell lineages to involve in adipogenesis, chrondrogenesis, and osteogenesis [13]. MSCs derived from various donors, including rat, mouse, rabbit, or humans, have been transplanted by intravenous (IV), intra-artery (IA), intracerebral (IC), or intracisternal routes into animals, from different time points (hours to months) after induction of stroke, and have shown to improve functional recovery during cerebral ischemia [112]. Following IV and IC injection, MSCs migrate to the ischemic boundary; however, few cells have been shown to survive, and long-term cell engraftment has not been detected with IV administration [96]. Another study stated that only 3 % of administered cells expressed neuronal markers in vivo [15], which argued with the concept that tissue replacement is likely to be a potential mechanism for this strategy. More studies support that trophic factors secreted by the MSCs in response to the local microenvironment stimulate endogenous neurogenesis, angiogenesis, and immunomodulation and further improve functional recovery. Higher levels of BDNF, NT3, and VEGF have been detected in the penumbra region 14 days after human MSC transplantation [5]. Increased VEGF and bFGF drive angiogenesis and facilitate regional blood flow [53]. In addition to secreting trophic factors, MSCs were also detected to influence astrocyte survival and astrocyte-related trophic factor expression after ischemic insult, by activating kinase pathways and protein functions [67]. Up to date, clinical reports also reveal that MSCs significantly improve patients' functional recovery without adverse side effects, probably through neuronal differentiation or secreting anti-inflammatory as well as neurotrophic factors [11].

9.1.1.2 HSC

Quiescent of CD34+ HSCs are able to migrate quickly from bone marrow to blood circulation in response to cerebral ischemia, which is induced by a wide array of chemokines and cytokines, including stromal cell-derived factor-1 (SDF-1) and granulocyte-colony-stimulating factor (G-CSF) [50]. During stroke, HSCs exit from bone marrow, migrate to the brain, adhere to the vascular wall, and cross over the blood–brain barrier (BBB), mediated by SDF-1/CXCR4 axis or G-CSF. This recruitment of HSCs from BM to stroke-induced lesion area has been employed in clinical protocols, for the creation of ample supply of HSCs for brain repair [10]. Both experimental and clinical studies have demonstrated the safety and feasibility of HSCs-based therapy for ischemic stroke. Intracerebral implantation of CD34+ HSCs promoted angiogenesis and neurogenesis and increased the local cortical blood flow in the ischemic hemisphere in ischemic rats [100] and mice [107]. A small clinical study demonstrated that by autologously IA injection of CD34+ cells into five stroke patients, all patients showed behavior recovery and infarct reduction, suggesting the potential of direct IA infusion of autologous CD34+ selected cells for the treatment of stroke [4].

9.1.1.3 EPC

In 1997, Asahara and coworkers first isolated Flk-1+/CD34+ cells from human peripheral blood, which were defined as EPCs. In that study they found that these cells could integrate into blood vessels when transplanted into a hind-limb ischemic mouse model [1]. EPCs were usually generated and maintained in bone marrow and could migrate into lesion region to help blood vessel remodeling and repair [71]. Recent studies showed that EPC transplantation could promote cerebral blood flow, reduce infarct volume and neuronal cell death, increase focal angiogenesis and neurogenesis, and improve neurobehavioral recovery after ischemia [26]. Grafted EPCs could either secrete neurotrophic factors, which is supported by the evidence that

EPC medium could also promote angiogenesis [144], or differentiate into endothelial cells to replace/repair injured ECs and integrate into endogenous blood vessels, which is detected by histological studies [26]. These results support that EPCs have great therapeutic potential for stroke, most possibly through both directly integrating into blood vessels and secreting trophic factors.

9.1.1.4 Very Small Embryonic-Like Cells

In 2006, Ratajczak's group first discovered a nonhematopoietic population that expresses neural lineage markers (GFAP, Nestin, Olig1, Olig2, Sox2, and Musashi-1) and resides in the nonhematopoietic CXCR4+/Sca-1+/lin-/CD45- BM mononuclear cell fraction, named as very small embryonic-like cells (VSELs) [90]. The number of circulating VSELs in PB increases in mice after experimental stroke [55] and in stroke patients [89], suggesting that VSELs residing in adult tissues or mobilized into PB are a potent source of adult tissue-derived stem cells that can be used for regenerative medicine, particularly for neural repair after stroke. Ratajczak et al. observed increased gene expression of both pluripotent and NSC markers in PB-borne nucleated cells in stroke patients, resembling what they previously noted in murine stroke model. Further analyses using computer tomography imaging revealed differences in VSEL mobilization between patients with posterior circulation infarcts and patients with partial anterior circulation infarcts [55]. In addition, the observation that murine VSELs are capable of differentiating into neurons, oligodendrocytes, and microglia further encourages us to use these cells as donor grafts for regeneration of a damaged CNS. However, a limitation for clinical application is the small number of VSELs that could be harvested, requiring ex vivo expansion strategy, especially to generate enough supply of VSELs for stroke therapy in clinical setting.

9.1.2 Mechanism of Death of Transplanted Stem Cells During Cerebral Ischemia

Although stem cell transplantation appears to be very promising for stroke, a number of problems remain unresolved and need specific attention in order to improve therapeutic efficacy for further successful clinical translation, including low survival and engraftment of transplanted cells in the brain subjected to multiple insults including ischemia, reactive oxygen species (ROS) generation, inflammatory response, apoptotic cascade activation, and so on.

Accumulating evidence demonstrates that less than 10 % of transplanted stem cells could survive in the lesion site after transplantation as they are exposed in hostile environment, and cell death is initiated via multiple mechanisms [48]. It's reported that more cells survive when they are transplanted into sham animals (no brain injury) compared to injured animals [3], indicating that factors in the lesion site induce death of the transplanted cells. These factors include but not limited to time after injury [47], distance from the transplantation site to the lesion site [86], state of the cells transplanted (differentiated or undifferentiated) [59], aging of the cells transplanted [108], host immune response [35], and phagocytic response of host [2]. Subsequent evidence shows that delivery time is the major determinant of the survival of transplanted stem cells. It is reported that NPC survival was significantly reduced following delayed cell delivery [20], which was mediated by the inflammatory milieu.

Cell death is initiated even prior to transplantation, explained by two main mechanisms: detachment of cells from adherent surface and the removal of growth factor, during the procedure of trypsinization and suspension. Inhibition of cell adhesion-induced cell death was first reported in 1994 by Frisch and Francis. They found that when epithelial cells were seeded in medium with the addition of soluble peptide-GRGDSP, which prevented cell attachment by blocking integrins, it resulted in increased apoptosis [28]. This kind of cell death is termed anoikis, which can be rescued by culturing cells on ECM-coated surfaces to promote cell adhesion. For example, oligodendrocyte progenitor cells cultured on glass coverslips coated with fibronectin or laminin showed greater viability compared to those cultured on non-coated surfaces [37]. In vitro study demonstrated that addition of laminin to neural progenitor cells increased the number of neurospheres and reduced cell death in comparison to control groups, while blocking the beta 1 integrin inhibited the effect of laminin, suggesting this is beta 1 integrin mediated [32]. One proposed explanation for detachment-induced cell death is that Bmf released from actin in terms of cytoskeleton stabilization is reduced after cell detachment. Bmf binds to Bcl-2 in mitochondria and neutralizes its antiapoptotic effect, which activates caspase-8, further releasing Bcl-2 from the mitochondria to induce cell death [29].

In addition to detachment-mediated cell death, removal of growth factors also induces apoptosis. Typically, c-Jun amino-terminal kinase (JNK) signaling pathway is activated when trophic support is removed, mediates c-Jun phosphorylation, thus induces the expression of proapoptotic factor-Bcl-2 family (DP5/Hrk). It further demonstrates that DP5 activates a proapoptotic member of the Bcl-2 family-Bax, causes mitochondrial damage, and releases cytochrome c, leading to the formation of apoptotic protease-activating factor 1 (Apaf-1)/caspase-9 complex, which activates caspase-3 resulting in cell apoptosis [138].

Therefore, cell–ECM interactions are reduced, and apoptosis is initiated even prior to transplantation when stem cells are trypsinized as single cells, cell survival is further reduced by needle insertion, and growth factors withdraw during the injection process, as well as hostile environment they confront in the lesion site after transplantation, given the generation of reactive oxygen species (ROS) and inflammatory response mediators in brain postischemia. It is highly accepted that cerebral ischemia caused excessive ROS would induce the apoptosis of the transplanted cells [12]. Our study showed that more than 80% of grafted cells died within 72 h after administration [110], and our in vitro studies also suggested that exposure of stem cells to culture conditions which mimic the hostile environment in vivo (such as oxygen–glucose deprivation and H2O2 stimulation) led to the apoptosis mediated by ROS [110].

9.1.3 *Strategies to Improve BMSC Survival*

Both basic studies and clinical evidence strongly support that BMSCs could serve as a promising restorative therapy for stroke. However, as stated above, high stem cell death rate is the main hurdle that hinders the therapy. Scientists in the field propose several strategies to conquer this challenge, including gene modification, preconditioning, and biomaterial-based methods.

9.1.3.1 Gene Modification

After cerebral ischemia, both intrinsic and extrinsic apoptosis pathways are activated [3]. More than 80 % of stem cells died after their transplantation, which is mainly caused by the activation of proapoptotic signals. Thus, downregulation of proapoptotic or upregulation of antiapoptotic cues by manipulating gene expression of stem cells posttransplantation may ameliorate the microenvironment and further enhance their survival. Indeed, overexpressing of Bcl-2 in embryonic stem cells (ESCs) increased their survival after injection into ischemic rat brain, as well as enhanced their neuronal differentiation, and improved functional outcome [123].

In addition to regulate apoptotic-related genes, amounting evidence shows that modification of trophic genes in stem cells also has significant impacts on their survival and therapeutic efficacy (Table 9.1). MSCs overexpressing BDNF or GDNF after injection into ischemic rats showed more cell survival, promoted functional recovery, and reduced ischemic damage at 7 and 14 days following MCAO, while rats that received CNTF- or NT3-transfected MSCs showed neither functional recovery nor ischemic damage reduction [57]. Liu et al. found that intravenously administered hMSCs overexpressing PIGF could accumulate in the ischemic lesions, further reduced lesion volume, enhanced angiogenesis, and elicited functional improvement [74]. FGF-2-modified MSCs with HSV-1 greatly reduced infarct volume and improved functional recovery at 14 days after stroke [40]. When surviving, a new apoptosis-inhibiting protein was overexpressed in MSCs and promoted MSCs' survival by 1.3-fold at 4 days and 3.4-fold higher at 14 days posttransplantation, which results in reduced infarct volume and improved neurological function [76].

Besides BMSCs, lots of studies from Dr. Kim's group showed that transplantation of human NSCs overexpressing BDNF [61], VEGF [60], or Akt-1 [61] could produce a two- to threefold increase in cell survival at 2 weeks and 8 weeks posttransplantation, as well as reduce infarct volume and improve functional recovery [171]. Transduction of NPCs with TAT-Hsp70 led to increased number of grafted NPCs, reduced BBB disruption, enhanced postischemic neurogenesis, and increased neurotrophic factor secretion [23].

During the last decade, microRNAs (miRs), a group of short RNA molecules that involve in posttranscriptional downregulation, have gained extensive attention in modulating cell survival. It is reported that miR-210 and miR-107 exert significant

Table 9.1 Enhancement of stem cell survival by gene modification

Ovexpressing genes	Stem cells	Transfecting agents	Stem cell fate	Therapeutic outcome
Bcl-2 [200]	Embryonic stem cells	Electroporation with Bcl-2 plasmid	Increased ES survival after injection	Into ischemic rat brain, as well as enhanced their neuronal differentiation and improved functional outcome
BDNF or GDNF [166]	MSCs	Adenovirus	More cells survival at 7 and 14 days following MCAO	Improved functional recovery and reduced ischemic damage at 7 and 14 days following MCAO
PIGF [175]	hMSCs	Adenovirus	More cells accumulate in the lesion area	Accumulate in the ischemic lesions, further reduce lesion volume, enhance angiogenesis, and elicit functional improvement
FGF-2 [160]	MSCs	HSV-1 vector	N/A	Reduced infarct volume and improved functional recovery
Ang-1 [184]	hMSC	Adenovirus	N/A	Rats receiving Ang-hMSCs exhibited comparable lesion reduction, improved functional recovery, and increased angiogenesis
Survivin [176]	MSCs	Lentiviral vector	Promote MSCs' survival by 1.3-fold at 4 days and 3.4-fold higher at 14 days posttransplantation	Reduced infarct volume and improved neurological function

(continued)

Table 9.1 (continued)

Ovexpressing genes	Stem cells	Transfecting agents	Stem cell fate	Therapeutic outcome
BDNF [61]	NSCs	Adenovirus	Threefold increase in cell survival at 2 weeks and 8 weeks post injection	Renewed angiogenesis, induce behavioral improvement in ICH animals
Akt-1 [61]	NSCs	Retroviral vector	50–100 % increased cell survival at 2 and 8 weeks posttransplantation	Induced behavioral improvement
Bcl-XL [170]	hNSCs	Retroviral vector	Number of hNSCs were 1.5-fold higher at 2 weeks and 10-fold higher at 7 weeks than controls posttransplantation	Improved locomotor scores and enhanced accuracy of hind-limb placement in a grid walk
HSP-70 [154]	NPCs; MSCs	TAT-Hsp70 protein transduction	Increased intracerebral numbers of grafted NPCs	Reduced BBB disruption, enhanced postischemic neurogenesis, and increased neurotrophic factor secretion; decreased apoptosis in the infarcted tissue and improved cardiac function
HSP 27 [180]		Lentiviral vector	Increased MSCs' survival in vitro and in vivo	
HGF [209]	MSC	HSV-1	N/A	Decreased apoptosis of neurons and reduced neurologic deficits and infarcts
CXCR4 [207]	MSC	Lentivirus	A significant increase in the number of eGFP-positive MSCs in the infarct areas	A reduction in the volume of the cerebral infarction and improved neurological function

(continued)

Table 9.1 (continued)

Ovexpressing genes	Stem cells	Transfecting agents	Stem cell fate	Therapeutic outcome
VEGF [169, 210]	NSC	Retroviral vector	2–3 fold increase in cell survival at 2 weeks and 8 weeks posttransplantation	Increased angiogenesis and behavioral recovery in mouse ICH model; improved focal angiogenesis and the Neurological Severity Scale score
	MSC	Plasmid transfection with lipofactamine	Improved cell viability	Enhanced the capillary formation in the infarction region and eventually attenuated left ventricular remodeling
		Facially amphipathic bile acid-modified polyethyleneimine (BA-PEI) conjugates [182]		

antiapoptotic effects in BMSCs by targeting caspase-8-associated protein-2 and programmed cell death-10 [81]. Pharmacological agents, including diazoxide [84], can induce protective miRs expression. Besides miRs, a recent investigation elucidated that preconditioning of MSCs with specific cell-free DNAs (cfDNAs) increased cell survival via Toll-like receptor 9 (TLR9) and translocation of nuclear factor-kappa B (NFkB) [54]. This evidence highlights the possibility that miRs and cfDNAs may be potential new targets to promote stem cell survival after transplantation.

Collectively, these exciting results suggest that gene modification is a promising strategy to increase cell survival after transplantation, and these enhanced cell survivals could contribute to reduced infarcts and improved behavioral recovery through neuronal differentiation and promoted trophic factor secretion.

9.1.3.2 Precondition-Based Method

Gene modification takes the risk that uncontrolled expression of introducing gene may have adverse effects and induce tumor formation on normal brain. Recent studies show that precondition strategy including hypoxia preconditioning, growth factor preconditioning, and antiapoptosis drug preconditioning could be a safe and efficient method [136]. Up to now, a number of sublethal insults including hypoxia [131], anoxia [119], hydrogen sulfide (H2S) [127], hydrogen peroxide (H2O2) [141], as well as growth factors, such as erythropoietin (EPO) [68], stromal-derived factor-1 (SDF-1) [139], insulin-like growth factor-1 (IGF-1) [78], heat shock proteins (HSPs) [117], or pharmacological agents such as melatonin [110], minocycline [94], isoflurane [52], and lipopolysaccharide (LPS) [132], have been tested in stem cells (Table 9.2).

Sublethal hypoxia preconditioning applied to stem cells have shown to activate protective signals including hypoxia-inducible factor-1 (HIF-1), growth factors, Akt, and ERK signals to further enhance their resistance to apoptosis/necrosis cues by increasing survival signals [30]. Dr. Wei's group has performed extensive studies related to hypoxia preconditioning. In these studies they demonstrated that transplantation of hypoxia preconditioning MSCs improves infarcted heart function [38] and ischemic brain function [124] recovery via enhanced survival of implanted cells and angiogenesis. Also they found hypoxic precondition reduced ES-NPCs apoptosis by 40–50 % in serum-free medium via upregulation of erythropoietin (EPO), Bcl-2, and HIF-1alpha [114].

One study from Dr. Yang's group demonstrated that melatonin pretreatment increased MSCs' survival and proangiogenic activity through Erk1/2 signaling pathway [110], which is consistent with other studies that melatonin treatment enhanced adipose-derived mesenchymal stem cells (ADMSCs) survival and therapy for lung ischemia injury [134] and reduced grafted eEPC apoptosis/necrosis as well as increased their outgrowth in injured kidney [87]. For minocycline, Sakata et al. showed that transplantation of minocycline-preconditioned NSCs protected their survival from ischemic reperfusion injury via upregulation of Nrf2 and Nrf2-regulated antioxidant genes, increased their paracrine factors releasing, attenuated infarct size, and improved neurological performance [94], and doxycycline has the similar protective effects [80]. Additionally, low LPS pretreatment was found to protect MSCs against oxidative stress-induced apoptosis and increase cell engraftment after transplantation into ischemic heart [132]. A recent study showed that EPO pretreatment could also suppress MSCs' apoptosis in response to hydrogen peroxide stimuli [25].

9.1.3.3 Biomaterial-Based Method

The development of biomaterials has evolved from the first-generation, material-based approach that focused on mechanical strength, durability, and biocompatibility to the third-generation, bio-functional materials that try to integrate biological cues to modulate cellular functions by modifying with extracellular matrix (ECM) related to signaling molecules. In recent years, biomaterials have been proven to be an effective strategy for regulating cellular behavior, including promoting cell survival, directing cell differentiation. Advances in biomaterials engineering enable promoting grafted cell survival and engraftment and have generated much attention in stroke therapy.

With regard to injecting stem cells which are encapsulated within biomaterials into ischemic brain, the infarct cavity is always an ideal location. First, it is more clinical relevant since the transplantation procedure is not initiated until infarct cav-

Table 9.2 Preconditioning treatment to improve stem cell survival

Triggers	Stem cells involved	Animal model	Stem cell fate	Outcome
Hypoxia	hESCs [155]; MSC [158, 161, 201]; ES-NPCs [195]	Myocardial infarction; MCAO	Increased neural precursor cell survival; engraftment of MSC was increased; cell survival was increased; promoted their survival, migration, and homing to the ischemic brain region; promote transplanted cell survival	Promoted neuronal differentiation; improvement in global, regional, and diastolic left ventricular functions; an increase in angiogenesis, as well as enhanced morphologic and functional benefits of stem cell therapy; reduced infarct volume and improved behavioral recovery; ES-NPCs exhibited extensive neuronal differentiation in the ischemic brain, accelerated and enhanced recovery of sensorimotor function
Anoxia	MSC [172, 199]	Myocardial infarction	Increased cell survival	Increased fractional shortening, ejection fraction, arteriole density and decreased infarct size; increased the capillary density and the fractional shortening and attenuated myocardial fibrosis
Hydrogen sulfide	MSCs [202]	Myocardial infarction	Improved the survival rate of the transplanted MSCs	Reduced the infarct size and increased left ventricular (LV) function
IL-6	NSCs [94]	MCAO	Protected the grafted neural stem cells from ischemic reperfusion injury	Attenuated infarct size, improved neurological performance and angiogenesis
Melatonin	MSCs [181, 194]; ADMSC [205]	MCAO; acute lung ischemia-reperfusion injury; ischemic kidney	Improved survival of MSCs; decreased apoptosis of ADMSC	Reduced infarct volume, enhanced angiogenesis, neurogenesis and functional recovery; protected the lung from ischemic injury; increased survival, paracrine activity, and efficiency of MSCs
Minocycline	NSCs [188]; OPC [190]	MCAO; in vitro	Minocycline preconditioning protected the grafted NSCs from ischemic reperfusion injury; reduced apoptosis in response to OGD	Attenuated infarct size and improved neurological performance
Doxycycline	NSCs [178]	In vitro	Decreased cell death and increased cell viability after oxygen–glucose deprivation–reoxygenation	Showed cryoprotective via induced the expression of Nrf2
BDNF	NSC [186]	MCAO	Promoted cell survival 1 week after transplantation	BDNF pretreatment of NSCs results in higher initial NSC engraftment and survival, increased neuroprotection, and greater functional recovery

ity is formed, which is already 2–3 weeks after the onset of stroke; second, cavity is adjacent to the highly plastic peri-infarct region, and injection of stem cells into the cavity shows to achieve best outcome. Third, injection into the cavity will not damage normal brain tissues. Although directly injecting stem cells into infarct cavity shows its merit in reduced infarct volume, enhanced behavioral recovery, and increased angiogenesis and neurogenesis, low cell survival is still a major problem that hinders its clinical application. For instance, only 8 % of the grafted NSPC transplanted cells survived 4 weeks posttransplantation in Mongolian gerbils after focal ischemia [43]. In another study, approximately 4 % of grafted NPCs survived at 2 weeks posttransplantation [145].

Previous experimental studies showed that using Matrigel, fibrin glue gels, particles, and other scaffolds as matrices could improve the survival of stem cells in the infarct cavity posttransplantation (Table 9.3). Matrigel is an extracellular matrix comprised of ECM proteins and growth factor mixtures, including collagen, laminin, epidermal growth factor (EGF), and fibroblast growth factor 2 (FGF-2). Jin et al. injected NPCs encapsulated with Matrigel into the infarct cavity in both young and aged rats. Compared to control group, more cells were detected at the infarct site, and best functional recovery was achieved in the NPCs+Matrigel group [45, 46]. However, Matrigel is derived from a mouse sarcoma that raises a serious concern for its clinical application.

The functions of biodegradable polymers such as PGA and PLGA are also extensively investigated in stem cell-based therapy in stroke. For instance, Park et al. implanted NSCs seeded on polyglycolic acid (PGA), a high biocompatible scaffold into the infarct cavity, and found infarct volume was greatly reduced as well as establishment of neuronal connections between exogenous transplanted NSCs and endogenous neurons [85]. Modo's group demonstrated PLGA could act as a structural support for NSCs in infarct cavity to improve cell survival and function [7]. In their further study, they loaded VEGF into the PLGA microparticles and transplanted NSCs which were seeded on the VEGF-releasing PLGA particles into the cavity. Their results showed that VEGF-releasing PLGA not only provides structural support but also attracts ECs into the cavity to induce neurovascular formation [8].

Hyaluronan (HA), a glycosaminoglycan that naturally and abundantly exist in the brain, could involve in brain development and influence cell adhesion, migration, angiogenesis, and axon growth. Thus it is reasonable to choose HA as protective matrices to encapsulate cells for transplantation into the brain to maintain a hydrated and porous environment [83]. Recently, experimental studies from Dr. Thomas Carmichael's group proved that hydrogel composed of cross-linked hyaluronan and heparin sulfate significantly promoted NPCs' survival after transplantation into the infarct cavity, accompanied by reduced inflammation [145]. In their further study, they proposed to modify hyaluronic acid hydrogel with cell adhesion peptide RGD and cross-linked with either MMP degradable peptides or non-MMP degradable peptides through a Michael Addition reaction to produce two hydrogel formulations with two different stiffness moduli (300 Pa in MMP HA and 1000 Pa in non-MMP HA). NPCs derived from induced pluripotent stem cells (iPS-NPC) were encapsulated in the hydrogel matrix and delivered to the infarct cavity of stroke

Table 9.3 Enhancement of stem cell survival by biomaterials

Biomaterials	Stem cells involved	Animal model	Stem cell fate	Outcome
PLGA microparticles [147]	MHP36 cells	MCAO	Cell survival was increased, and cells were differentiated into neurons	PLGA microparticles acted as a structural support for NSCs;
PLGA microparticles loaded with VEGF [148]	hNSCs			NSCs showed neuronal differentiation, and neurovascular unit was performed in the infarct cavity
Matrigel [45, 162]	Human fetal NPCs	MCAO	Enhance the survival of transplanted NPCs	Behavioral recovery was improved, and infarct volume was reduced
Collagen [135, 167, 187, 203]	Cardiomyoblasts	Myocardial infarction	Enhanced early survival of H9c2 cardiomyoblasts after transplantation into ischemic hearts	Left ventricular function was improved
	NSCs	MCAO	Increased cell survival and distribution	Reduced infarct volume, induced angiogenesis
	Marrow stromal cell-derived NPCs;	Excisional wound healing model	Remained longer viability	Improved motor behavior;
	BM-MSCs			significantly enhanced angiogenesis and VEGF
Hyaluronan [168, 183]	NPCs; iPS-NSCs	MCAO	Promoted NPCs' survival and neuronal differentiation after transplantation into the infarct cavity	Enhanced neurovascular unit formation and reduced inflammation
Fibrin glue [151]	iPS	MCAO	N/A	Improved the motor function, reduced infarct size, attenuated inflammation cytokines, and mediated neuroprotection
Collagen with bFGF in gelatin microspheres [179]	NS-MSCs	MCAO	Increased cell survival and proliferation	Significantly improved histological and
				Functional recovery in the rat stroke model

mice. They found that hydrogel system with MMP and RGD modification promoted neuronal differentiation of iPS-NPC and induced minimum inflammation [58].

9.1.4 Bone Marrow Stem Cell Mobilization in Stroke

9.1.4.1 Factors Mediating BMSC Homing

Migration and homing of administered cells to the ischemic regions are clinically relevant and very critical to their therapeutic efficacy. A detailed analysis of the biological responses to brain injury would not only give us insight into the mechanism of stem cell homing but also give us important clues about how we can improve their homing capacity. Now it is clear that following brain injury, homing molecular cues, including chemokines, growth factors, and adhesion molecules, originating from the inflammatory zone in the injured brain, are activated and upregulated to cause BMSC homing. Chemokines such as G-CSF and SDF-1 have been demonstrated to be an important stem cell homing mediator that mobilizes stem cells from bone marrow into the PB. G-CSF treatment enhances tissue regeneration and improves recovery after stroke by mobilizing BMSCs from bone marrow into peripheral blood [91]. Previous studies showed that subcutaneous injection of G-CSF for 5 days after cerebral ischemia promotes BMSC migration to the lesion area, reduces infarcts, and enhances functional recovery in stroke rats [101]. G-CSF treatment is also demonstrated to facilitate neurogenesis in SVZ by increasing the infiltration of BMSC into the brain [99]. BMSCs exert their benefits on cerebral ischemic injuries through promoting neuronal repair and recovery of brain function, which provides a basis for the development of a noninvasive autologous therapy for cerebral ischemia. Some pilot clinical trials demonstrated that G-CSF could mobilize BMSCs in patients after acute stroke safely and provide better neurological outcome compared to conventional treatment [101].

SDF-1 is another important homing factor, which is secreted primarily by bone marrow fibroblasts and is required for BMSC homing/retention in the bone marrow microenvironment. SDF-1 and its receptors CXCR4 and CXCR7 were found upregulated after early focal cerebral ischemia [121] and showed beneficial for the adhesion and migration of BMSCs both to bone marrow and to ischemic tissue through activation of specific integrin molecules. Given that CXCR4 and CXCR7 are present on bone marrow stem cells [14], upregulation of SDF-1 in the local ischemic damage after injury may be related to stem cell homing and engraftment toward the injured tissue. During cerebral ischemia, SDF-1 was found primarily co-localized with endothelial cells and closely interacted with infiltrated BMSCs from bone marrow in the ischemic penumbra region, suggesting that SDF-1 may mediate trafficking of transplanted BMSCs to ischemically damaged tissue. Indeed, overexpression of SDF-1 in ischemic tissues has recently been found to augment EPC-induced vasculogenesis in hind-limb ischemic mice, as well as enhanced recovery of blood perfusion, increased capillary density, and induced partial incorporation of EPCs into the microvessels [129]. Our previous studies have highlighted biphasic function of SDF-1 in stroke mice in a time-dependent manner. One study demonstrated that injection of CXCR4 inhibitor AMD3100 into ischemic mice during acute phase significantly suppressed inflammatory response and reduced blood–brain barrier disruption via inhibiting leukocyte migration and infiltration [39]; however, another study showed that overexpression of SDF-1 in mice brain during post-acute phase promoted neurovascular recovery, neurogenesis, and angiogenesis through enhancing migration of neural progenitor cells and endothelial cells, while AMD3100 reversed protective effects of SDF-1 [66].

In addition to chemokines, growth factors, inflammatory cytokines, and adhesion-related molecules also play important roles in stem cell homing. For instance, PDGF and VEGF are demonstrated to act as chemoattractants to induce migration of MSCs [105]; IL-6, (TGF)-β1, interleukin (IL)-1β, and tumor necrosis factor (TNF)-α stimulate chemotactic migration through matrix metalloproteinases (MMPs) secreted by the MSCs [18]. During the transmigration process of MSCs through the vascular endothelium, integrins and adhesion molecules are involved. Based on the fact that MSCs express α4β1 integrin and vascular cell adhesion molecule-1 (VCAM-1), it is proposed that MSCs roll along the vascular endothelium may share the same mechanism as white blood cells and HSCs to move through the blood vessels. Indeed, Ip et al. identified that β1 integrins are important for the intramyocardial traffic of MSCs by developing a functional genomics approach [42]. Moreover, the adhesion of rat MSCs to endothelial cells of microvessels is reduced by anti-VCAM-1 antibody [98].

9.1.4.2 Tracking of Grafted Stem Cells In Vivo

Different administration routes will result in different homing, distribution, and engraftment. Experimental studies demonstrated that intracerebral [], intra-arterial, intravenous, and intracisternal injection of MSCs result in reduced infarct volume and enhanced behavioral functional recovery, irrespective of pros and cons existing in each injection method [112]. Intracerebral injection delivers and had the highest cell retention in a desired location compared to other methods [111], but it also induces adverse effects involving seizures and transient motor function impairment given its invasive procedure. Intraventricular transplantation is less invasive but achieves less therapeutic efficacy as intraventricularly injected human NSCs into ischemic rat brain did not show improvement [102]. Intravenous delivery is safer and more feasible, but only few cells could localize to the infarct region [111]. Intra-arterial administration contributes to more cells retaining in the brain than intravenous delivery and is beneficial for behavioral recovery [63]. However, intra-arterial transplantation leads to high mortality (about 40 %) and morbidity due to cell accumulation and microemboli, especially when large-sized stem cells (e.g., MSCs) were transplanted intra-arterially [44], which is a major concern for its clinical translation.

In order to determine stem cell migration and in vivo distribution, noninvasive and real-time imaging modalities are developed in recent years. Several multifunctional nanoprobes with high MR sensitivity are developed by our group to label stem cells and allow us to longitudinally track them after injection by MRI in terms of its high spatial resolution. In one study we labeled MSCs [122] and NSCs [142] with high MR sensitivity fluorescent-magnetite-nanocluster (FMNC) and tracked them by MRI and fluorescent imaging after injection into the contralateral hemisphere of the ischemic mice brain. MSCs were detected to migrate toward the perifocal region of the ipsilateral hemisphere through the corpus callosum. We further developed a trifunctional nanoprobe by adding iodine-125 to superparamagnetic iron oxide nanoparticles, which allows us to quantitatively track MSCs injected into the brain by micro-SPECT/CT and MRI. Using this method we found 30 % of intracerebrally grafted MSCs migrated from the injection hemisphere to the lesion area, and intravenously injection induced more than 90 % of MSCs migrated and accumulated in the lung, while no cells were found in the brain (Fig. 9.1) [111]. However, one major limitation of SPIO-based imaging strategy is that survival and dead cells cannot be distinguished. Signals from survival and dead cells are all captured by MRI and micro-SPECT. To resolve this problem, bioluminescence imaging (BLI) is developed and widely used to track the migration and survival of transplanted cells which are modified with a firefly or Renilla luciferase (Luc) enzyme [143]. However, the spatial resolution and the penetration depth of BLI are limited, which hinder its clinical application at current stage.

Recently, radionuclide probes for PET imaging were designed, as 18F-fluorodeoxyglucose ([18F]-FDG) is the most popular one. Several studies have reported direct imaging of transplanted cells with 18F-FDG [9, 113]. To track survival of grafted cells, the herpes simplex virus type 1-derived thymidine kinase (HSV-1-tk), which could exclusively phosphorylate substrates composed of acycloguanosines, is employed and routinely used to monitor human ESCs and C17.2 NSCs in the rodent brain [106, 118].

9.1.5 Strategies to Improve Bone Marrow Stem Cell Homing

Stem cell homing is a multistep process involving cell attachment, adhesion to the vascular endothelium, and migration through the tissue stromal, which are mediated by different factors, including chemokines, growth factors, integrins, and adhesion molecules. Understanding the mechanism of homing could help us to develop novel strategies to improve their homing ability and further increase the therapeutic efficacy. In principle, those methods that used to increase stem cell survival could also apply to improving stem cell homing.

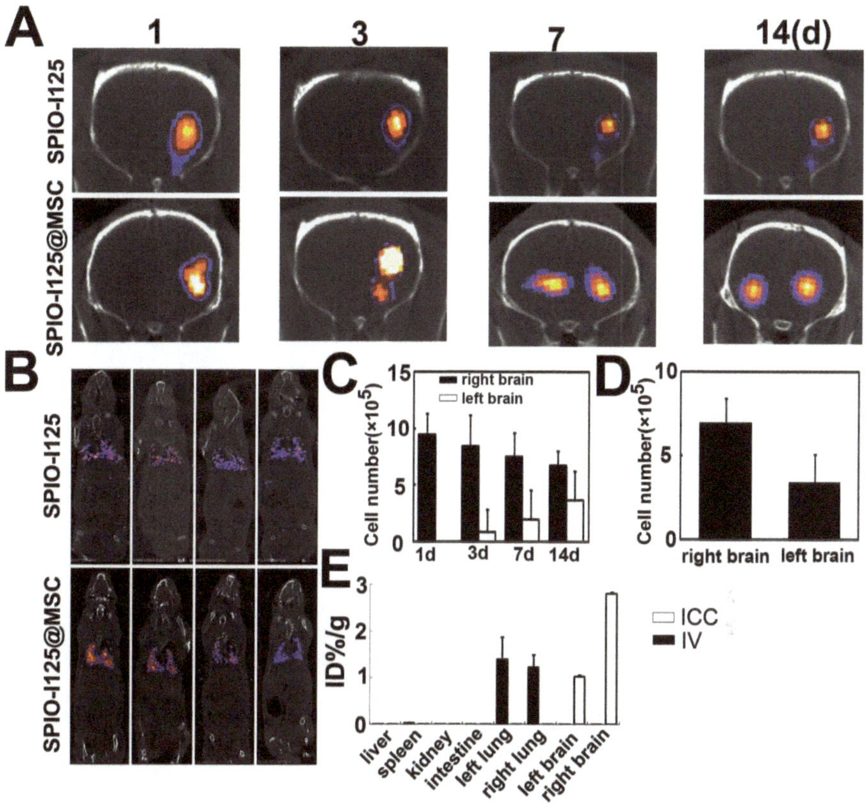

Fig. 9.1 SPECT/CT tracking of (125) I-fSiO4@SPIO-labeled MSCs in ischemic rats after IC and IV injection. (**a**, **b**) SPECT/CT imaging of labeled MSCs and particles alone in ischemic rats after IC (**a**) and IV injection (**b**). (**c**) The radioactivity detected in the right and left hemispheres accounting for the total transplanted dose at different time points after IC injection. (**d**) Ex vivo analysis of radioactivity in right and left hemispheres accounting for the total transplanted dose 14 days after IC injection. (**e**) Biodistribution of (125) I-fSiO4@SPIO-labeled MSCs at 14 days after IV or IC transplantation (Reprinted from Tang et al. [111], Copyright 2015, with permission from Wiley)

9.1.5.1 Homing Gene-Based Method

We and others demonstrated that genetic modification of the target tissue or the stem cells with homing genes is feasible to stimulate their homing ability and further improve behavioral recovery after stroke (Table 9.4). By stereotactic injection of adeno-associated virus (AAV) carrying SDF-1α gene into ischemic mice brain, Li et al. found that migration of endogenous neural stem cells and OPCs from subventricular zone to the peri-infarct region was enhanced and induced increased neurogenesis and oligodendrogenesis, reduced brain atrophy, as well as improved white matter and behavioral recovery [66, 69].

Table 9.4 Enhancement of stem cell homing by gene modification

Overexpressing genes	Stem cells	Animal model	Outcome
CXCR4 [207]	MSCs	Myocardial Infarction	Increased accumulation of BMSCs in the lesion area and an improvement in cardiac function
CCR1 [159]	MSC	Myocardium infarct	CCR1-MSCs accumulated in the infarcted myocardium at significantly higher levels. CCR1-MSC-injected hearts exhibited a significant reduction in infarct size, reduced cardiomyocyte apoptosis, and increased capillary density
ACE2 [150]	EPC	Cerebral ischemia	ACE2 overexpression improved the abilities of EPC migration and tube formation, reduced cerebral infarct volume and neurologic deficits, increased cerebral microvascular density and angiogenesis
HGF siRNA [149]	ASC	Hind-limb ischemia	Transduced ASC-shHGF secreted >80 % less HGF, which led to a reduced ability to promote survival, proliferation, and migration of mature and progenitor endothelial cells in vitro
IGF-1 [156]	MSC	Permanent coronary artery occlusion	IGF-1 transgene expression induced massive stem cell mobilization via SDF-1α signaling and culminated in extensive angiomyogenesis in the infarcted heart
GDNF [163]	NPC	Stroke	More NSPC-GDNF cells migrated toward the ischemic core, reduced infarct volume, and improved behavioral recovery
SCF [193]	NSPCs	Normal mice	Recombinant SCF induces potent NSPC migration in vitro and in vivo through the activation of c-kit on NSPCs
MicroRNA 9 [153]	hESC-derived neural progenitors	Stroke	hNPCs without miR-9 activity also showed enhanced migration

In addition to SDF-1 α, Yu and coworkers demonstrated its receptor CXCR4 also plays a pivotal role in stem cell homing. By transducing MSCs with CXCR4 by lentivirus and injecting them via the femoral vein following MCAO, they found that CXCR4 overexpression promoted MSCs' migration to the infarct region and enhanced neuroprotection via increased angiogenesis [137]. Besides stroke, MSCs overexpressing CXCR4 was also proved to migrate into the cardiac infarct area in a cardiac infarct animal model, leading to a significant improvement in cardiac function [17]. Additionally, growth factors also show their capacity to enhance stem cell migration posttransplantation. Haider et al. demonstrated that IGF-1 overexpression promoted MSC recruitment through paracrine activation of SDF-1α and enhanced myocardial repair [31]. When NPCs overexpressing GDNF were injected into ischemic rat brain, more cells were found accumulated in the lesion area [49]. For EPC it was reported that overexpression of angiotensin-converting enzyme 2 (ACE2) improved the EPC migration and tube formation, and injection of lentivirus-ACE2-transfected EPCs reduced cerebral infarct volume and neurological deficits, which was driven by eNOS [16].

Recently, miRNAs were demonstrated to play an important role in stem cell migration. One study from Delaloy et al. for the first time identified miR-9 as a novel regulator that coordinates the proliferation and migration of hNPCs. They found that hNPCs without miR-9 activity showed enhanced migration when transplanted into mouse embryonic or adult brains in a stroke mouse model [21]. Other miRNAs such as miR-10b and miR-204 have been also proven to play an important role in cell migration [41, 72].

9.1.5.2 Preconditioning-Based Method

As we discussed above, although overexpression of homing genes in both grafted stem cells and local brain tissues improves stem cell homing, several disadvantages exist in this strategy. For instance, uncontrolled expression of introducing genes raises the safety issue, and the risk of tumorigenicity such as leukemia also limits its application. Recently, upregulation of homing genes in MSCs under stress conditions including hypoxia has been confirmed, which may be mediated by HIF-1 alpha [24]. It is reported that hypoxia induces CXCR4 and CXCR7 expression in BMSCs via upregulated HIF-1α [75], and hypoxia preconditioning enhances migration of MSCs via increased expression of cMet [93], which hints at the possibility that hypoxia preconditioning could enhance mobilization of stem cells to lesion sites in ischemic brain. In addition to hypoxia preconditioning, H2O2 preconditioning could increase the migration of MSCs through upregulation of CXCR4 and activation of extracellular signal-regulated kinase (ERK) [65], and pretreatment of HSCs with SDF-1 or dextran sulfate enhances their homing to bone marrow, which is involved in several genes including CXCR4 and MMP-9 [33].

Accumulating evidence shows that pretreatment with growth factors also increases MSCs' mobilization (Table 9.5). In previous investigations, IGF-1 as well as VEGF increased MSC migratory responses via CXCR4 chemokine receptor signaling which is PI3/Akt dependent [70, 109]. Early studies have demonstrated that statins increased EPC number and function through activating the Akt/eNOS pathway [22]. Likewise, enhancement of eNOS enhancers improves the stem cell homing. In particular, pretreatment with eNOS enhancers significantly increased the homing of the intravenously infused EPCs or BMCs and led to increased exercise capacity in a hind-limb ischemia model [95].

Table 9.5 Preconditioning mediators to enhance stem cell homing

Triggers	Stem cells involved	Animal model	Stem cell fate	Outcome
Hypoxia	hESCs [155]; MSC [158, 161, 201]; ES-NPCs [195]	Myocardial infarction; MCAO	Increased neural precursor cell survival; engraftment of MSC was increased; cell survival was increased; promoted their survival, migration, and homing to the ischemic brain region; promote transplanted cell survival	Promoted neuronal differentiation; improvement in global, regional, and diastolic left ventricular functions; an increase in angiogenesis, as well as enhanced morphologic and functional benefits of stem cell therapy; reduced infarct volume and improved behavioral recovery; ES-NPCs exhibited extensive neuronal differentiation in the ischemic brain, accelerated and enhanced recovery of sensorimotor function
H2O2	MSC [172, 199]	Myocardial infarction	Increased cell survival	Increased fractional shortening, ejection fraction, arteriole density and decreased infarct size; increased the capillary density and the fractional shortening and attenuated myocardial fibrosis
Hydrogen sulfide	MSCs [202]	Myocardial infarction	Improved the survival rate of the transplanted MSCs	Reduced the infarct size and increased left ventricular (LV) function
IGF-1	NSCs [94]	MCAO	Protected the grafted neural stem cells from ischemic reperfusion injury	Attenuated infarct size, improved neurological performance and angiogenesis
VEGF	MSCs [181, 194]; ADMSC [205]	MCAO; acute lung ischemia-reperfusion injury; ischemic kidney	Improved survival of MSCs; decreased apoptosis of ADMSC	Reduced infarct volume, enhanced angiogenesis, neurogenesis, and functional recovery; protected the lung from ischemic injury; increased survival, paracrine activity, and efficiency of MSCs
Minocycline	NSCs [188]; OPC [190]	MCAO; in vitro	Minocycline preconditioning protected the grafted NSCs from ischemic reperfusion injury; reduced apoptosis in response to OGD	Attenuated infarct size and improved neurological performance

(continued)

Table 9.5 (continued)

Triggers	Stem cells involved	Animal model	Stem cell fate	Outcome
Doxycycline	NSCs [178]	In vitro	Decreased cell death and increased cell viability after oxygen–glucose deprivation–reoxygenation	Showed cryoprotective via induced the expression of Nrf2
BDNF	NSC [186]	MCAO	Promoted cell survival 1 week after transplantation	BDNF pretreatment of NSCs results in higher initial NSC engraftment and survival, increased neuroprotection, and greater functional recovery
Valporate and lithium [198]	MSC	MCAO	Priming with VPA or lithium increased the number of MSC homing to the cerebral infarcted regions, and copriming with VPA and lithium further enhanced this effect through VPA-induced CXCR4 overexpression and lithium-induced MMP-9 upregulation	Priming with VPA and/or lithium improved functional recovery, reduced brain infarct volume, and enhanced angiogenesis

9.1.5.3 Biomaterial-Based Method

With the rapid development of tissue engineering, many state-of-the-art biomaterials have been developed to combine stem cells to treat cerebrovascular diseases, with the ultimate goal of repairing organs and tissue. In past two decades, many protein-based, polysaccharide-based, polymer-based, peptide-based, and ceramic-based scaffolds that have been proven to promote the viability, differentiation, and migration of stem cells are well designed [125]. Both natural and synthetic biomaterials have been developed and combined with stem cell-based therapy to promote cell survival and migration posttransplantation (Table 9.6).

Fibrin gel is ranked as the first biomaterial to prevent bleeding and promote wound healing in terms of the abundance of fibrinogen, ease fabrication, controllable gelation time, and tunable mechanical property. Fibrin gel is able to exclusively enhance the migration of the transplanted cells toward the lesion boundary zone, even it disappears completely 4 weeks after transplantation [133]. In one study performed by Lee and coworkers, they designed a VEGF-releasing gel that could attract NSC migration [171]. It is also reported that PEGylated fibrin patch controlled the release of SDF-1α at the infarct site and increased the rate of c-kit+ stem

Table 9.6 Biomaterial-based method to enhance stem cell homing

Biomaterials	Stem cells/ homing factors involved	Animal model	Stem cell fate
Fibrin gel	BMSC [204]	Cortical injury	Fibrin matrix enhanced the retention of the transplanted cells within the lesion, migration toward the lesion boundary zone, and differentiation into the neurons and perivascular cells
	C17.2 cell line [171]	Myocardial ischemia	The cells migrated toward the fibrin gel, with the total migration distance of 102.4 ± 76.1 μm over 3 days
	(PEGylated) fibrin patch [208]		The myocardial recruitment of c-kit+ cells was significantly higher in the group treated with the SDF-1a PEGylated fibrin patch
Alginate microspheres	Bone marrow-derived progenitor cells [165]	Hind-limb ischemia	Increased mobilization of bone marrow-derived progenitor cells and also improved recruitment of angiogenic cells expressing CXCR4 from bone marrow and local tissue
	hMSCs [206]	Myocardial ischemia	RGD-modified alginate improved cell attachment and growth and increased angiogenic growth factor expression
starPEG-heparin hydrogels	EPCs [146]	In vitro	Higher migration rates were achieved
Gtn-HPA hydrogels and PCNs	NPCs [174]	In vitro	Gtn-HPA/SDF-1-PCN hydrogels promoted hemotactic recruitment to enhance infiltration of aNPCs by 3- to 45-fold relative to hydrogels that lacked SDF-1
Collagen microgel	hMSCs [197]	Hind-limb ischemia	Optimized hMSC embedded microgels were shown to induce vascular repair and functional improvement by increasing SDF-1 expression
HA	EPCs [177]	Myocardial ischemia	Induced continuous homing of EPCs and improved left ventricular function in a rat model of myocardial infarction
	SDF-1 [191]		Injection of biomimetic hydrogels containing SDF-1 and Ac-SDKP increased stem cell homing and significantly improved left ventricle function, increased angiogenesis, decreased infarct size and great

(continued)

Table 9.6 (continued)

Biomaterials	Stem cells/ homing factors involved	Animal model	Stem cell fate
PLGA	SDF-1 [152, 196]	In vitro	Released SDF-1α caused significant migration of MSCs throughout the duration of release from the microspheres
			Threefold increase of the host-derived stem cell migration at the interface for up to 2 weeks
PCL	MSCs [189]	Bone tissue engineering model	MSCs were shown to migrate within a polycaprolactone scaffold in response to SDF-1
PLEOF [157]	BMSCs	In vitro	The migration of BMS cells in response to time-released SDF-1alpha closely followed the protein release kinetics from the hydrogels
PUASM [164]	SDF-1	MCAO	Systemic administration of SDF-1α-loaded copolymer into ischemic rat resulted in enhanced angiogenesis and neurogenesis
SPIONs combined with exterior magnet	EPCs [173]	MCAO	SPION-labeled EPC homing was greatly increased in ischemic hemisphere with magnetic field treatment
	MSCs [185]	Balloon angioplasty in a rabbit model	Magnetic targeting of mesenchymal stem cells gives rise to a sixfold increase in cell retention following balloon angioplasty in a rabbit model
	hNSCs [192]		Magnet treated rats had a larger number and greater distribution of ferumoxide-labeled NSCs as compared with controls

cell recruitment and offered potential therapeutic benefits in the myocardium ischemic mice [140]. This body of work suggests that migration of stem cells can be monitored by fibrin scaffolds.

Recently, scaffolds fabricated from gelatin [130], collagen [88], alginate [36], and hyaluronic acid (HA) [128] have been developed for the controlled release of growth factors, which could provide homing signals to enhance stem cell migration. Kuraitis et al. found encapsulating SDF-1 into alginate microspheres led to increased mobilization of bone marrow-derived CXCR4+ progenitor cells and restoring perfusion to ischemic tissues via neovascularization [56]. Further studies demonstrated hMSCs encapsulated in RGD-modified alginate microspheres are capable of facilitating myocardial repair [135]. Baumann et al. reported that encapsulating SDF-1α with starPEG-heparin hydrogels enhanced migration of EPCs in vitro [6]. Lim et al.

developed a multifunctional biomaterial comprising injectable gelatin-hydroxyphenylpropionic acid (Gtn-HPA) hydrogels and dextran sulfate/chitosan polyelectrolyte complex nanoparticles (PCNs) to carry SDF-1 to promote infiltration of NPCs through MMP-9 [73]. In particular, an interesting study fabricated and optimized a shape-controlled 3D type-I collagen-based microgel platform to modulate SDF-1 expression of hMSCs, and hMSCs embedded in the microgels were shown to induce vascular repair and functional improvement in hind-limb ischemic mouse [116]. Currently, HA is gaining its popularity as a biomaterial for tissue regeneration [62]. By chemically modifying HA with hydroxyethyl methacrylate, controlled release of SDF-1 was achieved after its encapsulation into HA, and enhanced endothelial progenitor cell chemotaxis was identified [79]. It is also reported that loading SDF-1 and angiogenic peptides (Ac-SDKP) to HA-based hydrogel promoted regeneration of cardiac function through increasing stem cell homing and angiogenesis [103].

Poly lactic-co-glycolic acid (PLGA) is an FDA-approved polymer and the most attractive polymeric drug/protein carrier among those synthetic materials as its high biocompatibility, biodegradability, and tunable mechanical property. PLGA has been extensively designed for controlled release of small molecule drugs, proteins, and other macromolecules in commercial use and in research. Double-emulsion solvent extraction/evaporation is a routine technique to load proteins to biodegradable PLGA microspheres. Using this strategy, Cross et al. loaded SDF-lα into PLGA microspheres for releasing SDF-1α over 50 days without affecting its bioactivity, and significant migration of MSCs throughout the duration of release from the microspheres was observed [19]. Thevenot and colleagues fabricated PLGA salt-leached scaffolds to carry SDF-1 and implanted in the subcutaneous cavity of Balb/c mice. They found this strategy enhanced host-derived stem cell engraftment by threefold compared to conventional mini-osmotic pump delivery for up to 2 weeks with limited inflammatory response [115].

In addition to PLGA, polycaprolactone (PCL) and poly (lactide ethylene oxide fumarate) hydrogel (PLEOF) have also been used to achieve MSC recruitment. Schantz et al. have developed acellular PCL scaffolds that allowed sequential delivery of VEGF, SDF-1, and bone morphogenetic protein-6 (BMP-6) in the rat and increased MSCs infiltrating into the scaffold, with concomitant angiogenesis [97]. In another study, He et al. synthesized SDF-1-loaded PLEOF hydrogel with poly(l-lactide) (PLA) fractions. A pronounced burst release followed by a period of sustained release was achieved, and MSCs showed migration to SDF-1 in a dose-dependent manner [34]. Recently, Kim et al. synthesized a dual pH-sensitive copolymer-poly (urethane amino sulfamethazine) (PUASM)-based random copolymer for controlled release of SDF-1 in stroke. This copolymer showed high protein encapsulation efficiency at pH 7.4, and at pH 5.5, it could release protein rapidly. Systemic administration of SDF-1α-loaded copolymer into ischemic rat resulted in enhanced angiogenesis and neurogenesis [51].

Recent studies have highlighted the role of superparamagnetic iron oxide nanoparticles in targeted cell delivery. Experimental studies from Dr. Yang's lab showed that intravenous injection of SPION-labeled EPCs into ischemic mice and followed by magnetic field treatment promoted their migration to the infarcts, further reduced brain atrophic volume, and improved neurobehavioral outcomes [64]. Other studies with this method also showed that magnetic targeting of MSCs or hNSCs led to increased cell retention following their injection [92, 104]. An interesting study reported that small direct current (DC) electric fields induced significant directional migration of hNSCs toward the cathode independent of CXCR4 signal [27].

9.2 Conclusion

Bone marrow-derived stem cells have been demonstrated as promising sources of adult stem cells for regeneration and repair of neurological disorders, including ischemic stroke. On the other hand, many experimental studies make us recognize many fundamental questions related to the cell survival, homing, and engraftment that contribute to the limited efficacy of BM-derived stem cell transplantation in the clinic. We and other groups have proposed many strategies such as gene modification, preconditioning treatment, and biomaterial-based method to overcome these limitations. Strategies to improve cell survival and homing would enhance their therapeutic efficacy and strengthen the application potential of stem cell therapy. In summary, stem cell-based therapy for ischemic stroke in humans is still in its infancy. Further basic and translational studies are required before it becomes a scientifically proven strategy in clinical setting.

References

1. Asahara T, Murohara T, Sullivan A, Silver M, van der Zee R, Li T, Witzenbichler B, Schatteman G, Isner JM. Isolation of putative progenitor endothelial cells for angiogenesis. Science. 1997;275(5302):964–7.
2. Bacigaluppi M, Pluchino S, Peruzzotti-Jametti L, Kilic E, Kilic U, Salani G, Brambilla E, West MJ, Comi G, Martino G, Hermann DM. Delayed post-ischaemic neuroprotection following systemic neural stem cell transplantation involves multiple mechanisms. Brain J Neurol. 2009;132(Pt 8):2239–51. doi:10.1093/brain/awp174.
3. Bakshi A, Keck CA, Koshkin VS, LeBold DG, Siman R, Snyder EY, McIntosh TK. Caspase-mediated cell death predominates following engraftment of neural progenitor cells into traumatically injured rat brain. Brain Res. 2005;1065(1–2):8–19. doi:10.1016/j.brainres.2005.09.059.
4. Banerjee S, Bentley P, Hamady M, Marley S, Davis J, Shlebak A, Nicholls J, Williamson DA, Jensen SL, Gordon M, Habib N, Chataway J. Intra-arterial immunoselected CD34+ stem cells for acute ischemic stroke. Stem Cells Transl Med. 2014;3(11):1322–30. doi:10.5966/sctm.2013-0178.
5. Bao X, Wei J, Feng M, Lu S, Li G, Dou W, Ma W, Ma S, An Y, Qin C, Zhao RC, Wang R. Transplantation of human bone marrow-derived mesenchymal stem cells promotes behavioral recovery and endogenous neurogenesis after cerebral ischemia in rats. Brain Res. 2011;1367:103–13. doi:10.1016/j.brainres.2010.10.063.

6. Baumann L, Prokoph S, Gabriel C, Freudenberg U, Werner C, Beck-Sickinger AG. A novel, biased-like SDF-1 derivative acts synergistically with starPEG-based heparin hydrogels and improves eEPC migration in vitro. J Control Release. 2012;162(1):68–75. doi:10.1016/j.jconrel.2012.04.049.
7. Bible E, Chau DY, Alexander MR, Price J, Shakesheff KM, Modo M. The support of neural stem cells transplanted into stroke-induced brain cavities by PLGA particles. Biomaterials. 2009;30(16):2985–94. doi:10.1016/j.biomaterials.2009.02.012.
8. Bible E, Qutachi O, Chau DY, Alexander MR, Shakesheff KM, Modo M. Neo-vascularization of the stroke cavity by implantation of human neural stem cells on VEGF-releasing PLGA microparticles. Biomaterials. 2012;33(30):7435–46. doi:10.1016/j.biomaterials.2012.06.085.
9. Bonios M, Terrovitis J, Chang CY, Engles JM, Higuchi T, Lautamaki R, Yu J, Fox J, Pomper M, Wahl RL, Tsui BM, O'Rourke B, Bengel FM, Marban E, Abraham MR. Myocardial substrate and route of administration determine acute cardiac retention and lung bio-distribution of cardiosphere-derived cells. J Nucl Cardiol. 2011;18(3):443–50. doi:10.1007/s12350-011-9369-9.
10. Borlongan CV. Bone marrow stem cell mobilization in stroke: a 'bonehead' may be good after all! Leukemia. 2011;25(11):1674–86. doi:10.1038/leu.2011.167.
11. Borlongan CV, Glover LE, Tajiri N, Kaneko Y, Freeman TB. The great migration of bone marrow-derived stem cells toward the ischemic brain: therapeutic implications for stroke and other neurological disorders. Prog Neurobiol. 2011;95(2):213–28. doi:10.1016/j.pneurobio.2011.08.005.
12. Calio ML, Marinho DS, Ko GM, Ribeiro RR, Carbonel AF, Oyama LM, Ormanji M, Guirao TP, Calio PL, Reis LA, Simoes Mde J, Lisboa-Nascimento T, Ferreira AT, Bertoncini CR. Transplantation of bone marrow mesenchymal stem cells decreases oxidative stress, apoptosis, and hippocampal damage in brain of a spontaneous stroke model. Free Radic Biol Med. 2014;70:141–54. doi:10.1016/j.freeradbiomed.2014.01.024.
13. Caplan AI. Mesenchymal stem cells. J Orthop Res. 1991;9(5):641–50. doi:10.1002/jor.1100090504.
14. Cencioni C, Capogrossi MC, Napolitano M. The SDF-1/CXCR4 axis in stem cell preconditioning. Cardiovasc Res. 2012;94(3):400–7. doi:10.1093/cvr/cvs132.
15. Chen J, Li Y, Wang L, Zhang Z, Lu D, Lu M, Chopp M. Therapeutic benefit of intravenous administration of bone marrow stromal cells after cerebral ischemia in rats. Stroke J Cereb Circ. 2001;32(4):1005–11.
16. Chen J, Xiao X, Chen S, Zhang C, Chen J, Yi D, Shenoy V, Raizada MK, Zhao B, Chen Y. Angiotensin-converting enzyme 2 priming enhances the function of endothelial progenitor cells and their therapeutic efficacy. Hypertension. 2013;61(3):681–9. doi:10.1161/HYPERTENSIONAHA.111.00202.
17. Cheng Z, Ou L, Zhou X, Li F, Jia X, Zhang Y, Liu X, Li Y, Ward CA, Melo LG, Kong D. Targeted migration of mesenchymal stem cells modified with CXCR4 gene to infarcted myocardium improves cardiac performance. Mol Ther J Am Soc Gene Ther. 2008;16(3):571–9. doi:10.1038/sj.mt.6300374.
18. Coronel MF, Musolino PL, Villar MJ. Selective migration and engraftment of bone marrow mesenchymal stem cells in rat lumbar dorsal root ganglia after sciatic nerve constriction. Neurosci Lett. 2006;405(1–2):5–9. doi:10.1016/j.neulet.2006.06.018.
19. Cross DP, Wang C. Stromal-derived factor-1 alpha-loaded PLGA microspheres for stem cell recruitment. Pharm Res. 2011;28(10):2477–89. doi:10.1007/s11095-011-0474-x.
20. Darsalia V, Kallur T, Kokaia Z. Survival, migration and neuronal differentiation of human fetal striatal and cortical neural stem cells grafted in stroke-damaged rat striatum. Eur J Neurosci. 2007;26(3):605–14. doi:10.1111/j.1460-9568.2007.05702.x.
21. Delaloy C, Liu L, Lee JA, Su H, Shen F, Yang GY, Young WL, Ivey KN, Gao FB. MicroRNA-9 coordinates proliferation and migration of human embryonic stem cell-derived neural progenitors. Cell Stem Cell. 2010;6(4):323–35. doi:10.1016/j.stem.2010.02.015.

22. Dimmeler S, Aicher A, Vasa M, Mildner-Rihm C, Adler K, Tiemann M, Rutten H, Fichtlscherer S, Martin H, Zeiher AM. HMG-CoA reductase inhibitors (statins) increase endothelial progenitor cells via the PI 3-kinase/Akt pathway. J Clin Invest. 2001;108(3):391–7. doi:10.1172/JCI13152.
23. Doeppner TR, Ewert TA, Tonges L, Herz J, Zechariah A, ElAli A, Ludwig AK, Giebel B, Nagel F, Dietz GP, Weise J, Hermann DM, Bahr M. Transduction of neural precursor cells with TAT-heat shock protein 70 chaperone: therapeutic potential against ischemic stroke after intrastriatal and systemic transplantation. Stem Cells. 2012;30(6):1297–310. doi:10.1002/stem.1098.
24. Ejtehadifar M, Shamsasenjan K, Movassaghpour A, Akbarzadehlaleh P, Dehdilani N, Abbasi P, Molaeipour Z, Saleh M. The effect of hypoxia on mesenchymal stem cell biology. Adv Pharm Bull. 2015;5(2):141–9. 10.15171/apb.2015.021.
25. Ercan E, Bagla AG, Aksoy A, Gacar G, Unal ZS, Asgun HF, Karaoz E. In vitro protection of adipose tissue-derived mesenchymal stem cells by erythropoietin. Acta Histochem. 2014;116(1):117–25. doi:10.1016/j.acthis.2013.06.007.
26. Fan Y, Shen F, Frenzel T, Zhu W, Ye J, Liu J, Chen Y, Su H, Young WL, Yang GY. Endothelial progenitor cell transplantation improves long-term stroke outcome in mice. Ann Neurol. 2010;67(4):488–97. doi:10.1002/ana.21919.
27. Feng JF, Liu J, Zhang XZ, Zhang L, Jiang JY, Nolta J, Zhao M. Guided migration of neural stem cells derived from human embryonic stem cells by an electric field. Stem Cells. 2012;30(2):349–55. doi:10.1002/stem.779.
28. Frisch SM, Francis H. Disruption of epithelial cell-matrix interactions induces apoptosis. J Cell Biol. 1994;124(4):619–26.
29. Frisch SM, Screaton RA. Anoikis mechanisms. Curr Opin Cell Biol. 2001;13(5):555–62.
30. Haider H, Ashraf M. Preconditioning and stem cell survival. J Cardiovasc Transl Res. 2010;3(2):89–102. doi:10.1007/s12265-009-9161-2.
31. Haider H, Jiang S, Idris NM, Ashraf M. IGF-1-overexpressing mesenchymal stem cells accelerate bone marrow stem cell mobilization via paracrine activation of SDF-1alpha/CXCR4 signaling to promote myocardial repair. Circ Res. 2008;103(11):1300–8. doi:10.1161/CIRCRESAHA.108.186742.
32. Hall PE, Lathia JD, Caldwell MA, Ffrench-Constant C. Laminin enhances the growth of human neural stem cells in defined culture media. BMC Neurosci. 2008;9:71. doi:10.1186/1471-2202-9-71.
33. Hayakawa J, Migita M, Ueda T, Fukazawa R, Adachi K, Ooue Y, Hayakawa M, Shimada T, Fukunaga Y. Dextran sulfate and stromal cell derived factor-1 promote CXCR4 expression and improve bone marrow homing efficiency of infused hematopoietic stem cells. J Nippon Med Sch = Nippon Ika Daigaku Zasshi. 2009;76(4):198–208.
34. He X, Ma J, Jabbari E. Migration of marrow stromal cells in response to sustained release of stromal-derived factor-1alpha from poly(lactide ethylene oxide fumarate) hydrogels. Int J Pharm. 2010;390(2):107–16. doi:10.1016/j.ijpharm.2009.12.063.
35. Hill CE, Moon LD, Wood PM, Bunge MB. Labeled Schwann cell transplantation: cell loss, host Schwann cell replacement, and strategies to enhance survival. Glia. 2006;53(3):338–43. doi:10.1002/glia.20287.
36. Ho YC, Mi FL, Sung HW, Kuo PL. Heparin-functionalized chitosan-alginate scaffolds for controlled release of growth factor. Int J Pharm. 2009;376(1–2):69–75. doi:10.1016/j.ijpharm.2009.04.048.
37. Hu J, Deng L, Wang X, Xu XM. Effects of extracellular matrix molecules on the growth properties of oligodendrocyte progenitor cells in vitro. J Neurosci Res. 2009;87(13):2854–62. doi:10.1002/jnr.22111.
38. Hu X, Yu SP, Fraser JL, Lu Z, Ogle ME, Wang JA, Wei L. Transplantation of hypoxia-preconditioned mesenchymal stem cells improves infarcted heart function via enhanced survival of implanted cells and angiogenesis. J Thorac Cardiovasc Surg. 2008;135(4):799–808. doi:10.1016/j.jtcvs.2007.07.071.

39. Huang J, Li Y, Tang Y, Tang G, Yang GY, Wang Y. CXCR4 antagonist AMD3100 protects blood-brain barrier integrity and reduces inflammatory response after focal ischemia in mice. Stroke J Cereb Circ. 2013;44(1):190–7. doi:10.1161/STROKEAHA.112.670299.
40. Ikeda N, Nonoguchi N, Zhao MZ, Watanabe T, Kajimoto Y, Furutama D, Kimura F, Dezawa M, Coffin RS, Otsuki Y, Kuroiwa T, Miyatake S. Bone marrow stromal cells that enhanced fibroblast growth factor-2 secretion by herpes simplex virus vector improve neurological outcome after transient focal cerebral ischemia in rats. Stroke J Cereb Circ. 2005;36(12):2725–30. doi:10.1161/01.STR.0000190006.88896.d3.
41. Imam JS, Plyler JR, Bansal H, Prajapati S, Bansal S, Rebeles J, Chen HI, Chang YF, Panneerdoss S, Zoghi B, Buddavarapu KC, Broaddus R, Hornsby P, Tomlinson G, Dome J, Vadlamudi RK, Pertsemlidis A, Chen Y, Rao MK. Genomic loss of tumor suppressor miRNA-204 promotes cancer cell migration and invasion by activating AKT/mTOR/Rac1 signaling and actin reorganization. PLoS ONE. 2012;7(12), e52397. doi:10.1371/journal.pone.0052397.
42. Ip JE, Wu Y, Huang J, Zhang L, Pratt RE, Dzau VJ. Mesenchymal stem cells use integrin beta1 not CXC chemokine receptor 4 for myocardial migration and engraftment. Mol Biol Cell. 2007;18(8):2873–82. doi:10.1091/mbc.E07-02-0166.
43. Ishibashi S, Sakaguchi M, Kuroiwa T, Yamasaki M, Kanemura Y, Shizuko I, Shimazaki T, Onodera M, Okano H, Mizusawa H. Human neural stem/progenitor cells, expanded in long-term neurosphere culture, promote functional recovery after focal ischemia in Mongolian gerbils. J Neurosci Res. 2004;78(2):215–23. doi:10.1002/jnr.20246.
44. Janowski M, Lyczek A, Engels C, Xu J, Lukomska B, Bulte JW, Walczak P. Cell size and velocity of injection are major determinants of the safety of intracarotid stem cell transplantation. J Cereb Blood Flow Metab. 2013;33(6):921–7. doi:10.1038/jcbfm.2013.32.
45. Jin K, Mao X, Xie L, Galvan V, Lai B, Wang Y, Gorostiza O, Wang X, Greenberg DA. Transplantation of human neural precursor cells in Matrigel scaffolding improves outcome from focal cerebral ischemia after delayed postischemic treatment in rats. J Cereb Blood Flow Metab. 2010;30(3):534–44. doi:10.1038/jcbfm.2009.219.
46. Jin K, Mao X, Xie L, Greenberg RB, Peng B, Moore A, Greenberg MB, Greenberg DA. Delayed transplantation of human neural precursor cells improves outcome from focal cerebral ischemia in aged rats. Aging Cell. 2010;9(6):1076–83. doi: 10.1111/j.1474-9726.2010.00638.x.
47. Johann V, Schiefer J, Sass C, Mey J, Brook G, Kruttgen A, Schlangen C, Bernreuther C, Schachner M, Dihne M, Kosinski CM. Time of transplantation and cell preparation determine neural stem cell survival in a mouse model of Huntington's disease. Exp Brain Res. 2007;177(4):458–70. doi:10.1007/s00221-006-0689-y.
48. Kallur T, Darsalia V, Lindvall O, Kokaia Z. Human fetal cortical and striatal neural stem cells generate region-specific neurons in vitro and differentiate extensively to neurons after intrastriatal transplantation in neonatal rats. J Neurosci Res. 2006;84(8):1630–44. doi:10.1002/jnr.21066.
49. Kameda M, Shingo T, Takahashi K, Muraoka K, Kurozumi K, Yasuhara T, Maruo T, Tsuboi T, Uozumi T, Matsui T, Miyoshi Y, Hamada H, Date I. Adult neural stem and progenitor cells modified to secrete GDNF can protect, migrate and integrate after intracerebral transplantation in rats with transient forebrain ischemia. Eur J Neurosci. 2007;26(6):1462–78. doi:10.1111/j.1460-9568.2007.05776.x.
50. Kaplan RN, Psaila B, Lyden D. Niche-to-niche migration of bone-marrow-derived cells. Trends Mol Med. 2007;13(2):72–81. doi:10.1016/j.molmed.2006.12.003.
51. Kim DH, Seo YK, Thambi T, Moon GJ, Son JP, Li G, Park JH, Lee JH, Kim HH, Lee DS, Bang OY. Enhancing neurogenesis and angiogenesis with target delivery of stromal cell derived factor-1alpha using a dual ionic pH-sensitive copolymer. Biomaterials. 2015;61:115–25. doi:10.1016/j.biomaterials.2015.05.025.
52. Kim JH, Oh AY, Choi YM, Ku SY, Kim YY, Lee NJ, Sepac A, Bosnjak ZJ. Isoflurane decreases death of human embryonic stem cell-derived, transcriptional marker Nkx2.5(+) cardiac progenitor cells. Acta Anaesthesiol Scand. 2011;55(9):1124–31. doi:10.1111/j.1399-6576.2011.02509.x.

53. Kinnaird T, Stabile E, Burnett MS, Lee CW, Barr S, Fuchs S, Epstein SE. Marrow-derived stromal cells express genes encoding a broad spectrum of arteriogenic cytokines and promote in vitro and in vivo arteriogenesis through paracrine mechanisms. Circ Res. 2004;94(5):678–85. doi:10.1161/01.RES.0000118601.37875.AC.
54. Kostjuk S, Loseva P, Chvartatskaya O, Ershova E, Smirnova T, Malinovskaya E, Roginko O, Kuzmin V, Izhevskaia V, Baranova A, Ginter E, Veiko N. Extracellular GC-rich DNA activates TLR9- and NF-kB-dependent signaling pathways in human adipose-derived mesenchymal stem cells (haMSCs). Expert Opin Biol Ther. 2012;12 Suppl 1:S99–111. doi:10.1517/14712598.2012.690028.
55. Kucia M, Zhang YP, Reca R, Wysoczynski M, Machalinski B, Majka M, Ildstad ST, Ratajczak J, Shields CB, Ratajczak MZ. Cells enriched in markers of neural tissue-committed stem cells reside in the bone marrow and are mobilized into the peripheral blood following stroke. Leukemia. 2006;20(1):18–28. doi:10.1038/sj.leu.2404011.
56. Kuraitis D, Zhang P, Zhang Y, Padavan DT, McEwan K, Sofrenovic T, McKee D, Zhang J, Griffith M, Cao X, Musaro A, Ruel M, Suuronen EJ. A stromal cell-derived factor-1 releasing matrix enhances the progenitor cell response and blood vessel growth in ischaemic skeletal muscle. Eur Cells Mater. 2011;22:109–23.
57. Kurozumi K, Nakamura K, Tamiya T, Kawano Y, Ishii K, Kobune M, Hirai S, Uchida H, Sasaki K, Ito Y. Mesenchymal stem cells that produce neurotrophic factors reduce ischemic damage in the rat middle cerebral artery occlusion model. Mol Ther. 2005;11(1):96–104.
58. Lam J, Lowry WE, Carmichael ST, Segura T. Delivery of iPS-NPCs to the stroke cavity within a hyaluronic acid matrix promotes the differentiation of transplanted cells. Adv Funct Mater. 2014;24(44):7053–62. doi:10.1002/adfm.201401483.
59. Le Belle JE, Caldwell MA, Svendsen CN. Improving the survival of human CNS precursor-derived neurons after transplantation. J Neurosci Res. 2004;76(2):174–83. doi:10.1002/jnr.20035.
60. Lee HJ, Kim KS, Park IH, Kim SU. Human neural stem cells over-expressing VEGF provide neuroprotection, angiogenesis and functional recovery in mouse stroke model. PLoS ONE. 2007;2(1), e156. doi:10.1371/journal.pone.0000156.
61. Lee HJ, Kim MK, Kim HJ, Kim SU. Human neural stem cells genetically modified to overexpress Akt1 provide neuroprotection and functional improvement in mouse stroke model. PLoS ONE. 2009;4(5), e5586. doi:10.1371/journal.pone.0005586.
62. Lei Y, Gojgini S, Lam J, Segura T. The spreading, migration and proliferation of mouse mesenchymal stem cells cultured inside hyaluronic acid hydrogels. Biomaterials. 2011;32(1):39–47. doi:10.1016/j.biomaterials.2010.08.103.
63. Li L, Jiang Q, Qu CS, Ding GL, Li QJ, Wang SY, Lee JH, Lu M, Mahmood A, Chopp M. Transplantation of marrow stromal cells restores cerebral blood flow and reduces cerebral atrophy in rats with traumatic brain injury: in vivo MRI study. J Neurotrauma. 2011;28(4):535–45. doi:10.1089/neu.2010.1619.
64. Li Q, Tang G, Xue S, He X, Miao P, Li Y, Wang J, Xiong L, Wang Y, Zhang C, Yang GY. Silica-coated superparamagnetic iron oxide nanoparticles targeting of EPCs in ischemic brain injury. Biomaterials. 2013;34(21):4982–92. doi:10.1016/j.biomaterials.2013.03.030.
65. Li S, Deng Y, Feng J, Ye W. Oxidative preconditioning promotes bone marrow mesenchymal stem cells migration and prevents apoptosis. Cell Biol Int. 2009;33(3):411–8. doi:10.1016/j.cellbi.2009.01.012.
66. Li Y, Huang J, He X, Tang G, Tang YH, Liu Y, Lin X, Lu Y, Yang GY, Wang Y. Postacute stromal cell-derived factor-1alpha expression promotes neurovascular recovery in ischemic mice. Stroke J Cereb Circ. 2014;45(6):1822–9. doi:10.1161/STROKEAHA.114.005078.
67. Li Y, Liu Z, Xin H, Chopp M. The role of astrocytes in mediating exogenous cell-based restorative therapy for stroke. Glia. 2014;62(1):1–16. doi:10.1002/glia.22585.
68. Li Y, Lu Z, Keogh CL, Yu SP, Wei L. Erythropoietin-induced neurovascular protection, angiogenesis, and cerebral blood flow restoration after focal ischemia in mice. J Cereb Blood Flow Metab. 2007;27(5):1043–54. doi:10.1038/sj.jcbfm.9600417.

69. Li Y, Tang G, Liu Y, He X, Huang J, Lin X, Zhang Z, Yang GY, Wang Y. CXCL12 gene therapy ameliorates ischemia-induced white matter injury in mouse brain. Stem Cells Transl Med. 2015;4(10):1122–30. doi:10.5966/sctm.2015-0074.
70. Li Y, Yu X, Lin S, Li X, Zhang S, Song YH. Insulin-like growth factor 1 enhances the migratory capacity of mesenchymal stem cells. Biochem Biophys Res Commun. 2007;356(3):780–4. doi:10.1016/j.bbrc.2007.03.049.
71. Li YF, Ren LN, Guo G, Cannella LA, Chernaya V, Samuel S, Liu SX, Wang H, Yang XF. Endothelial progenitor cells in ischemic stroke: an exploration from hypothesis to therapy. J Hematol Oncol. 2015;8:33. doi:10.1186/s13045-015-0130-8.
72. Liao CG, Kong LM, Zhou P, Yang XL, Huang JG, Zhang HL, Lu N. miR-10b is overexpressed in hepatocellular carcinoma and promotes cell proliferation, migration and invasion through RhoC, uPAR and MMPs. J Transl Med. 2014;12:234. doi:10.1186/s12967-014-0234-x.
73. Lim TC, Rokkappanavar S, Toh WS, Wang LS, Kurisawa M, Spector M. Chemotactic recruitment of adult neural progenitor cells into multifunctional hydrogels providing sustained SDF-1alpha release and compatible structural support. FASEB J. 2013;27(3):1023–33. doi:10.1096/fj.12-221515.
74. Liu H, Honmou O, Harada K, Nakamura K, Houkin K, Hamada H, Kocsis JD. Neuroprotection by PlGF gene-modified human mesenchymal stem cells after cerebral ischaemia. Brain J Neurol. 2006;129(Pt 10):2734–45. doi:10.1093/brain/awl207.
75. Liu H, Xue W, Ge G, Luo X, Li Y, Xiang H, Ding X, Tian P, Tian X. Hypoxic preconditioning advances CXCR4 and CXCR7 expression by activating HIF-1alpha in MSCs. Biochem Biophys Res Commun. 2010;401(4):509–15. doi:10.1016/j.bbrc.2010.09.076.
76. Liu N, Zhang Y, Fan L, Yuan M, Du H, Cheng R, Liu D, Lin F. Effects of transplantation with bone marrow-derived mesenchymal stem cells modified by survivin on experimental stroke in rats. J Transl Med. 2011;9:105. doi:10.1186/1479-5876-9-105.
77. Liu X, Ye R, Yan T, Yu SP, Wei L, Xu G, Fan X, Jiang Y, Stetler RA, Liu G, Chen J. Cell based therapies for ischemic stroke: from basic science to bedside. Prog Neurobiol. 2014;115:92–115. doi:10.1016/j.pneurobio.2013.11.007.
78. Lu G, Ashraf M, Haider KH. Insulin-like growth factor-1 preconditioning accentuates intrinsic survival mechanism in stem cells to resist ischemic injury by orchestrating protein kinase calpha-erk1/2 activation. Antioxid Redox Signal. 2012;16(3):217–27. doi:10.1089/ars.2011.4112.
79. MacArthur Jr JW, Purcell BP, Shudo Y, Cohen JE, Fairman A, Trubelja A, Patel J, Hsiao P, Yang E, Lloyd K, Hiesinger W, Atluri P, Burdick JA, Woo YJ. Sustained release of engineered stromal cell-derived factor 1-alpha from injectable hydrogels effectively recruits endothelial progenitor cells and preserves ventricular function after myocardial infarction. Circulation. 2013;128(11 Suppl 1):S79–86. doi:10.1161/CIRCULATIONAHA.112.000343.
80. Malik YS, Sheikh MA, Zhu X. Doxycycline can stimulate cytoprotection in neural stem cells with oxygen-glucose deprivation-reoxygenation injury: a potential approach to enhance effectiveness of cell transplantation therapy. Biochem Biophys Res Commun. 2013;432(2):355–8. doi:10.1016/j.bbrc.2013.01.097.
81. Meng S, Cao J, Wang L, Zhou Q, Li Y, Shen C, Zhang X, Wang C. MicroRNA 107 partly inhibits endothelial progenitor cells differentiation via HIF-1beta. PLoS ONE. 2012;7(7), e40323. doi:10.1371/journal.pone.0040323.
82. Miller DJ, Simpson JR, Silver B. Safety of thrombolysis in acute ischemic stroke: a review of complications, risk factors, and newer technologies. Neurohospitalist. 2011;1(3):138–47. doi:10.1177/1941875211408731.
83. Moshayedi P, Carmichael ST. Hyaluronan, neural stem cells and tissue reconstruction after acute ischemic stroke. Biomatter. 2013;3(1):1–9. doi:10.4161/biom.23863.
84. Niagara MI, Haider H, Jiang S, Ashraf M. Pharmacologically preconditioned skeletal myoblasts are resistant to oxidative stress and promote angiomyogenesis via release of paracrine factors in the infarcted heart. Circ Res. 2007;100(4):545–55. doi:10.1161/01.RES.0000258460.41160.ef.

85. Park KI, Teng YD, Snyder EY. The injured brain interacts reciprocally with neural stem cells supported by scaffolds to reconstitute lost tissue. Nat Biotechnol. 2002;20(11):1111–7. doi:10.1038/nbt751.
86. Parr AM, Kulbatski I, Tator CH. Transplantation of adult rat spinal cord stem/progenitor cells for spinal cord injury. J Neurotrauma. 2007;24(5):835–45. doi:10.1089/neu.2006.3771.
87. Patschan D, Hildebrandt A, Rinneburger J, Wessels JT, Patschan S, Becker JU, Henze E, Kruger A, Muller GA. The hormone melatonin stimulates renoprotective effects of "early outgrowth" endothelial progenitor cells in acute ischemic kidney injury. Am J Physiol Ren Physiol. 2012;302(10):F1305–12. doi:10.1152/ajprenal.00445.2011.
88. Quinlan E, Lopez-Noriega A, Thompson EM, Hibbitts A, Cryan SA, O'Brien FJ. Controlled release of vascular endothelial growth factor from spray-dried alginate microparticles in collagen-hydroxyapatite scaffolds for promoting vascularization and bone repair. J Tissue Eng Regen Med. 2015. doi:10.1002/term.2013.
89. Ratajczak MZ, Kim CH, Wojakowski W, Janowska-Wieczorek A, Kucia M, Ratajczak J. Innate immunity as orchestrator of stem cell mobilization. Leukemia. 2010;24(10):1667–75. doi:10.1038/leu.2010.162.
90. Ratajczak MZ, Zuba-Surma E, Wojakowski W, Suszynska M, Mierzejewska K, Liu R, Ratajczak J, Shin DM, Kucia M. Very small embryonic-like stem cells (VSELs) represent a real challenge in stem cell biology: recent pros and cons in the midst of a lively debate. Leukemia. 2014;28(3):473–84. doi:10.1038/leu.2013.255.
91. Rennert RC, Sorkin M, Garg RK, Gurtner GC. Stem cell recruitment after injury: lessons for regenerative medicine. Regen Med. 2012;7(6):833–50. doi:10.2217/rme.12.82.
92. Riegler J, Liew A, Hynes SO, Ortega D, O'Brien T, Day RM, Richards T, Sharif F, Pankhurst QA, Lythgoe MF. Superparamagnetic iron oxide nanoparticle targeting of MSCs in vascular injury. Biomaterials. 2013;34(8):1987–94. doi:10.1016/j.biomaterials.2012.11.040.
93. Rosova I, Dao M, Capoccia B, Link D, Nolta JA. Hypoxic preconditioning results in increased motility and improved therapeutic potential of human mesenchymal stem cells. Stem Cells. 2008;26(8):2173–82. doi:10.1634/stemcells.2007-1104.
94. Sakata H, Niizuma K, Yoshioka H, Kim GS, Jung JE, Katsu M, Narasimhan P, Maier CM, Nishiyama Y, Chan PH. Minocycline-preconditioned neural stem cells enhance neuroprotection after ischemic stroke in rats. J Neurosci Off J Soc Neurosci. 2012;32(10):3462–73. doi:10.1523/JNEUROSCI.5686-11.2012.
95. Sasaki K, Heeschen C, Aicher A, Ziebart T, Honold J, Urbich C, Rossig L, Koehl U, Koyanagi M, Mohamed A, Brandes RP, Martin H, Zeiher AM, Dimmeler S. Ex vivo pretreatment of bone marrow mononuclear cells with endothelial NO synthase enhancer AVE9488 enhances their functional activity for cell therapy. Proc Natl Acad Sci U S A. 2006;103(39):14537–41. doi:10.1073/pnas.0604144103.
96. Savitz SI, Dinsmore JH, Wechsler LR, Rosenbaum DM, Caplan LR. Cell therapy for stroke. NeuroRx J Am Soc Exp Neurother. 2004;1(4):406–14. doi:10.1602/neurorx.1.4.406.
97. Schantz JT, Chim H, Whiteman M. Cell guidance in tissue engineering: SDF-1 mediates site-directed homing of mesenchymal stem cells within three-dimensional polycaprolactone scaffolds. Tissue Eng. 2007;13(11):2615–24. doi:10.1089/ten.2006.0438.
98. Segers VF, Van Riet I, Andries LJ, Lemmens K, Demolder MJ, De Becker AJ, Kockx MM, De Keulenaer GW. Mesenchymal stem cell adhesion to cardiac microvascular endothelium: activators and mechanisms. Am J Physiol Heart Circ Physiol. 2006;290(4):H1370–7. doi:10.1152/ajpheart.00523.2005.
99. Sehara Y, Hayashi T, Deguchi K, Zhang H, Tsuchiya A, Yamashita T, Lukic V, Nagai M, Kamiya T, Abe K. Potentiation of neurogenesis and angiogenesis by G-CSF after focal cerebral ischemia in rats. Brain Res. 2007;1151:142–9. doi:10.1016/j.brainres.2007.01.149.
100. Shyu WC, Lin SZ, Chiang MF, Su CY, Li H. Intracerebral peripheral blood stem cell (CD34+) implantation induces neuroplasticity by enhancing beta1 integrin-mediated angiogenesis in chronic stroke rats. J Neurosci Off J Soc Neurosci. 2006;26(13):3444–53. doi:10.1523/JNEUROSCI.5165-05.2006.

101. Shyu WC, Lin SZ, Lee CC, Liu DD, Li H. Granulocyte colony-stimulating factor for acute ischemic stroke: a randomized controlled trial. CMAJ Can Med Assoc J = J Assoc Med Can. 2006;174(7):927–33. doi:10.1503/cmaj.051322.
102. Smith EJ, Stroemer RP, Gorenkova N, Nakajima M, Crum WR, Tang E, Stevanato L, Sinden JD, Modo M. Implantation site and lesion topology determine efficacy of a human neural stem cell line in a rat model of chronic stroke. Stem Cells. 2012;30(4):785–96. doi:10.1002/stem.1024.
103. Song M, Jang H, Lee J, Kim JH, Kim SH, Sun K, Park Y. Regeneration of chronic myocardial infarction by injectable hydrogels containing stem cell homing factor SDF-1 and angiogenic peptide Ac-SDKP. Biomaterials. 2014;35(8):2436–45. doi:10.1016/j.biomaterials.2013.12.011.
104. Song M, Kim YJ, Kim YH, Roh J, Kim SU, Yoon BW. Using a neodymium magnet to target delivery of ferumoxide-labeled human neural stem cells in a rat model of focal cerebral ischemia. Hum Gene Ther. 2010;21(5):603–10. doi:10.1089/hum.2009.144.
105. Spaeth E, Klopp A, Dembinski J, Andreeff M, Marini F. Inflammation and tumor microenvironments: defining the migratory itinerary of mesenchymal stem cells. Gene Ther. 2008;15(10):730–8. doi:10.1038/gt.2008.39.
106. Sun N, Lee A, Wu JC. Long term non-invasive imaging of embryonic stem cells using reporter genes. Nat Protoc. 2009;4(8):1192–201. doi:10.1038/nprot.2009.100.
107. Taguchi A, Soma T, Tanaka H, Kanda T, Nishimura H, Yoshikawa H, Tsukamoto Y, Iso H, Fujimori Y, Stern DM, Naritomi H, Matsuyama T. Administration of CD34+ cells after stroke enhances neurogenesis via angiogenesis in a mouse model. J Clin Invest. 2004;114(3):330–8. doi:10.1172/JCI20622.
108. Takahashi K, Yasuhara T, Shingo T, Muraoka K, Kameda M, Takeuchi A, Yano A, Kurozumi K, Agari T, Miyoshi Y, Kinugasa K, Date I. Embryonic neural stem cells transplanted in middle cerebral artery occlusion model of rats demonstrated potent therapeutic effects, compared to adult neural stem cells. Brain Res. 2008;1234:172–82. doi:10.1016/j.brainres.2008.07.086.
109. Tang J, Wang J, Kong X, Yang J, Guo L, Zheng F, Zhang L, Huang Y, Wan Y. Vascular endothelial growth factor promotes cardiac stem cell migration via the PI3K/Akt pathway. Exp Cell Res. 2009;315(20):3521–31. doi:10.1016/j.yexcr.2009.09.026.
110. Tang Y, Cai B, Yuan F, He X, Lin X, Wang J, Wang Y, Yang GY. Melatonin pretreatment improves the survival and function of transplanted mesenchymal stem cells after focal cerebral ischemia. Cell Transplant. 2014;23(10):1279–91. doi:10.3727/096368913X667510.
111. Tang Y, Zhang C, Wang J, Lin X, Zhang L, Yang Y, Wang Y, Zhang Z, Bulte JW, Yang GY. MRI/SPECT/fluorescent tri-modal probe for evaluating the homing and therapeutic efficacy of transplanted mesenchymal stem cells in a rat ischemic stroke model. Adv Funct Mater. 2015;25(7):1024–34. doi:10.1002/adfm.201402930.
112. Tang YH, Ma YY, Zhang ZJ, Wang YT, Yang GY. Opportunities and challenges: stem cell-based therapy for the treatment of ischemic stroke. CNS Neurosci Ther. 2015;21(4):337–47. doi:10.1111/cns.12386.
113. Terrovitis J, Lautamaki R, Bonios M, Fox J, Engles JM, Yu J, Leppo MK, Pomper MG, Wahl RL, Seidel J, Tsui BM, Bengel FM, Abraham MR, Marban E. Noninvasive quantification and optimization of acute cell retention by in vivo positron emission tomography after intramyocardial cardiac-derived stem cell delivery. J Am Coll Cardiol. 2009;54(17):1619–26. doi:10.1016/j.jacc.2009.04.097.
114. Theus MH, Wei L, Cui L, Francis K, Hu X, Keogh C, Yu SP. In vitro hypoxic preconditioning of embryonic stem cells as a strategy of promoting cell survival and functional benefits after transplantation into the ischemic rat brain. Exp Neurol. 2008;210(2):656–70. doi:10.1016/j.expneurol.2007.12.020.
115. Thevenot PT, Nair AM, Shen J, Lotfi P, Ko CY, Tang L. The effect of incorporation of SDF-1alpha into PLGA scaffolds on stem cell recruitment and the inflammatory response. Biomaterials. 2010;31(14):3997–4008. doi:10.1016/j.biomaterials.2010.01.144.

116. Thomas D, Fontana G, Chen X, Sanz-Nogues C, Zeugolis DI, Dockery P, O'Brien T, Pandit A. A shape-controlled tuneable microgel platform to modulate angiogenic paracrine responses in stem cells. Biomaterials. 2014;35(31):8757–66. doi:10.1016/j.biomaterials.2014.06.053.
117. Tilkorn DJ, Davies EM, Keramidaris E, Dingle AM, Gerrand YW, Taylor CJ, Han XL, Palmer JA, Penington AJ, Mitchell CA, Morrison WA, Dusting GJ, Mitchell GM. The in vitro preconditioning of myoblasts to enhance subsequent survival in an in vivo tissue engineering chamber model. Biomaterials. 2012;33(15):3868–79. doi:10.1016/j.biomaterials.2012.02.006.
118. Waerzeggers Y, Klein M, Miletic H, Himmelreich U, Li H, Monfared P, Herrlinger U, Hoehn M, Coenen HH, Weller M, Winkeler A, Jacobs AH. Multimodal imaging of neural progenitor cell fate in rodents. Mol Imaging. 2008;7(2):77–91.
119. Wang JA, He A, Hu X, Jiang Y, Sun Y, Jiang J, Gui C, Wang Y, Chen H. Anoxic preconditioning: a way to enhance the cardioprotection of mesenchymal stem cells. Int J Cardiol. 2009;133(3):410–2. doi:10.1016/j.ijcard.2007.11.096.
120. Wang L, Lin Z, Shao B, Zhuge Q, Jin K. Therapeutic applications of bone marrow-derived stem cells in ischemic stroke. Neurol Res. 2013;35(5):470–8. doi:10.1179/1743132813Y.0000000210.
121. Wang Y, Huang J, Li Y, Yang GY. Roles of chemokine CXCL12 and its receptors in ischemic stroke. Curr Drug Targets. 2012;13(2):166–72.
122. Wang Y, Xu F, Zhang C, Lei D, Tang Y, Xu H, Zhang Z, Lu H, Du X, Yang GY. High MR sensitive fluorescent magnetite nanocluster for stem cell tracking in ischemic mouse brain. Nanomed Nanotechnol Biol Med. 2011;7(6):1009–19. doi:10.1016/j.nano.2011.03.006.
123. Wei L, Cui L, Snider BJ, Rivkin M, Yu SS, Lee CS, Adams LD, Gottlieb DI, Johnson Jr EM, Yu SP, Choi DW. Transplantation of embryonic stem cells overexpressing Bcl-2 promotes functional recovery after transient cerebral ischemia. Neurobiol Dis. 2005;19(1–2):183–93. doi:10.1016/j.nbd.2004.12.016.
124. Wei L, Fraser JL, Lu ZY, Hu X, Yu SP. Transplantation of hypoxia preconditioned bone marrow mesenchymal stem cells enhances angiogenesis and neurogenesis after cerebral ischemia in rats. Neurobiol Dis. 2012;46(3):635–45. doi:10.1016/j.nbd.2012.03.002.
125. Willerth SM, Sakiyama-Elbert SE. Combining stem cells and biomaterial scaffolds for constructing tissues and cell delivery. In: StemBook. Cambridge, MA. 2008. doi:10.3824/stembook.1.1.1.
126. Writing Group M, Lloyd-Jones D, Adams RJ, Brown TM, Carnethon M, Dai S, De Simone G, Ferguson TB, Ford E, Furie K, Gillespie C, Go A, Greenlund K, Haase N, Hailpern S, Ho PM, Howard V, Kissela B, Kittner S, Lackland D, Lisabeth L, Marelli A, McDermott MM, Meigs J, Mozaffarian D, Mussolino M, Nichol G, Roger VL, Rosamond W, Sacco R, Sorlie P, Roger VL, Thom T, Wasserthiel-Smoller S, Wong ND, Wylie-Rosett J, American Heart Association Statistics C, Stroke Statistics S. Heart disease and stroke statistics – 2010 update: a report from the American Heart Association. Circulation. 2010;121(7):e46–215. doi:10.1161/CIRCULATIONAHA.109.192667.
127. Xie X, Sun A, Zhu W, Huang Z, Hu X, Jia J, Zou Y, Ge J. Transplantation of mesenchymal stem cells preconditioned with hydrogen sulfide enhances repair of myocardial infarction in rats. Tohoku J Exp Med. 2012;226(1):29–36.
128. Xu X, Jha AK, Duncan RL, Jia X. Heparin-decorated, hyaluronic acid-based hydrogel particles for the controlled release of bone morphogenetic protein 2. Acta Biomater. 2011;7(8):3050–9. doi:10.1016/j.actbio.2011.04.018.
129. Yamaguchi J, Kusano KF, Masuo O, Kawamoto A, Silver M, Murasawa S, Bosch-Marce M, Masuda H, Losordo DW, Isner JM, Asahara T. Stromal cell-derived factor-1 effects on ex vivo expanded endothelial progenitor cell recruitment for ischemic neovascularization. Circulation. 2003;107(9):1322–8.
130. Yamamoto M, Ikada Y, Tabata Y. Controlled release of growth factors based on biodegradation of gelatin hydrogel. J Biomater Sci Polym Ed. 2001;12(1):77–88.

131. Yan F, Yao Y, Chen L, Li Y, Sheng Z, Ma G. Hypoxic preconditioning improves survival of cardiac progenitor cells: role of stromal cell derived factor-1alpha-CXCR4 axis. PLoS ONE. 2012;7(7), e37948. doi:10.1371/journal.pone.0037948.
132. Yao Y, Zhang F, Wang L, Zhang G, Wang Z, Chen J, Gao X. Lipopolysaccharide preconditioning enhances the efficacy of mesenchymal stem cells transplantation in a rat model of acute myocardial infarction. J Biomed Sci. 2009;16:74. doi:10.1186/1423-0127-16-74.
133. Yasuda H, Kuroda S, Shichinohe H, Kamei S, Kawamura R, Iwasaki Y. Effect of biodegradable fibrin scaffold on survival, migration, and differentiation of transplanted bone marrow stromal cells after cortical injury in rats. J Neurosurg. 2010;112(2):336–44. doi:10.3171/2009.2.JNS08495.
134. Yip HK, Chang YC, Wallace CG, Chang LT, Tsai TH, Chen YL, Chang HW, Leu S, Zhen YY, Tsai CY, Yeh KH, Sun CK, Yen CH. Melatonin treatment improves adipose-derived mesenchymal stem cell therapy for acute lung ischemia-reperfusion injury. J Pineal Res. 2013;54(2):207–21. doi:10.1111/jpi.12020.
135. Yu J, Du KT, Fang Q, Gu Y, Mihardja SS, Sievers RE, Wu JC, Lee RJ. The use of human mesenchymal stem cells encapsulated in RGD modified alginate microspheres in the repair of myocardial infarction in the rat. Biomaterials. 2010;31(27):7012–20. doi:10.1016/j.biomaterials.2010.05.078.
136. Yu SP, Wei Z, Wei L. Preconditioning strategy in stem cell transplantation therapy. Transl Stroke Res. 2013;4(1):76–88. doi:10.1007/s12975-012-0251-0.
137. Yu X, Chen D, Zhang Y, Wu X, Huang Z, Zhou H, Zhang Y, Zhang Z. Overexpression of CXCR4 in mesenchymal stem cells promotes migration, neuroprotection and angiogenesis in a rat model of stroke. J Neurol Sci. 2012;316(1–2):141–9. doi:10.1016/j.jns.2012.01.001.
138. Yuan J, Yankner BA. Apoptosis in the nervous system. Nature. 2000;407(6805):802–9. doi:10.1038/35037739.
139. Zemani F, Silvestre JS, Fauvel-Lafeve F, Bruel A, Vilar J, Bieche I, Laurendeau I, Galy-Fauroux I, Fischer AM, Boisson-Vidal C. Ex vivo priming of endothelial progenitor cells with SDF-1 before transplantation could increase their proangiogenic potential. Arterioscler Thromb Vasc Biol. 2008;28(4):644–50. doi:10.1161/ATVBAHA.107.160044.
140. Zhang G, Nakamura Y, Wang X, Hu Q, Suggs LJ, Zhang J. Controlled release of stromal cell-derived factor-1 alpha in situ increases c-kit+ cell homing to the infarcted heart. Tissue Eng. 2007;13(8):2063–71. doi:10.1089/ten.2006.0013.
141. Zhang J, Chen GH, Wang YW, Zhao J, Duan HF, Liao LM, Zhang XZ, Chen YD, Chen H. Hydrogen peroxide preconditioning enhances the therapeutic efficacy of Wharton's Jelly mesenchymal stem cells after myocardial infarction. Chin Med J. 2012;125(19):3472–8.
142. Zhang L, Wang Y, Tang Y, Jiao Z, Xie C, Zhang H, Gu P, Wei X, Yang GY, Gu H, Zhang C. High MRI performance fluorescent mesoporous silica-coated magnetic nanoparticles for tracking neural progenitor cells in an ischemic mouse model. Nanoscale. 2013;5(10):4506–16. doi:10.1039/c3nr00119a.
143. Zhang SJ, Wu JC. Comparison of imaging techniques for tracking cardiac stem cell therapy. J Nucl Med. 2007;48(12):1916–9. doi:10.2967/jnumed.107.043299.
144. Zhao YH, Yuan B, Chen J, Feng DH, Zhao B, Qin C, Chen YF. Endothelial progenitor cells: therapeutic perspective for ischemic stroke. CNS Neurosci Ther. 2013;19(2):67–75. doi:10.1111/cns.12040.
145. Zhong J, Chan A, Morad L, Kornblum HI, Fan G, Carmichael ST. Hydrogel matrix to support stem cell survival after brain transplantation in stroke. Neurorehabil Neural Repair. 2010;24(7):636–44. doi:10.1177/1545968310361958.
146. Baumann L, Prokoph S, Gabriel C, Freudenberg U, Werner C, Beck-Sickinger AG. A novel, biased-like SDF-1 derivative acts synergistically with starPEG-based heparin hydrogels and improves eEPC migration in vitro. J Control Release. 2012;162(1):68–75. doi:10.1016/j.jconrel.2012.04.049.
147. Bible E, Chau DY, Alexander MR, Price J, Shakesheff KM, Modo M. The support of neural stem cells transplanted into stroke-induced brain cavities by PLGA particles. Biomaterials. 2009;30(16):2985–94. doi:10.1016/j.biomaterials.2009.02.012.

148. Bible E, Qutachi O, Chau DY, Alexander MR, Shakesheff KM, Modo M. Neo-vascularization of the stroke cavity by implantation of human neural stem cells on VEGF-releasing PLGA microparticles. Biomaterials. 2012;33(30):7435–46. doi:10.1016/j.biomaterials.2012.06.085.
149. Cai L, Johnstone BH, Cook TG, Liang Z, Traktuev D, Cornetta K, Ingram DA, Rosen ED, March KL. Suppression of hepatocyte growth factor production impairs the ability of adipose-derived stem cells to promote ischemic tissue revascularization. Stem Cells. 2007;25(12):3234–43. doi:10.1634/stemcells.2007-0388.
150. Chen J, Xiao X, Chen S, Zhang C, Chen J, Yi D, Shenoy V, Raizada MK, Zhao B, Chen Y. Angiotensin-converting enzyme 2 priming enhances the function of endothelial progenitor cells and their therapeutic efficacy. Hypertension. 2013;61(3):681–9. doi:10.1161/HYPERTENSIONAHA.111.00202.
151. Chen SJ, Chang CM, Tsai SK, Chang YL, Chou SJ, Huang SS, Tai LK, Chen YC, Ku HH, Li HY, Chiou SH. Functional improvement of focal cerebral ischemia injury by subdural transplantation of induced pluripotent stem cells with fibrin glue. Stem Cells Dev. 2010;19(11):1757–67. doi:10.1089/scd.2009.0452.
152. Cross DP, Wang C. Stromal-derived factor-1 alpha-loaded PLGA microspheres for stem cell recruitment. Pharm Res. 2011;28(10):2477–89. doi:10.1007/s11095-011-0474-x.
153. Delaloy C, Liu L, Lee JA, Su H, Shen F, Yang GY, Young WL, Ivey KN, Gao FB. MicroRNA-9 coordinates proliferation and migration of human embryonic stem cell-derived neural progenitors. Cell Stem Cell. 2010;6(4):323–35. doi:10.1016/j.stem.2010.02.015.
154. Doeppner TR, Ewert TA, Tonges L, Herz J, Zechariah A, ElAli A, Ludwig AK, Giebel B, Nagel F, Dietz GP, Weise J, Hermann DM, Bahr M. Transduction of neural precursor cells with TAT-heat shock protein 70 chaperone: therapeutic potential against ischemic stroke after intrastriatal and systemic transplantation. Stem Cells. 2012;30(6):1297–310. doi:10.1002/stem.1098.
155. Francis KR, Wei L. Human embryonic stem cell neural differentiation and enhanced cell survival promoted by hypoxic preconditioning. Cell Death Dis. 2010;1, e22. doi:10.1038/cddis.2009.22.
156. Haider H, Jiang S, Idris NM, Ashraf M. IGF-1-overexpressing mesenchymal stem cells accelerate bone marrow stem cell mobilization via paracrine activation of SDF-1alpha/CXCR4 signaling to promote myocardial repair. Circ Res. 2008;103(11):1300–8. doi:10.1161/CIRCRESAHA.108.186742.
157. He X, Ma J, Jabbari E. Migration of marrow stromal cells in response to sustained release of stromal-derived factor-1alpha from poly(lactide ethylene oxide fumarate) hydrogels. Int J Pharm. 2010;390(2):107–16. doi:10.1016/j.ijpharm.2009.12.063.
158. Hu X, Yu SP, Fraser JL, Lu Z, Ogle ME, Wang JA, Wei L. Transplantation of hypoxia-preconditioned mesenchymal stem cells improves infarcted heart function via enhanced survival of implanted cells and angiogenesis. J Thorac Cardiovasc Surg. 2008;135(4):799–808. doi:10.1016/j.jtcvs.2007.07.071.
159. Huang J, Zhang Z, Guo J, Ni A, Deb A, Zhang L, Mirotsou M, Pratt RE, Dzau VJ. Genetic modification of mesenchymal stem cells overexpressing CCR1 increases cell viability, migration, engraftment, and capillary density in the injured myocardium. Circ Res. 2010;106(11):1753–62. doi:10.1161/CIRCRESAHA.109.196030.
160. Ikeda N, Nonoguchi N, Zhao MZ, Watanabe T, Kajimoto Y, Furutama D, Kimura F, Dezawa M, Coffin RS, Otsuki Y, Kuroiwa T, Miyatake S. Bone marrow stromal cells that enhanced fibroblast growth factor-2 secretion by herpes simplex virus vector improve neurological outcome after transient focal cerebral ischemia in rats. Stroke J Cereb Circ. 2005;36(12):2725–30. doi:10.1161/01.STR.0000190006.88896.d3.
161. Jaussaud J, Biais M, Calderon J, Chevaleyre J, Duchez P, Ivanovic Z, Couffinhal T, Barandon L. Hypoxia-preconditioned mesenchymal stromal cells improve cardiac function in a swine model of chronic myocardial ischaemia. Eur J Cardiothorac Surg. 2013;43(5):1050–7. doi:10.1093/ejcts/ezs549.

162. Jin K, Mao X, Xie L, Galvan V, Lai B, Wang Y, Gorostiza O, Wang X, Greenberg DA. Transplantation of human neural precursor cells in Matrigel scaffolding improves outcome from focal cerebral ischemia after delayed postischemic treatment in rats. J Cereb Blood Flow Metab. 2010;30(3):534–44. doi:10.1038/jcbfm.2009.219.
163. Kameda M, Shingo T, Takahashi K, Muraoka K, Kurozumi K, Yasuhara T, Maruo T, Tsuboi T, Uozumi T, Matsui T, Miyoshi Y, Hamada H, Date I. Adult neural stem and progenitor cells modified to secrete GDNF can protect, migrate and integrate after intracerebral transplantation in rats with transient forebrain ischemia. Eur J Neurosci. 2007;26(6):1462–78. doi:10.1111/j.1460-9568.2007.05776.x.
164. Kim DH, Seo YK, Thambi T, Moon GJ, Son JP, Li G, Park JH, Lee JH, Kim HH, Lee DS, Bang OY. Enhancing neurogenesis and angiogenesis with target delivery of stromal cell derived factor-1alpha using a dual ionic pH-sensitive copolymer. Biomaterials. 2015;61:115–25. doi:10.1016/j.biomaterials.2015.05.025.
165. Kuraitis D, Zhang P, Zhang Y, Padavan DT, McEwan K, Sofrenovic T, McKee D, Zhang J, Griffith M, Cao X, Musaro A, Ruel M, Suuronen EJ. A stromal cell-derived factor-1 releasing matrix enhances the progenitor cell response and blood vessel growth in ischaemic skeletal muscle. Eur Cells Mater. 2011;22:109–23.
166. Kurozumi K, Nakamura K, Tamiya T, Kawano Y, Ishii K, Kobune M, Hirai S, Uchida H, Sasaki K, Ito Y, Kato K, Honmou O, Houkin K, Date I, Hamada H. Mesenchymal stem cells that produce neurotrophic factors reduce ischemic damage in the rat middle cerebral artery occlusion model. Mol Ther J Am Soc Gene Ther. 2005;11(1):96–104. doi:10.1016/j.ymthe.2004.09.020.
167. Kutschka I, Chen IY, Kofidis T, Arai T, von Degenfeld G, Sheikh AY, Hendry SL, Pearl J, Hoyt G, Sista R, Yang PC, Blau HM, Gambhir SS, Robbins RC. Collagen matrices enhance survival of transplanted cardiomyoblasts and contribute to functional improvement of ischemic rat hearts. Circulation. 2006;114(1 Suppl):I167–73. doi:10.1161/CIRCULATIONAHA.105.001297.
168. Lam J, Lowry WE, Carmichael ST, Segura T. Delivery of iPS-NPCs to the stroke cavity within a hyaluronic acid matrix promotes the differentiation of transplanted cells. Adv Funct Mater. 2014;24(44):7053–62. doi:10.1002/adfm.201401483.
169. Lee HJ, Kim KS, Park IH, Kim SU. Human neural stem cells over-expressing VEGF provide neuroprotection, angiogenesis and functional recovery in mouse stroke model. PLoS ONE. 2007;2(1), e156. doi:10.1371/journal.pone.0000156.
170. Lee HJ, Kim MK, Kim HJ, Kim SU. Human neural stem cells genetically modified to overexpress Akt1 provide neuroprotection and functional improvement in mouse stroke model. PLoS ONE. 2009;4(5), e5586. doi:10.1371/journal.pone.0005586.
171. Lee YB, Polio S, Lee W, Dai G, Menon L, Carroll RS, Yoo SS. Bio-printing of collagen and VEGF-releasing fibrin gel scaffolds for neural stem cell culture. Exp Neurol. 2010;223(2):645–52. doi:10.1016/j.expneurol.2010.02.014.
172. Li JH, Zhang N, Wang JA. Improved anti-apoptotic and anti-remodeling potency of bone marrow mesenchymal stem cells by anoxic pre-conditioning in diabetic cardiomyopathy. J Endocrinol Investig. 2008;31(2):103–10. doi:10.1007/BF03345575.
173. Li Q, Tang G, Xue S, He X, Miao P, Li Y, Wang J, Xiong L, Wang Y, Zhang C, Yang GY. Silica-coated superparamagnetic iron oxide nanoparticles targeting of EPCs in ischemic brain injury. Biomaterials. 2013;34(21):4982–92. doi:10.1016/j.biomaterials.2013.03.030.
174. Lim TC, Rokkappanavar S, Toh WS, Wang LS, Kurisawa M, Spector M. Chemotactic recruitment of adult neural progenitor cells into multifunctional hydrogels providing sustained SDF-1alpha release and compatible structural support. FASEB J. 2013;27(3):1023–33. doi:10.1096/fj.12-221515.
175. Liu H, Honmou O, Harada K, Nakamura K, Houkin K, Hamada H, Kocsis JD. Neuroprotection by PlGF gene-modified human mesenchymal stem cells after cerebral ischaemia. Brain J Neurol. 2006;129(Pt 10):2734–45. doi:10.1093/brain/awl207.

176. Liu N, Zhang Y, Fan L, Yuan M, Du H, Cheng R, Liu D, Lin F. Effects of transplantation with bone marrow-derived mesenchymal stem cells modified by survivin on experimental stroke in rats. J Transl Med. 2011;9:105. doi:10.1186/1479-5876-9-105.
177. MacArthur Jr JW, Purcell BP, Shudo Y, Cohen JE, Fairman A, Trubelja A, Patel J, Hsiao P, Yang E, Lloyd K, Hiesinger W, Atluri P, Burdick JA, Woo YJ. Sustained release of engineered stromal cell-derived factor 1-alpha from injectable hydrogels effectively recruits endothelial progenitor cells and preserves ventricular function after myocardial infarction. Circulation. 2013;128(11 Suppl 1):S79–86. doi:10.1161/CIRCULATIONAHA.112.000343.
178. Malik YS, Sheikh MA, Zhu X. Doxycycline can stimulate cytoprotection in neural stem cells with oxygen-glucose deprivation-reoxygenation injury: a potential approach to enhance effectiveness of cell transplantation therapy. Biochem Biophys Res Commun. 2013;432(2):355–8. doi:10.1016/j.bbrc.2013.01.097.
179. Matsuse D, Kitada M, Ogura F, Wakao S, Kohama M, Kira J, Tabata Y, Dezawa M. Combined transplantation of bone marrow stromal cell-derived neural progenitor cells with a collagen sponge and basic fibroblast growth factor releasing microspheres enhances recovery after cerebral ischemia in rats. Tissue Eng A. 2011;17(15–16):1993–2004. doi:10.1089/ten.TEA.2010.0585.
180. McGinley LM, McMahon J, Stocca A, Duffy A, Flynn A, O'Toole D, O'Brien T. Mesenchymal stem cell survival in the infarcted heart is enhanced by lentivirus vector-mediated heat shock protein 27 expression. Hum Gene Ther. 2013;24(10):840–51. doi:10.1089/hum.2011.009.
181. Mias C, Trouche E, Seguelas MH, Calcagno F, Dignat-George F, Sabatier F, Piercecchi-Marti MD, Daniel L, Bianchi P, Calise D, Bourin P, Parini A, Cussac D. Ex vivo pretreatment with melatonin improves survival, proangiogenic/mitogenic activity, and efficiency of mesenchymal stem cells injected into ischemic kidney. Stem Cells. 2008;26(7):1749–57. doi:10.1634/stemcells.2007-1000.
182. Moon HH, Joo MK, Mok H, Lee M, Hwang KC, Kim SW, Jeong JH, Choi D, Kim SH. MSC-based VEGF gene therapy in rat myocardial infarction model using facial amphipathic bile acid-conjugated polyethyleneimine. Biomaterials. 2014;35(5):1744–54. doi:10.1016/j.biomaterials.2013.11.019.
183. Moshayedi P, Carmichael ST. Hyaluronan, neural stem cells and tissue reconstruction after acute ischemic stroke. Biomatter. 2013;3(1):1–9. doi:10.4161/biom.23863.
184. Onda T, Honmou O, Harada K, Houkin K, Hamada H, Kocsis JD. Therapeutic benefits by human mesenchymal stem cells (hMSCs) and Ang-1 gene-modified hMSCs after cerebral ischemia. J Cereb Blood Flow Metab. 2008;28(2):329–40. doi:10.1038/sj.jcbfm.9600527.
185. Riegler J, Liew A, Hynes SO, Ortega D, O'Brien T, Day RM, Richards T, Sharif F, Pankhurst QA, Lythgoe MF. Superparamagnetic iron oxide nanoparticle targeting of MSCs in vascular injury. Biomaterials. 2013;34(8):1987–94. doi:10.1016/j.biomaterials.2012.11.040.
186. Rosenblum S, Smith TN, Wang N, Chua JY, Westbroek E, Wang K, Guzman R. BDNF pretreatment of human embryonic-derived neural stem cells improves cell survival and functional recovery after transplantation in hypoxic-ischemic stroke. Cell Transplant. 2015;24(12):2449–61. doi:10.3727/096368914X679354.
187. Rustad KC, Wong VW, Sorkin M, Glotzbach JP, Major MR, Rajadas J, Longaker MT, Gurtner GC. Enhancement of mesenchymal stem cell angiogenic capacity and stemness by a biomimetic hydrogel scaffold. Biomaterials. 2012;33(1):80–90. doi:10.1016/j.biomaterials.2011.09.041.
188. Sakata H, Narasimhan P, Niizuma K, Maier CM, Wakai T, Chan PH. Interleukin 6-preconditioned neural stem cells reduce ischaemic injury in stroke mice. Brain J Neurol. 2012;135(Pt 11):3298–310. doi:10.1093/brain/aws259.
189. Schantz JT, Chim H, Whiteman M. Cell guidance in tissue engineering: SDF-1 mediates site-directed homing of mesenchymal stem cells within three-dimensional polycaprolactone scaffolds. Tissue Eng. 2007;13(11):2615–24. doi:10.1089/ten.2006.0438.
190. Schmitz T, Endesfelder S, Chew LJ, Zaak I, Buhrer C. Minocycline protects oligodendroglial precursor cells against injury caused by oxygen-glucose deprivation. J Neurosci Res. 2012;90(5):933–44. doi:10.1002/jnr.22824.

191. Song M, Jang H, Lee J, Kim JH, Kim SH, Sun K, Park Y. Regeneration of chronic myocardial infarction by injectable hydrogels containing stem cell homing factor SDF-1 and angiogenic peptide Ac-SDKP. Biomaterials. 2014;35(8):2436–45. doi:10.1016/j.biomaterials.2013.12.011.
192. Song M, Kim YJ, Kim YH, Roh J, Kim SU, Yoon BW. Using a neodymium magnet to target delivery of ferumoxide-labeled human neural stem cells in a rat model of focal cerebral ischemia. Hum Gene Ther. 2010;21(5):603–10. doi:10.1089/hum.2009.144.
193. Sun L, Lee J, Fine HA. Neuronally expressed stem cell factor induces neural stem cell migration to areas of brain injury. J Clin Invest. 2004;113(9):1364–74. doi:10.1172/JCI20001.
194. Tang Y, Cai B, Yuan F, He X, Lin X, Wang J, Wang Y, Yang GY. Melatonin pretreatment improves the survival and function of transplanted mesenchymal stem cells after focal cerebral ischemia. Cell Transplant. 2014;23(10):1279–91. doi:10.3727/096368913X667510.
195. Theus MH, Wei L, Cui L, Francis K, Hu X, Keogh C, Yu SP. In vitro hypoxic preconditioning of embryonic stem cells as a strategy of promoting cell survival and functional benefits after transplantation into the ischemic rat brain. Exp Neurol. 2008;210(2):656–70. doi:10.1016/j.expneurol.2007.12.020.
196. Thevenot PT, Nair AM, Shen J, Lotfi P, Ko CY, Tang L. The effect of incorporation of SDF-1alpha into PLGA scaffolds on stem cell recruitment and the inflammatory response. Biomaterials. 2010;31(14):3997–4008. doi:10.1016/j.biomaterials.2010.01.144.
197. Thomas D, Fontana G, Chen X, Sanz-Nogues C, Zeugolis DI, Dockery P, O'Brien T, Pandit A. A shape-controlled tuneable microgel platform to modulate angiogenic paracrine responses in stem cells. Biomaterials. 2014;35(31):8757–66. doi:10.1016/j.biomaterials.2014.06.053.
198. Tsai LK, Wang Z, Munasinghe J, Leng Y, Leeds P, Chuang DM. Mesenchymal stem cells primed with valproate and lithium robustly migrate to infarcted regions and facilitate recovery in a stroke model. Stroke J Cereb Circ. 2011;42(10):2932–9. doi:10.1161/STROKEAHA.110.612788.
199. Wang JA, He A, Hu X, Jiang Y, Sun Y, Jiang J, Gui C, Wang Y, Chen H. Anoxic preconditioning: a way to enhance the cardioprotection of mesenchymal stem cells. Int J Cardiol. 2009;133(3):410–2. doi:10.1016/j.ijcard.2007.11.096.
200. Wei L, Cui L, Snider BJ, Rivkin M, Yu SS, Lee CS, Adams LD, Gottlieb DI, Johnson Jr EM, Yu SP, Choi DW. Transplantation of embryonic stem cells overexpressing Bcl-2 promotes functional recovery after transient cerebral ischemia. Neurobiol Dis. 2005;19(1–2):183–93. doi:10.1016/j.nbd.2004.12.016.
201. Wei N, Yu SP, Gu X, Taylor TM, Song D, Liu XF, Wei L. Delayed intranasal delivery of hypoxic-preconditioned bone marrow mesenchymal stem cells enhanced cell homing and therapeutic benefits after ischemic stroke in mice. Cell Transplant. 2013;22(6):977–91. doi:10.3727/096368912X657251.
202. Xie X, Sun A, Zhu W, Huang Z, Hu X, Jia J, Zou Y, Ge J. Transplantation of mesenchymal stem cells preconditioned with hydrogen sulfide enhances repair of myocardial infarction in rats. Tohoku J Exp Med. 2012;226(1):29–36.
203. Yan F, Yue W, Zhang YL, Mao GC, Gao K, Zuo ZX, Zhang YJ, Lu H. Chitosan-collagen porous scaffold and bone marrow mesenchymal stem cell transplantation for ischemic stroke. Neural Regen Res. 2015;10(9):1421–6. doi:10.4103/1673-5374.163466.
204. Yasuda H, Kuroda S, Shichinohe H, Kamei S, Kawamura R, Iwasaki Y. Effect of biodegradable fibrin scaffold on survival, migration, and differentiation of transplanted bone marrow stromal cells after cortical injury in rats. J Neurosurg. 2010;112(2):336–44. doi:10.3171/2009.2.JNS08495.
205. Yip HK, Chang YC, Wallace CG, Chang LT, Tsai TH, Chen YL, Chang HW, Leu S, Zhen YY, Tsai CY, Yeh KH, Sun CK, Yen CH. Melatonin treatment improves adipose-derived mesenchymal stem cell therapy for acute lung ischemia-reperfusion injury. J Pineal Res. 2013;54(2):207–21. doi:10.1111/jpi.12020.
206. Yu H, Cao B, Feng M, Zhou Q, Sun X, Wu S, Jin S, Liu H, Lianhong J. Combinated transplantation of neural stem cells and collagen type I promote functional recovery after cerebral ischemia in rats. Anat Rec. 2010;293(5):911–7. doi:10.1002/ar.20941.

207. Yu X, Chen D, Zhang Y, Wu X, Huang Z, Zhou H, Zhang Y, Zhang Z. Overexpression of CXCR4 in mesenchymal stem cells promotes migration, neuroprotection and angiogenesis in a rat model of stroke. J Neurol Sci. 2012;316(1–2):141–9. doi:10.1016/j.jns.2012.01.001.
208. Zhang G, Nakamura Y, Wang X, Hu Q, Suggs LJ, Zhang J. Controlled release of stromal cell-derived factor-1 alpha in situ increases c-kit+ cell homing to the infarcted heart. Tissue Eng. 2007;13(8):2063–71. doi:10.1089/ten.2006.0013.
209. Zhao MZ, Nonoguchi N, Ikeda N, Watanabe T, Furutama D, Miyazawa D, Funakoshi H, Kajimoto Y, Nakamura T, Dezawa M, Shibata MA, Otsuki Y, Coffin RS, Liu WD, Kuroiwa T, Miyatake S. Novel therapeutic strategy for stroke in rats by bone marrow stromal cells and ex vivo HGF gene transfer with HSV-1 vector. J Cereb Blood Flow Metab. 2006;26(9):1176–88. doi:10.1038/sj.jcbfm.9600273.
210. Zhu W, Mao Y, Zhao Y, Zhou LF, Wang Y, Zhu JH, Zhu Y, Yang GY. Transplantation of vascular endothelial growth factor-transfected neural stem cells into the rat brain provides neuroprotection after transient focal cerebral ischemia. Neurosurgery. 2005;57(2):325–33. discussion 325–333.Fig. 9.1 (continued)

Chapter 10
Clinical Studies of Bone Marrow-Derived Stem Cell Therapy in Stroke Patients

Francisco Moniche, Irene Escudero, Elena Zapata, Fernando Mancha, Ángela Vega-Salvatierra, Blanca Pardo, and Joan Montaner

Abstract Stroke is the leading cause of long-term disability in adults and the third cause of mortality worldwide. In the very acute phase of stroke, thrombolytics and endovascular thrombectomy can reduce stroke disability; however, only a small minority of patients receive these treatments. Once the neurological deficits are established, there are few options for recovery. In recent years, extensive cell therapy preclinical research has demonstrated a neurorestorative effect after cerebral ischemia. In cerebral ischemia animal models, bone marrow-derived stem cells improve neurological outcomes even in the long term, increasing brain plasticity and enhancing recovery mainly due to secretion of growth factors and cytokines.

In the bone marrow, different types of cells have been used for cell therapy in stroke. The first type of cells used for stroke and the most extensive studied in preclinical research are mesenchymal stem cells (MSCs). In recent years some other cells have been studied for stroke therapy with promising results, such as bone marrow mononuclear cells (BM-MNCs), hematopoietic stem cells (HSCs), and multipotent adult progenitor cells (MAPCs). Several phase I and II clinical trials have been published to date with these stem cells, which have already demonstrated the feasibility and safety of this therapy in the stroke setting. An increasing number of

F. Moniche (✉)
Stroke Unit, Neurology Department, Instituto de Biomedicina de Sevilla-IBiS, Hospital Universitario Virgen del Rocío, Avda. Manuel Siurot, s/n., 41013 Seville, Spain

Instituto de Biomedicina de Sevilla-IBiS, Hospital Universitario Virgen del Rocío, Avda. Manuel Siurot, s/n., 41013 Seville, Spain
e-mail: pmoniche@gmail.com

I. Escudero • E. Zapata • B. Pardo
Stroke Unit, Neurology Department, Instituto de Biomedicina de Sevilla-IBiS, Hospital Universitario Virgen del Rocío, Avda. Manuel Siurot, s/n., 41013 Seville, Spain

F. Mancha • Á. Vega-Salvatierra • J. Montaner
Instituto de Biomedicina de Sevilla-IBiS, Hospital Universitario Virgen del Rocío, Avda. Manuel Siurot, s/n., 41013 Seville, Spain

K. Jin et al. (eds.), *Bone Marrow Stem Cell Therapy for Stroke*,
DOI 10.1007/978-981-10-2929-5_10

clinical trials, mainly with bone marrow MSCs and BM-MNCs, are ongoing to further assess the best dose, route, and timing of this therapy and to elucidate the efficacy cell therapy in stroke.

Keywords Patients • Bone marrow-derived stem cells • Therapy • Stroke

10.1 Introduction

The World Health Organization (WHO) defined stroke as a "rapidly developing clinical signs of focal (or global) disturbance of cerebral function, with symptoms lasting 24 h or longer or leading to death, with no apparent cause other than of vascular origin" [1].

Annually, 15 million people worldwide suffer a stroke and, in the United States alone, a person dies every 3 min due to stroke. In high-income countries, stroke is the third most common cause of death, only after coronary heart disease and cancer. Stroke is also the main cause of acquired adult disability [2]. Of those who survive a stroke, five million people annually are left permanently disabled, placing a burden on family and society. Only 41 % are independent 6 months after a stroke.

The ischemic stroke represents 80–85 % of strokes and the incidence is 150–200 patients/100,000 per year [3]. Stroke prevalence is about 2 % of the population of >20 years; however, it increases to up to 6–7 % in older adults (>65 years). Projections show that by 2030, an additional 3.4 million people in the United States aged ≥18 years will have had a stroke, a 20.5 % increase in prevalence from 2012 [4]. In the EU, the cost attributed to stroke was €64.1 billion in 2010, mainly due to the high costs of long-term special care and rehabilitation [5]. Moreover, the socio-economic burden of stroke is expected to increase due to aging of the population and the rise in diabetes and obesity, which are reaching an epidemic level.

The currently available therapies of acute stroke target rapid vessel recanalization, since, without restoration of cerebral blood flow, hypoperfused cerebral tissue in the penumbral region progresses to cellular death that ultimately expands the necrotic core lesion. Nowadays, thrombolytics (i.e., tissue plasminogen activator or tPA) and endovascular thrombectomy are the main therapies for restoring normal perfusion in acute ischemic stroke. However, tPA has important limitations, with a narrow therapeutic window of 4.5 h, which means that less than 5–10 % of ischemic stroke patients receive this treatment. Moreover, recanalization rates after the administration of tPA are low and prevent disability in only 55 patients per 1000 people treated, without reducing mortality [6]. Recently, endovascular thrombectomy has demonstrated efficacy in several clinical trials in those patients with a large vessel occlusion [7]. Although the treatment approach of acute ischemic stroke is rapidly evolving [8], recanalization therapies are only administered to a minority of acute stroke patients.

Another approach to improve outcomes in stroke is the administration of neuroprotective drugs. Neuroprotective treatment aims to reduce the damage of stroke, but as most of the injury occurs in the first 24–48 h, these therapies must be administrated soon after stroke onset, and to date, no drug has been demonstrated to ameliorate the disability or mortality after stroke [9, 10].

In recent years, many studies have shown that once the stroke is established, profound neurorestorative processes are induced in brain tissue in response to focal cerebral ischemia [11]. Although these processes are insufficient to restore neurological function, neurorestorative treatments with pharmacological or cell-based therapies could stimulate and amplify these endogenous mechanisms in stroke patients. This approach has the major advantage of a wider therapeutic window, as neurorestorative therapies can be instituted during the recovery phase of the stroke, and promotes the remodeling of brain tissue. This makes the treatment available to a much larger number of stroke patients.

Also, neurorestorative treatments target not only the ischemic and "penumbra" tissue (hypoperfused tissue) but also viable brain tissue with normal perfusion stimulating neuronal plasticity and neurological recovery [11]. However, until now, neurorestorative drugs targeting single steps in the cascade of cerebral ischemia have failed to improve neurological deficits, probably related to stroke complexity, with necrosis, apoptosis, inflammation, and remodeling occurring as a continuum.

Stem cell therapy, such as transplantation of bone marrow stem cells, represents one of the most exciting fields in regenerative medicine and has emerged as an attractive approach for the treatment of stroke. These stem cells are believed to exert multiple therapeutic actions. They might target simultaneously several processes by releasing different factors inducing neuroprotection and brain remodeling and modulating the post-ischemic inflammatory response [12–15]. Extensive basic research has been done during the last two decades in cell therapy and stroke animal models, but we are still in the first steps in the clinical research with stroke patients.

The potential of cell-based therapy relies on several *key properties*: (1) their capacity to differentiate into several cell lineages, (2) their immunomodulatory properties, (3) their ex vivo expansion potential, (4) their ability to secrete factors to regulate biological functions such as proliferation and differentiation over a broad target of cells, and (5) their ability to home to damaged tissues.

10.2 Bone Marrow Cell Therapy and Clinical Trials

To date, there are many different cells being investigated for stroke, both in preclinical studies and in clinical trials such as embryonic stem cells, neural stem cells, adipose-derived stem cells, induced pluripotent stem cells (iPS), and stem cells obtained from bone marrow, umbilical cord, and amniotic or placental tissue. However, in this chapter, we will focus on bone marrow cell therapy as most of preclinical and clinical studies have used bone marrow stem cells [16, 17].

There are several advantages of using bone marrow stem cells as a cell therapy for stroke. First, the efficacy and reproducible benefits of these cells have been demonstrated in several laboratories and in different animal stroke models. Second, bone marrow stem cells are adult cells and therefore do not have ethical problems, unlike fetal and embryonic cells. Third, it has been proven reliable to use bone marrow stem cells in different time periods of stroke, even in the acute stroke phase. Finally, although some studies have been done with allogenic cells, bone marrow stem cells allow autologous administration avoiding the possibility of rejection.

In the bone marrow, different types of cells have been used for cell therapy in stroke. The first type of cells used for stroke and the most extensive studied in preclinical research are mesenchymal stem cells (MSCs). In recent years, some other cells have been studied for stroke therapy with promising results, such as bone marrow mononuclear cells (BM-MNCs), hematopoietic stem cells (HSCs), and multipotent adult progenitor cells (MAPCs).

10.2.1 Mesenchymal Stem Cells

Mesenchymal stem cells (MSCs or marrow stromal cells) are one of the first types of cells that have been studied for ischemic stroke. MSCs have a considerable therapeutic potential that has generated markedly increasing interest in a wide variety of biomedical disciplines. Extensive preclinical studies with MSCs have made this therapy a very promising cell-based approach for stroke.

MSCs are multipotent adult stem cells defined as those cells which have three characteristics: (a) must be plastic adherent when maintained in standard culture conditions; (b) must express CD105, CD73, and CD90 and lack expression of CD45, CD34, CD14, or CD11b, CD79alpha, or CD19 and HLA-DR surface molecules; and (c) must at least differentiate to osteoblasts, adipocytes, and chondroblasts in vitro [18]. The lack of expression of HLA-DR (class II major histocompatibility complex) gives them an immunoprivileged status, and also their relative ease of isolation from bone marrow makes these cells a good candidate for cell therapy in different illness such as stroke.

The safety of MSCs has been analyzed in a recent meta-analysis of clinical trials with more than 1000 patients in different clinical conditions that included ischemic stroke, Crohn's disease, myocardial infarction, cardiomyopathy, and graft versus host disease. An excellent safety profile of MSCs was demonstrated, as there was no association between MSC treatment and acute infusional toxicity, organ system complications, infection, death, or malignancy [19]. However, larger controlled clinical trials are required before defining a definitive safety profile of MSCs.

Regarding efficacy outcomes, MSCs have demonstrated efficacy in clinical trials in other conditions such as graft versus host disease and are under study in autoimmune diseases (i.e., Crohn's disease, multiple sclerosis, and type 1 diabetes) and different models of ischemia (i.e., stroke, ischemic cardiac diseases, and limb ischemia) [20–22].

In animal stroke models, injection of MSCs resulted in very large and favorable effects on neurological outcomes. In a recent meta-analysis of preclinical studies, MSCs improved consistently multiple outcome measures with very large effect sizes. These data are robust across species studied, administration route, dose, and presence of comorbidities [23]. Furthermore, MSCs attenuated tissue damage and migrated into the ischemic boundary zone accompanied by reduced neuronal apoptosis and enhanced neoangiogenesis and synaptogenesis [24, 25].

The first clinical trial published with bone marrow stem cells in stroke patients was done by Bang et al. which used autologous MSCs [26] (Table 10.1). In this placebo-controlled phase I/II trial of 30 patients with chronic stroke, 5 patients were treated with MSCs and 25 were controls. Those patients in the treated group received two doses of IV autologous mesenchymal stromal cells at 4–5 weeks and 7–9 weeks from the onset of symptoms. This method was reportedly safe and feasible in the short term. In 2010 Lee et al. published the long-term follow-up, with 52 patients finally included (16 in MSC group and 36 in control group). No MSC-related adverse events were reported, with significant improved neurologic recovery in those patients receiving cellular therapy compared to controls (the proportion of patients with modified Rankin scale score 0–3 increased in the MSC group, $p=0.046$) [27].

Other pilot trials have been published using MSCs in stroke. Honmou et al. reported IV MSCs transplantation in 12 patients with chronic ischemic stroke [28]. No adverse events were described from transplantation, and interestingly, a reduction of more than 20% of infarction volume was observed in magnetic resonance imaging at 1 week after cell injection.

In another trial, Bhasin et al. published a trial including 40 chronic stroke patients [29]. Of these, 14 patients received intravenous BM-MNCs, 6 patients received intravenous MSCs, and 20 were control patients. During follow-up, stem cell transplantation was reported to be safe, and there was a significant improvement at 6 months in the Barthel index when the whole stem cell group was compared to the control group, although there was no difference in Rankin scale or Fugl-Meyer scale.

In spite of these previous experiences with MSCs, there are several disadvantages of using MSCs in stroke patients (Table 10.2):

1. MSCs require several weeks of cell culture to obtain sufficient quantity of cells for transplantation, not allowing the autologous injection of MSCs in the acute-subacute phase of stroke.
2. The large size of cells (13–19 μm) could also lead to pulmonary entrapment when administered by intravenous injection or even to microvascular occlusions and new cerebral infarctions in the intra-arterial (IA) route [30].

To date, there is no published data of allogenic transplantation of bone marrow MSCs or with a different route than intravenous but a very recent interim report of an open-label single-arm study of surgical transplantation of modified bone marrow-derived mesenchymal stem cells [31]. In this interim analysis of 16 patients that have completed 12 months of follow-up, authors describe significant improvement in NIHSS (National Institute of Health Stroke Scale) (mean decrease 2.00 [95%

Table 10.1 Published clinical trials with bone marrow-derived stem cell therapy in stroke patients

Cell population	Author	No. of patient treated (controls)	Design	Route	Cell dose	Time window	Follow-up (month)	Adverse events
MSCs								
Autologous	Bang (2005) [26]	5 (25)	Randomized, observer-blinded phase I/II	Intravenous	5×10^7	4–5 weeks	12	None reported
Autologous	Lee (2010) [27]	16 (36)	Observer-blinded phase II	Intravenous	5×10^7	5 weeks	60	None study-related reported
Autologous	Honmou (2011) [28]	12 (0)	Open-label phase I	Intravenous	$6–16 \times 10^7$	36–133 days	12	Mild fever, nausea, appetite loss
Autologous	Bhasin (2013) [29]	6 (6)	Open-label phase I/II	Intravenous	$5–6 \times 10^7$	3–24 months	6	None study-related reported
Allogenic	Steinberg (2016) [31]	18 (0)	Open-label phase I/II	Intraparenchymal	$2.5–10 \times 10^6$	6–60 months	12	Headache, nausea, subdural hematoma (2 of 18)
BM-MNCs								
Autologous	Suarez-Monteagudo (2009) [41]	5 (0)	Open-label phase I	Intraparenchymal	$1.4–5.5 \times 10^7$	1–10 years	12	Headache, drowsiness, nausea, fever
Autologous	Barbosa (2010) [60]	6 (0)	Open-label phase I	Intra-arterial	$12.5–50 \times 10^7$	<90 days	6	None study-related reported
Autologous	Battistella (2011) [43]	6 (0)	Open-label phase I	Intra-arterial	$10–50 \times 10^7$	2–3 months	6	Seizures (2 of 6)
Autologous	Savitz (2011) [46]	10 (0)	Open-label phase I	Intravenous	$7–10 \times 10^6$/kg	24–72 h	6	None study-related reported

Autologous	Friedrich (2012) [44]	20 (0)	Open-label phase I	Intra-arterial	2.2×10^7	3–7 days	6	None study-related reported
Autologous	Moniche (2012) [47]	10 (10)	Observer-blinded phase I/II	Intra-arterial	16×10^7	5–9 days	6	Seizures (2 of 10)
Autologous	Prasad (2012) [49]	11 (0)	Open-label phase I	Intravenous	$0.2–18 \times 10^7$	7–30 days	6	One reinfarction
Autologous	Li (2013) [42]	60 (40)	Observer-blinded phase I/II (hemorrhagic stroke)	Intraparenchymal	$0.2–2 \times 10^7$	5–7 days	6	Fever, one unspecified pulmonary tumor
Autologous	Rosado-de-Castro (2013) [45]	12 (0)	Open-label phase I	Intravenous ($n=5$), intra-arterial ($n=7$)	$10–50 \times 10^7$	19–89 days	6	Seizures (7 of 12)
Autologous	Prasad (2014) [50]	85 (35)	Blinded randomized phase II	Intravenous	28×10^7	18.5 days	12	None reported
Autologous	Sharma (2014) [59]	24 (0)	Open-label phase I/II	Intrathecal	1×10^6/kg	40 months	30	None reported
CD34+								
Autologous	Banerjee (2014) [16]	5 (0)	Open-label phase I	Intra-arterial	$1.2–2.5 \times 10^6$	<7 days	6	None study-related reported
MAPC								
Allogenic	Hess (public presentation)	65 (61)	Blinded phase II	Intravenous	40×10^7	24–48 h	6	None study-related reported

Table 10.2 Comparative of bone marrow mononuclear cells and bone marrow mesenchymal cells

Type of cell	Advantages	Disadvantages
BM-MNCs	Consistent beneficial effect and an excellent safety profile in animal models	Not allow allogenic transplantation without immunosuppressive drugs
	Excellent safety profile in pilot clinical trials in stroke	Variability in the number of cells obtained after bone marrow harvest
	Prepared for administration within hours	
	No tumor formation	
BM-MSCs	Very large and favorable effects in stroke models	Require cell culture (several weeks)
	Immunoprivileged status, allow allogenic transplantation	No allow autologous administration in acute stroke patients
	Excellent safety profile in other clinical conditions and pilot stroke trials	Large size of cells that could lead to microvascular occlusions or pulmonary entrapment
	No tumor formation	

confidence interval, −2.7 to −1.3; $P<0.001$]) and Fugl-Meyer scale (mean increase 19.20 [95 % confidence interval, 11.4–27.0; $P<0.001$]). Patients included had a chronic ischemic stroke (mean 22 months from stroke onset) and received a stereotactic injection of allogenic MSCs (*SB623 cells*) in the peri-infarct area. Serious adverse events were unrelated or unlikely to be related to cell treatment. Postsurgery headache was the most common adverse event that was probably or definitely related to the procedure, experienced by 77.8 % of patients, and subdural hematoma and epileptic seizure were detected in two patients (11 %).

Although the immunoprivileged status of MSCs makes a rejection of allogenic transplantation very unlikely, its safety has to be proven in stroke patients. However, previous reports of allogenic use of MSCs in other conditions described no acute infusional toxicity [19]. Regarding the route, the probably main reason for the absence of MSC clinical trials using intra-arterial route is due to the potential of arterial embolism that has been described with animal stroke models, even in mammalians. Lu et al. [30] described an intra-arterial MSC transplantation in a canine stroke model with the development of new infarctions 24 h after transplantation in 16 % of dogs, probably due to impeded cerebral blood flow [32].

Due to the large and favorable effects in preclinical studies and in spite of the disadvantages described, MSCs are still one of the best candidates for cell therapy in stroke patients, and several clinical trials are currently ongoing.

10.2.2 Hematopoietic Stem Cells

Other bone marrow stem cells that have been investigated in animal stroke models are hematopoietic stem cells (HSCs). These cells express CD34 (CD34+ cell) and can be also found in peripheral blood and umbilical cord blood. Preclinical studies of CD34+ cells have shown significant benefits in animal stroke models, with evidence of functional improvement as well as reduced infarct volume [33]. In a preclinical study, intravenous CD34+ cell transplantation resulted in increased perilesional angiogenesis and subsequent neurogenesis in mice at 48 h post-stroke [34]. There are also other evidences of neurogenesis and angiogenesis induced by CD34+ cells in subacute stroke, with cells transplanted expressing neuronal, glial, and vascular endothelial cell markers [35]. There are some preliminary clinical trials demonstrating the safety of autologous CD34+ peripheral blood stem cells [36, 37]. Bone marrow hematopoietic stem cells have only been used in a pilot open-label clinical trial of five stroke patients [38]. CD34+ cells were collected from the bone marrow of the subjects before being delivered by catheter angiography into the ipsilateral middle cerebral artery within 7 days from stroke onset. No safety issues were described and all patients show improvement of neurological deficit during follow-up, although no comparison was done with a control group. Authors found a nonsignificant reduction in the mean lesion volume from inclusion to day 180, with no new lesions in MRI (edema, hemorrhage, or tumor). To date, very few clinical trials are currently ongoing testing bone marrow CD34+ cells in stroke patients.

10.2.3 Bone Marrow Mononuclear Cells (BM-MNCs)

BM-MNCs are one of the most studied types of cells for use as stroke therapy. BM-MNCs are composed of a mixture of myeloid, lymphoid, erythroid, and stem cell populations, which includes HSCs, MSCs, and endothelial progenitor cells. The main advantage over other types of cell therapy is that autologous transplantation is feasible, even in the acute phase of stroke, as they are isolated from bone marrow and prepared for administration within hours. As MSCs, BM-MNCs have been extensively studied in animal models demonstrating a consistent beneficial effect and an excellent safety profile. Several biological effects such as attenuation of neuronal death, modulating microglia, reducing pro-inflammatory responses, increasing neoangiogenesis, and promoting proliferation of endogenous neural stem cells have been invoked [12, 39, 40]. However, few clinical studies have assessed the safety and efficacy of BM-MNC transplantation in stroke patients.

The first trial published with BM-MNCs was an open-label trial with five stroke patients treated with intraparenchymal route by Suarez-Monteagudo et al. [41]. Patients included had a chronic stroke from 1 to 10 years from onset and authors describe an excellent tolerance of procedure and with no important adverse events derived from surgery or transplant. After this study, only Li et al. have published a clinical trial using intraparenchymal route, although not including ischemic stroke but intracerebral hemorrhage (ICH) [42]. In this study, autologous BM-MNCs were injected to the perihemorrhage area in the base ganglia through an intracranial drainage tube 6 days after ICH. Surgical drainage of ICH was performed in every patient within the first day from ICH onset, and after 5 days those patients who consent to be treated with BM-MNCs were included in the study group ($n=60$), and those who rejected cell therapy were the control group ($n=40$). Both groups had similar baseline characteristics and similar NIHSS (National Institute of Health Stroke Scale) after surgery, but authors describe a significant improvement in Barthel and NIHSS scores in the study group 6 months after inclusion (57.39 ± 23.51 in study group vs 46.90 ± 20.29 in control group, $P<0.01$ in Barthel scale and 10.09 ± 8.86 vs 14.35 ± 10.14, $P<0.01$ in NIHSS).

Since the trial published by Suarez-Monteagudo et al. in 2009, several clinical trials have been published with BM-MNCs. Most of them used less invasive routes as intravenous or intra-arterial injection.

A Brazilian trial published by Battistella et al. included six patients treated with intra-arterial BM-MNCs with a time window of 2–3 months from stroke onset [43]. There was no worsening immediately after the procedure or during follow-up period. At the 180-day follow-up evaluation, there was a slight improvement in NIHSS (range −1 to −8 points). Although BM-MNC transplantation was safe in these patients, there is less evidence from the animal studies to suggest that BM-MNCs could be effective in this time window. The same group later published another trial, including 20 patients with moderate to severe middle cerebral artery (MCA) ischemic stroke in a time window of 3–7 days [44], showing no procedure-related adverse events, with 40 % of good clinical outcomes at 6 months.

In 2013, this group (Rosado de Castro et al.) compared IV vs. IA routes in BM-MNC transplantation in 12 stroke patients, demonstrating that with the IV route more cells were trapped in lungs after injection than IA injection. However, they found similar rates of brain homing between both routes [45]. Remarkably, all of the intravenous patients suffered seizures during the follow-up period. Authors hypothesize that the infused cells could modify excitability in the perilesional regions, generating seizures, which should be evaluated further in future clinical trials.

In another trial evaluating the test, feasibility, and safety of autologous BM-MNC infusion in patients with acute ischemic stroke, Savitz et al. [46] included ten patients with a time window of 24–72 h after stroke onset treated with intravenous BM-MNCs. This methodology is supported by a preclinical study in which rats with middle cerebral artery occlusion performed better on neurologic tests with IV mononuclear cells infused up to 72 h, compared with 1 week from stroke onset. There were no study-related severe adverse events. However, of the ten patients included, two of them required hemicraniectomy due to malignant middle cerebral artery infarction after transplantation. In the efficacy analysis, there was a trend toward better outcomes in BM-MNC patients when compared to 79 historical controls who met the NIHSS inclusion criteria.

Our group performed a pilot single-blind (outcomes assessor) phase I/II controlled clinical trial in patients with subacute MCA ischemic stroke [47]. The aim was to assess the safety, feasibility, and clinical effects of autologous intra-arterial BM-MNC transplantation. Twenty patients (ten cases and ten controls) with severe ischemic stroke in the middle cerebral artery territory within 5–9 days from stroke onset were included. The primary outcome was safety and feasibility of the procedure. Secondary outcomes were the improvement in neurological function assessed by modified Rankin scale, Barthel index, and NIHSS.

All were severely disabled at inclusion (mean NIHSS score of 15.6 in BM-MNC group vs. 15.0 in control group, $p=0.82$). BM-MNC transplantation was done at 6.4 ± 1.3 days after stroke onset. A mean 1.59×10^8 BM-MNCs ($\pm 1.21 \times 10^8$) were intra-arterially injected. Rate of infusion through microcatheter was 0.5–1 mL/min, as rates of up to 2 mL/min do not seem to produce cell damage nor the use of heparin or iodine contrast [48].

There were no adverse events related to BM-MNC transplantation. No significant hemodynamic or respiratory changes occurred during the bone marrow harvest or the intra-arterial BM-MNC injection. DWI-MRI did not show new ischemic lesions in the active group after transplantation. During follow-up, two BM-MNC-treated patients had an isolated partial seizure. No deaths or stroke recurrence were observed during the follow-up period, and the 6-month MRI also showed no tumor formation in either group. There were no significant differences in neurological function compared to the control group. At 6 months, a greater nonsignificant proportion of BM-MNC-treated patients had an mRankin (modified Rankin) ≤ 2 (20 %) than the control group (0 %) ($p=0.47$). No differences were found in the Barthel index ($p=0.80$) or in NIHSS scores compared to the control group ($p=0.43$).

Prasad et al. also reported a trial with 11 stroke patients within 7–30 days from stroke onset [49]. Patients received IV BM-MNC transplantation and were followed up for a year, with no detection of tumor formation or other adverse events related to cell therapy.

The same group published in 2014 the biggest trial to date with BM-MNCs using intravenous route, including 120 patients in a phase II trial [50]. Fifty-eight patients were treated with BM-MNCs and 60 patients were controls. Patients with subacute ischemic stroke between 7 and 30 days were included in the study. A randomization was done in a 1:1 ratio and a single intravenous infusion of autologous BM-MNCs was performed in experimental group with a mean of 280.75 million BM-MNCs at median of 18.5 days after stroke onset. During follow-up, 8.4 % patients died and Kaplan-Meier survival curve showed no differences between both groups. Adverse events and serious adverse events were also comparable between the two arms.

In the efficacy analysis, there were no significant differences between BM-MNC arm and control arm in the Barthel index score (63.1 versus 63.6; $p=0.92$), modified Rankin scale shift analysis ($p=0.53$) or score >3 (47.5 % versus 49.2 %; $p=0.85$), NIHSS score (6.3 versus 7.0; $p=0.53$), or change in infarct volume (−11.1 versus −7.36; $P=0.63$) at day 180. Authors concluded that with the methods and timing used, the intravenous injection of BM-MNCs is safe, but there is no beneficial effect on stroke outcome.

Several other trials are ongoing testing different time windows, doses, and routes, which will give more light about the possible efficacy of BM-MNCs in stroke.

One of the *disadvantages* of BM-MNCs is that the mixture of cells (i.e., myeloid, erythroid, lymphoid, and stem cell populations) makes not possible to perform an allogenic BM-MNC transplantation without immunosuppressive drugs due to rejection. Another issue is the variability in the number of cells obtained after a bone marrow harvest, with a variability in final dose of cells injected when a standardized volume of bone marrow is harvested. In our previous trial, a volume of 50 mL of bone marrow leads to doses as different as $0 \cdot 33$ and $4 \cdot 96 \times 10^6$/kg.

On the other hand, the presence of different populations of cells within the mononuclear fraction of bone marrow could be an *advantage* and seems to be beneficial, as not only stem cells contribute to improved outcomes after stroke [51]. In a recent paper, Yang et al. showed that, in a mouse stroke model, both myeloid cells and stem cell populations are important cell types that reduce inflammation and subsequent infarct maturation. The stem cell subpopulation within BM-MNCs is critical for the therapeutic effect in post-stroke recovery. However, myeloid cells (granulocytes and monocytes) seem to modify also pro-inflammatory cytokines and regulate the microglia decreasing the neurotoxic effect and improving neuron survival rates leading to improve stroke outcomes [52].

10.2.4 Human Multipotent Adult Progenitor Cells (MAPCs®)

Recently, multipotent adult progenitor cells (MAPCs), a subpopulation of stem cells isolated from bone marrow, have been described and characterized. Human MAPCs are multipotent stem cells that have been shown to differentiate into various mesodermal cell types, with a remarkable proliferative capacity in culture. In particular, their vascular potential in vitro and in vivo has been demonstrated which make them an attractive candidate for novel cell-based treatment of ischemic diseases. Moreover MAPCs are also immunoprivileged. In a recent preclinical study comparing MSC and MAPC, the latter compared favorably with hMSC and provides a greater beneficial effect as indicated by the increase in angiogenesis, SVZ cell proliferation, and decreased inflammatory response providing an attractive new source of allogenic source of cells for stroke [53]. With data not yet published, Hess et al. (Table 10.1) have communicated the safety and feasibility of intravenous MAPC therapy in acute stroke patients.

10.3 Timing, Route, and Dose of Bone Marrow Stem Cell Transplantation

10.3.1 Time Window

The *optimal time window* for stem cell therapy is not well known. In the stroke rat model, this time window seems to be wide, even up to 1 month after cerebral infarction, but only rats receiving bone marrow stem cells 7 days after MCA occlusion exhibit decreased ischemic lesion volume [54]. However, some groups have demonstrated that an earlier transplantation results in better neurological recovery, especially when MSC or BM-MNC injection is performed during the first week after stroke or even in the first 72 h [55, 56]. Therefore, it is plausible that an earlier treatment could produce a greater effect on inflammation, apoptosis, and remodeling after stroke. In line with this preclinical evidence, our group described that when

BM-MNCs are administered intra-arterially in subacute MCA stroke patients, they seem to induce changes in serum levels of cytokines and growth factors (i.e., GM-CSF, PDGF-BB, and MMP-2) even 3 months after transplantation, which seem to be associated with better functional outcomes in stroke patients [57].

On the other hand, stem cell transplantation in the acute stroke phase (i.e., within 72 h) could be challenging, as these patients are usually neurologically unstable and prone to deteriorate. Also, this short window needs extensive logistical efforts to perform an autologous bone marrow cell injection in a hospitalized stroke patient. An allogenic transplantation would probably be more feasible in this setting although the usual complications of patients in this early stage of stroke could make more difficult to evaluate safety issues of transplantation.

Although there is no much evidence from preclinical studies to perform a clinical trial with stem cells in the chronic phase of stroke, several trials are treating patients with MSCs or BM-MNCs and stable deficit from chronic strokes [31, 41].

10.3.2 Route of Delivery

Based on animal models of stroke, it is not clear which route of delivery is preferable. Although intravenous (IV) stem cell delivery is increasingly used in clinical trials, IV injection leads to an initial random dispersion of cells throughout the body, and recent data suggest that the majority of the stem cells administered are trapped in filter organs such as the lungs, liver, and spleen, with a therapeutically questionable number of cells reaching the ischemic brain [58]. In contrast, intra-arterial cell delivery provides the opportunity to target the entire ischemic lesion enabling exposure of cells to chemoattractant signals (originating from the lesion). Other routes are being tested such as the intrathecal route [59] or the report of Steinberg et al. [31] using intraparenchymal route with exciting preliminary results, but with some serious adverse events (i.e., subdural hematoma and pneumocephalus).

However, similar to prior animal experiments, clinical trials with IV or IA injection of bone marrow stem cells also have found cells sequestered in the spleen, lung, liver, and kidney [55, 60]. This fact raises the question of whether cells need brain homing to produce the beneficial effects or the cytokine and growth factor secretion is enough to improve stroke outcomes. Although paracrine mechanisms are now the leading hypotheses to explain how cell therapies may enhance stroke recovery [16], it seems critical to expose cells to the ischemic environment to stimulate growth factor production [61].

10.3.3 Cell Dose

A wide range of number of cells has been used for transplantation in animal stroke models and in clinical trials. While in preclinical studies there is strong evidence that a higher dose of cells increases the probability of a good neurological outcome

[23, 55], the optimal number of cells to be transplanted for ischemic stroke is unknown. This raises the question of whether a higher dose of stem cells would produce a greater effect in recovery in stroke patients, but to date clinical data regarding dose is scarce.

Our group [47] found that although no significant correlation between the functional status and the amount of transplanted BM-MNCs was detected, there was a trend toward a better outcome when higher numbers of CD34+ cells were injected. In the three follow-up evaluations, a trend to positive correlation with Barthel index and negative correlations with mRankin scale and NIHSS was found, especially in the Barthel index at 1 month after transplantation ($r=0.57$, $p=0.09$). These data may support the hypothesis that a higher number of cells could lead to better outcomes.

Taguchi et al. [62] evaluated in a clinical trial two different doses of BM-MNCs administered intravenously in stroke patients after 7–10 days of stroke onset (250×10^6 and 340×10^6 cells in the lower and higher dose groups, respectively), and although it was a phase I/IIa clinical trial not designed to test efficacy, authors described a trend toward improved neurological outcomes in those patients receiving the higher dose of bone marrow cells.

On the other side, Prasad et al. [50] published a phase II trial including 120 stroke patients with 58 of them being treated with intravenous injection of BM-MNCs, showing no relationship between cell dose and outcomes.

Also, in a meta-analysis of cell-based therapies for treating stroke patients [63], authors found that stem cell therapy was more effective with higher dose of cells and also when intra-arterial route was used.

In a recent pooling data of two different clinical trials with BM-MNCs [64], a higher dose of autologous BM-MNC was related to better outcome in stroke patients. In this paper, 22 patients were analyzed and intra-arterial route was used in 77.3 % and intravenous in 22.7 % of patients. A higher number of cells injected were associated with better outcomes at 6 months ($p=0.015$). Also, a strong negative correlation was found between cell dose and disability when intravenous patients were excluded from analysis ($r=-0.63$, $p=0.006$), pointing to the hypothesis that the combination of higher number of cells and intra-arterial route could be a key factor to improve neurological outcomes in stroke patients. This pooling data showed that the optimal threshold of transplanted cells is probably around 310×10^6 BM-MNCs in order to obtain good functional outcome with high probability among treated stroke patients. However, further clinical data is needed and dose-finding clinical trials are ongoing in ischemic stroke patients [31, 65].

10.4 Conclusions

As no effective neuroprotective or neurorestorative drug has demonstrated efficacy for ischemic stroke, new therapeutic strategies such as cell therapies to enhance neurological recovery after stroke are urgently needed.

Data from preclinical and clinical studies with stem cells in stroke strengthens the notion that stem cells could increase brain plasticity and improve stroke recovery. Extensive preclinical studies have demonstrated large and favorable effects of different types of bone marrow stem cells in stroke.

Several phase I and II clinical trials have been published to date with bone marrow stem cells that have already demonstrated the feasibility and safety of this therapy in the stroke setting. An increasing number of clinical trials, mainly with bone marrow MSCs and BM-MNCs, are ongoing to further assess the best dose, route, and timing of this therapy and to elucidate the efficacy of cell therapy in stroke.

Funding No disclosure information. This study is funded by grants from Junta de Andalucía [TCRM 0001/2006], Fundación Mutua Madrileña, and Instituto de Salud Carlos III [PI15/01197].

References

1. WHO MONICA Project Investigators. The World Health Organization MONICA Project (Monitoring trends and determinants in cardiovascular disease). J Clin Epidemiol. 1988;41:105–14.
2. World Health Organization. The Atlas of disease cardiovascular and stroke. http://www.who.int/cardiovascular_diseases/resources/atlas/en/
3. Sudlow CLM, Warlow CP, for the International stroke incidence collaboration. Comparable studies of the incidence of stroke and its pathological types. Results from a International collaboration. Stroke. 1997;28:491–9.
4. Ovbiagele B, Goldstein LB, Higashida RT, Howard VJ, Johnston SC, Khavjou OA, Lackland DT, Lichtman JH, Mohl S, Sacco RL, Saver JL, Trogdon JG, on behalf of the American Heart Association Advocacy Coordinating Committee and Stroke Council. Forecasting the future of stroke in the United States: a policy statement from the American Heart Association and American Stroke Association. Stroke. 2013;44:2361–75.
5. Gustavsson A, Svensson M, et al. Cost of disorders of the brain in Europe 2010. Eur Neuropsychopharmacol. 2011;21:718–79.
6. Wardlaw JM, Murray V, Berge E, del Zoppo G, Sandercock P, Lindley RL, Cohen G. Recombinant tissue plasminogen activator for acute ischaemic stroke: an updated systematic review and meta-analysis. Lancet. 2012;379:2364–72.
7. Goyal M, Menon BK, van Zwam WH, et al, for the HERMES collaborators. Endovascular thrombectomy after large-vessel ischaemic stroke: a meta-analysis of individual patient data from five randomised trials. Lancet 2016; published online Feb 18. http://dx.doi.org/10.1016/S0140-6736(16)00163-X.
8. Muir KW, White P. HERMES: messenger for stroke interventional treatment. Lancet. 2016 Feb 18. [Epub ahead of print].
9. Lees KR, Zivin JA, Ashwood T, et al. NXY-059 for acute ischemic stroke. N Engl J Med. 2006;354:588–600.
10. Dávalos A, Alvarez-Sabín J, Castillo J, et al. Citicoline in the treatment of acute ischaemic stroke: an international, randomised, multicentre, placebo-controlled study (ICTUS trial). Lancet. 2012;380:349–57.
11. Hermann DM, Chopp M. Promoting brain remodelling and plasticity for stroke recovery: therapeutic promise and potential pitfalls of clinical translation. Lancet Neurol. 2012;11:369–80.
12. Brenneman M, Sharma S, Harting M, Strong R, Cox Jr CS, Aronowski J, Grotta JC, Savitz SI. Autologous bone marrow mononuclear cells enhance recovery after acute ischemic stroke in young and middle-aged rats. J Cereb Blood Flow Metab. 2010;30:140–9.

13. Fujita Y, Ihara M, Ushiki T, Hirai H, Kizaka-Kondoh S, Hiraoka M, Ito H, Takahashi R. Early protective effect of bone marrow mononuclear cells against ischemic white matter damage through augmentation of cerebral blood flow. Stroke. 2010;41:2938–43.
14. Zhang ZG, Chopp M. Neurorestorative therapies for stroke: underlying mechanisms and translation to the clinic. Lancet Neurol. 2009;8:491–500.
15. Yoo SW, Chang DY, Lee HS, Kim GH, Park JS, Ryu BY, et al. Immune suppression following mesenchymal stem cell transplantation in the ischemic brain is mediated by TGF-β. Neurobiol Dis. 2013;58:249–57.
16. Banerjee S, Williamson DA, Habib N, Chataway J. The potential benefit of stem cell therapy after stroke: an update. Vasc Health Risk Manag. 2012;8:569–80.
17. Rosado-de-Castro PH, Pimentel-Coelho PM, da Fonseca LM, de Freitas GR, Mendez-Otero R. The rise of cell therapy trials for stroke: review of published and registered studies. Stem Cells Dev. 2013;22:2095–111.
18. Dominici M, Le Blanc K, Mueller I, et al. Minimal criteria for defining multipotent mesenchymal stromal cells: the International Society for Cellular Therapy position statement. Cytotherapy. 2006;8:315–7.
19. Lalu MM, McIntyre L, Pugliese C, et al. Safety of cell therapy with mesenchymal stromal cells (SafeCell): a systematic review and meta-analysis of clinical trials. PLoS One. 2012;7:e47559.
20. Gupta PK, Chullikana A, Parakh R, Desai S, Das A, Gottipamula S, Krishnamurthy S, Anthony N, Pherwani A, Majumdar AS. A double blind randomized placebo controlled phase I/II study assessing the safety and efficacy of allogeneic bone marrow derived mesenchymal stem cell in critical limb ischemia. J Transl Med. 2013;11:143.
21. Chou SH, Lin SZ, Kuo WW, Pai P, Lin JY, Lai CH, Kuo CH, Lin KH, Tsai FJ, Huang CY. Mesenchymal stem cell insights: prospects in cardiovascular therapy. Cell Transplant. 2014;23:513–29.
22. De Miguel MP, Fuentes-Julián S, Blázquez-Martínez A, Pascual CY, Aller MA, Arias J, Arnalich-Montiel F. Immunosuppressive properties of mesenchymal stem cells: advances and applications. Curr Mol Med. 2012;12:574–91.
23. Vu Q, Xie K, Eckert M, Zhao W, Cramer SC. Meta-analysis of preclinical studies of mesenchymal stromal cells for ischemic stroke. Neurology. 2014;82:1277–86.
24. Chen J, Li Y, Katakowski M, Chen X, Wang L, Lu D, Lu M, Gautam SC, Chopp M. Intravenous bone marrow stromal cell therapy reduces apoptosis and promotes endogenous cell proliferation after stroke in female rat. J Neurosci Res. 2003;73:778–86.
25. Chen J, Zhang ZG, Li Y, Wang L, Xu YX, Gautam SC, Lu M, Zhu Z, Chopp M. Intravenous administration of human bone marrow stromal cells induces angiogenesis in the ischemic boundary zone after stroke in rats. Circ Res. 2003;92:692–9.
26. Bang OY, Lee JS, Lee PH, Lee G. Autologous mesenchymal stem cell transplantation in stroke patients. Ann Neurol. 2005;57:874–82.
27. Lee JS, Hong JM, Moon GJ, Lee PH, Ahn YH. OY Bang and STARTING collaborators. A long-term follow-up study of intravenous autologous mesenchymal stem cell transplantation in patients with ischemic stroke. Stem Cells. 2010;28:1099–106.
28. Honmou O, Houkin K, Matsunaga T, Niitsu Y, Ishiai S, Onodera R, Waxman SG, Kocsis JD. Intravenous administration of auto serum-expanded autologous mesenchymal stem cells in stroke. Brain. 2011;134:1790–807.
29. Bhasin A, Srivastava MV, Mohanty S, Bhatia R, Kumaran SS, Bose S. Stem cell therapy: a clinical trial of stroke. Clin Neurol Neurosurg. 2013;115:1003–8.
30. Lu SS, Liu S, Zu QQ, Xu XQ, Yu J, Wang JW, Zhang Y, Shi HB. In vivo MR imaging of intraarterially delivered magnetically labeled mesenchymal stem cells in a canine stroke model. PLoS One. 2013;8:e54963.
31. Steinberg GK, Kondziolka D, Wechsler LR, Lunsford LD, Coburn ML, Billigen JB, Kim AS, Johnson JN, Bates D, King B, Case C, McGrogan M, Yankee EW, Schwartz NE. Clinical outcomes of transplanted modified bone marrow-derived mesenchymal stem cells in stroke: a phase 1/2a study. Stroke. 2016;47(7):1817–24.

32. Walczak P, Zhang J, Gilad AA, Kedziorek DA, Ruiz-Cabello J, Young RG, Pittenger MF, van Zijl PC, Huang J, Bulte JW. Dual-modality monitoring of targeted intraarterial delivery of mesenchymal stem cells after transient ischemia. Stroke. 2008;39:1569–74.
33. Schwarting S, Litwak S, Hao W, Bähr M, Weise J, Neumann H. Hematopoietic stem cells reduce postischemic inflammation and ameliorate ischemic brain injury. Stroke. 2008;39:2867–75.
34. Taguchi A, Soma T, Tanaka H, et al. Administration of CD34+ cells after stroke enhances neurogenesis via angiogenesis in a mouse model. J Clin Invest. 2004;114:330–8.
35. Shyu WC, Lin SZ, Chiang MF, Su CY, Li H. Intracerebral peripheral blood stem cell (CD34+) implantation induces neuroplasticity by enhancing beta1 integrin-mediated angiogenesis in chronic stroke rats. J Neurosci. 2006;26:3444–53.
36. Chen DC, Lin SZ, Fan JR, Lin CH, Lee W, Lin CC, Liu YJ, Tsai CH, Chen JC, Cho DY, Lee CC, Shyu WC. Intracerebral implantation of autologous peripheral blood stem cells in stroke patients: a randomized phase II study. Cell Transplant. 2014;23:1599.
37. England TJ, Abaei M, Auer DP, Lowe J, Jones DR, Sare G, Walker M, Bath PM. Granulocyte-colony stimulating factor for mobilizing bone marrow stem cells in subacute stroke: the stem cell trial of recovery enhancement after stroke 2 randomized controlled trial. Stroke. 2012;43:405–11.
38. Banerjee S, Bentley P, Hamady M, Marley S, Davis J, Shlebak A, Nicholls J, Williamson DA, Jensen SL, Gordon M, Habib N, Chataway J. Intra-arterial immunoselected CD34+ stem cells for acute ischemic stroke. Stem Cells Transl Med. 2014;3:1322–30.
39. Nakano-Doi A, Nakagomi T, Fujikawa M, Nakagomi N, Kubo S, Lu S, et al. Bone marrow mononuclear cells promote proliferation of endogenous neural stem cells through vascular niches after cerebral infarction. Stem Cells. 2010;28:1292–302.
40. Sharma S, Yang B, Strong R, Xi X, Brenneman M, Grotta JC, et al. Bone marrow mononuclear cells protect neurons and modulate microglia in cell culture models of ischemic stroke. J Neurosci Res. 2010;88:2869–76.
41. Suárez-Monteagudo C, Hernández-Ramírez P, Alvarez-González L, García-Maeso I, de la Cuétara-Bernal K, Castillo-Díaz L, et al. Autologous bone marrow stem cell neurotransplantation in stroke patients. An open study. Restor Neurol Neurosci. 2009;27:151–61.
42. Li ZM, Zhang ZT, Guo CJ, Geng FY, Qiang F, Wang LX. Autologous bone marrow mononuclear cell implantation for intracerebral hemorrhage-a prospective clinical observation. Clin Neurol Neurosurg. 2013;115:72–6.
43. Battistella V, de Freitas GR, da Fonseca LM, Mercante D, Gutfilen B, Goldenberg RC, Dias JV, Kasai-Brunswick TH, Wajnberg E, Rosado-de-Castro PH, Alves-Leon SV, Mendez-Otero R, Andre C. Safety of autologous bone marrow mononuclear cell transplantation in patients with nonacute ischemic stroke. Regen Med. 2011;6:45–52.
44. Friedrich MA, Martins MP, Araújo MD, Klamt C, Vedolin L, Garicochea B, Raupp EF, Sartori El Ammar J, Machado DC, Costa JC, Nogueira RG, Rosado-de-Castro PH, Mendez-Otero R, Freitas GR. Intra-arterial infusion of autologous bone marrow mononuclear cells in patients with moderate to severe middle cerebral artery acute ischemic stroke. Cell Transplant. 2012;21 Suppl 1:S13–21.
45. Rosado de Castro PH, Schmidt FR, Battistella V, Lopes de Souza SA, Gutfilen B, Goldenberg RC, Kasai-Brunswick TH, Vairo L, Silva RM, et al. Biodistribution of bone marrow mononuclear cells after intra-arterial or intravenous transplantation in subacute stroke patients. Regen Med. 2013;8:145–55.
46. Savitz SI, Misra V, Kasam M, Juneja H, Cox Jr CS, Alderman S, Aisiku I, Kar S, Gee A, Grotta JC. Intravenous autologous bone marrow mononuclear cells for ischemic stroke. Ann Neurol. 2011;70:59–69.
47. Moniche F, Gonzalez A, Gonzalez-Marcos JR, Carmona M, Piñero P, Espigado I, Garcia-Solis D, Cayuela A, Montaner J, Boada C, Rosell A, Jimenez MD, Mayol A, Gil-Peralta A. Intra-arterial bone marrow mononuclear cells in ischemic stroke. A pilot clinical trial. Stroke. 2012;43:2242–4.

48. El Khoury R, Misra V, Sharma S, Cox CS, Walker P, Grotta JC, et al. The effect of transcatheter injections on cell viability and cytokine release of mononuclear cells. AJNR Am J Neuroradiol. 2010;31:1488–92.
49. Prasad K, Mohanty S, Bhatia R, Srivastava MV, Garg A, Srivastava A, Goyal V, Tripathi M, Kumar A, Bal C, Vij A, Mishra NK. Autologous intravenous bone marrow mononuclear cell therapy for patients with subacute ischaemic stroke: a pilot study. Indian J Med Res. 2012;136:221–8.
50. Prasad K, Sharma A, Garg A, Mohanty S, Bhatnagar S, Johri S, Singh KK, Nair V, Sarkar RS, Gorthi SP, Hassan KM, Prabhakar S, Marwaha N, Khandelwal N, Misra UK, Kalita J, Nityanand S, InveST Study Group. Intravenous autologous bone marrow mononuclear stem cell therapy for ischemic stroke: a multicentric, randomized trial. Stroke. 2014;45:3618–24.
51. Yang B, Parsha K, Schaar K, Xi X, Aronowski J, Savitz SI. Various cell populations within the mononuclear fraction of bone marrow contribute to the beneficial effects of autologous bone marrow cell therapy in a rodent stroke model. Transl Stroke Res. 2016 Mar 20. [Epub ahead of print].
52. Womble TA, Green S, Shahaduzzaman M, Grieco J, Sanberg PR, Pennypacker KR, Willing AE. Monocytes are essential for the neuroprotective effect of human cord blood cells following middle cerebral artery occlusion in rat. Mol Cell Neurosci. 2014;59:76–84.
53. Mora-Lee S, Sirerol-Piquer MS, Gutiérrez-Pérez M, Gomez-Pinedo U, Roobrouck VD, López T, Casado-Nieto M, Abizanda G, Rabena MT, Verfaille C, Prósper F, García-Verdugo JM. Therapeutic effects of hMAPC and hMSC transplantation after stroke in mice. PLoS One. 2012;7(8):e43683.
54. Komatsu K, Honmou O, Suzuki J, Houkin K, Hamada H, Kocsis JD. Therapeutic time window of mesenchymal stem cells derived from bone marrow after cerebral ischemia. Brain Res. 2010;1334:84–92.
55. Yang B, Strong R, Sharma S, Brenneman M, Mallikarjunarao K, Xi X, Grotta JC, Aronowski J, Savitz SI. Therapeutic time window and dose response of autologous bone marrow mononuclear cells for ischemic stroke. J Neurosci Res. 2011;89:833–9.
56. de Vasconcelos Dos Santos A, da Costa RJ, Diaz Paredes B, Moraes L, Jasmin, Giraldi-Guimarães A, Mendez-Otero R. Therapeutic window for treatment of cortical ischemia with bone marrow-derived cells in rats. Brain Res. 2010;1306:149–58.
57. Moniche F, Montaner J, Gonzalez-Marcos JR, Carmona M, Piñero P, Espigado I, Cayuela A, Escudero I, de la Torre-Laviana FJ, Boada C, Rosell A, Mayol A, Jimenez MD, Gil-Peralta A, Gonzalez A. Intra-arterial bone marrow mononuclear cell transplantation correlates with GM-CSF, PDGF-BB, and MMP-2 serum levels in stroke patients: results from a clinical trial. Cell Transplant. 2014;23 Suppl 1:S57–64.
58. Fischer UM, Harting MT, Jimenez F, Monzon-Posadas WO, Xue H, Savitz SI, et al. Pulmonary passage is a major obstacle for intravenous stem cell delivery: the pulmonary first-pass effect. Stem Cells Dev. 2009;18:683–92.
59. Sharma A, Sane H, Gokulchandran N, Khopkar D, Paranjape A, Sundaram J, Gandhi S, Badhe P. Autologous bone marrow mononuclear cells intrathecal transplantation in chronic stroke. Stroke Res Treat. 2014;2014:234095.
60. Barbosa da Fonseca LM, Gutfilen B, Rosado de Castro PH, Battistella V, Goldenberg RC, Kasai-Brunswick T, Chagas CL, Wajnberg E, Maiolino A, Salles Xavier S, Andre C, Mendez-Otero R, de Freitas GR. Migration and homing of bone-marrow mononuclear cells in chronic ischemic stroke after intra-arterial injection. Exp Neurol. 2010;221:122–8.
61. Chen X, Li Y, Wang L, Katakowski M, Zhang L, Chen J, Xu Y, Gautam SC, Chopp M. Ischemic rat brain extracts induce human marrow stromal cell growth factor production. Neuropathology. 2002;22:275–9.
62. Taguchi A, Sakai C, Soma T, Kasahara Y, Stern DM, Kajimoto K, Ihara M, Daimon T, Yamahara K, Doi K, Kohara N, Nishimura H, Matsuyama T, Naritomi H, Sakai N, Nagatsuka K. Intravenous autologous bone marrow mononuclear cell transplantation for stroke: phase1/2a clinical trial in a homogeneous group of stroke patients. Stem Cells Dev. 2015;24:2207–18.

63. Jeong H, Yim HW, Cho YS, Kim YI, Jeong SN, Kim HB, Oh IH. Efficacy and safety of stem cell therapies for patients with stroke: a systematic review and single arm meta-analysis. Int J Stem Cells. 2014;7:63–9.
64. Moniche F, Rosado-de-Castro PH, Escudero I, Zapata E, de la Torre Laviana FJ, Mendez-Otero R, Carmona M, Piñero P, Bustamante A, Lebrato L, Cabezas JA, Gonzalez A, de Freitas GR, Montaner J. Increasing dose of autologous bone marrow mononuclear cells (BM-MNCs) transplantation is related with stroke outcome. Results from a pooled analysis of two clinical trials. Stem Cells Int. 2016;2016:8657173.
65. Moniche F, Escudero I, Zapata-Arriaza E, Usero-Ruiz M, Prieto-León M, de la Torre J, Gamero MA, Tamayo JA, Ochoa-Sepúlveda JJ, Maestre J, Carmona M, Piñero P, Calderón-Cabrera C, Jimenez MD, Gonzalez A, Montaner J. Intra-arterial bone marrow mononuclear cells (BM-MNCs) transplantation in acute ischemic stroke (IBIS trial): protocol of a phase II, randomized, dose-finding, controlled multicenter trial. Int J Stroke. 2015;10:1149–52.

Chapter 11
Bone Marrow Stem Cell Delivery Methods, Routes, Time, Efficacy, and Safety

Lijie Huang, Jianjing Yang, Mark Nyanzu, Felix Siaw-Debrah, and Qichuan Zhuge

Abstract Stem cell transplantation offers an exciting new therapeutic avenue for stroke, as many studies have demonstrated favorable results in animal models with various cell types. Several early phase I and II clinical trials are now underway with promising outcomes. However, cell transplantation for stroke is still in its infancy with many issues that need to be addressed in order to achieve full potential as a therapy. Among the major hurdles for a successful clinical translation is determining the optimal conditions of transplantation for stroke. In this chapter, we review the impact of implanted cell number, delivery sites of cells, and transplantation time on the stroke outcome. In addition, we also discuss the efficacy and safety of bone marrow stem cell transplantation for stroke treatment.

Keywords Transplantation • Stroke • Optimal condition • Time • Cell number

11.1 Introduction

Stem cell transplantation offers an exciting new therapeutic avenue for stroke, as many studies have demonstrated favorable results in animal models with various cell types. Several early phase I and II clinical trials are now underway with promising outcomes. Therefore, the potential therapeutic impact of stem cell transplantation on regeneration of damaged brain tissue opens up enormous possibilities. If successful, millions of stroke survivors with disability may benefit. However, cell transplantation for stroke is still in its infancy with many issues that need to be addressed in order to achieve full potential as a therapy. Among the major hurdles

L. Huang • J. Yang • M. Nyanzu • F. Siaw-Debrah • Q. Zhuge (✉)
Department of Neurosurgery, First Affiliated Hospital, Wenzhou Medical University, Wenzhou, China

Zhejiang Provincial Key Laboratory of Aging and Neurological Disorder Research, First Affiliated Hospital, Wenzhou Medical University, Wenzhou, China
e-mail: zhugeqichuan@vip.163.com

K. Jin et al. (eds.), *Bone Marrow Stem Cell Therapy for Stroke*,
DOI 10.1007/978-981-10-2929-5_11

for a successful clinical translation is determining the optimal conditions of transplantation for stroke. As different groups used different protocols or conditions, the results may not compare well with each other, which makes it difficult to determine the best conditions for stem cell therapy following a stroke. The optimal conditions, including the best cell type, the cell number, the timing of transplantation, the route and site of delivery, and the stroke model, are highly important. All of those conditions still need further study. In the following parts, we will summary recent studies, which are associated with BMSC delivery method, time point, efficacy and safety.

11.2 Delivery Methods and Routes

Bone marrow stem cell (BMSC) transplantation is a promising therapy for some kind of diseases like traumatic spinal cord injury (SCI) and degenerative conditions of the central nervous system (CNS). The number of transplanted cells in the brain depends on the effectiveness of the transplantation. BMSC transplantation has been investigated and explored in animal models to determine its therapeutic effects for disorders such as SCI and brain ischemia [1]. BMSCs produce different trophic factors (e.g., brain-derived neurotrophic factor (BDNF), vascular endothelial growth factor (VEGF), nerve growth factor (NGF), and hepatocellular growth factor (HGF)) and cytokines. BMSCs do not only activate endogenous restorative responses such as angiogenesis, synaptogenesis, and neurogenesis, they also have a negative effect on the death of the brain cells in the ischemic boundary zone [2]. Prior to the administration of BMSCs, it's required to culture them for a certain amount of time [3]. BMSCs have the protective potentials to repair the tissue [4]. Also BMSCs have the ability to cross the blood–brain barrier (BBB) and migrate into the brain parenchyma [5]. In order to achieve greater efficacy in terms of neuroprotection, BMSCs should be modified and injected intracerebrally [6]. The neuroprotective effects of BMSCs may result from their behavior as small molecular factories [7], although the basic mechanisms are still not comprehensive. Different routes of administration BMSCs exhibit different effects in treating disorders.

11.2.1 Intravenous and Intra-arterial Routes

When BMSCs are intra-arterially administered, larger numbers of the cell are able to reach their target tissues such as the brain parenchyma as opposed to the intravenous (IV) routes. The cells are thus able to confer their neuroprotective functions efficiently through the intra-arterial (IA) routes. Protective factors from the marrow cells provide the neuroprotection after transplantation. Injection of BMSCs into the ipsilateral carotid artery after transient middle cerebral artery occlusion (MCAO) results in many BMSCs in the ischemic hemisphere [1]. Transplantation of BMSCs after brain ischemia results in up to 21 % of the cells in the MCA territory [2]. Administration of BMSCs after brain ischemia decreases infarct volume and increases transplanted cells in the brain when infused in intravenously [8]. Following transplantations of BMSCs, analysis of brain sections can be performed by fluorescence photography. BMSCs transplanted by IA routes appear to express larger numbers of PKH26-positive cells in an ischemic hemisphere compared to the IV routes. IA transplantation is an effective route in reducing infarct volumes unlike the IV routes. Rotarod score test of an IA-transplanted BMSCs shows improved and higher motor function than the IV-transplanted ones. BMSCs thus improve functional outcome when administered intra-arterially following ischemia [3, 9]. Reports indicate that adverse effects after IA administration of BMSC are often minimal and insignificant in patients with stroke [10–12]. IA routes however require a cerebral angiography which is an invasive procedure with its associated risk factors such as new strokes. The risk is nevertheless very minimal with research indicating the possibility of a new stroke is about 0.14 % and other complications is close to 2 % [13]. Other notable risk factors following BMSC transplantation intra-arterially involve embolism and occlusion of brain vessels [14]. IV routes thus might be safer than IA routes to some extent. That's not convinced that IA routes seem to be an effective and a superior route than the IV routes. Further research and validation need to be performed to elucidate that one route is more functional than the other [15]. Stem cell delivery to the injured spinal cord has to overcome several arterial branches to maximize efficacy. Hence, a highly selective and technically challenging cannulation is required. One advantage of intravenous stem cell delivery is comparatively the least invasive approach and has been investigated in several studies [16]. After intravenous injection, cellular homing occurs into the pathological CNS tissues. However, IV route is still less efficient when measured with other approaches such as intra-cerebrospinal fluid (CSF). Additional problems associated with intravenous stem cell delivery include reliance on injury-mediated opening of the BBB to allow cell access to the CNS parenchyma (or the need for additional drugs such as lipopolysaccharide to open the BBB) [17]. First-pass effects affect BMSCs and trap them in extra tissues such as the liver and lungs exposing them to longer periods of immunity and reticuloendothelial cells after injection into the bloodstream. Very few to no cells usually present within the injured spinal segments that receive BMSCs intravenously indicates that the effectiveness of cell therapy might not be necessarily related to the number of cells reaching the brain parenchyma [18].

11.2.2 Intrathecal/Lumbar Puncture Routes

Lumbar puncture (LP) is a minimally invasive way of cell and drug delivery, and BMSC may be well suited for LP transplantation because of their responsiveness to signals from the injured CNS [17]. Clinical and theoretical studies show that LP delivery of stem cells is extremely attractive. LP is performed at the L3–4 level, far away from the cervical or thoracic spinal cord, which is the region most commonly effected by SCI. This makes LP delivery of stem cells relatively safe and unlikely to worsen compromised patients as a direct result of the intervention. Stem cells can be injected directly into the lesion site; however, the additional trauma from intramedullary transplantation can further compromise injured tissue and impede clinical outcomes. A potential and effective alternative is intrathecal transplantation via LP [19]. Additional advantages of LP delivery are related to several factors: (1) they are far more superior to IV route because they are injected directly into the CSF without encountering the BBB; (2) the CSF transports the cells to injured tissues without encountering first effect degradation by the liver or lungs; and (3) because the transplanted cells are delivered away from the hostile environment of the injured tissue, they are given a greater opportunity to survive and migrate to the injured site. Cell transplantation via LP may be relevant for conditions such as multiple sclerosis with widely disseminated lesions making intramedullary transplantation impractical [20]. Neuronal progenitor cells provide neuroprotective functions for injured spinal cord after subacute transplantation directly into the cord [21]. Proliferation of BMSC occurs both in injured and uninjured spinal cord after LP transplantation during the early stages of the transplantation; however, the number of proliferating cells decreases with time. LP-transplanted BMSCs are distributed in the intrathecal space, along the length of the spinal cord, and a few will migrate to the lesion cavity. This implies that BMSC can reach the injured spinal cord using minimally invasive method of LP injection of cells into the lumbar intrathecal space. Prior to transplantation of the cells, extensive incision and disruption of the dura need to be done to allow for transplanted cells to migrate freely into the injured spinal cord [22].

Swelling of the injured cord together with the meningeal reaction causes intrathecally injected cells to be attracted to the cord parenchyma. Collagenous matrix could be implanted to serve as an adhesive substrate for the cells to adhere and attach. The substrate also serves as a barrier for the subarachnoid space to prevent the passage of the intrathecal cells [23].

11.2.3 Cerebral Injections

A major unresolved problem in the context of SCI is the delivery of cells to an already compromised spinal cord without causing further damage. Most investigators have undertaken direct injection into the injured spinal [24]. Although this is acceptable in animal experiments, its extrapolation to humans may be difficult because a major neurosurgical operation will be required. This difficulty in translation will limit clinical trials, at least initially, to patients with complete SCIs in whom further deterioration cannot occur but in whom significant benefit from transplantation therapies is also least likely [25]. Another problem associated with direct parenchymal injections is the likelihood of damaging spared spinal tissues with the injecting needle. It is a well-known principle of neurosurgery that injured tissues do not tolerate operative manipulation as well as normal tissues, because of the presence of edema, altered blood flow, and injury-related cytokines. Finally, direct injection of cells into the parenchyma does not allow suitable delivery of multiple therapeutic doses because of its invasive nature and because injecting cells into multifocal diseases presents many logistical and technical challenges. The direct delivery of stem cells into the CSF has also been explored [26], and intraventricular injection has been the favored delivery method [27]. This technique, however, is too invasive for clinical applicability, which makes its transplantation challenging. Investigators have demonstrated that neurosphere-derived stem cells delivered into the ventricular CSF can reach the injured tissue in a spinal contusion model [28]. BMSCs are more appropriate because of their evidenced therapeutic effect, their availability, and the possibility of an autologous model in humans. Injection of cells into the lumbar CSF via an indwelling cannula has been shown to be effective for delivering embryonic germ cell derivatives. Considerably more cells will be detected in the injured tissues after both intrathecal and intraventricular delivery. Transplanting cells into the CSF leads to more successful grafting when injection is via an intrathecal or intraventricular rather than intravenous route. The number of cells within the injured spinal cord tissues increased with passage of time. Few cells are present at the early days after transplantation; however, many more cells will be recorded as the time increases after transplantation.

11.3 Effective Times for BMSC Transplantation

Bone marrow stem cell transplantation has over the years been a promising field for the treatment of various disorders like ischemic stroke, hematologic diseases, cardiac disorders like cardiac infarctions, etc. The efficacy of treatment not only depends on the route of cells but also on the cell dose and the time of delivery [29]. For example, de Vasconcelos Dos Santos et al. highlighted the benefit of using BMSCs in the treatment of a thermocoagulation-induced ischemic rat model at different therapeutic windows. They concluded in their experiment that BMSCs might be an efficient treatment protocol for stroke only in the acute/subacute phase of the disease since they were unable to decrease glial scarring significantly [30]. Another example was a recent study which clearly showed that IA administration of 1×10^7 BMSCs immediately after reperfusion is much more effective in delivering BMSCs to the brain than IV administration. In addition, the larger number BMSCs are transplanted in the brain during the early stage of reperfusion, the better protective effect may be presented. The study therefore suggested further understanding into the dose–response influence and therapeutic time window for efficient BMSC delivery to the ischemic site [31].

BMSC transplantation like any other treatment option has an optimum time to which to get adequate outcome. There is therefore the need to optimize treatment by taking advantage of the best time for optimum outcome and reduce disease progress. During ischemia, a series of inflammatory response is initiated which is mediated by many transcription factors of which nuclear factor-kB is a key factor. When hypoxia or ischemia occurs, a cascade of signal transductions is triggered, causing nuclear factor-kB inhibitor IkB phosphorylation degradation and activation of nuclear factor-kB to enter nuclei and stimulate target gene transcription. These processes eventually trigger a positive feedback which leads to an overwhelming inflammatory response. This secondary inflammatory response if not controlled will accelerate and cause further cerebral ischemia/reperfusion injury. A study performed recently indicated that nuclear factor-kB DNA-binding activity is exponentially enhanced within 6–12 h after ischemia, whereas there is gradual decrease between 24 and 72 h. This indicates that NF-kB translocation occurs in a time-dependent manner after cerebral ischemia [32].

In an experiment to test the hypothesis that IV administration of BMSCs could lead to improvement of functional recovery after MCAO for 45 min in the rat and to determine specific time windows for efficacy. Iihoshi et al. injected rats intravenously with transfected mononuclear cells at 3, 6, 12, 24, and 72 h after MCAO. The ischemic lesion was histologically analyzed at 14 days. It was noted that there was no lesion detected at 3 h transplantation after lesion induction. Lesions were however detected from 6 h post-lesion group and progressively increase at times 12, 24, and 72 h. Infused LacZ(+) bone marrow cells are implanted extensively in and around the ischemic site, with immunohistochemistry studies indicating some amount of differentiation of neuronal and glial cells. Behavioral testing (Morris water maze and treadmill stress test) also indicated improved functional recovery in the transplanted group. These findings further stress the need to intervene as fast as possible and also suggest that IV administration of autologous mononuclear cells from the bone marrow could help improve functional outcome [3]. Other experiments suggest other time windows. Of notable example is the administration of allogeneic human umbilical cord blood MSCs (hUCB-MSC) by LP 3 days after stroke, which was stated to be a valuable method for efficient cell delivery and therapy in stroke model in rats [33, 34].

11.4 Efficacy and Safety of BMSC Transplantation

According to therapy purpose, proper BMSC delivery method should be selected. The delivery efficacy and safety are two critical factors that determine the application of delivery method. When comparing those different delivery methods, every route has some apparently advantages and shortages.

Using IV infusion to deliver BMSC is the simplest and safest method. In animal models, engraftment was demonstrated [35]. However, IV infusion has low cell retention rate; the efficacy is pretty low. Also, BMSC cannot diffuse to specific sites for therapeutic effect. The amount of BMSC arrive the target organ may not efficiently repair the primary injury by differentiation. But, studies demonstrated that recovery improved after BMSC administration should be partially owed to the inflammation milieu improvement by trophic factors and inflammation factors, which are secreted by transplanted BMSC. Comparing to IV infusion, local IA infusion has higher efficacy. The cell distribution after IA infusion of BMSC is much better than IV infusion [36]. IA infusion of BMSC can be applied for heart diseases. IA infusion has some shortage, which may cause microembolism or ischemia during infusion. And sometimes these shortages are fatal. Direct route still is the highest efficacy delivery method, which could deliver a maximum amount of cells to intended area [37]. Safety would be essential for patients; due to the big invasion, the direct route is limited to some certain situation. LP route is a novel minimally invasive method for delivery of BMSC, which can be applied for many kinds of CNS disease. The important thing is that researches already have demonstrated that BMSC could pass through BBB after LP administration [38, 39]. LP route delivered BMSC able to survive and accumulate and can exhibit the function therapy in CNS disease [22, 40]. The problem is the efficacy of LP still kind of low. Therefore, optimal transplantation technique should be developed to serve maximally safe and efficacy results.

In spite of the restriction of the delivery method, there still have some methods to improving delivery efficacy. Preconditions could improve the survival of BMSC, which including hypoxia and pharmacological treatment. In myocardial infarction model, hypoxia preconditioning can increase the expression of pro-survival and proangiogenic factors including hypoxia-inducible factor 1, angiopoietin-1, vascular endothelial growth factor and its receptor, Flk-1, erythropoietin, Bcl-2, and Bcl-xL. Cell death of hypoxic stem cells and caspase-3 activation in these cells were significantly lower. Transplantation of hypoxic BMSCs after myocardial infarction results in an increase in angiogenesis, as well as enhanced morphologic and functional benefits of stem cell therapy [41]. Indeed, in intracerebral hemorrhage (ICH) model, BMSCs pretreated with hypoxia preconditioning can significantly improve behavioral performance, and increase neurogenesis compared with the vehicle group after ICH [42]. Alternatively, many kinds of pharmacological treatment also could enhance mesenchymal stem cell survival. Trimetazidine (TMZ) preconditioning increases the survival rate of BMSCs through upregulation of HIF1-α in rat myocardial injury model [43]. Hypoxia-inducible factor 1-α (HIF1-α) prolyl hydroxylase inhibitor dimethyloxalylglycine (DMOG) also can upregulate expression of survival and angiogenic factors including HIF1-α, vascular endothelial growth factor, glucose transporter 1, and phospho-Akt, which enhance BMSC survival and therapeutic efficacy after transplantation [44]. Noiseux et al. demonstrated that oxytocin treatment can evoke MSC protection through both intrinsic pathways and secretion of cytoprotective factors [45]. TGF-α stimulates MSC VEGF production in part via a p38 MAPK-dependent mechanism, and preconditioning MSCs with TGF-α could enhance their ability to protect myocardium injury [46]. Tadalafil could increase Bcl2/Bax during the early phase and transcriptional upregulation of PKG-I by STAT3 during the late phase which promotes stem cell protection against ischemic injury [47].

In clinical trials, the efficacy and safety of BMSC transplant for many kinds of diseases have been studied. In stroke patients, the BMSC treatment safety appeared to be safe up to 1 year [48–50]. No significant abnormal EEG/seizures are observed in those patients. Also BMSC transplant treatment in animal stroke model indicated that it has beneficial effects compared to controls [51]. There is another study published by Prasad et al. which showed that intravenous infusion of BMSC doesn't have beneficial effects of treatment on stroke outcome [50]. It is not possible to evaluate efficacy outcome as only one randomized controlled study was available. There still have 15 ongoing clinical trials in phase I or II [52]. After these trials finish, we can gain more insights into the therapeutic potential of BMSC transplant. In diabetes mellitus, study showed that stem cell transplantation can be a safe and effective approach for therapy [53]. In degenerative diseases of the retina, initial data from early stage clinical trials suggest that short-term safety objectives can be met [54]. However, the question of efficacy will require additional time and testing to be adequately resolved. In spinal cord injury, based on short–medium terms following up, stem cell transplantation appears to be safe and valid in patients and more effective in chronic and complete injury [55]. Nonetheless, prospective, randomized trials in larger cohorts are still needed. In acute myocardial infarction, there is insufficient evidence for a beneficial effect of cell therapy for patients [56]. Further adequately powered trials are needed, and until then the efficacy of this intervention remains unproven.

In summary, base on different disease characters, we can select proper delivery method to increase the safety and efficacy. Preconditions via hypoxia or pharmacological treatments also can improve BMSC survival and enhance the efficacy. Indeed, abundance of researches had demonstrated that BMSCs have beneficial effect on many kinds of disease models. But, for clinical application, the effects of BMSC still need to be confirmed in the following clinical trials. BMSC transplantation therapy is a promising approach for curing so many difficult diseases.

References

1. Ohta T, Kikuta K, Imamura H, Takagi Y, Nishimura M, Arakawa Y, et al. Administration of ex vivo-expanded bone marrow-derived endothelial progenitor cells attenuates focal cerebral ischemia-reperfusion injury in rats. Neurosurgery. 2006;59:679–86. discussion 679–686.
2. Li Y, Chen J, Chen XG, Wang L, Gautam SC, Xu YX, et al. Human marrow stromal cell therapy for stroke in rat: neurotrophins and functional recovery. Neurology. 2002;59:514–23.
3. Iihoshi S, Honmou O, Houkin K, Hashi K, Kocsis JD. A therapeutic window for intravenous administration of autologous bone marrow after cerebral ischemia in adult rats. Brain Res. 2004;1007:1–9.
4. Zhang S, Ge J, Sun A, Xu D, Qian J, Lin J, et al. Comparison of various kinds of bone marrow stem cells for the repair of infarcted myocardium: single clonally purified non-hematopoietic mesenchymal stem cells serve as a superior source. J Cell Biochem. 2006;99:1132–47.
5. Brazelton TR, Rossi FM, Keshet GI, Blau HM. From marrow to brain: expression of neuronal phenotypes in adult mice. Science. 2000;290:1775–9.
6. Kurozumi K, Nakamura K, Tamiya T, Kawano Y, Ishii K, Kobune M, et al. Mesenchymal stem cells that produce neurotrophic factors reduce ischemic damage in the rat middle cerebral artery occlusion model. Mol Ther. 2005;11:96–104.
7. Chopp M, Li Y. Treatment of neural injury with marrow stromal cells. Lancet Neurol. 2002;1:92–100.
8. Honma T, Honmou O, Iihoshi S, Harada K, Houkin K, Hamada H, et al. Intravenous infusion of immortalized human mesenchymal stem cells protects against injury in a cerebral ischemia model in adult rat. Exp Neurol. 2006;199:56–66.
9. Giraldi-Guimardes A, Rezende-Lima M, Bruno FP, Mendez-Otero R. Treatment with bone marrow mononuclear cells induces functional recovery and decreases neurodegeneration after sensorimotor cortical ischemia in rats. Brain Res. 2009;1266:108–20.

10. Battistella V, de Freitas GR, da Fonseca LM, Mercante D, Gutfilen B, Goldenberg RC, et al. Safety of autologous bone marrow mononuclear cell transplantation in patients with nonacute ischemic stroke. Regen Med. 2011;6:45–52.
11. Mendez-Otero R, de Freitas GR, Andre C, de Mendonca ML, Friedrich M, Oliveira-Filho J. Potential roles of bone marrow stem cells in stroke therapy. Regen Med. 2007;2:417–23.
12. Mendonca ML, Freitas GR, Silva SA, Manfrim A, Falcao CH, Gonzales C, et al. safety of intra-arterial autologous bone marrow mononuclear cell transplantation for acute ischemic stroke. Arq Bras Cardiol. 2006;86:52–5.
13. Kaufmann TJ, Huston 3rd J, Mandrekar JN, Schleck CD, Thielen KR, Kallmes DF. Complications of diagnostic cerebral angiography: evaluation of 19,826 consecutive patients. Radiology. 2007;243:812–9.
14. Walczak P, Zhang J, Gilad AA, Kedziorek DA, Ruiz-Cabello J, Young RG, et al. Dual-modality monitoring of targeted intraarterial delivery of mesenchymal stem cells after transient ischemia. Stroke. 2008;39:1569–74.
15. Brenneman M, Sharma S, Harting M, Strong R, Cox Jr CS, Aronowski J, et al. Autologous bone marrow mononuclear cells enhance recovery after acute ischemic stroke in young and middle-aged rats. J Cereb Blood Flow Metab. 2010;30:140–9.
16. Garbuzova-Davis S, Willing AE, Zigova T, Saporta S, Justen EB, Lane JC, et al. Intravenous administration of human umbilical cord blood cells in a mouse model of amyotrophic lateral sclerosis: distribution, migration, and differentiation. J Hematother Stem Cell Res. 2003;12:255–70.
17. Pluchino S, Quattrini A, Brambilla E, Gritti A, Salani G, Dina G, et al. Injection of adult neurospheres induces recovery in a chronic model of multiple sclerosis. Nature. 2003;422:688–94.
18. Borlongan CV, Hadman M, Sanberg CD, Sanberg PR. Central nervous system entry of peripherally injected umbilical cord blood cells is not required for neuroprotection in stroke. Stroke. 2004;35:2385–9.
19. Tator CH. Review of treatment trials in human spinal cord injury: issues, difficulties, and recommendations. Neurosurgery. 2006;59:957–82. discussion 982–957.
20. Weiss S, Dunne C, Hewson J, Wohl C, Wheatley M, Peterson AC, et al. Multipotent cns stem cells are present in the adult mammalian spinal cord and ventricular neuroaxis. J Neurosci. 1996;16:7599–609.
21. Parr AM, Kulbatski I, Zahir T, Wang X, Yue C, Keating A, et al. Transplanted adult spinal cord-derived neural stem/progenitor cells promote early functional recovery after rat spinal cord injury. Neuroscience. 2008;155:760–70.
22. Bakshi A, Barshinger AL, Swanger SA, Madhavani V, Shumsky JS, Neuhuber B, et al. Lumbar puncture delivery of bone marrow stromal cells in spinal cord contusion: a novel method for minimally invasive cell transplantation. J Neurotrauma. 2006;23:55–65.
23. Lepore AC, Bakshi A, Swanger SA, Rao MS, Fischer I. Neural precursor cells can be delivered into the injured cervical spinal cord by intrathecal injection at the lumbar cord. Brain Res. 2005;1045:206–16.
24. Hofstetter CP, Schwarz EJ, Hess D, Widenfalk J, El Manira A, Prockop DJ, et al. Marrow stromal cells form guiding strands in the injured spinal cord and promote recovery. Proc Natl Acad Sci U S A. 2002;99:2199–204.
25. Burns AS, Lee BS, Ditunno Jr JF, Tessler A. Patient selection for clinical trials: the reliability of the early spinal cord injury examination. J Neurotrauma. 2003;20:477–82.
26. Snyder EY, Taylor RM, Wolfe JH. Neural progenitor cell engraftment corrects lysosomal storage throughout the mps vii mouse brain. Nature. 1995;374:367–70.
27. Yandava BD, Billinghurst LL, Snyder EY. "Global" cell replacement is feasible via neural stem cell transplantation: evidence from the dysmyelinated shiverer mouse brain. Proc Natl Acad Sci U S A. 1999;96:7029–34.

28. Wu S, Suzuki Y, Noda T, Bai H, Kitada M, Kataoka K, et al. Immunohistochemical and electron microscopic study of invasion and differentiation in spinal cord lesion of neural stem cells grafted through cerebrospinal fluid in rat. J Neurosci Res. 2002;69:940–5.
29. Chen J, Li Y, Wang L, Lu M, Zhang X, Chopp M. Therapeutic benefit of intracerebral transplantation of bone marrow stromal cells after cerebral ischemia in rats. J Neurol Sci. 2001;189:49–57.
30. de Vasconcelos Dos Santos A, da Costa Reis J, Diaz Paredes B, Moraes L, Jasmin, Giraldi-Guimaraes A, et al. Therapeutic window for treatment of cortical ischemia with bone marrow-derived cells in rats. Brain Res. 2010;1306:149–58.
31. Kamiya N, Ueda M, Igarashi H, Nishiyama Y, Suda S, Inaba T, et al. Intra-arterial transplantation of bone marrow mononuclear cells immediately after reperfusion decreases brain injury after focal ischemia in rats. Life Sci. 2008;83:433–7.
32. Barta M, Jakubicka I. Collection of sperm from foxes using electroejaculation under halothane anesthesia. Vet Med (Praha). 1989;34:637–40.
33. Lim JY, Jeong CH, Jun JA, Kim SM, Ryu CH, Hou Y, et al. Therapeutic effects of human umbilical cord blood-derived mesenchymal stem cells after intrathecal administration by lumbar puncture in a rat model of cerebral ischemia. Stem Cell Res Ther. 2011;2:38.
34. Stutzmann JM, Mary V, Wahl F, Grosjean-Piot O, Uzan A, Pratt J. Neuroprotective profile of enoxaparin, a low molecular weight heparin, in in vivo models of cerebral ischemia or traumatic brain injury in rats: a review. CNS Drug Rev. 2002;8:1–30.
35. Liechty KW, MacKenzie TC, Shaaban AF, Radu A, Moseley AM, Deans R, et al. Human mesenchymal stem cells engraft and demonstrate site-specific differentiation after in utero transplantation in sheep. Nat Med. 2000;6:1282–6.
36. Forest VF, Tirouvanziam AM, Perigaud C, Fernandes S, Fusellier MS, Desfontis JC, et al. Cell distribution after intracoronary bone marrow stem cell delivery in damaged and undamaged myocardium: implications for clinical trials. Stem Cell Res Ther. 2010;1:4.
37. Stamm C, Nasseri B, Choi YH, Hetzer R. Cell therapy for heart disease: great expectations, as yet unmet. Heart Lung Circ. 2009;18:245–56.
38. Osaka M, Honmou O, Murakami T, Nonaka T, Houkin K, Hamada H, et al. Intravenous administration of mesenchymal stem cells derived from bone marrow after contusive spinal cord injury improves functional outcome. Brain Res. 2010;1343:226–35.
39. Akiyama Y, Radtke C, Honmou O, Kocsis JD. Remyelination of the spinal cord following intravenous delivery of bone marrow cells. Glia. 2002;39:229–36.
40. Bakshi A, Hunter C, Swanger S, Lepore A, Fischer I. Minimally invasive delivery of stem cells for spinal cord injury: advantages of the lumbar puncture technique. J Neurosurg Spine. 2004;1:330–7.
41. Hu X, Yu SP, Fraser JL, Lu Z, Ogle ME, Wang JA, et al. Transplantation of hypoxia-preconditioned mesenchymal stem cells improves infarcted heart function via enhanced survival of implanted cells and angiogenesis. J Thorac Cardiovasc Surg. 2008;135:799–808.
42. Sun J, Wei ZZ, Gu X, Zhang JY, Zhang Y, Li J, et al. Intranasal delivery of hypoxia-preconditioned bone marrow-derived mesenchymal stem cells enhanced regenerative effects after intracerebral hemorrhagic stroke in mice. Exp Neurol. 2015;272:78–87.
43. Hu X, Yang J, Wang Y, Zhang Y, Ii M, Shen Z, et al. Mesenchymal stem cells preconditioned with trimetazidine promote neovascularization of hearts under hypoxia/reoxygenation injury. Int J Clin Exp Med. 2015;8:16991–7005.
44. Liu XB, Wang JA, Ji XY, Yu SP, Wei L. Preconditioning of bone marrow mesenchymal stem cells by prolyl hydroxylase inhibition enhances cell survival and angiogenesis in vitro and after transplantation into the ischemic heart of rats. Stem Cell Res Ther. 2014;5:111.
45. Noiseux N, Borie M, Desnoyers A, Menaouar A, Stevens LM, Mansour S, et al. Preconditioning of stem cells by oxytocin to improve their therapeutic potential. Endocrinology. 2012;153:5361–72.

46. Herrmann JL, Wang Y, Abarbanell AM, Weil BR, Tan J, Meldrum DR. Preconditioning mesenchymal stem cells with transforming growth factor-alpha improves mesenchymal stem cell-mediated cardioprotection. Shock. 2010;33:24–30.
47. Kumar S, Ashraf M. Tadalafil, a phosphodiesterase inhibitor protects stem cells over longer period against hypoxia/reoxygenation injury through stat3/pkg-i signaling. Stem Cells Dev. 2015;24:1332–41.
48. Kumar A, Prasad M, Jali VP, Pandit AK, Misra S, Kumar P, et al. Bone marrow mononuclear cell therapy in ischaemic stroke: a systematic review. Acta Neurol Scand. 2016. in press.
49. Moniche F, Gonzalez A, Gonzalez-Marcos JR, Carmona M, Pinero P, Espigado I, et al. Intra-arterial bone marrow mononuclear cells in ischemic stroke: a pilot clinical trial. Stroke. 2012;43:2242–4.
50. Prasad K, Sharma A, Garg A, Mohanty S, Bhatnagar S, Johri S, et al. Intravenous autologous bone marrow mononuclear stem cell therapy for ischemic stroke: a multicentric, randomized trial. Stroke. 2014;45:3618–24.
51. Vahidy FS, Rahbar MH, Zhu H, Rowan PJ, Bambhroliya AB, Savitz SI. Systematic review and meta-analysis of bone marrow-derived mononuclear cells in animal models of ischemic stroke. Stroke. 2016;47:1632–9.
52. Taguchi A, Sakai C, Soma T, Kasahara Y, Stern DM, Kajimoto K, et al. Intravenous autologous bone marrow mononuclear cell transplantation for stroke: phase1/2a clinical trial in a homogeneous group of stroke patients. Stem Cells Dev. 2015;24:2207–18.
53. El-Badawy A, El-Badri N. Clinical efficacy of stem cell therapy for diabetes mellitus: a meta-analysis. PLoS One. 2016;11:e0151938.
54. Klassen H. Stem cells in clinical trials for treatment of retinal degeneration. Expert Opin Biol Ther. 2016;16:7–14.
55. Li XC, Zhong CF, Deng GB, Liang RW, Huang CM. Efficacy and safety of bone marrow-derived cell transplantation for spinal cord injury: a systematic review and meta-analysis of clinical trials. Clin Transplant. 2015;29:786–95.
56. Fisher SA, Zhang H, Doree C, Mathur A, Martin-Rendon E. Stem cell treatment for acute myocardial infarction. Cochrane Database Syst Rev. 2015:CD006536.

Chapter 12
Potential Mechanisms of Transplanted Cell-Mediated Recovery After Stroke

Brian Wang, Andrew Hui, and Kunlin Jin

Abstract The underlying mechanisms of transplanted cell-mediated recovery after stroke have received increased attention over the last decade. In this chapter, we review some of the well-known mechanisms in detail and what the more recent findings and relevant clinical trials are suggesting. We also highlight pertinent reports in order to inform the reader of current opportunities and challenges with regard to the following potential mechanisms: (1) functional integration with the existing brain circuitry; (2) neuroprotection through trophic factor secretion; (3) increased formation of new vasculature; (4) structural and functional changes to the brain; and (5) recruitment of endogenous progenitors.

Keywords Stroke • Cell transplantation • Neuroprotection • Angiogenesis • Plasticity

12.1 Introduction

Stroke is the leading cause of adult disability worldwide. Yet, there is only one FDA-approved drug for the treatment of stroke where only 2–4 % of stroke patients can benefit from it due to its short therapeutic window [1]. Therefore, one must look to the post-acute phase of stroke, which has a wider treatment window, and, even more so, therapies that seek to prevent, restore, and repair the damage caused after an episode of stroke. This is where cell transplantation presents itself as an exciting and innovative approach for the treatment of stroke. Several studies have shown

B. Wang
Center for Neuroscience Discovery, Institute for Healthy Aging, University of North Texas Health Science Center, Fort Worth, TX 76107, USA

A. Hui
Texas College of Osteopathic Medicine, University of North Texas Health Science Center, Fort Worth, TX 76107, USA

K. Jin (✉)
Department of Pharmacology Neuroscience, University of North Texas Health Science Center, Fort Worth, TX, USA
e-mail: kunlin.jin@unthsc.edu

K. Jin et al. (eds.), *Bone Marrow Stem Cell Therapy for Stroke*,
DOI 10.1007/978-981-10-2929-5_12

favorable in vivo results using cells from various sources. Similarly, data from a number of published clinical trials also support the notion of cell transplantation after stroke, and these have been succinctly compiled and reviewed (see [2, 3]). Although the outcomes seem promising, cell transplantation for stroke therapy is still in its stages of infancy that is not without its own unresolved issues. Many cite the following as conditions that need to be met for the successful application of cell transplantation in the clinic (see Chap. 11 for a vivid discussion): optimal timing of cell transplantation, optimal route of administration and site of cell delivery, and optimal number of cells to be transplanted. Of course, these will hardly be rectified any time soon with the myriad of studies differing in their research approach. What might be of help is attempting to understand the probable mechanisms that underlie the phenomenon of transplanted cell-mediated recovery after stroke. Researchers could perhaps obtain insight into the perplexing problems that arise from cell transplantation to further inform them of their future studies.

This chapter endeavors to examine the potential mechanisms in detail and to offer fresh perspectives with regard to this topic. We hope this will not only spark extensive conversations and collaborations to propel the field of cell transplantation forward but will also promote the application of this knowledge to stroke recovery and possibly other diseases of the human body requiring this new lease of life.

12.2 Potential Mechanisms of Transplanted Cell-Mediated Stroke Recovery

Included in the list of potential mechanisms are functional integration with the existing brain circuitry, neuroprotection through trophic factor secretion, increased formation of new vasculature, structural and functional changes to the brain, and recruitment of endogenous progenitors. We detail here recent studies and relevant clinical trials to inform the reader of current opportunities and challenges with regard to mechanisms underlying stroke recovery after cell transplantation.

12.2.1 Functional Integration with the Existing Brain Circuitry

The hope of cell replacement therapy is to restore the brain to its previous normal working condition. When stroke occurs, brain cells in the damaged region die. These are being removed by circulating monocytes that mature into macrophages at the site of injury [4], which leaves an open circuit in the brain that manifests as deficits in the functioning of the body depending on where the stroke developed. It was found that transplanted cells have the potential to form these lost connections by differentiating into what is required to close the circuit again. What is important in functional integration is that the differentiated cells show structural characteristics analogous to different native neuronal subtypes or glial cells, express synaptic proteins, form synapses with the existing circuitry, and demonstrate synaptic input through electrophysical recordings.

There is evidence that neural stem cells (NSCs) transplanted into the hippocampus of transient global ischemia rats could replace lost circuitry [5]. Although only 1–3 % of the transplanted NSCs survived (out of the 7.5×10^4 being grafted), these NSCs were found to have neurite-like structures in the pyramidal layer of the hippocampus and also express synaptobrevin, suggesting that the transplanted NSCs were able to differentiate into functional neurons and form synapses.

One study that grafted murine embryonic stem cell (mESC)-derived precursors into a rat model of endothelin-1 middle cerebral artery occlusion (MCAO) demonstrated that these mESC-derived precursors were able to differentiate in large numbers into mature neurons with varying neurotransmitter subtypes and supporting glial cells [6]. The group reported the presence of cholinergic, GABAergic, and serotonergic neurons, as well as striatal neurons expressing substance P and DARPP32 at 4 weeks after transplantation using immunohistochemistry. The presynaptic vesicle marker, SV2, was also found to be expressed by the transplanted cells, suggesting their functional maturation. Furthermore, 27.7 % of the transplanted cells exhibited action potentials and voltage-gated Na^+ and K^+ currents, confirming their neuronal phenotype. Unexpectedly, spontaneous excitatory postsynaptic potentials (EPSPs) and excitatory postsynaptic currents (EPSCs) were seen in these neurons between 4 and 7 weeks after grafting, thus establishing their ability to receive synaptic input.

With the advent of human induced pluripotent stem cells (iPSCs), there is a distinct possibility of using patient-specific cells for autologous transplantation [7–9]. However, whether these cells could survive in the long run and differentiate to neurons to promote functional recovery in the ischemic brain are unclear. Lindvall and Kokaia's group was the first to demonstrate that grafting human fibroblast-derived iPSCs is an efficient and safe approach for neuronal replacement in the ischemic brain [7]. They found that functional recovery was seen from just 4 days after grafting the human iPSCs into the striatum of the ischemic rats, while at 5 months after transplantation, the differentiated neurons from the iPSCs were seen to develop axonal projections to the global pallidus and display electrophysiological properties of matured neurons that received synaptic input from the host neurons. This points to the possibility that other mechanisms other than neuronal replacement could be responsible for this early recovery, which makes it more apparent that neuronal replacement may not play a critical role in transplanted cell-mediated recovery in the post-acute phase of stroke. Could it perhaps serve an important function in the chronic stage of stroke that is still yet to be discovered? To truly assess whether or not the transplanted iPSCs or any transplanted cell types are functionally integrated, longer time points are needed to observe further behavioral improvements. Another study showed that grafted iPSC-derived neuronal progenitors that were fated to possess a cortical phenotype in vitro developed into mature and functional cortical neurons, which integrated into the ischemic brains of rats and consequently promoted functional recovery [9]. Interestingly, the groups that were transplanted with cells showed behavioral improvements at 2 months but not at 1 week or 1 month, once again supporting the notion that another mechanism that alters the microenvironment by the transplanted cells such as immunomodulation [10] could be at play. In fact, it was demonstrated that when neuroinflammation prevailed, grafted cells remained undifferentiated. This could be attributed to the presence of soluble mediators such as BMP-4 and noggin that were secreted by reactive astrocytes, $CD45^+$ cells, and inflamed endothelial cells [11]. Therefore, one can hypothesize that when inflammation-associated cues begin to decrease, grafted cells will then be able to differentiate into functionally mature cell types. Taking it a step further, could this signal a paradigm change with regard to when and how cell transplantation should be performed? Instead of grafting cells early in the stroke recovery phase where inflammation is exceedingly active, studies can perhaps use anti-inflammation strategies to combat the overactive inflammation then proceed with cell transplantation to hasten the process of differentiation and functional integration to achieve behavioral improvement quickly.

A very recent study using what is called multilineage-differentiating stress-enduring (Muse) cells was assessed for fate and function in a rat model of transient MCAO [12]. Muse cells are a distinct stem cell population with self-renewing capabilities and can differentiate into cells that represent all three germ layers (ectoderm, mesoderm, and endoderm) from just a single cell. They are also stress tolerant and were discovered in mesenchymal tissues such as the bone marrow and cultured fibroblasts [13] and have been shown to be non-tumorigenic [13–15]. Muse cells were injected into the ischemic rat's cortex at 2 days after transient MCAO, while the authors observed the behavior and effects of the grafted Muse cells for the following 84 days. Ischemic rats that received Muse cells showed significant improvements in neurological and motor functions at days 70 and 84 compared with control. Despite the fact that infarct volume remained the same, the transplanted Muse cells demonstrated their abilities to increase the survival of the host cells in the peri-infarct area. The grafted Muse cells were able to survive in the ischemic rat brain for a period of 84 days and surprisingly, differentiated with high ratio into neuronal cells (~60 %) and with moderate ratio into oligodendrocytes (~25 %), while glial fibrillary acidic protein-positive ($GFAP^+$) cells were scarce. Of note, the Muse cells displayed presence of synaptophysin, were able to integrate into the sensory-motor cortex with neurites that extended into the cervical spinal cord, and exhibited the normalization of somatosensory evoked potentials (SEPs).

12.2.2 Neuroprotection Through Trophic Factor Secretion

It has been shown time and again that within the first 2 days of cell transplantation after stroke, there often is a reduction of infarct volume and apoptosis in the peri-infarct tissue [16]. This would suggest a form of protection that is mediated directly or otherwise by the transplanted cells to aid in the recovery process. A plausible mechanism for neuroprotection many studies have found is through transplanted cell-mediated release (and/or stimulation) of endogenous trophic factors such as brain-derived neurotrophic factor (BDNF), fibroblast growth factor-2 (FGF-2), and vascular endothelial growth factors (VEGFs).

The essentiality of BDNF cannot be understated. It has been shown to sustain physiological processes in the normal adult brain such as learning and memory by modulating dendritic spine morphology and branching [17, 18] and synaptic plasticity [19]. In the poststroke brain, BDNF is involved in protective and reparative processes including neurogenesis, motor learning, and the downregulation of anti-apoptotic proteins [20, 21]. To date, there are only two stroke studies – one in the neonate rat [22] and the other in an adult rat [23] – that have transplanted BDNF-secreting mesenchymal stem cells (BDNF-MSCs) which were genetically modified to overexpress BDNF using an adenovirus construct. van Velthoven et al. [22] delivered rat BDNF-MSCs via the intranasal route to ischemic P10 rats and found significant reductions in infarct size, gray matter loss (both $p<0.01$ when compared with vehicle) and white matter loss ($p<0.05$ vs. vehicle) at 28 days post-MCAO. The effects of BDNF-MSCs, however, did not differ significantly from those rats transplanted with MSCs only. Although BDNF-MSCs promoted a significant reduction in motor deficits at 14 days, the effect was seemingly temporary as there was no noticeable difference at 28 days post-MCAO when compared with the MSCs-only group. Moreover, the BDNF-MSC and MSC groups saw significant increases in SVZ cell proliferation at 28 days post-MCAO, suggesting that there are indeed active endogenous repair processes taking place. Taken together, the data indicate that BDNF-MSCs could confer similar levels of benefits than MSCs alone. This observation can be attributed to the use of a non-integrating adenoviral construct wherein the resulting forced overexpression of BDNF is only temporary as the vector is subsequently removed from the cell. Administering additional BDNF-MSCs at later time points or forcing sustained BDNF secretion may be critical to induce longer-lasting effects. In a different study, Jeong et al. [23] transplanted (at 3 days post-MCAO) a recombinant replication-deficient adenoviral vector with a protein transduction domain that encodes for BDNF to produce a consistently high level of BDNF secretion from human bone marrow-derived MSCs (hBM-MSCs). At 14 days after MCAO, there not only was significant proliferation and differentiation of NSCs and functional recovery but also substantial reduction of apoptotic cell death and infarct volume when comparing the hBM-MSCs with vehicle and MSCs-only groups, suggesting that BDNF could indeed be neuroprotective if its expression could be consistently upregulated for a period of time. Longer-term studies, however, are needed to support the constant upregulation of BDNF as a potential therapeutic for stroke recovery.

Another trophic factor that can be released by transplanted cells after stroke is FGF-2; it is known to regulate stem cell proliferation, differentiation, and migration [24–26] as well as induce neuronal sprouting [27]. Injection of neutralizing antibodies to FGF-2 in newborn rats saw a 60 % decrease in DNA synthesis in the hippocampus just after 4 h, suggesting the importance of endogenous FGF-2 in brain development [28]. Further, centrally administering FGF-2 in the adult rat at 1 day after permanent proximal MCAO demonstrated an increase in the rate and degree of functional recovery after 4 weeks when compared with control [29]. Even though

FGF-2 is a very well-studied factor, reports with regard to transplanted cell release of FGF-2 after stroke are scarce. Ikeda et al. [30] demonstrated the effects of functional recovery after MCAO in adult Wistar rats by transplanting MSCs with the FGF-2 gene transferred by a replication-incompetent herpes simplex virus type 1 (HSV-1) artificially fused with an interleukin-2 secretory signal sequence. At 3 and 7 days after MCAO, only FGF-2 production in the FGF-2-MSCs group was substantially increased ($p<0.05$) compared with the sham and MSCs-only groups. Interestingly, the level of FGF-2 was maintained in the ipsilateral hemisphere of the MSCs-only group, while it was shown to decrease over time in the sham group, suggesting that continuous FGF-2 secretion by the transplanted cells is a critical factor in determining the extent of damage caused by an episode of stroke. At day 14 after MCAO, only the FGF-2-MSCs group had a significant reduction in infarct volume ($p<0.05$) compared with the other groups. Furthermore, at 21 days after MCAO, the FGF-2-MSCs group exhibited significant neurological improvement ($p<0.05$) compared with the MSCs-only and sham groups. Although the results indicate that transplantation of FGF-2-MSCs with the HSV-1 vector is an efficacious treatment modality for stroke, it would be interesting to examine the long-term effects of FGF-2-MSCs in aged rats since the aged systemic milieu is vastly different from that of the young. On a different note, one could also consider transplanting NSPCs transduced with FGF-2 as these have been shown to associate with the vasculature, which seems to be critical for NSPCs to remain in a proliferative and undifferentiated state, thereby improving their survivability and integration in an ischemic environment [31].

Last but certainly not the least, the family of VEGFs including VEGF-A, VEGF-B, and placental growth factor is critical for the proper functioning and development of the nervous and circulatory systems [32]. Apart from its involvement in the different phases of vascular development, VEGFs also possess direct trophic and neuroprotective effects. For these reasons, it is not surprising that VEGFs are germane to the topic of stroke recovery. In a rat model of global cerebral ischemia where the hippocampus is the main area of damage, VEGF-A mRNA was demonstrated to be induced in neurons at 12 h after reperfusion, with a peak of 1 day, and in astrocytes at 1 day after reperfusion. Interestingly, the VEGF-A signal seemingly disappeared after 3 days [33]. By the same token, a 3-day intraventricular infusion of VEGF-A starting at 24 h after MCAO in rats revealed behavioral improvements lasting for a minimum of 2 months as well as decreased infarct volume by almost a third at 30 days after MCAO [34, 35], indicating that the timing of VEGF-A administration is crucial to achieving maximum recovery from stroke. With regard to conferring neuroprotection by way of secreting VEGF from transplanted cells, a few reports agree that a possible mechanism is more often than not through vasculogenesis, which is to be expected [36, 37]. However, it is becoming increasingly apparent that VEGF is more than a one-trick pony. One group transplanted neuroepithelial-like stem cells that were derived from human iPSCs expressing VEGF into the striatum and cortex of MCAO mice and rats and discovered that although VEGF was increased in astrocytes and blood vessels in the peri-infarct area, there was no evidence to suggest that the higher VEGF levels correlated with VEGF-induced angiogenesis and vascularization [7]. This phenomenon was confirmed in another study they performed [8]. Rather than arguing against a significant role VEGF may or may not have from cell transplantation, this could suggest possible non-angiogenic roles for VEGF. In fact, higher VEGF levels from grafted cells at early time points, e.g., day 14 after stroke, could be involved with promoting endogenous plasticity [38] as well as suppressing inflammation [39].

12.2.3 Increased Formation of New Vasculature

A large indicator of whether tissue can be repaired is whether there is adequate blood flow near the penumbra. Vasculature allows for transmission of metabolites necessary for living tissue, the removal of damaged tissue, and the creation of a new location for existing NSCs to both migrate toward or new NSCs to be generated from [40]. The current treatment model consisting of thrombolytic therapy helps replenish the lack of blood flow during a stroke, but this treatment can sometimes be limiting due to a very small treatment window of 4.5 h. If there was a way to create collateral circulation after a stroke event, this could greatly expand the window for recovery from hours to days [41]. Recent studies provide evidence that transplanted cells can lead to increased angiogenesis, which could serve as a possible repair mechanism for stroke patients. Some of the cell types that lead to neovascularization posttransplantation include bone marrow stem/progenitor cell, adipose tissue-derived stem/progenitor cell, embryonic stem cell-derived cells, and cord blood and peripheral blood stem/progenitor cells [41]. It has been reported that these transplanted cells can lead to the increase of endogenous factors such as vascular endothelial growth factor-A/B (VEGF-A/B), transforming growth factor (TGF), TGF-B, fibroblast growth factor (FGF), stromal cell-derived factor-1 (SDF-1), brain-derived neurotrophic factor (BDNF), hepatocyte growth factor (HGF), and insulin-like growth factor-1 (IGF-1) [42]. As such, the mechanisms that involve these molecules are exciting new avenues that may help serve as possible stroke treatment models. These mechanisms are still being explored, and more data must show whether the new vasculature created via the transplantation of stem cells is physiologic or dysfunctional.

Recent research has been seeking to identify a clear mechanistic model for whether BMSCs promote angiogenesis through factors such as VEGF [43]. In one study, conditioned medium bone marrow mesenchymal stem cells were shown to operate via the PI3K-Akt pathway in response to hypoxia to decrease apoptosis while enhancing angiogenesis [44]. Another study suggested that endothelial nitric oxide synthase (eNOS) plays a critical role in whether transplanted BMSCs can serve to regain physiological function [45]. Mice were infused with BMSCs along with Nω-nitro-L-arginine methyl ester (L-NAME), an eNOS inhibitor. The mice injected with L-NAME were analyzed against control, revealing that the L-NAME group failed to increase angiogenesis and neurogenesis, thus indicating that the eNOS/BDNF pathway is an important regulator in whether angiogenesis will occur in poststroke states. Moreover, the data is unclear whether this is an effective treatment for all types of patients. One study showed that BMSCs mitigated stroke neurological impairment by promoting angiogenesis, neurogenesis, and synaptogenesis in aged mice, while another study showed that BMSC treatment of type 1 diabetic mice contributed to greater amounts of dysfunctional angiogenesis and increased the risk of cerebral hemorrhaging [46]. Continuing research seems to underscore that the process of angiogenesis does occur; however, there is still much to be learned about the mechanistic causality of BMSC transplantation as it relates to functional vs. dysfunctional angiogenesis. This field is particularly beneficial because BMSCs avoid ethical barriers and is currently in the process of ongoing clinical trials to determine treatment feasibility and safety [47]. Results and analyses are still underway and pending.

In addition to BMSCs, adipose tissue-derived stem cells (ADSCs) are another source of possible transplanted cells that can help increase neovascularization via VEGF and HGF release. These factors help increase endothelial cell survivability, migration, and vessel formation [48]. What makes ADSCs great candidates are that they are abundant and relatively easy to extract without the need of undergoing invasive surgery. There are already clinical trials underway including a phase IIa study that involves the administration of adipose-derived mesenchymal stem cells within 2 weeks of a stroke event. Another clinical study is analyzing the safety of direct injection of human adipose-derived stem cells into the brains of chronic nonhemorrhagic stroke patients (ClinicalTrials.gov Identifier, NCT02813512).

Recent research has provided some clarity regarding the mechanism of action that causes ADSCs to promote neovascularization. One study analyzed TGF-B1 expression in mice and found significantly greater amounts of TGF-B1 in the MCAO + ADSC group when compared to control [49]. This significant increase in expression of TGF-B1 suggests the MCAO + ADSC group was able to create new vessels. In another report, the following proteins were found to be significantly upregulated in rats containing transplanted ADSCs: Notch1, DII4, and Hes1 [50]. This data seems to suggest that ADSCs can increase angiogenesis via the Notch1-DII4 pathway.

Transplanted human cord has also been used as a way to obtain neovascularization via endothelial colony-forming cell (ECFC) separation. In one study, there was evidence of increased CD31 expression along the ischemic border when compared to the control group, which suggests the presence of new endothelial cells [51]. In another study regarding Wharton's jelly mesenchymal stem cells (WJ-MSCs), there were increased levels of the transcription factor FOXF1, which is important in the process of neovascularization [52]. While this evidence is encouraging, the exact mechanism of angiogenesis is still unknown after grafting. Hopefully, more studies will discover a model for how angiogenesis occurs when using human cord blood cells.

As we find more potential sources for transplanted stem cells to promote neovascularization, there needs to be an increased focus on how these sources compare and contrast with one another after transplantation. In at least one study, stem cells derived from the bone marrow were evaluated against Wharton's jelly-derived stem cells. This was performed by analyzing the aforementioned stem cells via cell migration analysis as well as tube formation assays on human microvascular endothelial cell (HMEC) lines [52]. It was found that WJ-MSCs were able to generate longer tube length when compared to BMSCs and that the HMECs showed migration preference for WJ-MSCs. Increased amounts of chemokine ligand 5 (CXCL5) were also secreted by WJ-MSCs as well. CXCL5 has been shown to be an important mediator in angiogenesis [53], which seems to suggest that WJ-MSCs may have better neovascularization potential than BMSCs. Whether this also directly translates to physiologically better outcomes remains to be seen. Ethical barriers may also play a deciding role in which of these transplanted cell sources become a mainstay in future treatment.

12.2.4 Structural and Functional Changes in the Brain

Thus far we have reviewed a number of ways that the body attempts to recover after an ischemic event through anatomical remodeling including angiogenesis and recovery of neural networks. This section serves to explain other potential areas where brain plasticity can be conducive to treatment after stem cell transplantation.

One of the ways stem cell transplantation can affect plasticity after a stroke is through its effects on microRNA (miRNA) expression. This process is usually performed by the miRNA's role in the regulation of gene expression at both the level of transcription and translation. It has been shown that injected stem cells upregulate endogenous brain cells and stimulate miRNAs through the release of microvesicles. These microvesicles contain modified miRNAs that can exert effects including the stimulation of tissue repair processes such as inflammation, neurogenesis, angiogenesis, and the body's response in a hypoxic environment [54]. An example of an important miRNA expression is miR-133b, which has been shown to increase total neurite length and branch number after being exposed to brain extracts of MCAO-induced mice for 72 h [54]. This process occurs through the transmission of miR-133b from BMSCs to both neurons and astrocytes. Interestingly, the release of microvesicles has a bidirectional effect between stem cells and damaged cells by having both an effect on inducing differentiating BMSCs and repair processes [54]. This bidirectional feedback seems to suggest that a large amount of modulation occurs through the microvesicle release between BMSCs and damaged cells. A similar treatment process model is currently being evaluated to treat brain tumors through the use of BMSC exerting its effects on miRNA. Specifically, BMSCs are being used with miR-146b as a vehicle to help minimize neoplastic proliferation by downregulating VEGF [54, 55]. With such treatments in other types of pathology being treated through similar processes, it is important to consider how other miRNAs may be able to influence areas of concern and to continue research into the exact mechanisms.

In addition to BMSC transplantation, human neural precursor cells have been studied for their effects on plasticity after stroke. In one study, dendrites were Golgi stained and analyzed when comparing human NPC treatment mice against vehicle controls [38]. It was found that those that underwent the human NPC treatment had better outcomes and that the changes in dendritic branching were more proliferative in areas that were closer to the human NPCs. Results also showed that human NPC transplantation increased overall dendritic and axon lengths [38]. A quantitative PCR analysis was performed to reveal which factors were expressed by the human NPCs to clarify what mediators were directly involved. Four of the following factors were involved: VEGF, thrombospondin 1, thrombospondin 2, and secreted protein acidic and rich in cysteine (SPARC). Of these, only SPARC had no effects on dendrites or axons when neutralized. When VEGF, thrombospondin 1, and thrombospondin 2 were neutralized by selective antibodies and compared with controls, it was shown that there were reductions in both dendritic branching and length [38]. While these three factors play a role in stem cell effects on brain plasticity during a stroke, the exact mechanism is still unclear.

12.2.5 Recruitment of Endogenous Progenitors

Another prospective route to treat strokes through the use of transplanted cells includes the recruitment of endogenous progenitor cells. While a healthy individual will have NSPCs primarily located at the SVZ regions of the lateral ventricles and SGZ of the dentate gyrus in the hippocampus, evidence has shown that the body will send signals that cause endogenous NSPCs to migrate toward infarct regions of the brain following ischemic injury. The presence of NSPCs in the penumbra is encouraging because they may offer the possibility of treatment through the reparation of damaged neurons or recruitment of cells that can differentiate into mature neurons [56]. Additionally, this also suggests that there is a pathway where neurogenic NSPCs can flow through to reach infarct regions. For simplicity, we shall refer to this pathway as a biobridge [57]. It has been shown that this biobridge can be stimulated by several chemotactic factors following the transplantation of stem cells such as bone marrow stem cells (BMSCs) [58]. The fact that a biobridge exists after stroke events is not enough. Major obstacles exist, causing most of the migrating NSPCs to die out. Some of these obstacles that affect the biobridge include enzymatic degradation, which causes the loss of structural integrity and deteriorating blood-brain barrier (BBB) resulting in dysfunctional anatomical connections. Even if NSPCs were able to navigate these obstacles, radical oxygen species await newly arriving NSPCs to disrupt their survival capabilities [56].

A good treatment model will circumvent disruptive factors by having a good understanding of the mechanisms leading to migration and survival of NSPCs. While we still have much to learn about the mechanisms, several studies have provided evidence for different factors released by transplanted stem cells that play a role in regulating the migration, proliferation, and differentiation of NSPCs in the

CNS. Here we will discuss research findings related to factors derived from stem cells including stromal cell-derived factor-1 (SDF-1), BDNF, and MMP-9.

Stromal cell-derived factor-1 (SDF-1, also known as CXCL12) released by BMSCs has been shown to cause migration, proliferation, and differentiation of NSPCs as well [58, 59]. SDF-1 binds to chemokine receptor type 4 (CXCR4) and CXCR7 receptors, which are shown to help recruit multiple types of stem/progenitor cells. While CXCR7 receptors are mainly expressed in mature myelin sheath [60], CXCR4 receptors provide the signal to regulate NSPCs, endothelial progenitor cells, hematopoietic stem cells, and endothelial progenitor cells. In addition, CXCR4 has been revealed to play a role in the differentiation of oligodendrocyte progenitor cells (OPCs) leading to re-myelination after stroke-induced white matter damage [60]. This finding was suggested after a study analyzed mice coadministered with AMD3100, an inhibitor of the SDF-1/CXCR4 signaling pathway. Mice treated with AMD3100 had much less myelin sheath integrity than mice that showed expression of CXCR4. Furthermore, the time window of treatment using CXCR4 gene therapy seems to be promising. Research has shown that mice treated an entire week after the stroke event still received therapeutic benefit [60]. Comparatively, current treatment has a short window of only a few hours. Future clinical studies will elucidate whether such treatments are safe and viable options.

Several research studies have shown that SDF-1 can be modulated due to certain factors. However, the exact mechanisms still seem to be lacking. For instance, upregulation of the SDF-1 pathway has been found to be stimulated by decreased oxygen levels and increased amounts of hypoxia-inducible factor-1 (HIF-1) [61]. While it can be suggested then that HIF-1 may increase adhesion, cell proliferation, and survival, it is unclear how and if HIF-1 directly impacts SDF-1. C-kit was found to be another regulator in a study where decreased recruitment of endogenous progenitors was seen after inhibiting c-kit [62]. Additionally, eNOS has been shown to upregulate SDF-1 through a cGMP-dependent process in ischemic mice tissue and plays a role in causing progenitor cells to adhere onto endothelium through ICAM-1- and CXCR4-dependent pathways [62]. Research data does not seem to draw clear indications about whether these factors integrate with one another or have indirect impact with one another.

Another key player in recruiting endogenous progenitors is BDNF, which has been shown to increase axonal growth and synaptic plasticity after stroke events. In one study, BDNF was shown to have a larger presence in the ipsilateral hemisphere of the mouse brain following middle cerebral artery occlusion (MCAO) while showing improvements in functional recovery [45]. This suggests that BDNF-secreting mesenchymal stem cells may be beneficial in treatment. The exact pathways in which BDNF causes migration of stem cells require further research.

MMP-9 is another protein stimulated by stem cell transplantation and has been shown to be of benefit to the biobridge. In one experiment, migration of stem cells correlated with a ninefold increase in MMP-9. Inhibitors of MMP-9 were injected into the mice so that they could be analyzed for neurovascular effects. What was found was that MMP inhibition correlated with dysfunctional neurovascular remodeling as well as significantly decreased amounts of SVZ neurogenic migration to infarct regions [56]. It can be concluded from this study that MMP plays an important role as to why there are increased amounts of endogenous stem cells in damaged regions.

As mentioned earlier, an arising issue that was discovered in the efficacy of stem cell transplantation has been due to the environment and pathway. Current research is underway to create an encasement scheme to help facilitate the delivery of transplanted stem cells to create a stable environment that allows for safe migration. Some of these schemes attempt to replicate a pathway similar to glial tubes consisting of seeded astrocytes [63]. It has also been shown that these bioscaffolds are able to protect the stem cells from oxidative environments [62]. More experimentation is currently being performed to verify the safety and efficacy of these models. Due to the promising results of research, we are continually learning more about how the transplantation of stem cells will recruit endogenous progenitor cells in the clinical setting ClinicalTrials.gov Identifier: (NCT02448641). Within the next few years, we hope that the clinical trials can provide us with new treatments that will produce a larger treatment window.

12.3 Conclusion

The existing research data and clinical trials continue to show that cell transplantation is an avenue of therapeutic for stroke that is worth investing in. While it is promising to see results, we continue to encourage researchers and scholars to collaborate with one another to take extra steps to provide mechanistic proof in order to extend the field of knowledge not only in breadth but also in depth. It is our hope that the many outstanding questions regarding the individual mechanisms highlighted above would provide the inspiration for challenging existing notions in the decades to come.

Acknowledgments This work was supported by the American Heart Association Predoctoral Fellowship (Award #: 15PRE25700416 to BW), the National Heart, Lung, and Blood Institute of the National Institutes of Health (Award #: R25HL125447 to Dr. J. K. Vishwanatha), and RO1 AG021980, R21NS084148, and R01NS057186 (to KJ). The content is solely the responsibility of the authors and does not necessarily represent the official views of the American Heart Association or the National Institutes of Health.

References

1. Molina CA. Reperfusion therapies for acute ischemic stroke: current pharmacological and mechanical approaches. Stroke. 2011;42(1 Suppl):S16–9. doi:10.1161/STROKEAHA.110.598763.
2. Rosado-de-Castro PH, Pimentel-Coelho PM, da Fonseca LM, de Freitas GR, Mendez-Otero R. The rise of cell therapy trials for stroke: review of published and registered studies. Stem Cells Dev. 2013;22(15):2095–111. doi:10.1089/scd.2013.0089.

3. Liu X, Ye R, Yan T, Yu SP, Wei L, Xu G, Fan X, Jiang Y, Stetler RA, Liu G, Chen J. Cell based therapies for ischemic stroke: from basic science to bedside. Prog Neurobiol. 2014;115:92–115. doi:10.1016/j.pneurobio.2013.11.007.
4. Wattananit S, Tornero D, Graubardt N, Memanishvili T, Monni E, Tatarishvili J, Miskinyte G, Ge R, Ahlenius H, Lindvall O, Schwartz M, Kokaia Z. Monocyte-derived macrophages contribute to spontaneous long-term functional recovery after stroke in mice. J Neurosci. 2016;36(15):4182–95. doi:10.1523/JNEUROSCI.4317-15.2016.
5. Toda H, Takahashi J, Iwakami N, Kimura T, Hoki S, Mozumi-Kitamura K, Ono S, Hashimoto N. Grafting neural stem cells improved the impaired spatial recognition in ischemic rats. Neurosci Lett. 2001;316(1):9–12.
6. Buhnemann C, Scholz A, Bernreuther C, Malik CY, Braun H, Schachner M, Reymann KG, Dihne M. Neuronal differentiation of transplanted embryonic stem cell-derived precursors in stroke lesions of adult rats. Brain. 2006;129(Pt 12):3238–48. doi:10.1093/brain/awl261.
7. Oki K, Tatarishvili J, Wood J, Koch P, Wattananit S, Mine Y, Monni E, Tornero D, Ahlenius H, Ladewig J, Brustle O, Lindvall O, Kokaia Z. Human-induced pluripotent stem cells form functional neurons and improve recovery after grafting in stroke-damaged brain. Stem Cells. 2012;30(6):1120–33. doi:10.1002/stem.1104.
8. Tatarishvili J, Oki K, Monni E, Koch P, Memanishvili T, Buga AM, Verma V, Popa-Wagner A, Brustle O, Lindvall O, Kokaia Z. Human induced pluripotent stem cells improve recovery in stroke-injured aged rats. Restor Neurol Neurosci. 2014;32(4):547–58. doi:10.3233/RNN-140404.
9. Tornero D, Wattananit S, Gronning Madsen M, Koch P, Wood J, Tatarishvili J, Mine Y, Ge R, Monni E, Devaraju K, Hevner RF, Brustle O, Lindvall O, Kokaia Z. Human induced pluripotent stem cell-derived cortical neurons integrate in stroke-injured cortex and improve functional recovery. Brain. 2013;136(Pt 12):3561–77. doi:10.1093/brain/awt278.
10. Wang B, Jin K. Current perspectives on the link between neuroinflammation and neurogenesis. Metab Brain Dis. 2015;30(2):355–65. doi:10.1007/s11011-014-9523-6.
11. Pluchino S, Zanotti L, Rossi B, Brambilla E, Ottoboni L, Salani G, Martinello M, Cattalini A, Bergami A, Furlan R, Comi G, Constantin G, Martino G. Neurosphere-derived multipotent precursors promote neuroprotection by an immunomodulatory mechanism. Nature. 2005;436(7048):266–71. doi:10.1038/nature03889.
12. Uchida H, Morita T, Niizuma K, Kushida Y, Kuroda Y, Wakao S, Sakata H, Matsuzaka Y, Mushiake H, Tominaga T, Borlongan CV, Dezawa M. Transplantation of unique subpopulation of fibroblasts, muse cells, ameliorates experimental stroke possibly via robust neuronal differentiation. Stem Cells. 2016;34(1):160–73. doi:10.1002/stem.2206.
13. Kuroda Y, Kitada M, Wakao S, Nishikawa K, Tanimura Y, Makinoshima H, Goda M, Akashi H, Inutsuka A, Niwa A, Shigemoto T, Nabeshima Y, Nakahata T, Nabeshima Y, Fujiyoshi Y, Dezawa M. Unique multipotent cells in adult human mesenchymal cell populations. Proc Natl Acad Sci U S A. 2010;107(19):8639–43. doi:10.1073/pnas.0911647107.
14. Kuroda Y, Wakao S, Kitada M, Murakami T, Nojima M, Dezawa M. Isolation, culture and evaluation of multilineage-differentiating stress-enduring (Muse) cells. Nat Protoc. 2013;8(7):1391–415. doi:10.1038/nprot.2013.076.
15. Wakao S, Kitada M, Kuroda Y, Shigemoto T, Matsuse D, Akashi H, Tanimura Y, Tsuchiyama K, Kikuchi T, Goda M, Nakahata T, Fujiyoshi Y, Dezawa M. Multilineage-differentiating stress-enduring (Muse) cells are a primary source of induced pluripotent stem cells in human fibroblasts. Proc Natl Acad Sci U S A. 2011;108(24):9875–80. doi:10.1073/pnas.1100816108.
16. Lindvall O, Kokaia Z. Stem cell research in stroke: how far from the clinic? Stroke. 2011;42(8):2369–75. doi:10.1161/STROKEAHA.110.599654.
17. Shimojo M, Courchet J, Pieraut S, Torabi-Rander N, Sando 3rd R, Polleux F, Maximov A. SNAREs controlling vesicular release of BDNF and development of callosal axons. Cell Rep. 2015;11(7):1054–66. doi:10.1016/j.celrep.2015.04.032.
18. Guo J, Ji Y, Ding Y, Jiang W, Sun Y, Lu B, Nagappan G. BDNF pro-peptide regulates dendritic spines via caspase-3. Cell Death Dis. 2016;7, e2264. doi:10.1038/cddis.2016.166.

19. Glerup S, Bolcho U, Molgaard S, Boggild S, Vaegter CB, Smith AH, Nieto-Gonzalez JL, Ovesen PL, Pedersen LF, Fjorback AN, Kjolby M, Login H, Holm MM, Andersen OM, Nyengaard JR, Willnow TE, Jensen K, Nykjaer A. SorCS2 is required for BDNF-dependent plasticity in the hippocampus. Mol Psychiatry. 2016. doi:10.1038/mp.2016.108.
20. Schabitz WR, Steigleder T, Cooper-Kuhn CM, Schwab S, Sommer C, Schneider A, Kuhn HG. Intravenous brain-derived neurotrophic factor enhances poststroke sensorimotor recovery andstimulatesneurogenesis.Stroke.2007;38(7):2165–72.doi:10.1161/STROKEAHA.106.477331.
21. Ploughman M, Windle V, MacLellan CL, White N, Dore JJ, Corbett D. Brain-derived neurotrophic factor contributes to recovery of skilled reaching after focal ischemia in rats. Stroke. 2009;40(4):1490–5. doi:10.1161/STROKEAHA.108.531806.
22. van Velthoven CT, Sheldon RA, Kavelaars A, Derugin N, Vexler ZS, Willemen HL, Maas M, Heijnen CJ, Ferriero DM. Mesenchymal stem cell transplantation attenuates brain injury after neonatal stroke. Stroke. 2013;44(5):1426–32. doi:10.1161/STROKEAHA.111.000326.
23. Jeong CH, Kim SM, Lim JY, Ryu CH, Jun JA, Jeun SS. Mesenchymal stem cells expressing brain-derived neurotrophic factor enhance endogenous neurogenesis in an ischemic stroke model. Biomed Res Int. 2014;2014:129145. doi:10.1155/2014/129145.
24. Zheng W, Nowakowski RS, Vaccarino FM. Fibroblast growth factor 2 is required for maintaining the neural stem cell pool in the mouse brain subventricular zone. Dev Neurosci. 2004;26(2–4):181–96. doi:10.1159/000082136.
25. Wagner JP, Black IB, DiCicco-Bloom E. Stimulation of neonatal and adult brain neurogenesis by subcutaneous injection of basic fibroblast growth factor. J Neurosci. 1999;19(14):6006–16.
26. Kuhn HG, Winkler J, Kempermann G, Thal LJ, Gage FH. Epidermal growth factor and fibroblast growth factor-2 have different effects on neural progenitors in the adult rat brain. J Neurosci. 1997;17(15):5820–9.
27. Kawamata T, Dietrich WD, Schallert T, Gotts JE, Cocke RR, Benowitz LI, Finklestein SP. Intracisternal basic fibroblast growth factor enhances functional recovery and up-regulates the expression of a molecular marker of neuronal sprouting following focal cerebral infarction. Proc Natl Acad Sci U S A. 1997;94(15):8179–84.
28. Tao Y, Black IB, DiCicco-Bloom E. In vivo neurogenesis is inhibited by neutralizing antibodies to basic fibroblast growth factor. J Neurobiol. 1997;33(3):289–96.
29. Kawamata T, Alexis NE, Dietrich WD, Finklestein SP. Intracisternal basic fibroblast growth factor (bFGF) enhances behavioral recovery following focal cerebral infarction in the rat. J Cereb Blood Flow Metab. 1996;16(4):542–7. doi:10.1097/00004647-199607000-00003.
30. Ikeda N, Nonoguchi N, Zhao MZ, Watanabe T, Kajimoto Y, Furutama D, Kimura F, Dezawa M, Coffin RS, Otsuki Y, Kuroiwa T, Miyatake S. Bone marrow stromal cells that enhanced fibroblast growth factor-2 secretion by herpes simplex virus vector improve neurological outcome after transient focal cerebral ischemia in rats. Stroke. 2005;36(12):2725–30. doi:10.1161/01.STR.0000190006.88896.d3.
31. Jenny B, Kanemitsu M, Tsupykov O, Potter G, Salmon P, Zgraggen E, Gascon E, Skibo G, Dayer AG, Kiss JZ. Fibroblast growth factor-2 overexpression in transplanted neural progenitors promotes perivascular cluster formation with a neurogenic potential. Stem Cells. 2009;27(6):1309–17. doi:10.1002/stem.46.
32. Greenberg DA, Jin K. Vascular endothelial growth factors (VEGFs) and stroke. Cell Mol Life Sci. 2013;70(10):1753–61. doi:10.1007/s00018-013-1282-8.
33. Lee MY, Ju WK, Cha JH, Son BC, Chun MH, Kang JK, Park CK. Expression of vascular endothelial growth factor mRNA following transient forebrain ischemia in rats. Neurosci Lett. 1999;265(2):107–10.
34. Wang Y, Galvan V, Gorostiza O, Ataie M, Jin K, Greenberg DA. Vascular endothelial growth factor improves recovery of sensorimotor and cognitive deficits after focal cerebral ischemia in the rat. Brain Res. 2006;1115(1):186–93. doi:10.1016/j.brainres.2006.07.060.
35. Sun Y, Jin K, Xie L, Childs J, Mao XO, Logvinova A, Greenberg DA. VEGF-induced neuroprotection, neurogenesis, and angiogenesis after focal cerebral ischemia. J Clin Invest. 2003;111(12):1843–51. doi:10.1172/JCI17977.

36. Wang Y, Kilic E, Kilic U, Weber B, Bassetti CL, Marti HH, Hermann DM. VEGF overexpression induces post-ischaemic neuroprotection, but facilitates haemodynamic steal phenomena. Brain. 2005;128(Pt 1):52–63. doi:10.1093/brain/awh325.
37. Ishikawa H, Tajiri N, Shinozuka K, Vasconcellos J, Kaneko Y, Lee HJ, Mimura O, Dezawa M, Kim SU, Borlongan CV. Vasculogenesis in experimental stroke after human cerebral endothelial cell transplantation. Stroke. 2013;44(12):3473–81. doi:10.1161/STROKEAHA.113.001943.
38. Andres RH, Horie N, Slikker W, Keren-Gill H, Zhan K, Sun G, Manley NC, Pereira MP, Sheikh LA, McMillan EL, Schaar BT, Svendsen CN, Bliss TM, Steinberg GK. Human neural stem cells enhance structural plasticity and axonal transport in the ischaemic brain. Brain. 2011;134(Pt 6):1777–89. doi:10.1093/brain/awr094.
39. Horie N, Pereira MP, Niizuma K, Sun G, Keren-Gill H, Encarnacion A, Shamloo M, Hamilton SA, Jiang K, Huhn S, Palmer TD, Bliss TM, Steinberg GK. Transplanted stem cell-secreted vascular endothelial growth factor effects poststroke recovery, inflammation, and vascular repair. Stem Cells. 2011;29(2):274–85. doi:10.1002/stem.584.
40. Ergul A, Alhusban A, Fagan SC. Angiogenesis: a harmonized target for recovery after stroke. Stroke. 2012;43(8):2270–4. doi:10.1161/STROKEAHA.111.642710.
41. Bliss T, Guzman R, Daadi M, Steinberg GK. Cell transplantation therapy for stroke. Stroke. 2007;38(2 Suppl):817–26. doi:10.1161/01.STR.0000247888.25985.62.
42. Gutierrez-Fernandez M, Fuentes B, Rodriguez-Frutos B, Ramos-Cejudo J, Vallejo-Cremades MT, Diez-Tejedor E. Trophic factors and cell therapy to stimulate brain repair after ischaemic stroke. J Cell Mol Med. 2012;16(10):2280–90. doi:10.1111/j.1582-4934.2012.01575.x.
43. Yang Z, Cai X, Xu A, Xu F, Liang Q. Bone marrow stromal cell transplantation through tail vein injection promotes angiogenesis and vascular endothelial growth factor expression in cerebral infarct area in rats. Cytotherapy. 2015;17(9):1200–12. doi:10.1016/j.jcyt.2015.06.005.
44. Hung SC, Pochampally RR, Chen SC, Hsu SC, Prockop DJ. Angiogenic effects of human multipotent stromal cell conditioned medium activate the PI3K-Akt pathway in hypoxic endothelial cells to inhibit apoptosis, increase survival, and stimulate angiogenesis. Stem Cells. 2007;25(9):2363–70. doi:10.1634/stemcells.2006-0686.
45. Bai YY, Peng XG, Wang LS, Li ZH, Wang YC, Lu CQ, Ding J, Li PC, Zhao Z, Ju SH. Bone marrow endothelial progenitor cell transplantation after ischemic stroke: an investigation into its possible mechanism. CNS Neurosci Ther. 2015;21(11):877–86. doi:10.1111/cns.12447.
46. Ye X, Hu J, Cui G. Therapy effects of bone marrow stromal cells on ischemic stroke. Oxid Med Cell Longev. 2016;2016:7682960. doi:10.1155/2016/7682960.
47. Loubinoux I, Demain B, Davoust C, Plas B, Vaysse L. Stem cells and motor recovery after stroke. Ann Phys Rehabil Med. 2014;57(8):499–508. doi:10.1016/j.rehab.2014.08.008.
48. Gimble JM, Katz AJ, Bunnell BA. Adipose-derived stem cells for regenerative medicine. Circ Res. 2007;100(9):1249–60. doi:10.1161/01.RES.0000265074.83288.09.
49. Huang H, Liu N, Wang JH, Zhang YX, Du HW, Chen RH, Huang HP. The effects of adipose-derived stem cells transplantation on the expression of TGF-beta1 in rat brain after cerebral ischemia. Xi Bao Yu Fen Zi Mian Yi Xue Za Zhi. 2011;27(8):872–5.
50. Lin FF, Liu N, Huang H, Chen AZ, Liu DS, Lin XH. Effects of the transplantation of adipose-derived stem cell on the expression of Notch1-Dll4 signaling pathway in brain of rats with focal cerebral ischemia. Zhonghua Yi Xue Za Zhi. 2011;91(33):2348–52.
51. Ding J, Zhao Z, Wang C, Wang CX, Li PC, Qian C, Teng GJ. Bioluminescence imaging of transplanted human endothelial colony-forming cells in an ischemic mouse model. Brain Res. 2016;1642:209–18. doi:10.1016/j.brainres.2016.03.045.
52. Hsieh JY, Wang HW, Chang SJ, Liao KH, Lee IH, Lin WS, Wu CH, Lin WY, Cheng SM. Mesenchymal stem cells from human umbilical cord express preferentially secreted factors related to neuroprotection, neurogenesis, and angiogenesis. PLoS ONE. 2013;8(8), e72604. doi:10.1371/journal.pone.0072604.
53. Rowland KJ, Diaz-Miron J, Guo J, Erwin CR, Mei J, Worthen GS, Warner BW. CXCL5 is required for angiogenesis, but not structural adaptation after small bowel resection. J Pediatr Surg. 2014;49(6):976–80. doi:10.1016/j.jpedsurg.2014.01.034. discussion 980.
54. Chen J, Venkat P, Zacharek A, Chopp M. Neurorestorative therapy for stroke. Front Hum Neurosci. 2014;8:382. doi:10.3389/fnhum.2014.00382.

55. Zhang Z, Zhang Y, Sun XX, Ma X, Chen ZN. microRNA-146a inhibits cancer metastasis by downregulating VEGF through dual pathways in hepatocellular carcinoma. Mol Cancer. 2015;14:5. doi:10.1186/1476-4598-14-5.
56. Sullivan R, Duncan K, Dailey T, Kaneko Y, Tajiri N, Borlongan CV. A possible new focus for stroke treatment – migrating stem cells. Expert Opin Biol Ther. 2015;15(7):949–58. doi:10.15 17/14712598.2015.1043264.
57. Duncan K, Gonzales-Portillo GS, Acosta SA, Kaneko Y, Borlongan CV, Tajiri N. Stem cell-paved biobridges facilitate stem transplant and host brain cell interactions for stroke therapy. Brain Res. 2015;1623:160–5. doi:10.1016/j.brainres.2015.03.007.
58. Pirzad Jahromi G, Shabanzadeh Pirsaraei A, Sadr SS, Kaka G, Jafari M, Seidi S, Charish J. Multipotent bone marrow stromal cell therapy promotes endogenous cell proliferation following ischemic stroke. Clin Exp Pharmacol Physiol. 2015;42(11):1158–67. doi:10.1111/1440-1681.12466.
59. Jablonska A, Lukomska B. Stroke induced brain changes: implications for stem cell transplantation. Acta Neurobiol Exp (Wars). 2011;71(1):74–85.
60. Li Y, Tang G, Liu Y, He X, Huang J, Lin X, Zhang Z, Yang GY, Wang Y. CXCL12 gene therapy ameliorates ischemia-induced white matter injury in mouse brain. Stem Cells Transl Med. 2015;4(10):1122–30. doi:10.5966/sctm.2015-0074.
61. Cheng M, Qin G. Progenitor cell mobilization and recruitment: SDF-1, CXCR4, alpha4-integrin, and c-kit. Prog Mol Biol Transl Sci. 2012;111:243–64. doi:10.1016/B978-0-12-398459-3.00011-3.
62. Rennert RC, Sorkin M, Garg RK, Gurtner GC. Stem cell recruitment after injury: lessons for regenerative medicine. Regen Med. 2012;7(6):833–50. doi:10.2217/rme.12.82.
63. Winter CC, Katiyar KS, Hernandez NS, Song YJ, Struzyna LA, Harris JP, Cullen DK. Transplantable living scaffolds comprised of micro-tissue engineered aligned astrocyte networks to facilitate central nervous system regeneration. Acta Biomater. 2016;38:44–58. doi:10.1016/j.actbio.2016.04.021.

Chapter 13
Bone Marrow Stem Cell-Stimulating Factors and Brain Recovery After Stroke

Li-Ru Zhao and Suning Ping

Abstract Stroke is a major cause of death and long-term neurological disability in adults worldwide. In the United States alone, stroke presents a serious public health problem, and it has created heavy public and personal financial burdens. By contrast to the severity of stroke in public health, the treatment of stroke is very limited. Currently, a clot-dissolving drug (rtPA) is the only treatment available for ischemic stroke. The majority of stroke patients are not able to receive this treatment due to the narrow therapeutic window: 4.5 h after stroke onset. Developing new treatment that fits the majority of stroke patients is a huge challenge for stroke research. Over the past decade, numerous studies have shown the therapeutic potential of stem cell factor (SCF) and granulocyte-colony-stimulating factor (G-CSF), the two essential hematopoietic growth factors, in acute, subacute, and chronic stroke. In this chapter, we have reviewed the biological function of SCF and G-CSF in both the hematopoietic system and central nervous system, summarized the progress of SCF and G-CSF research in adult ischemic stroke in both basic and clinical studies, and discussed the directions for future studies.

Keywords Stroke • SCF • G-CSF • Neuroprotection • Neurorestoration • Cerebral ischemia

Abbreviations

AKT	Serine/threonine kinase or protein kinase B
Bcl-2	B-cell lymphoma 2
BDNF	Brain-derived neurotrophic factor
BFU-E	Erythroid burst-forming units
BrdU	Bromodeoxyuridine (5-bromo-2′-deoxyuridine)

L.-R. Zhao (✉) • S. Ping
Department of Neurosurgery, State University of New York, Upstate Medical University, 750 E. Adams Street, Syracuse, NY 13210, USA
e-mail: ZHAOL@upstate.edu; ZLRLUND@gmail.com

K. Jin et al. (eds.), *Bone Marrow Stem Cell Therapy for Stroke*,
DOI 10.1007/978-981-10-2929-5_13

CFU-GM	Granulocyte-macrophage colonies
CNS	Central nervous system
ERK	Extracellular signal-regulated kinase
FDA	Food and Drug Administration
G-CSF	Granulocyte-colony-stimulating factor
G-CSFR	Granulocyte-colony-stimulating factor receptor
GFP	Green fluorescent protein
HMGB1	High mobility group protein B1
HPCs	Hematopoietic progenitor cells
HSCs	Hematopoietic stem cells
IL-1	Interleukin-1
i.p.	Intraperitoneal
i.v.	Intravenous
LTP	Long-term potentiation
MEK	Mitogen-activated protein/extracellular signal-regulated kinase
MRI	Magnetic resonance imaging
M-type	Mushroom type
NF-kB	Nuclear factor kappa beta
NMDA	N-Methyl-D-aspartate
NPCs	Neural progenitor cells
NSCs	Neural stem cells
PI3K	Phosphatidylinositol-3 kinase
PSD-95	Postsynaptic density protein 95
rtPA	Recombinant tissue plasminogen activator
s.c.	Subcutaneous
SCF	Stem cell factor
SGZ	Subgranular zone
SHRs	Spontaneously hypertensive rats
STAT3	Signal transducer and activator of transcription 3
SVZ	Subventricular zone
TNF	Tumor necrosis factor
U-type	Uncertain type
YFP	Yellow fluorescent protein

13.1 Introduction

Stroke is the fifth leading cause of death in the United States, and it remains the number one cause of long-term disability in the world. Over the past two decades, major advances have been made in the understanding of the pathophysiology of stroke, while there has not been much progress in the development of stroke treatment [26].

Of all strokes, 87 % are ischemic [61]. Most stroke studies, therefore, target ischemic stroke. Although great efforts have been made in developing treatments for ischemic stroke, only one drug, recombinant tissue plasminogen activator (rtPA) for thrombolysis, has been approved by the Food and Drug Administration (FDA) for treatment of acute ischemic stroke. This therapeutic approach must be initiated within 4.5 h after stroke onset (1995; [31]). Because of the narrow time window for treatment and the potential risk of intracerebral hemorrhage, in fact, only 1–3 % of stroke patients are able to receive this treatment [90]. As a result, more than 97 % of stroke patients lack a specific treatment. Thus, developing new therapeutic strategies to save a patient's life and improve their functional recovery is a major challenge for stroke research.

Stem cell factor (SCF) and granulocyte-colony-stimulating factor (G-CSF) are the essential hematopoietic growth factors that govern the growth, survival, differentiation, and mobilization of bone marrow stem cells [91, 102, 103]. Since 2003, the therapeutic effects of SCF and G-CSF on neuroprotection and neurorestoration in ischemic stroke have been frequently studied. Here we have briefly reviewed current understanding for the biological function of SCF and G-CSF in both the hematopoietic system and central nervous system (CNS), summarized the progress of SCF and G-CSF research in adult ischemic stroke, and discussed the directions for future studies.

13.2 The Origin and Biological Function of the Stem Cell Factor and Granulocyte-Colony-Stimulating Factor

SCF and G-CSF are the key members of the hematopoietic growth factor family and play important roles in regulating hematopoietic stem cell (HSC) proliferation, differentiation, and mobilization [20, 30, 55, 91, 102]. Since the discovery of SCF and G-CSF, great advancements have been made in understanding of the biological function of SCF and G-CSF and in developing pharmaceutical intervention of SCF and G-CSF to treat hematopoietic diseases and to repopulate and mobilize HSCs. Accumulating evidence has shown that SCF and G-CSF are not only crucially involved in the hematopoietic system [9, 53], but they also play a role in the CNS [95].

13.2.1 The Origin and Biological Function of SCF and G-CSF in the Hematopoietic System

SCF exists as a dimeric glycoprotein with a molecular weight of approximately 45 kDa [55]. SCF is classified into two forms: a membrane-bound form and a soluble form. Both the soluble and the transmembrane forms of SCF are biologically active. SCF is produced by the endothelial cells and fibroblasts [50]. Bone marrow stromal cells also produce SCF, which regulates the hematopoietic cell development in endocrine and paracrine manners. It has been revealed that HSCs and hematopoietic progenitor cells (HPCs) contain SCF mRNA; therefore, the growth and survival of HSCs/HPCs may also be regulated by autocrine synthesis of SCF [9, 67]. The production of SCF is increased by inflammatory stimuli such as interleukin-1 (IL-1) or tumor necrosis factor (TNF) [9].

C-kit has been demonstrated to be the receptor of SCF [6]. C-kit encodes a transmembrane tyrosine protein kinase receptor, and it has been found to express in HSCs/HPCs [2, 54, 89]. SCF/C-kit binding is a key process for SCF to regulate hematopoiesis. SCF acts directly on an enriched population containing HSCs/HPCs to accelerate their entry into the cell cycle [48] and enhance HSC/HPC expansion and survival in vitro [5, 92]. SCF is also crucially involved in the generation of white blood cells and red blood cells. It has been demonstrated that SCF is a key player for stimulating CD34-positive HSCs to form granulocyte-macrophage colonies (CFU-GM) and macroscopic erythroid burst-forming units (BFU-E) [5].

G-CSF is an approximately 24 kDa hydrophobic glycoprotein containing a neuraminic acid moiety, which regulates biological activity of G-CSF [16]. There are two recombinant forms of G-CSF, one is a non-glycosylated form and the other is a glycosylated form. Both of the two forms have similar biological activities and bioavailability when administered subcutaneously or intravenously. Many cells produce G-CSF after appropriate stimulation. Monocytes are the most prominent source of G-CSF [88]. Mesothelial cells, fibroblasts, and endothelial cells have also been found to produce G-CSF [44, 47, 101].

The receptor of G-CSF (CD114; G-CSFR) is a typical cytokine receptor with one transmembrane domain, an intracellular signal transduction domain, and homo-oligomerizes upon ligand binding [16]. G-CSF receptors are expressed on HSCs/HPCs [54] and mature neutrophilic granulocytes, monocytes, and platelets [7, 77]. The function of G-CSF has been demonstrated to play a vital role in directing the commitment of HSCs/HPCs to common myeloid lineage [68].

Numerous studies have determined the effects of the SCF and G-CSF combination (SCF+G-CSF) on hematopoiesis and HSC/HPC mobilization. It has been shown that SCF+G-CSF synergistically enhances the proliferation, differentiation, and survival of HSCs/HPCs [19]. The synergistic effect of SCF+G-CSF in hematopoiesis may be partially regulated by phosphatidylinositol-3 kinase (PI3K) and ERK signaling [19]. SCF+G-CSF has also been demonstrated to have synergistic effects in the mobilization of HSCs/HPCs from the bone marrow to the bloodstream in both laboratory animals and humans [3, 8, 20, 35, 84].

13.2.2 The Biological Function of SCF and G-CSF in the Central Nervous System

In addition to the effects of SCF and G-CSF in the hematopoietic system, increasing evidence shows that SCF and G-CSF also play a role in the CNS. Receptors for SCF and G-CSF have been found to express in the brain [38, 52], particularly in the neural stem cells/neural progenitor cells (NSCs/NPCs) [39, 64, 73, 96], and in cerebral neurons [73, 96] of adult mice and rats. It has been demonstrated that both SCF and G-CSF can pass through the blood-brain barrier [73, 100]. These findings suggest that hematopoietic growth factors, SCF and G-CSF, may have biological function in the CNS.

The role of SCF and G-CSF in directing NSCs/NPCs to give rise to neurons has been illustrated in both in vitro and in vivo studies. In cultured NSCs/NPCs, G-CSF has been shown to promote differentiation of NSCs/NPCs into neurons in a dose-dependent manner [73]. Infusing SCF into the cerebrolateral ventricle results in increases of newborn neurons in the neurogenic region, the subventricular zone [39]. When adding SCF and G-CSF during the proliferating stage of NSCs/NPCs, SCF in combination with G-CSF (SCF+G-CSF) shows a dual function in directing cell cycle arrest and promoting neuronal fate commitment through the enhancement of neurogenin 1 activity [64]. Together, these studies reveal that SCF and G-CSF are involved in neurogenesis.

Numerous in vitro and in vivo studies have examined the contribution of SCF and G-CSF in neuronal survival and neuronal plasticity. SCF selectively enhanced the survival of cultured embryonic chick dorsal root ganglia neurons [11]. SCF protects cultured neurons from apoptosis through the regulation of MEK/ERK or PI3K/AKT/NF-kB/Bcl-2 pathways [17]. Using cultured cortical neurons, G-CSF has been demonstrated to counteract programmed neuron death via PI3K mediation [73]. SCF acts as a neurotrophic factor supporting neuron survival during the development of the peripheral nervous system [11, 37]. SCF enhances neurite outgrowth in embryonic dorsal root ganglia [36, 37]. SCF+G-CSF synergistically promotes neurite outgrowth and network formation of cultured cortical neurons through PI3K/AKT/NF-kB/BDNF pathway [82]. Mice deficient in either SCF [60] or C-kit [43] display impaired long-term potentiation (LTP) and spatial learning and memory. G-CSF knockout mice show cognitive impairments, LTP reduction, and impairments in neural network formation in the hippocampus [18]. Collectively, these research data suggest that SCF and G-CSF, the two hematopoietic growth factors, act as neurotrophic factors to regulate neuron survival and neural plasticity. These findings provide insights into the potential role of SCF and G-CSF on neuroprotection and neurorestoration in the treatment of stroke.

13.3 The Role of SCF and G-CSF on Neuroprotection and Neurorestoration in Adult Ischemic Stroke

After a stroke, brain tissue that is located in and outside the infarct area undergoes significant changes including primary neuron loss, secondary neuron loss, neuroinflammation, neuron functional reorganization, neural network rewiring, and blood vessel regeneration. Based on the pathological progression and timing poststroke, stroke is classified into three clinical phases: the acute, subacute, and chronic phase. The duration and pathological severity of the three phases vary between individuals and depend on the infarction size, infarct location, cerebrovascular collateral response, patient's age, and medical comorbidities. Generally, the acute phase of stroke is the first 48 h after stroke, and the subacute phase of stroke is the period between 48 h and 6 weeks or to 3 months poststroke, whereas the chronic phase starts 3–6 months after stroke [33, 42, 63]. The primary neuron loss in the infarct core and penumbra zone [87] occurs during the acute phase of stroke, and the secondary neuron loss outside the infarct area mainly happens in the subacute phase of stroke [33]. In contrast to the pathological features of neuron loss in the acute and subacute phases [63], in the chronic phase, a stroke patient's neurological status becomes relatively stable, and the surviving neurons establish new networks in an effort to take over the function of the dead neurons [10, 13, 85, 95].

As stated above, the pathological features of the three phases of stroke are different. Therefore, the therapeutic strategies for each phase should be specific to the pathological alterations. The challenge of the specific treatment for each phase of stroke, however, is that the precise boundary among the three phases is difficult to identify and distinguish. It is often seen that some targeting molecules, such as NMDA receptor, matrix metalloproteinases, and intracellular mediator HMGB1, may have neuroprotective benefits in the acute phase of stroke but they may also risk negatively influencing the process of brain repair in the later recovery phase [59]. By contrast to the targeting molecules, increasing evidence has shown that administration of SCF and G-CSF in any of the stroke phases appears to be beneficial. Systemic administration of SCF and G-CSF in the acute or subacute phase of experimental stroke displays neuroprotective benefits; when administering during the chronic phase, SCF and G-CSF show neurorestorative effects in enhancing brain recovery.

13.3.1 The Effects of SCF and G-CSF in Acute Stroke

The effects of SCF and G-CSF in acute stroke have been extensively investigated. There is a large body of publications studying the role of SCF and G-CSF in neuroprotection in the phase of acute stroke in both animal models and stroke patients.

13.3.1.1 The Effects of SCF in Acute Stroke

C-kit, the receptor for SCF, has been shown to be increased in the neurogenic regions (the subventricular zone (SVZ) and the subgranular zone (SGZ)) of adult rats 24 h after cerebral ischemia [39]. Intraventricular delivery of SCF for 3 days post-cerebral ischemia [39] or subcutaneous daily injections of SCF during the period of 3 h and 7 days after induction of cortical ischemia [96] result in increases in the number of BrdU-labeled neural progenitor cells in the SVZ in rat models of focal cerebral ischemia. Intraventricular delivery of SCF for 3 days after focal cerebral ischemia enhances neurogenesis in the neurogenic regions [39]. Subcutaneous daily injection of SCF beginning at 3 h and ending 7 days after cerebral cortical ischemia shows a robust improvement in sensory motor function 1 week posttreatment. The SCF-induced functional improvement lasts more than 10 weeks after treatment, and the infarction size is reduced by SCF treatment [96]. These findings suggest neuroprotective and neuroregenerative effectiveness of SCF treatment in the acute phase of stroke.

13.3.1.2 The Effects of G-CSF in Acute Stroke

In comparison to SCF, G-CSF has been extensively studied in acute stroke.

Several studies have revealed the role of endogenous G-CSF in neuroprotection after stroke. Using rat models of focal cerebral ischemia, Schneider and coworkers [73] reported that both G-CSF and its receptor, G-CSFR, were widely expressed in the neurons throughout the brain. Two and 6 h after focal cerebral ischemia, G-CSF and G-CSFR were strongly increased in the neurons adjacent to the infarct area. In addition to rodents, G-CSFR is also robustly expressed in the peri-infarct neurons of human brain in the acute phase of ischemic stroke [34]. In G-CSF-deficient mice, the infarct volume is increased, and cerebral ischemia-induced neurological deficits are exacerbated as compared to wild-type mice. Systemic injections of G-CSF to the mice lacking of G-CSF before and 2 days after focal cerebral ischemia prevent G-CSF deficiency-induced enlarged infarction size and worsened neurological outcome [76]. These studies suggest that endogenous G-CSF and G-CSFR in neurons play an important role in neuroprotection.

The efficacy of exogenous administration of G-CSF in the acute phase of stroke has been largely examined in rat models [49, 72–74, 78, 80, 96] or mouse models (C57BL mice) of focal cerebral ischemia [27, 46, 79]. G-CSF treatment is initiated at the time points ranging from 30 min to 48 h after induction of ischemia with either single injection [27, 46, 73, 74, 79] or daily injections up to 10 days post-ischemia [49, 72, 73, 78, 80, 96]. Systemic administration of G-CSF (s.c., i.v., or i.p.) with treatment dosages of 50 μg/kg or 60 μg/kg shows beneficial effects in reducing infarction size and ameliorating neurological deficits [27, 46, 49, 72–74, 78–80, 96]. These research findings indicate that exogenous administration of G-CSF in the acute phase of stroke protects neurons from ischemic injury.

Understanding the mechanism underlying G-CSF treatment-induced neuroprotection in the acute phase of cerebral ischemia remains incomplete. Accumulating evidence shows that G-CSF treatment in the acute phase of focal cerebral ischemia reduces the disruption of the blood-brain barrier [49], reduces brain edema [28], decreases glutamate release in the infarcted striatum [32], suppresses pro-inflammatory cytokines and inflammatory mediators in the peri-ischemic areas [28, 74], and inhibits peripheral inflammatory cell infiltration to the ischemic hemisphere [49]. Anti-apoptosis may be involved in G-CSF-induced neuroprotection in acute ischemic stroke. Activation of STAT3/Bcl2 signaling, which plays a role in inhibiting apoptosis, is increased in the ipsilesional hemisphere by G-CSF treatment in the acute phase of cerebral ischemia [46, 70, 80]. G-CSF treatment in acute ischemic stroke reduces the number of cleaved caspase-3 expressing neurons in the injured cerebral cortex [80]. In addition, BrdU-labeled proliferating cells are increased in the ipsilesional hemisphere [78] or peri-infarct areas [96] after G-CSF treatment in acute cerebral ischemia. Neurogenesis and angiogenesis in the ipsilesional hemisphere are also enhanced by G-CSF treatment in acute ischemic stroke [49, 75]. However, these studies demonstrate the correlation between G-CSF-induced neuroprotection and G-CSF treatment-related cellular and molecular events such as anti-inflammation, anti-apoptosis, and pro-angiogenesis. The causal link mechanism of G-CSF-induced neuroprotection remains to be determined.

Controversial results have been reported regarding the neuroprotective role of G-CSF in acute stroke. Using a mouse model of permanent occlusion of the middle cerebral artery in CB-17 mice, Taguchi and colleagues [83] reported a negative result of G-CSF in acute ischemic stroke. Subcutaneous (s.c.) delivery of G-CSF (0.5, 5, 50, or 250 μg/kg) beginning at 1 h or 24 h poststroke and continuing up to 3 days or 7 days resulted in no change in infarction size at 3 days poststroke but increases in brain atrophy at 35 days poststroke in all tested doses except for the lowest one (0.5 μg/kg). The G-CSF treatment also impaired neurobehavioral function and increased infiltration of inflammatory cells (CD11b expressing cells and F4/80 positive cells) into the peri-infarct area. This negative study raises a caution flag on the neuroprotective effectiveness of G-CSF in acute stroke.

Two studies of meta-analysis have assessed the effects of G-CSF on acute stroke using animal models of focal cerebral ischemia. These studies have revealed that (1) G-CSF treatment initiating within 6 h or later than 6 h post-ischemic induction reduces infarction size and enhances functional recovery and the effectiveness of treatment within 6 h poststroke shows a dose-dependent manner [56] and (2) G-CSF treatment reduces motor impairments and death, while G-CSF reduces infarction size in transient but not in permanent models of focal ischemic stroke [22].

Clinical trials of G-CSF intervention for acute stroke have been carried out. In a phase IIa study, intravenous delivery of G-CSF to acute ischemic patients at doses of 30–180 μg/kg over the course of 3 days was reported to be safe and well tolerated [71]. In a small patient-sized phase I clinical trial, G-CSF was also proven safe and well tolerated in acute ischemic stroke patients, and G-CSF intervention showed improvement of neurological function [58]. However, the positive results do not display in a large phase IIb trial [69]. In this multinational, multicenter, randomized, and placebo-controlled trial (NCT00927836), G-CSF (135 μg/kg, i.v. over 72 h) was administered within 9 h post-ischemic stroke to the patients with an infarct location in the middle cerebral artery territory. However, G-CSF intervention failed to reduce the infarct size and improve functional outcome.

13.3.1.3 The Effects of SCF+G-CSF in Acute Stroke

Using a cerebral cortical ischemia model of stroke in spontaneously hypertensive rats (SHRs), in addition to SCF or G-CSF alone treatment [96], we have determined the therapeutic effects of SCF+G-CSF in acute stroke [96]. In this study, SCF+G-CSF was subcutaneously injected for 7 days beginning at 3 h post-ischemia. Similar to treatments of SCF or G-CSF alone, the combined SCF+G-CSF treatment led to reduction of infarction size. The BrdU-labeled cells in the peri-infarct area were also increased by SCF+G-CSF treatment. In contrast to the robust effects of SCF alone treatment in somatosensory and motor function recovery at 1 week posttreatment, SCF+G-CSF-treated rats did not show functional improvement at this early time point. However, the SCF+G-CSF-treated rats displayed a delayed but long-lasting improvement of somatosensory and motor function, which was detected at 4, 7 and 10 weeks after treatment. Toth and coworkers [86] examined the efficacy of SCF+G-CSF treatment in acute stroke using a cerebral cortical ischemia model of stroke in C57BL mice. SCF+G-CSF treatment, which initiated at immediately after ischemia followed by daily injections (i.p.) for 5 days, reduced infarct volume, promoted homing of BrdU/GFP co-expressing bone marrow-derived progenitor cells into the ischemic brain, increased $CD31^+/GFP^+$ endothelial cells in the ipsilesional hemisphere, and increased angiogenesis. Using the same mouse model of ischemic stroke as stated above [86], SCF+G-CSF was administered daily (s.c.) during the period of 1–10 days after induction of cerebral ischemia. In addition to reducing infarction size, SCF+G-CSF treatment increased bone marrow-derived neurons in the peri-infarct area and improved spatial learning and memory [45]. These studies shed light on the involvement of bone marrow-derived progenitor cells in SCF+G-CSF-induced beneficial effects in acute ischemic stroke.

13.3.2 The Effects of SCF and G-CSF in Subacute Stroke

There are even fewer studies targeting the therapeutic effects of SCF and G-CSF in subacute stroke than the studies of acute stroke. Using a rat model of transient focal ischemia, Lee and coworkers [49] determined the efficacy of G-CSF in subacute stroke. G-CSF was injected daily (i.p.) for 3 days beginning at 4 or 7 days post-ischemia. G-CSF treatment initiated at 4 days post-ischemia showed better functional improvement and greater reduction in hemispheric atrophy when compared to the treatment that was initiated at 7 days post-ischemia. Using a mouse (C57BL) model of cerebral cortical ischemia, Kawada and colleagues [45] examined the effects of SCF+G-CSF in both acute stroke (treatment during 1–10 days post-ischemia) and subacute stroke (treatment during 11–20 days post-ischemia). They observed that the treatment in both acute and subacute stroke reduced infarction size; however, SCF+G-CSF treatment in the subacute phase of cerebral ischemia showed better improvement in functional recovery and greater increases in bone marrow-derived neurons in the ipsi-infarct hemisphere and in NPC proliferation in the neurogenic region (the SVZ). In addition, they also found that SCF+G-CSF displayed a synergistic effect in promoting proliferation of NPCs in the SVZ as compared to SCF or G-CSF alone treatment. A study from the same group of Kawada showed that SCF+G-CSF treatment in the subacute phase of cerebral ischemia upregulated IL-10, an anti-inflammatory cytokine, on a much greater scale than the treatment in the acute phase [57]. Overall, it appears that the optimal timing for treatment is quite different between the treatment of G-CSF alone and SCF+G-CSF combination treatment: for G-CSF, the earlier the better, while for SCF+G-CSF, the later the better. The mechanism underlying this timing sensitivity for G-CSF or SCF+G-CSF treatment remains largely unknown. It provides an insightful notion, however, in directing future therapeutic studies for stroke: treatment efficacy depends on the timing of the intervention.

A phase IIb single-center, randomized, and placebo-controlled clinical trial for assessing the safety of G-CSF in subacute stroke has been completed. In this clinical study, G-CSF (10 μg/kg) or placebo was given subcutaneously for 5 days to 60 stroke patients. The treatment was initiated 3–30 days after stroke. G-CSF treatment showed a trend toward the reduction of the infarct volume in magnetic resonance imaging (MRI) results but did not cause significant adverse effects as compared to placebo controls, suggesting that G-CSF is safe when administered in the subacute phase of stroke [23].

13.3.3 The Effects of SCF and G-CSF in Chronic Stroke

Over the past decade, our research team has demonstrated the efficacy and possible mechanisms of SCF and G-CSF on stroke recovery in the chronic phase of stroke using rodent models of cerebral cortical ischemia.

SCF and G-CSF intervention in chronic stroke has been tested and validated to be effective when administered 3.5–6 months after cerebral cortical ischemia in both SHRs, C57BL mice, or transgenic mice with C57BL genetic background [13, 14, 51, 65, 66, 99].

First, we determined the efficacy of SCF and G-CSF treatment on stroke recovery in the chronic phase using a cerebral cortical ischemia model in SHRs.

The rationale for using SHRs is that hypertension is the most important risk factor for stroke in humans [41]. Chronic hypertension causes extensive changes in the cerebrovascular bed [4, 40]. Occlusion of the middle cerebral artery distal to the striatal branch and/or of the ipsilateral common carotid artery in SHRs leads to a more consistent and larger infarction in the cortex than in normotensive rats because of inadequate blood flow through collateral vessels [4, 12, 21, 29, 81, 94, 96–99]. In addition to the consistent infarction, this model also induces permanent deficits in somatosensorimotor function that last up to the chronic phase of stroke. Further, this model has no problem for long-term survival [62, 81, 94, 96–99]. Using the cortical ischemia model in SHRs, SCF (200 μg/kg), G-CSF (50 μg/kg), or SCF+G-CSF was subcutaneously administered for 7 days beginning at 3.5 months post-ischemic stroke. We found that only the SCF+G-CSF combination treatment led to a stable and long-term (17 weeks) improvement in somatosensory motor function. SCF alone treatment resulted in functional improvement but the improvement did not present as stable as the SCF+G-CSF combination treatment. G-CSF alone treatment did not show functional benefits. In addition, field-evoked potentials further validated the neurobehavioral findings and revealed that a normal pattern of somatosensory pathways was reestablished by SCF+G-CSF treatment. In addition to the functional improvement, the infarct cavity was reduced in SCF+G-CSF-treated SHRs [99], suggesting that neural regeneration may be involved in brain repair by SCF+G-CSF treatment in the chronic phase. This study provides first evidence that functional restoration in the chronic phase of stroke is possible through the SCF+G-CSF combination treatment.

We have also assessed the safety and effectiveness of the SCF+G-CSF combination treatment on stroke recovery in the chronic phase using aged animals [65], because stroke has the highest incidence in those over the age of 60 [61]. Aged male SHRs (11–13 months) and C57BL mice (16–18 months) were subjected to focal cerebral cortical ischemia. These ages of SHRs and C57BL mice are equivalent to 61–72 years in humans based on their differences in average lifespan [65]. Six dosages were examined examined in the chronic phase of ischemic stroke in the aged SHRs and C57BL mice: (1) SCF+G-CSF at 200 μg/kg for SCF and 50 μg/kg for G-CSF, (2) SCF+G-CSF at 100 μg/kg for SCF and 25 μg/kg for G-CSF, (3) SCF+G-CSF at 50 μg/kg for SCF and 25 μg/kg for G-CSF, (4) SCF+G-CSF at 20 μg/kg for SCF and 10 μg/kg for G-CSF, (5) SCF+G-CSF at 10 μg/kg for SCF and 5 μg/kg for G-CSF, and (6) SCF+G-CSF at 5 μg/kg for SCF and 2.5 μg/kg for G-CSF. Subcutaneous injections of SCF+G-CSF were given for 5 days beginning at 3–4 months after induction of cerebral ischemia. We observed that all six tested dosages did not cause either acute or chronic toxicity to the livers and kidneys, demonstrating that SCF+G-CSF treatment for chronic stroke is safe for the aged population. When determining the effects of SCF+G-CSF in mobilizing bone marrow stem cells into the blood, the three higher dosages of SCF+G-CSF showed significant elevation of C-kit-expressing stem cells in the blood. In a somatosensory motor testing (limb placement test), two higher dosages of SCF+G-CSF (100/25 μg/kg, and 50/25 μg/kg) led to stable and long-term functional improvement. The intermediate dose of SCF+G-CSF (20/10 μg/kg) showed a short-term improvement, whereas the two lower dosages did not improve somatosensory motor function in the chronic phase of stroke in aged SHRs. These findings suggest that the SCF+G-CSF combination treatment for chronic stroke recovery is a safe and effective approach for the aged population. SCF+G-CSF combination treatment in chronic stroke mobilizes bone marrow stem cells and improves functional recovery in a dose-dependent manner.

We have carried out several mechanistic studies to understand how SCF+G-CSF combination treatment repairs a stroke-damaged brain in the chronic phase. Using a bone marrow transplantation approach to track bone marrow-derived cells, our study revealed that bone marrow-derived endothelial cells and bone marrow-derived neurons were involved in SCF+G-CSF-enhanced angiogenesis and neurogenesis in the brain of chronic stroke [66]. To determine the effects of SCF+G-CSF in regulating dynamics of synaptic circuits in the chronic phase of experimental stroke, we used two-photon microscopy to scan the brain area adjacent to the infarct cavity before and after SCF+G-CSF treatment in aged Thy-1-YFPH mice (C57BL background) [13] (Fig. 13.1). The Thy-1-YFPH mice express yellow fluorescent protein (YFP) only in the layer V pyramidal neurons [24]. Before treatment, the number of

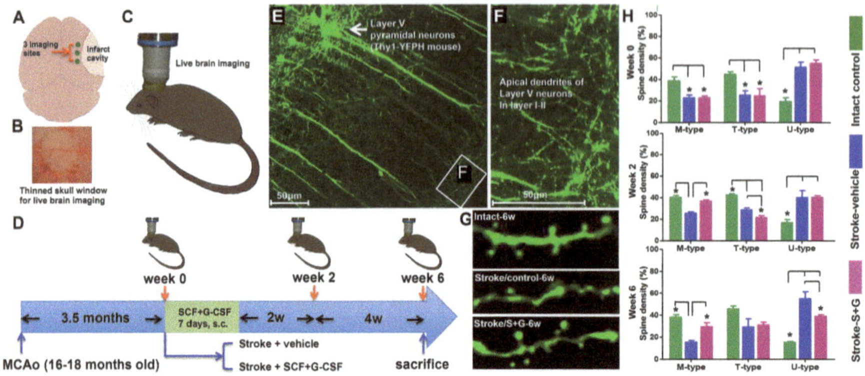

Fig. 13.1 Rebuilding synaptic circuits in the peri-infarct cavity cortex are enhanced by SCF+G-CSF treatment in the chronic phase of experimental stroke in aged mice. (**a**) Schematic diagram showing the three imaging sites adjacent to the infarct cavity. (**b**) Schematic diagram showing a thinned skull window that was prepared for live brain imaging. (**c**) Schematic diagram of the live brain imaging. (**d**) Schematic diagram of experimental design. (**e**) Layer V pyramidal neurons in the cortex of Thy-1-YFPH mice. Note that *yellow* fluorescent protein (YFP) is only expressed in the layer V pyramidal neurons. Boxed area is the layer I–II of cortex enlarged in the panel F. (**f**) The apical dendrites and dendritic spines of the layer V pyramidal neurons distributing in the layer I–II of cortex of Thy-1-YFPH mice. (**g**) Representative live brain images of the apical dendrites and dendritic spines in the brains of intact controls, stroke-vehicle controls, and stroke-SCF+G-CSF-treated mice at 6 weeks posttreatment. Note that the majority of dendritic spines in the cortex of intact controls are mushroom type of spines (M-type), while the majority of dendritic spines in the cortex of stroke-vehicle controls are uncertain type of spines (U-type). The mushroom type of spines appears to be increased in the cortex of stroke-SCF+G-CSF-treated mice. (**h**) Dynamics of dendritic spines before and after treatment. Note that M-type spines and thin-type of spines (T-type) are reduced in the two stroke groups, while the U-type spines are increased in the two stroke groups before treatment. These findings suggest that synaptic degeneration or reduced synaptic circuits occurs in the peri-infarct cavity cortex in the chronic phase of experimental stroke. Two weeks after treatment, stroke-SCF+G-CSF-treated mice show increases in M-type spines and decreases in T-type spines, suggesting the reestablishment of synaptic circuits in the peri-infarct cavity cortex. Six weeks after treatment, M-type spines are increased, and the U-type spines are reduced in the stroke mice treated with SCF+G-CSF. These data indicate that SCF+G-CSF treatment rebuilds synaptic circuits in the peri-infarct cavity cortex. This figure summarizes the results of the study published elsewhere [13]

mushroom-type (M-type) spines in the layer V pyramidal neurons was reduced, and the uncertain-type (U-type) spines, which cannot build synapses with other neurons, were increased in the stroke mice. This observation suggests that reduced synaptic circuits occur in the peri-infarct cavity cortex in chronic stroke brain. Six weeks after SCF+G-CSF treatment, however, the M-type spines were significantly increased, and the U-type spines were significantly reduced in the layer V pyramidal neurons adjacent to the infarct cavity. In addition to the two-photon live brain imaging, immunohistochemistry data showed significant increases of postsynaptic density protein 95 (PSD-95) puncta and dendritic branches in the peri-infarct cavity cortex 6 weeks after SCF+G-CSF treatment. These findings suggest that SCF+G-CSF treatment in the chronic phase of stroke enhances synaptic network regeneration in the peri-infarct cavity cortex (Fig. 13.2). This study advances the current knowledge of stroke recovery: an aged brain damaged by ischemic stroke is reparable by a pharmaceutical approach, SCF+G-CSF.

Our follow-up studies have clarified that neural network rewiring in the peri-infarct cavity cortex is required for SCF+G-CSF-enhanced functional improvement in the chronic phase of experimental stroke. In an in vitro study, we have demonstrated that SCF+G-CSF synergistically enhances neurite outgrowth and neural network formation through NF-kB mediation [82]. In an in vivo study [14], we sought to use an approach for blocking the NF-kB-mediated neural network rewiring to determine the causal link between the SCF+G-CSF-promoted neural network regeneration and SCF+G-CSF-enhanced functional improvement in the chronic phase of experimental stroke. In this study, NF-kB inhibitor was infused into the lateral ventricle in the contralesional cortex before and during the 7-day subcutaneous injections of SCF+G-CSF. Motor function was evaluated before treatment as well as 2 and 6 weeks after treatment. Our data revealed that SCF+G-CSF treatment in the chronic phase of stroke increased axonal sprouting, synaptogenesis, and angiogenesis specifically in the peri-infarct cavity cortex but not in the contralesional cortex. NF-kB inhibitor completely abolished the SCF+G-CSF-increased axonal sprouting, synaptogenesis and angiogenesis in the peri-infarct cavity cortex 10 weeks after treatment. In addition, the SCF+G-CSF-improved motor function at 2 and 6 weeks posttreatment was also eliminated by NF-kB inhibitor. This study demonstrates a key role of neural network rewiring in the peri-infarct cavity cortex in the SCF+G-CSF-enhanced motor function recovery in the chronic phase of stroke.

To further identify whether the SCF+G-CSF-enhanced synaptic network rewiring and the SCF+G-CSF-enhanced motor function recovery occurred simultaneously, we carried out an independent study. In this study [15], using a combination approach through live brain imaging, whole brain imaging, molecular manipulation, synaptic and vascular assessments, and motor function examination, we further validated our findings that the SCF+G-CSF-enhanced motor function recovery in the chronic phase of stroke was linked to neural network rewiring in the peri-infarct cavity cortex. Thy-1-YFPH mice were also used for this study. SCF+G-CSF treatment was initiated at 6 months post-experimental stroke. Similar to the earlier study [14], infusion of NF-kB inhibitor was used for blocking the SCF+G-CSF-enhanced neural network rewiring in the peri-infarct cavity cortex.

A previous study reported that motor activity in a Rota-Rod could modify dendritic spine formation [93]. To prevent altering dendritic spines by repeated motor function tests with a Rota-Rod, the chronic stroke mice without behavioral tests were used for live brain imaging and whole brain imaging to identify SCF+G-CSF per se induced remodeling of synaptic circuits in the cortex adjacent to the infarct cavity and/or in the contralesional cortex and to determine whether the synaptic circuit rewiring in the peri-infarct cavity cortex simultaneously happens when motor function is improved by SCF+G-CSF treatment. Our findings showed that SCF+G-CSF treatment at 6 months poststroke improved motor function recovery. SCF+G-

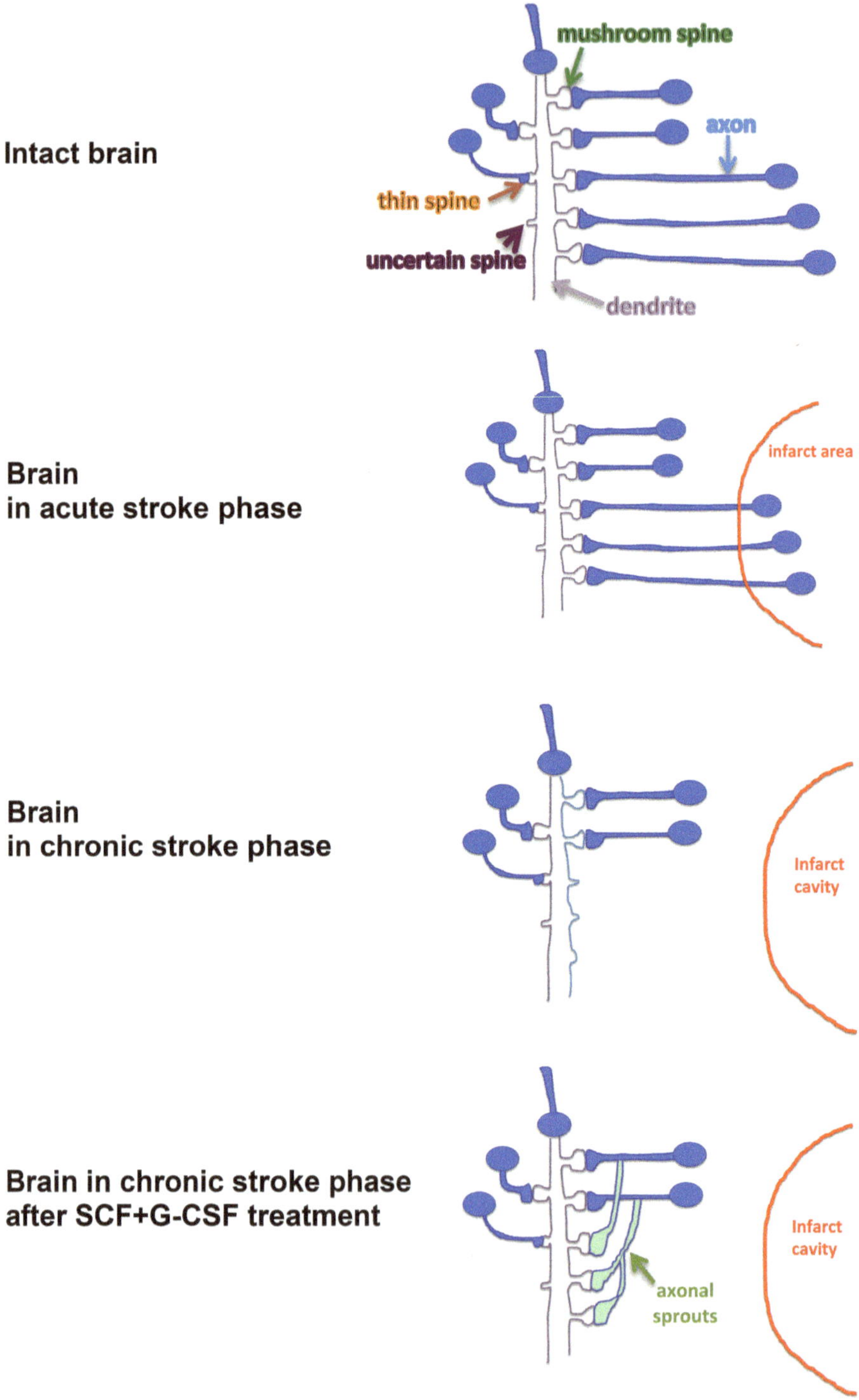

Intact brain
mushroom spine
axon
thin spine
uncertain spine
dendrite
Brain
in acute stroke phase
infarct area
Brain
in chronic stroke phase
Infarct
cavity
Brain in chronic stroke phase
after SCF+G-CSF treatment
Infarct
cavity
axonal
sprouts

CSF promoted mushroom spine formation, enlarged postsynaptic membrane size, and increased postsynaptic PSD-95 accumulation and blood vessel density in the peri-infarct cavity cortex but not in the contralesional cortex. When two-photon live brain imaging showed SCF+G-CSF-enhanced synaptic circuit regeneration in the peri-infarct cavity cortex 2 and 6 weeks posttreatment, motor functional improvement was also seen in the SCF+G-CSF-treated mice 2 and 6 weeks posttreatment.

Once the SCF+G-CSF-increased synaptic network regeneration in the peri-infarct cavity cortex was blocked by NF-kB inhibitor, the SCF+G-CSF-improved motor function was also eliminated. This study has further confirmed that the enhanced neural network formation in the peri-infarct cavity cortex via NF-kB regulation is crucially involved in the SCF+G-CSF-improved motor function in chronic stroke.

A double-blinded, randomized, and placebo-controlled clinical trial for examining the safety and efficacy of G-CSF in chronic stroke has been conducted. Forty-one ischemic stroke patients (>4 months after stroke) were included in this trial. G-CSF (10 μg/kg, s.c.) was given for 10 days. The results showed that the G-CSF treatment was safe to the chronic stroke patients, whereas the improved functional outcome was not seen in G-CSF-treated patients. Authors discussed including more patients in future studies to increase the power of statistical analysis [25].

13.4 Concluding Remarks

Unlike rtPA therapy that has a limited therapeutic window within 4.5 h after ischemic stroke onset, SCF and G-CSF appear to have broad therapeutic potential for acute, subacute, and chronic stroke according to the basic studies using animal models of focal ischemic stroke. The majority of these studies used one-dose and one-time treatment, and some of the studies examined different dosages and treatment time points. Based on the findings, the optimal time for G-CSF treatment appears to be the acute phase or earlier subacute phase of stroke; for SCF, the best treatment time may be the acute phase of stroke; and for SCF+G-CSF combination treatment, the optimal treatment time appears to be the later stage – the subacute and chronic

Fig. 13.2 Schematic diagram of synaptic networks in different conditions. In intact brain, the majority of dendritic spines are the mushroom-type spines. Thin-type spines are the flexible spines that either grow into large mushroom spines or shrink/disappear in response to microenvironment changes. The uncertain-type spines are the spines under degeneration, and this type of spine cannot form synaptic connections with other neurons. During the acute phase of stroke, neurons in the infarct area die due to lack of blood supply. As a result, the post-synapses of the dead neurons undergo degeneration (mushroom spines shrink to uncertain-type spines). SCF+G-CSF treatment in the chronic phase of experimental stroke promotes axonal sprouting and dendritic branching and enhances mushroom spine formation and synaptogenesis. Thus, the SCF+G-CSF treatment enhances rebuilding of synaptic circuits and neural networks in the peri-infarct cavity cortex in the chronic phase of experimental stroke. The schematic diagram shown in this figure is based on our previous studies published elsewhere ([13, 14], 2016)

phases of experimental stroke. However, it remains largely unknown why the treatments of SCF and G-CSF alone or combination treatment have different optimal timings and how SCF and G-CSF alone or combination treatment protects neurons from ischemic injury or restores/repairs neuron function after stroke.

Before moving to clinical trials, several crucial questions need to be addressed: Do we use the most clinically relevant animal models of stroke to determine the therapeutic effects? Do we clarify the precise pathological features for acute, subacute, and chronic phases of stroke? Do we demonstrate the optimal intervention timing, dosage, and delivery route for SCF and G-CSF alone or combination treatment? Do we validate the research findings using different animal models especially in nonhuman primates?

Although clinical trials have proven the safety of using G-CSF in treating acute, subacute, and chronic stroke patients, the efficacy of G-CSF in functional recovery has not yet been demonstrated positively. It is worth noting that keeping the infarct type/size and treatment time point uniform is relatively easier in basic science research using animal models than clinical studies using stroke patients. The variation in infarction size and location, cerebrovascular collateral response, patient's age, sex, race, and medical comorbidities as well as differences in intervention timing may cause robust increases in standard deviation of research data. As a result, significant increases of sample sizes (number of stroke patients) are required for reaching the levels of statistical difference in clinical trials.

Overall, SCF and G-CSF research brings new hope for developing a new treatment for stroke as these hematopoietic growth factors show therapeutic potential for acute, subacute, and chronic stroke. Many open questions, however, need to be addressed in the future for both basic research and clinical trials.

Acknowledgments The authors would like to thank Sandy McGillis for her assistance in artworks for the live brain imaging and in editing the manuscript. This work was partially supported by the National Institutes of Health and National Institute of Neurological Disorders and Stroke (NINDS)(R01 NS060911) in the United States.

References

1. The National Institute of Neurological Disorders and Stroke rt-PA Stroke Study Group. (1995) Tissue plasminogen activator for acute ischemic stroke. N Engl J Med. 333:1581–1587.
2. Andre C, d'Auriol L, Lacombe C, Gisselbrecht S, Galibert F. c-kit mRNA expression in human and murine hematopoietic cell lines. Oncogene. 1989;4:1047–9.
3. Andrews RG, Briddell RA, Knitter GH, Rowley SD, Appelbaum FR, McNiece IK. Rapid engraftment by peripheral blood progenitor cells mobilized by recombinant human stem cell factor and recombinant human granulocyte colony-stimulating factor in nonhuman primates. Blood. 1995;85:15–20.
4. Barone FC, Price WJ, White RF, Willette RN, Feuerstein GZ. Genetic hypertension and increased susceptibility to cerebral ischemia. Neurosci Biobehav Rev. 1992;16:219–33.

5. Bernstein ID, Andrews RG, Zsebo KM. Recombinant human stem cell factor enhances the formation of colonies by CD34+ and CD34+lin- cells, and the generation of colony-forming cell progeny from CD34+lin- cells cultured with interleukin-3, granulocyte colony-stimulating factor, or granulocyte-macrophage colony-stimulating factor. Blood. 1991;77:2316–21.
6. Besmer P, Murphy JE, George PC, Qiu FH, Bergold PJ, Lederman L, Snyder Jr HW, Brodeur D, Zuckerman EE, Hardy WD. A new acute transforming feline retrovirus and relationship of its oncogene v-kit with the protein kinase gene family. Nature. 1986;320:415–21.
7. Boneberg EM, Hareng L, Gantner F, Wendel A, Hartung T. Human monocytes express functional receptors for granulocyte colony-stimulating factor that mediate suppression of monokines and interferon-gamma. Blood. 2000;95:270–6.
8. Briddell RA, Hartley CA, Smith KA, McNiece IK. Recombinant rat stem cell factor synergizes with recombinant human granulocyte colony-stimulating factor in vivo in mice to mobilize peripheral blood progenitor cells that have enhanced repopulating potential. Blood. 1993;82:1720–3.
9. Broudy VC. Stem cell factor and hematopoiesis. Blood. 1997;90:1345–64.
10. Carmichael ST. Brain excitability in stroke: the yin and yang of stroke progression. Arch Neurol. 2012;69:161–7.
11. Carnahan JF, Patel DR, Miller JA. Stem cell factor is a neurotrophic factor for neural crest-derived chick sensory neurons. J Neurosci: Off J Soc Neurosci. 1994;14:1433–40.
12. Coyle P. Different susceptibilities to cerebral infarction in spontaneously hypertensive (SHR) and normotensive Sprague-Dawley rats. Stroke. 1986;17:520–5.
13. Cui L, Murikinati SR, Wang D, Zhang X, Duan WM, Zhao LR. Reestablishing neuronal networks in the aged brain by stem cell factor and granulocyte-colony stimulating factor in a mouse model of chronic stroke. PLoS ONE. 2013;8, e64684.
14. Cui L, Duchamp NS, Boston DJ, Ren X, Zhang X, Hu H, Zhao LR. NF-kappaB is involved in brain repair by stem cell factor and granulocyte-colony stimulating factor in chronic stroke. Exp Neurol. 2015;263:17–27.
15. Cui L, Wang D, McGillis S, Kyle M, Zhao LR. Repairing the brain by SCF+G-CSF treatment at 6 months post-experimental stroke: mechanistic determination of the causal link between neurovascular regeneration and motor functional recovery. ASN Neuro. 2016;8(4).
16. Demetri GD, Griffin JD. Granulocyte colony-stimulating factor and its receptor. Blood. 1991;78:2791–808.
17. Dhandapani KM, Wade FM, Wakade C, Mahesh VB, Brann DW. Neuroprotection by stem cell factor in rat cortical neurons involves AKT and NFkappaB. J Neurochem. 2005;95:9–19.
18. Diederich K, Sevimli S, Dorr H, Kosters E, Hoppen M, Lewejohann L, Klocke R, Minnerup J, Knecht S, Nikol S, Sachser N, Schneider A, Gorji A, Sommer C, Schabitz WR. The role of granulocyte-colony stimulating factor (G-CSF) in the healthy brain: a characterization of G-CSF-deficient mice. J Neurosci. 2009;29:11572–81.
19. Duarte RF, Frank DA. SCF and G-CSF lead to the synergistic induction of proliferation and gene expression through complementary signaling pathways. Blood. 2000;96:3422–30.
20. Duarte RF, Franf DA. The synergy between stem cell factor (SCF) and granulocyte colony-stimulating factor (G-CSF): molecular basis and clinical relevance. Leuk Lymphoma. 2002;43:1179–87.
21. Duverger D, MacKenzie ET. The quantification of cerebral infarction following focal ischemia in the rat: influence of strain, arterial pressure, blood glucose concentration, and age. J Cereb Blood Flow Metab. 1988;8:449–61.
22. England TJ, Gibson CL, Bath PM. Granulocyte-colony stimulating factor in experimental stroke and its effects on infarct size and functional outcome: a systematic review. Brain Res Rev. 2009;62:71–82.
23. England TJ, Abaei M, Auer DP, Lowe J, Jones DR, Sare G, Walker M, Bath PM. Granulocyte-colony stimulating factor for mobilizing bone marrow stem cells in subacute stroke: the stem

cell trial of recovery enhancement after stroke 2 randomized controlled trial. Stroke. 2012;43:405–11.
24. Feng G, Mellor RH, Bernstein M, Keller-Peck C, Nguyen QT, Wallace M, Nerbonne JM, Lichtman JW, Sanes JR. Imaging neuronal subsets in transgenic mice expressing multiple spectral variants of GFP. Neuron. 2000;28:41–51.
25. Floel A, Warnecke T, Duning T, Lating Y, Uhlenbrock J, Schneider A, Vogt G, Laage R, Koch W, Knecht S, Schabitz WR. Granulocyte-colony stimulating factor (G-CSF) in stroke patients with concomitant vascular disease – a randomized controlled trial. PLoS ONE. 2011;6, e19767.
26. George PM, Steinberg GK. Novel stroke therapeutics: unraveling stroke pathophysiology and its impact on clinical treatments. Neuron. 2015;87:297–309.
27. Gibson CL, Bath PM, Murphy SP. G-CSF reduces infarct volume and improves functional outcome after transient focal cerebral ischemia in mice. J Cereb Blood Flow Metab. 2005;25:431–9.
28. Gibson CL, Jones NC, Prior MJ, Bath PM, Murphy SP. G-CSF suppresses edema formation and reduces interleukin-1beta expression after cerebral ischemia in mice. J Neuropathol Exp Neurol. 2005;64:763–9.
29. Grabowski M, Nordborg C, Brundin P, Johansson BB. Middle cerebral artery occlusion in the hypertensive and normotensive rat: a study of histopathology and behaviour. J Hypertens. 1988;6:405–11.
30. Greenbaum AM, Link DC. Mechanisms of G-CSF-mediated hematopoietic stem and progenitor mobilization. Leukemia. 2011;25:211–7.
31. Hacke W, Kaste M, Bluhmki E, Brozman M, Davalos A, Guidetti D, Larrue V, Lees KR, Medeghri Z, Machnig T, Schneider D, von Kummer R, Wahlgren N, Toni D, Investigators E. Thrombolysis with alteplase 3 to 4.5 hours after acute ischemic stroke. N Engl J Med. 2008;359:1317–29.
32. Han JL, Blank T, Schwab S, Kollmar R. Inhibited glutamate release by granulocyte-colony stimulating factor after experimental stroke. Neurosci Lett. 2008;432:167–9.
33. Hara H, Harada K, Sukamoto T. Chronological atrophy after transient middle cerebral artery occlusion in rats. Brain Res. 1993;618:251–60.
34. Hasselblatt M, Jeibmann A, Riesmeier B, Maintz D, Schabitz WR. Granulocyte-colony stimulating factor (G-CSF) and G-CSF receptor expression in human ischemic stroke. Acta Neuropathol. 2007;113:45–51.
35. Hess DA, Levac KD, Karanu FN, Rosu-Myles M, White MJ, Gallacher L, Murdoch B, Keeney M, Ottowski P, Foley R, Chin-Yee I, Bhatia M. Functional analysis of human hematopoietic repopulating cells mobilized with granulocyte colony-stimulating factor alone versus granulocyte colony-stimulating factor in combination with stem cell factor. Blood. 2002;100:869–78.
36. Hirata T, Kasugai T, Morii E, Hirota S, Nomura S, Fujisawa H, Kitamura Y. Characterization of c-kit-positive neurons in the dorsal root ganglion of mouse. Brain Res Dev Brain Res. 1995;85:201–11.
37. Hirata T, Morii E, Morimoto M, Kasugai T, Tsujimura T, Hirota S, Kanakura Y, Nomura S, Kitamura Y. Stem cell factor induces outgrowth of c-kit-positive neurites and supports the survival of c-kit-positive neurons in dorsal root ganglia of mouse embryos. Development. 1993;119:49–56.
38. Hirota S, Ito A, Morii E, Wanaka A, Tohyama M, Kitamura Y, Nomura S. Localization of mRNA for c-kit receptor and its ligand in the brain of adult rats: an analysis using in situ hybridization histochemistry. Brain Res Mol Brain Res. 1992;15:47–54.
39. Jin K, Mao XO, Sun Y, Xie L, Greenberg DA. Stem cell factor stimulates neurogenesis in vitro and in vivo. J Clin Invest. 2002;110:311–9.
40. Johansson BB. Cerebral vascular bed in hypertension and consequences for the brain. Hypertension. 1984;6:III81–6.
41. Johansson BB, Auer LM, Sayama I. Reaction of pial arteries and veins to hypercapnia in hypertensive and normotensive rats. Stroke. 1985;16:320–3.

42. Kang DW, Latour LL, Chalela JA, Dambrosia JA, Warach S. Early and late recurrence of ischemic lesion on MRI: evidence for a prolonged stroke-prone state? Neurology. 2004;63:2261–5.
43. Katafuchi T, Li AJ, Hirota S, Kitamura Y, Hori T. Impairment of spatial learning and hippocampal synaptic potentiation in c-kit mutant rats. Learn Mem. 2000;7:383–92.
44. Kaushansky K, Lin N, Adamson JW. Interleukin 1 stimulates fibroblasts to synthesize granulocyte-macrophage and granulocyte colony-stimulating factors. Mechanism for the hematopoietic response to inflammation. J Clin Invest. 1988;81:92–7.
45. Kawada H, Takizawa S, Takanashi T, Morita Y, Fujita J, Fukuda K, Takagi S, Okano H, Ando K, Hotta T. Administration of hematopoietic cytokines in the subacute phase after cerebral infarction is effective for functional recovery facilitating proliferation of intrinsic neural stem/progenitor cells and transition of bone marrow-derived neuronal cells. Circulation. 2006;113:701–10.
46. Komine-Kobayashi M, Zhang N, Liu M, Tanaka R, Hara H, Osaka A, Mochizuki H, Mizuno Y, Urabe T. Neuroprotective effect of recombinant human granulocyte colony-stimulating factor in transient focal ischemia of mice. J Cereb Blood Flow Metab. 2006;26:402–13.
47. Lanfrancone L, Boraschi D, Ghiara P, Falini B, Grignani F, Peri G, Mantovani A, Pelicci PG. Human peritoneal mesothelial cells produce many cytokines (granulocyte colony-stimulating factor [CSF], granulocyte-monocyte-CSF, macrophage-CSF, interleukin-1 [IL-1], and IL-6) and are activated and stimulated to grow by IL-1. Blood. 1992;80:2835–42.
48. Leary AG, Zeng HQ, Clark SC, Ogawa M. Growth factor requirements for survival in G0 and entry into the cell cycle of primitive human hematopoietic progenitors. Proc Natl Acad Sci U S A. 1992;89:4013–7.
49. Lee ST, Chu K, Jung KH, Ko SY, Kim EH, Sinn DI, Lee YS, Lo EH, Kim M, Roh JK. Granulocyte colony-stimulating factor enhances angiogenesis after focal cerebral ischemia. Brain Res. 2005;1058:120–8.
50. Lennartsson J, Ronnstrand L. Stem cell factor receptor/c-Kit: from basic science to clinical implications. Physiol Rev. 2012;92:1619–49.
51. Liu Y, Popescu M, Longo S, Gao M, Wang D, McGillis S, Zhao LR. Fibrinogen reduction and motor function improvement by hematopoietic growth factor treatment in chronic stroke in aged mice: a treatment frequency study. Cell Transplant. 2016;25:729–34.
52. Manova K, Bachvarova RF, Huang EJ, Sanchez S, Pronovost SM, Velazquez E, McGuire B, Besmer P. c-kit receptor and ligand expression in postnatal development of the mouse cerebellum suggests a function for c-kit in inhibitory interneurons. J Neurosci. 1992;12:4663–76.
53. Martino M, Laszlo D, Lanza F. Long-active granulocyte colony-stimulating factor for peripheral blood hematopoietic progenitor cell mobilization. Expert Opin Biol Ther. 2014;14:757–72.
54. McKinstry WJ, Li CL, Rasko JE, Nicola NA, Johnson GR, Metcalf D. Cytokine receptor expression on hematopoietic stem and progenitor cells. Blood. 1997;89:65–71.
55. McNiece IK, Briddell RA. Stem cell factor. J Leukoc Biol. 1995;58:14–22.
56. Minnerup J, Heidrich J, Wellmann J, Rogalewski A, Schneider A, Schabitz WR. Meta-analysis of the efficacy of granulocyte-colony stimulating factor in animal models of focal cerebral ischemia. Stroke. 2008;39:1855–61.
57. Morita Y, Takizawa S, Kamiguchi H, Uesugi T, Kawada H, Takagi S. Administration of hematopoietic cytokines increases the expression of anti-inflammatory cytokine (IL-10) mRNA in the subacute phase after stroke. Neurosci Res. 2007;58:356–60.
58. Moriya Y, Mizuma A, Uesugi T, Ohnuki Y, Nagata E, Takahashi W, Kobayashi H, Kawada H, Ando K, Takagi S, Takizawa S. Phase I study of intravenous low-dose granulocyte colony-stimulating factor in acute and subacute ischemic stroke. J Stroke Cerebrovasc Dis. 2013;22:1088–97.
59. Moskowitz MA, Lo EH, Iadecola C. The science of stroke: mechanisms in search of treatments. Neuron. 2010;67:181–98.

60. Motro B, Wojtowicz JM, Bernstein A, van der Kooy D. Steel mutant mice are deficient in hippocampal learning but not long-term potentiation. Proc Natl Acad Sci U S A. 1996;93:1808–13.
61. Mozaffarian D, et al. Heart disease and stroke statistics-2016 update: a report from the American Heart Association. Circulation. 2016;133:e38–360.
62. Ohlsson AL, Johansson BB. Environment influences functional outcome of cerebral infarction in rats. Stroke. 1995;26:644–9.
63. Parsons MW, Li T, Barber PA, Yang Q, Darby DG, Desmond PM, Gerraty RP, Tress BM, Davis SM. Combined (1)H MR spectroscopy and diffusion-weighted MRI improves the prediction of stroke outcome. Neurology. 2000;55:498–505.
64. Piao CS, Li B, Zhang LJ, Zhao LR. Stem cell factor and granulocyte colony-stimulating factor promote neuronal lineage commitment of neural stem cells. Differentiation. 2012;83:17–25.
65. Piao CS, Gonzalez-Toledo ME, Gu X, Zhao LR. The combination of stem cell factor and granulocyte-colony stimulating factor for chronic stroke treatment in aged animals. Exp Transl Stroke Med. 2012;4:25.
66. Piao CS, Gonzalez-Toledo ME, Xue YQ, Duan WM, Terao S, Granger DN, Kelley RE, Zhao LR. The role of stem cell factor and granulocyte-colony stimulating factor in brain repair during chronic stroke. J Cereb Blood Flow Metab. 2009;29:759–70.
67. Ratajczak MZ, Kuczynski WI, Sokol DL, Moore JS, Pletcher Jr CH, Gewirtz AM. Expression and physiologic significance of Kit ligand and stem cell tyrosine kinase-1 receptor ligand in normal human CD34+, c-Kit+ marrow cells. Blood. 1995;86:2161–7.
68. Richards MK, Liu F, Iwasaki H, Akashi K, Link DC. Pivotal role of granulocyte colony-stimulating factor in the development of progenitors in the common myeloid pathway. Blood. 2003;102:3562–8.
69. Ringelstein EB, et al. Granulocyte colony-stimulating factor in patients with acute ischemic stroke: results of the AX200 for Ischemic Stroke trial. Stroke. 2013;44:2681–7.
70. Schabitz WR, Kollmar R, Schwaninger M, Juettler E, Bardutzky J, Scholzke MN, Sommer C, Schwab S. Neuroprotective effect of granulocyte colony-stimulating factor after focal cerebral ischemia. Stroke. 2003;34:745–51.
71. Schabitz WR, Laage R, Vogt G, Koch W, Kollmar R, Schwab S, Schneider D, Hamann GF, Rosenkranz M, Veltkamp R, Fiebach JB, Hacke W, Grotta JC, Fisher M, Schneider A. AXIS: a trial of intravenous granulocyte colony-stimulating factor in acute ischemic stroke. Stroke. 2010;41:2545–51.
72. Schneider A, Wysocki R, Pitzer C, Kruger C, Laage R, Schwab S, Bach A, Schabitz WR. An extended window of opportunity for G-CSF treatment in cerebral ischemia. BMC Biol. 2006;4:36.
73. Schneider A, Kruger C, Steigleder T, Weber D, Pitzer C, Laage R, Aronowski J, Maurer MH, Gassler N, Mier W, Hasselblatt M, Kollmar R, Schwab S, Sommer C, Bach A, Kuhn HG, Schabitz WR. The hematopoietic factor G-CSF is a neuronal ligand that counteracts programmed cell death and drives neurogenesis. J Clin Invest. 2005;115:2083–98.
74. Sehara Y, Hayashi T, Deguchi K, Zhang H, Tsuchiya A, Yamashita T, Lukic V, Nagai M, Kamiya T, Abe K. Decreased focal inflammatory response by G-CSF may improve stroke outcome after transient middle cerebral artery occlusion in rats. J Neurosci Res. 2007;85:2167–74.
75. Sehara Y, Hayashi T, Deguchi K, Zhang H, Tsuchiya A, Yamashita T, Lukic V, Nagai M, Kamiya T, Abe K. G-CSF enhances stem cell proliferation in rat hippocampus after transient middle cerebral artery occlusion. Neurosci Lett. 2007;418:248–52.
76. Sevimli S, Diederich K, Strecker JK, Schilling M, Klocke R, Nikol S, Kirsch F, Schneider A, Schabitz WR. Endogenous brain protection by granulocyte-colony stimulating factor after ischemic stroke. Exp Neurol. 2009;217:328–35.
77. Shimoda K, Okamura S, Harada N, Kondo S, Okamura T, Niho Y. Identification of a functional receptor for granulocyte colony-stimulating factor on platelets. J Clin Invest. 1993;91:1310–3.

78. Shyu WC, Lin SZ, Yang HI, Tzeng YS, Pang CY, Yen PS, Li H. Functional recovery of stroke rats induced by granulocyte colony-stimulating factor-stimulated stem cells. Circulation. 2004;110:1847–54.
79. Six I, Gasan G, Mura E, Bordet R. Beneficial effect of pharmacological mobilization of bone marrow in experimental cerebral ischemia. Eur J Pharmacol. 2003;458:327–8.
80. Solaroglu I, Tsubokawa T, Cahill J, Zhang JH. Anti-apoptotic effect of granulocyte-colony stimulating factor after focal cerebral ischemia in the rat. Neuroscience. 2006;143:965–74.
81. Stroemer RP, Kent TA, Hulsebosch CE. Neocortical neural sprouting, synaptogenesis, and behavioral recovery after neocortical infarction in rats. Stroke. 1995;26:2135–44.
82. Su Y, Cui L, Piao C, Li B, Zhao LR. The effects of hematopoietic growth factors on neurite outgrowth. PLoS ONE. 2013;8, e75562.
83. Taguchi A, Wen Z, Myojin K, Yoshihara T, Nakagomi T, Nakayama D, Tanaka H, Soma T, Stern DM, Naritomi H, Matsuyama T. Granulocyte colony-stimulating factor has a negative effect on stroke outcome in a murine model. Eur J Neurosci. 2007;26:126–33.
84. To LB, Bashford J, Durrant S, MacMillan J, Schwarer AP, Prince HM, Gibson J, Lewis I, Swart B, Marty J, Rawling T, Ashman L, Charles S, Cohen B. Successful mobilization of peripheral blood stem cells after addition of ancestim (stem cell factor) in patients who had failed a prior mobilization with filgrastim (granulocyte colony-stimulating factor) alone or with chemotherapy plus filgrastim. Bone Marrow Transplant. 2003;31:371–8.
85. Tombari D, Loubinoux I, Pariente J, Gerdelat A, Albucher JF, Tardy J, Cassol E, Chollet F. A longitudinal fMRI study: in recovering and then in clinically stable sub-cortical stroke patients. Neuroimage. 2004;23:827–39.
86. Toth ZE, Leker RR, Shahar T, Pastorino S, Szalayova I, Asemenew B, Key S, Parmelee A, Mayer B, Nemeth K, Bratincsak A, Mezey E. The combination of granulocyte colony-stimulating factor and stem cell factor significantly increases the number of bone marrow-derived endothelial cells in brains of mice following cerebral ischemia. Blood. 2008;111:5544–52.
87. Touzani O, Roussel S, MacKenzie ET. The ischaemic penumbra. Curr Opin Neurol. 2001;14:83–8.
88. Vellenga E, Rambaldi A, Ernst TJ, Ostapovicz D, Griffin JD. Independent regulation of M-CSF and G-CSF gene expression in human monocytes. Blood. 1988;71:1529–32.
89. Wang C, Curtis JE, Geissler EN, McCulloch EA, Minden MD. The expression of the proto-oncogene C-kit in the blast cells of acute myeloblastic leukemia. Leukemia. 1989;3:699–702.
90. Warlow C, Sudlow C, Dennis M, Wardlaw J, Sandercock P. Stroke. Lancet. 2003;362:1211–24.
91. Welte K, Platzer E, Lu L, Gabrilove JL, Levi E, Mertelsmann R, Moore MA. Purification and biochemical characterization of human pluripotent hematopoietic colony-stimulating factor. Proc Natl Acad Sci U S A. 1985;82:1526–30.
92. Wineman JP, Nishikawa S, Muller-Sieburg CE. Maintenance of high levels of pluripotent hematopoietic stem cells in vitro: effect of stromal cells and c-kit. Blood. 1993;81:365–72.
93. Yang G, Pan F, Gan WB. Stably maintained dendritic spines are associated with lifelong memories. Nature. 2009;462:920–4.
94. Zhao LR, Mattsson B, Johansson BB. Environmental influence on brain-derived neurotrophic factor messenger RNA expression after middle cerebral artery occlusion in spontaneously hypertensive rats. Neuroscience. 2000;97:177–84.
95. Zhao LR, Piao CS, Murikinati SR, Gonzalez-Toledo ME. The role of stem cell factor and granulocyte-colony stimulating factor in treatment of stroke. Recent Pat CNS Drug Discov. 2013;8:2–12.
96. Zhao LR, Singhal S, Duan WM, Mehta J, Kessler JA. Brain repair by hematopoietic growth factors in a rat model of stroke. Stroke. 2007;38:2584–91.
97. Zhao LR, Risedal A, Wojcik A, Hejzlar J, Johansson BB, Kokaia Z. Enriched environment influences brain-derived neurotrophic factor levels in rat forebrain after focal stroke. Neurosci Lett. 2001;305:169–72.

98. Zhao LR, Duan WM, Reyes M, Keene CD, Verfaillie CM, Low WC. Human bone marrow stem cells exhibit neural phenotypes and ameliorate neurological deficits after grafting into the ischemic brain of rats. Exp Neurol. 2002;174:11–20.
99. Zhao LR, Berra HH, Duan WM, Singhal S, Mehta J, Apkarian AV, Kessler JA. Beneficial effects of hematopoietic growth factor therapy in chronic ischemic stroke in rats. Stroke. 2007;38:2804–11.
100. Zhao LR, Navalitloha Y, Singhal S, Mehta J, Piao CS, Guo WP, Kessler JA, Groothuis DR. Hematopoietic growth factors pass through the blood-brain barrier in intact rats. Exp Neurol. 2007;204:569–73.
101. Zsebo KM, Yuschenkoff VN, Schiffer S, Chang D, McCall E, Dinarello CA, Brown MA, Altrock B, Bagby Jr GC. Vascular endothelial cells and granulopoiesis: interleukin-1 stimulates release of G-CSF and GM-CSF. Blood. 1988;71:99–103.
102. Zsebo KM, Wypych J, McNiece IK, Lu HS, Smith KA, Karkare SB, Sachdev RK, Yuschenkoff VN, Birkett NC, Williams LR, et al. Identification, purification, and biological characterization of hematopoietic stem cell factor from buffalo rat liver – conditioned medium. Cell. 1990;63:195–201.
103. Zsebo KM, Williams DA, Geissler EN, Broudy VC, Martin FH, Atkins HL, Hsu RY, Birkett NC, Okino KH, Murdock DC, et al. Stem cell factor is encoded at the Sl locus of the mouse and is the ligand for the c-kit tyrosine kinase receptor. Cell. 1990;63:213–24.

Chapter 14
Imaging and Tracking Transplanted Bone Marrow Stem Cells After Stroke

Hongbin Han, Weifang Zhang, Lan Yuan, Junhao Yan, Wei Wang, Chunyan Shi, and Yunqian Li

Abstract Stem cells provide a promising therapy for treating stroke. Several imaging techniques provide the possibility of the visual representation of biological processes of transplanted stem cells in vivo. Further means of monitoring the transplanted cells safely, noninvasively, and longitudinally would contribute to the understanding of the underlying stem cell therapeutic mechanism. In this chapter, we describe the state-of-the-art methods of monitoring transplanted stem cells in vivo, including magnetic resonance imaging (MRI), radionuclide imaging, and optical imaging. Their principles vary and each has advantages and drawbacks. Presently, no single technique is excellent through ideal criteria; the combination of multiple imaging modalities is thus an attractive strategy.

H. Han (✉) • W. Wang
Department of Radiology, Peking University Third Hospital,
49 North Garden Road, Haidian District, Beijing 100191, China

Beijing Key Laboratory of Magnetic Resonance Imaging Device and Technique,
Beijing 100191, China
e-mail: hanhongbin@bjmu.edu.cn; medimawei@bjmu.edu.cn

W. Zhang
Nuclear Medicine Department, Peking University Third Hospital,
49 Huayuanbei Road, Haidian District, Beijing, China, 100083
e-mail: 13681358745@126.com

L. Yuan • Y. Li
Medical and Health Analysis Center, Peking University,
38 Xueyuan Road, Haidian District, Beijing, China, 100083
e-mail: yuan_lan@bjmu.edu.cn; 806900948@qq.com

J. Yan
Department of Anatomy and Embryology, School of Basic Medical Sciences, Peking University, 38 Xueyuan Road, Haidian District, Beijing, China, 100083
e-mail: yjh@bjmu.edu.cn

C. Shi
Department of Radiology, Beijing An Zhen Hospital, Capital Medical University, Beijing Institute of Heart, Lung and Blood Vessel Disease,
Anzhen Road 2th, Chaoyang district, Beijing, China
e-mail: scy0518@126.com

K. Jin et al. (eds.), *Bone Marrow Stem Cell Therapy for Stroke*,
DOI 10.1007/978-981-10-2929-5_14

Keywords Stem cells • Stroke • Imaging • Magnetic resonance imaging • Radionuclide imaging • Optical imaging

14.1 Introduction

In spite of considerable advances in stroke management, current acute treatments have only mild effects and moderately restore its lost function [1, 2]. Therefore, stroke remains a major cause of disability and needs much more efficient therapeutic approaches. Cell transplantation is a novel treatment method in many fields of medicine, such as the cell replacement in the ischemic region to prevent further stroke disability [3]. In the last decade, evidence of neurogenesis in the adult human brain has provided the basic scientific hypothesis of cell transplantation therapy for various neurological disorders, including Parkinson's disease, multiple sclerosis, and stroke. Initial animal and experimental studies have identified a significant benefit of stem cell transplantation in neuroregeneration and improvement in neural function [4, 5]. Stroke is one of the most severe neurological disorder, which has been selected for a pioneering trial in the clinical application of stem cells [6, 7]. Furthermore, several studies demonstrated the feasibility of stem cell-based therapy for the restoration of lost brain function and improvement of the clinical outcome in stroke patients [2]. Several experimental and clinical researches have introduced different types of stem cells for transplantation in stroke [8, 9]. Moreover, clarification of optimal cell dosing, route of transplantation, cell delivery methods, and in vivo cell imaging techniques is needed to ensure safety, efficacy, expected outcome, and more success of potential stem cell transplantation trials [10].

Nowadays, radionuclide, fluorescence, and MRI are suitable candidates for human nervous system cellular imaging [11–13]. Among them, positron emission tomography (PET) is more sensitive to low concentrations of contrast agents. However, it has some limitations such as low spatial resolution, radiation exposure, and short-term signal production. Another technique is fluorescence, which is a sensitive method with some distinct advantages in small animal models, but it is not feasible for human whole-body visualization because of the limited penetration depth and low spatial resolution [12]. Although a high spatial resolution can be provided by other methods like micro-CT, this technique is not always suitable for in vivo human studies, and it needs to be optimized for better cell detection throughout the whole body. With respect to the full commitment of clinical studies and trials to patients' safety, radiation and radioactive exposures are important limitations of CT, PET, and single-photon emission computed tomography (SPECT). Therefore, MRI is a superior method for cell tracking and imaging. Several stem cell tracking studies have been performed using MRI, and the most popular area is neurological diseases (Table 14.1) [14].

Table 14.1 Comparison of several in vivo cell imaging modalities

Modality	MRI	Radionuclide	Fluorescence
Radiation injury	No	Yes	No
Acquisition time	Microseconds to hours	Minutes	Minutes
Depth	No limit	No limit	<1 cm
Resolution	10–100 μm	1–2 mm	2–3 mm
Longitudinal cell tracking	+++	+	+++
Sensitivity	+++	+++	+
Clinical application	Yes	Yes	No
Quantification of cell number	Possible	Yes	No
Assessment of cell viability	Possible	Yes	Yes

Reprinted and adapted with permission from J Res Med Sci. Ref. [34]

14.2 Magnetic Resonance Imaging (MRI)

Hongbin Han, Wei Wang and Chunyan Shi

MRI is a promising noninvasive in vivo stem cell tracking technique, and its primary advantage is the safety and clinical applicability without radiation and radioactive exposures. Furthermore, the living environment of brain cells can be quantitatively measured using a novel tracer-based MRI [15, 16]. Therefore, it is of tremendous value to dynamically monitor transplantation, migration, and therapeutic response over a long period of time. Based on the radio wave, MRI has no limit to tissue depth penetration. MRI can display three-dimensional information of the brains; meanwhile, arbitrary section images can be obtained. MRI can provide detailed information on host organs with superior soft tissue contrast and excellent spatial resolution that can reach below 50 μm and even single-cell detection under ideal conditions [17]. MRI also can provide multiparameter information, which are associated with relaxation time, proton density, flow, or biochemistry.

14.2.1 Concepts and Principles

MRI has been used as a medical diagnostic tool for more than 30 years, and the first image was published in 1973 by Lauterbur who received the 2003 Nobel Prize in Physiology or Medicine [18]. The principle of MRI is based on radio frequency emitted when selective atomic nuclei are placed in an external magnetic field, in which hydrogen atoms are most commonly used. The radio-frequency signal can be detected by the extracorporeal coils. A variety of parameters of the body's tissue can then be calculated and images are generated. Presently, two strategies are reported for tracking stem cell using MRI. Firstly, stem cells can be tracked using exogenous metal-based compounds and endogenous genetic-based agents, both of which can modify the relaxation time of radiowaves of the surrounding hydrogen protons.

Secondly, stem cells can also be labeled with selective atoms such as fluorine-19, which emits a characteristic radio frequency. The modified radio-frequency signals can be detected by MR and generate contrast in images; therefore, stem cells can be distinguished from the background once implanted in the brain.

14.2.2 Methods and Tracers

Metallic tracers are primarily developed among cell tracking strategies. The first studies of MR cell tracking in the brain were reported in 1992, in which transplanted cells were labeled and tracked using superparamagnetic iron oxide (SPIO) [19]. Over the years, a variety of metal-based tracers have been developed and have more efficiency in labeling and tracking the implanted cells. According to the effect on relaxation time or signal density, these metal-based tracers can be classified into positive and negative groups. Positive tracers can decrease the longitudinal/T_1 relaxation time and increase the MRI signal density on the T_1-weighted image. Presently, positive tracers mainly involve gadolinium (Gd^{3+}) and manganese (Mn^{3+}) chelates, while negative tracers can decrease the transverse/T_2 relaxation time and the MRI signal density on the T_1- or T_2^*-weighted image. Negative tracers are primarily superparamagnetic iron nanoparticles (SPIOs).

14.2.3 Application and Limitation

14.2.3.1 Positive Tracers

Positive tracers are paramagnetic agents, which have one or more unpaired electrons. Gd^{3+} chelates are the most effective paramagnetic agents, owing to their seven unpaired electrons. Gadolinium-diethylenetriaminepentaacetic acid (Gd-DTPA) is the most common contrast agent in clinical application. Tracers for stem cell labeling include Gd-chelated dextran nanoparticles and Gd-DOTA-peptide complexes and Gadolinium (Gd^{3+})-doped mesoporous silica nanoparticles [20]. However, the problem using gadolinium to track stem cells is the risk of toxicity on cells and tissues. Free ionic gadolinium is highly toxic; Gd-DTPA, an FDA-approved agent, can elicit nephrogenic systemic fibrosis [21]. Presently, because of the low sensitivity of Gd-based agents and low uptake of Gd^{3+} chelates by targeted cells, enough Gd^{3+} chelates are necessary for labeling cells to generate efficient contrast on images, which may increase the risk of toxicity to cells and host tissue [18]. For example, in a rat stroke model, transplants of stem cells labeled with gadolinium-rhodamine dextran resulted in a significant increase in lesion size compared to the groups with fluorescent dye labeling [22]. The exact mechanism of gadolinium toxicity is not known, but a possibility is that free Gd^{3+} can influence cell metabolism of the labeled cells through distorting the activity of K^+/Cl^- membrane cotransporter [23]. It is also

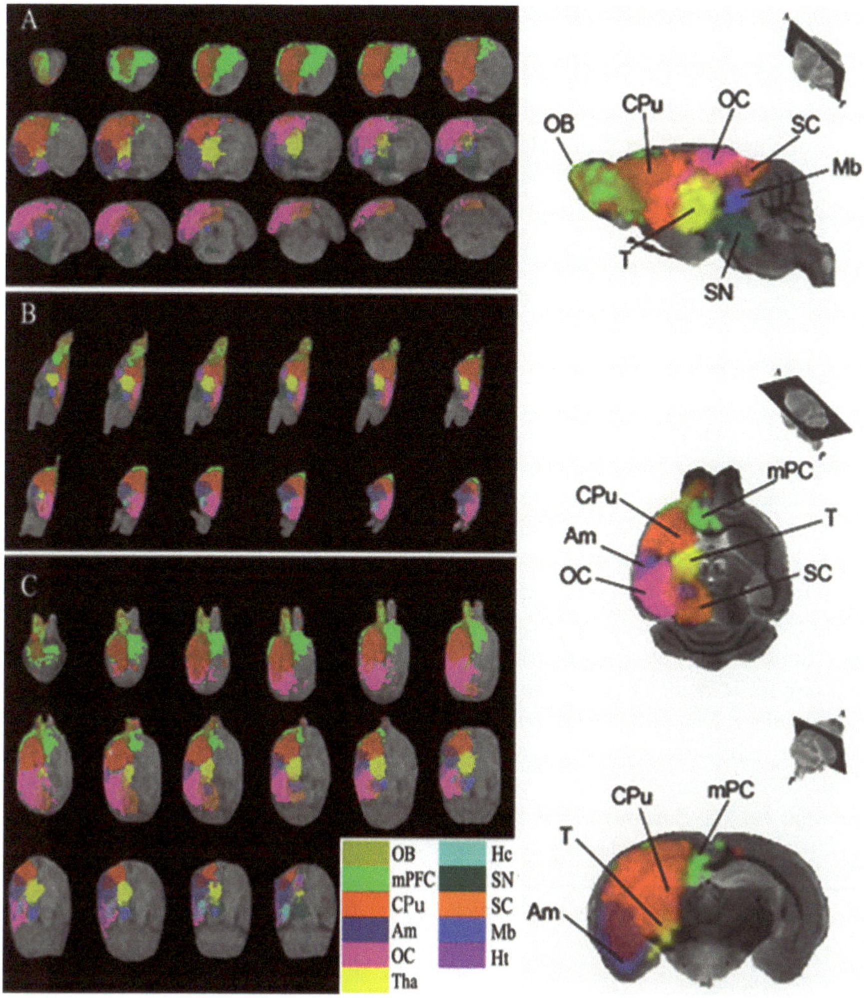

Fig. 14.1 Color map of the brain extracellular space divisions based on the brain interstitial space flow. Each colorized area represents the $Volume_{max}$ of the traced interstitial space flow with 11 different start points: coronal (**a**), sagittal (**b**), and transverse views (**c**) (area 1, *OB* olfactory bulb; area 2, *mPFC*, medial prefrontal cortex; area 3, *Cn*, caudate putamen; area 4, *Am*, amygdaloid nucleus; area 5, *Oc*, occipital cortex; area 6, *Tha*, thalamus; area 7, *Hc*, hippocampus; area 8, *SN*, substantia nigra; area 9, *SC*, superior colliculus; area 10, *Mb*, midbrain; area 11, *Ht*, hypothalamus)

unclear how these Gd^{3+} are metabolized in labeled cells. Recently, chelated gadolinium was employed to trace the interstitial or extracellular space in the deep brain [15] where in it provides the living environment of both the host and the implanted brain cells (Fig. 14.1). The new discovery of transportation barrier in the brain interstitial space has great implications for the site of implantation [16, 24].

Manganese is another paramagnetic tracer for positive T_1 contrast of labeled cells, owing to their five unpaired electrons. It was used as the first MR contrast agent in 1978. The behavior and metabolism of Mn^{2+} ions in the cell is similar with those of calcium (Ca^{2+}) ions and both enter cells through ligand- or voltage-gated Ca^{2+} ion channels. Therefore, Mn^{2+}-enhanced MRI has been performed to investigate neuronal activity and to visual neuronal connectivity. However, regarding the potential cytotoxicity, a few reports exist concerning the use of manganese-based agents to stem cell tracking. Silica-coated MnO nanoparticles have been developed for the labeling and MR tracking of mesenchymal stem cells [18].

14.2.3.2 Negative Tracers

Negative tracers are superparamagnetic agents. They can cause magnetic resonance signal dephasing related to local magnetic field inhomogeneity induced in water molecules near superparamagnetic particles (susceptibility effect) and shorten transverse relaxation times of hydrogen protons. As a result, labeled stem cells produce dark spots in T_1- or T_2^*-weighted images. Superparamagnetic iron oxide (SPIO) particles with various diameters are the most common negative tracers. SPIO is typically composed of a Fe_3O_4 or Fe_2O_3 core with varying chemical coatings, such as dextran, glycosaminoglycans, starch, polyethylene glycol, siloxane, and polylactic acid. They can be categorized according to size. Ultrasmall superparamagnetic iron oxide particles (USPIO) measure less than 50 nm in diameter. SPIO particle size varies from 50 to 200 nm, and micronized paramagnetic iron oxide (MPIO) particles measure more than 1 μm in diameter. SPIOs were first reported for the purpose of labeling and tracking transplanted rat brain cells in 1992–1993 and demonstrating cell migration in vivo in 2001. Among them, ferumoxide, an iron oxide nanoparticle coated with dextran, was approved by the FDA in 1995 as a MRI enhancement contrast agent for the clinical diagnosis of liver lesions. Since then, ferumoxides have been widely applied in cell labeling, although it is no longer being made due to the lack of sales. Presently, no SPIO has gained FDA approval for labeling stem cells. Other metal-based negative tracers such as bimetallic ferrite nanoparticles (e.g., $CoFe_2O_4$, $MnFe_2O_4$, and $NiFe_2O_4$), hybrid magnetic nanoparticles, and iron-platinum-based nanoparticles have the potential for high biocompatibility, sensitivity, and detectability, but are still in the early trial stages and only utilized in the lab.

Advantages of SPIO over conventional paramagnetic tracers are as follows:

Firstly, SPIO has a much stronger MR relaxivity and higher sensitivity. Gadolinium chelates can only be detected in the micromolar range, while SPIO is sensitive in the nanomolar range for visualizing single-cell tracking and is suitable for molecular imaging [25].

Secondly, the concentration level of SPIO for labeling stem cells has negligible cytotoxicity. Transfection agents have been used in conjunction with SPIO for labeling stem cells, which may be toxic at certain levels either through cell death or prevention of cell division. Currently, some alternative coatings have been developed to facilitate the administration of SPIO to cells without the use of a transfection agent, which significantly decreases the local toxicity to cells due to iron crossing both nuclear and mitochondrial membranes [26].

Thirdly, iron-containing SPIO shows an acceptable level of biocompatibility in part due to the body's innate ability to metabolize naturally occurring iron in the form of ferritin [27].

While SPIO is the most common tracer for magnetic cell labeling, there are some limitations:

Firstly, SPIO can produce the blooming effect, that is, "dark spots" in images are beyond the volume occupied by SPIO. The effect is helpful in increasing the visibility of the labeled stem cells however, it interferes with visualizing host anatomical structures and provides false locations of the labeled cells [28].

Secondly, with stem cell division and proliferation, the ratio of cells labeled with SPIO decreases. As a result, longitudinal tracking will result in gradual signal loss [29].

Thirdly, iron released from apoptotic or lytic cells can be internalized by macrophages in nearby tissues, resulting in signal not associated with the target cells [30].

Fourthly, besides SPIO, many factors, including bleeds, air, and hemorrhage, can present false-positive hypointensities on T_1 and T_2^* images [17].

Fifthly, SPIO has limited capability in reporting cell functionality and viability and is insufficient to attain comprehensive information of stem cell post transplantation. Histology has shown that hypointensity persists even after labeled cell death. This highlights the necessity of some external modality to confirm that the labeled cells are viable and unimpaired [28].

Sixthly, quantifying labeled cells is challenging. In the relaxometric-based experiments, relaxivity changes depend on physical properties of internalized SPIO during degradation. Signal intensity is not directly proportional to the concentration of SPIO and cannot be related back to cell numbers.

Lastly, SPIOs are generally considered as non-toxic. However, due to the complexity and variety of components of SPIOs, their toxicity profiles are still controversial. Ramaswamy reported successful labeling of mesenchymal stem cells (MSCs) using Feridex® without impairing cell chondrogenesis differentiation, while Bulte reported that chondrogenic differentiation of MSCs is inhibited after labeling with Feridex® [29].

14.2.4 Conclusion and Future Perspectives

Safety is paramount in the implementation of tracers for tracking stem cells. The ideal magnetic tracer must be nontoxic to cells, biodegradable, effective in low doses, chemically stable in the body, not affect stem cell function, must not transfer harmful by-products nor affect stem cell function [27]. Therefore, it is necessary to fully characterize tracer effects on stem cells (i.e., cytotoxicity), including viability, differentiation, migration/homing, distribution, and engraftment [29]. Cytotoxicity is related to the potential for organismal migration, penetration, and accumulation of metal. There are several developing techniques for eliminating cytotoxicity. Firstly, the use of a tracer based on ^{19}F rather than metal-based tracers, allows for quantitative cell tracking. In ^{19}F MR images, the background signal is lacking and an accurate, unambiguous detection of labeled cells can be achieved. The relationship between the concentration of ^{19}F and signal intensity is directly proportional and linear, and the labeled cell numbers can be quantified directly from the acquired images. Unfortunately, ^{19}F-based tracking MRI is poorly sensitive, and the minimum amount of cells that can be detected ranges from 10^3 to 10^5 cells [18]. Secondly, developing a high relaxivity tracer is necessary to achieve high cellular magnetization, so that less metal is required to obtain the same contrast. Indeed, given sufficient magnetization, in vivo single-cell detection is possible, while tracers in very low doses may have negligible side effects [31].

An ideal magnetic tracer should allow for longitudinal tracking of implanted stem cells for follow-up of its viability and function. Longitudinal tracking using a tracer is necessary for studying the efficiency of stem cell therapy. Conventional magnetic tracers are unsuitable for longitudinal tracking due to signal loss resulting from stem cell division and proliferation. Recently, novel methods using reporter genes are reported for use in longitudinal tracking. Reporter genes are DNA sequences that are introduced to the stem cell genome. The reporter genes encode reporter proteins like ferritin and tyrosinase. These reporter proteins accumulate intracellular iron, which allows for a paramagnetic effect to be detected on MRI. As the reporter genes are stably expressed in the host cell genome, the tracer will not be lost or diluted even after cell division. Furthermore, the signal emitted from reporter genes is based on the viability of the host cells. Thus, reporter genes can be used simultaneously for longitudinal tracking and viability detection [18, 20, 32–34].

The ideal magnetic tracer-based imaging for stroke should also be characterized as follows: Use of imaging in clinical trials is strongly encouraged to provide as much information as possible to assess vascular/structural lesions, infarct size, cell viability, location, the success and safety of implantation, and inflammation. Imaging should also be used to monitor safety and recovery and, when possible, investigate mechanisms of action and provide information on surrogate markers of treatment effect. Imaging measures might also be useful to help stratify patients at baseline [35]. Because no single technique fufills all these criteria, the combination of multiple imaging modalities is an attractive strategy. Multimodal imaging can exploit the strengths and overcome limitations and has been developed rapidly over the last decade [26]. Gadolinium and SPIO may be combined with the fluorescent compound allowing their detection by histology in vitro and by MRI in vivo [36, 37]. PET for high sensitivity and specificity coupled with MRI for high spatial and temporal resolution has been applied for cell tracking following ischemic stroke [17]. Magnetic nanoparticles can be conjugated to identify molecular sensors that detect stimuli associated with cell viability and functions, such as chemicals secreted during cell differentiation, physical contact with neighboring cells during stem cell engraftment, intercellular pH changes during cell death, and certain molecules related with stem cell differentiation. The interaction between the sensors and the stimuli generates detectable changes of signal in the magnetic nanoparticle that can be captured by MRI [29, 38].

14.3 Radionuclide Imaging

Weifang Zhang

Among the different imaging techniques available, nuclear medicine has become one of the most employed techniques, due to its favorable characteristics, such as the availability of different radiopharmaceuticals, high sensitivity, good tissue penetration, and translation to clinical applications [39]. Because of their exquisite picomolar (10^{-11} to 10^{-12} mol/l) sensitivity, radionuclide imaging modalities are able to detect trace quantities of radioisotopes for studying biological processes in living subjects. Radiopharmaceutical cell labeling has been used for decades to systemically monitor cells in nuclear medicine studies, such as labeled leukocyte scintigraphy for the detection of infectious and inflammatory diseases. Besides its high sensitivity, radionuclide imaging has several advantages. It is highly quantitative, which means it is dynamic and kinetic, observing the biodistribution of implanted cells in vivo. In fact, radionuclide imaging is the sole direct labeling technique used thus far in human studies, involving both autologous bone marrow-derived stem cells and peripheral hematopoietic stem cells. Technological developments of both PET and SPECT have further facilitated the implementation of specialized systems for small animal imaging with much greater spatial resolution (1–2 mm) to emerge as one of the most powerful tools and dramatically advance the field of in vivo cell tracking.

14.3.1 Concepts and Principles

A variety of labeling methods with radionuclides have been created and used to study cell distribution in the body. Technetium-99m (^{99m}Tc) is currently the most used radionuclide in the world due to favorable properties, such as its decay by gamma emission with an energy of 140 keV and a 6 h half-life, optimum physical characteristics for SPECT, allowing images for up to 24 h after injection. ^{99m}Tc has wider availability and lower cost compared with other radionuclides, with good resolution and low radiation dose to the patient and to the labeled cells [40]. Radionuclide indium-111 (^{111}In) may also be used for cell labeling in SPECT, which allows cell tracking for up to 96 h but results in lower-resolution images and a higher radiation dose to the patient and labeled cells. In addition, different studies have indicated that Auger electrons of ^{111}In-oxine labeling affect cellular integrity and lead to cytotoxicity of stem cells.

The radionuclide fluorine-18 (^{18}F) has a half-life of approximately 110 min and is the most commonly used in positron emission tomography (PET) and hybrid PET/CT, mainly in the radiopharmaceutical ^{18}F-fluorodeoxyglucose (^{18}F-FDG). ^{18}F-FDG allows cell labeling and tracking for a few hours. ^{18}F can also be incorporated into a modified thymidine analog, 3′-deoxy-3′-^{18}F-fluorothymidine (^{18}F-FLT), which is phosphorylated by thymidine kinase but not incorporated into DNA, thus sequestering ^{18}F-FLT within the cell. ^{18}F-FLT has been used to visualize neural stem cell mobilization in the mouse hippocampus. Unlike SPECT scans, PET has a two- to threefold higher spatial resolution than SPECT (3–6 mm versus 10–15 mm) and allows quantification of standardized uptake values, which may be used to compare the response to different therapies. Zirconium-89 and Cu-64 are other promising radionuclides for cell labeling in PET that possess longer half-lives (78.4 h and 12.7 h) and may allow cell tracking for several days to weeks.

Tracking cells with SPECT and PET may be separated to two strategies: direct and indirect. Direct tracking is that stem cells are labelled with a radiotracer in vitro prior to transplantation in vivo, while indirect. Indirect cell tracking may be carried out via reporter gene/probe systems.

14.3.2 *Methods and Tracers*

In general, radioisotopes with a relatively long decay half-life are used to track cells for a period of several hours or even days. These can be done with radiopharmaceuticals such as ^{99m}Tc-hexamethylpropyleneamine oxime (^{99m}Tc-HMPAO) or ^{111}In-oxine for SPECT and ^{18}F-FDG for PET. The isotope is carried into the cells via a lipophilic chelator, which governs the initial extraction of the tracer into the cells. Once inside the cells, a trapping mechanism reduces the lipophilicity of the molecule, and the isotope is retained. After a short incubation period, the cells are washed to remove any unbound activity and are injected into the host.

Both SPECT and PET imagings offer visualization of radiolabeled cells or tissue structures with high spatial resolution. However, because of the short half-lives of most radioisotopes used, one cannot track the long-term survival and fate of these cells. Another major concern is that direct radionuclide labeling cannot accurately distinguish live from apoptotic cells because radioisotopes will remain active even after cell death or may leak from dead cells into surrounding non target cells.

The concept of reporter gene for PET/SPECT detection involves the construction of a deoxyribonucleic acid sequence coding a specific reporter. A vector delivery system is used to carry the reporter gene into cells of interest. The expression of the reporter gene product in the cells, when exposed to the corresponding radiolabeled reporter probes, combined specifically with the probes, “turns on” the imaging signal in the probe and entraps the activated probe inside the cells. Finally, imaging and recording of the signal are achieved by use of appropriate devices, such as PET, SPECT, or multimodal device [41, 42]. Because accumulation of the reporter probe requires expression of the reporter gene and activity of the reporter-gene product, the imaging signal will be dependent on viability of the therapeutic cells. This approach could track transplanted cells in the human body as long as they remain alive and would also be suitable for assessing the proliferative capacity of the transplanted cells. This is in contrast with direct labeling techniques and provides a more specific readout. Additionally, imaging can be performed repeatedly and is not limited by radioactive decay of the initial label load.

Reporter gene/probe systems have traditionally been divided in three groups, according to the way that the protein product of the reporter gene interacts with the reporter probe and causes its accumulation on the surface or inside the cells: (1) reporter genes that encode enzymes to phosphorylate specific reporter probes leading to their entrapment; (2) reporter genes that encode protein receptors, which in turn bind to specific reporter probes; and (3) reporter genes that encode cell membrane transporters that accelerate the accumulation of reporter probes in the cells. The best known and most widely used reporter gene for both PET and SPECT imaging is the herpes simplex virus type 1 thymidine kinase gene (*HSV1-tk*). A lentivirus may be used to deliver a reporter gene for the expression of herpes simplex virus-truncated TK that catalyzes a reaction leading to the accumulation of the probe. Herpes thymidine kinase can phosphorylate two main classes of exogenously administered reporter probes: pyrimidine analog derivatives, such as 2′-fluoro-2′-deoxy-β-D-arabinofuranosyl-5-iodouracil (FIAU), and acycloguanosine derivatives, such as 9-(4-fluoro-3-hydroxymethyl-butyl)guanine (FHBG). When herpes thymidine kinase phosphorylates its reporter probe, not only is the probe retained within viable cells but signal amplification also occurs when the enzyme reacts with multiple substrates. To facilitate either PET or SPECT imaging, reporter probes can be synthesized with the appropriate isotope e.g., positron-emitting fluorine-18 and iodine-124 or gamma-emitting iodine-123, iodine-125, and iodine-131 [43].

Although reporter gene imaging has become very widespread in preclinical studies, few studies have shown a proof-of-principle application for cell therapy in the human [44] (Fig. 14.2). Reporter gene imaging requires genetic modification of transplanted cells, which increases the regulatory complexity for their approval and poses additional risks for mutagenesis. Furthermore, the radioactive tracers would expose the patient to higher radiation doses. However, in the case of MSCs, it has been shown that free radical scavenging and DNA repair mechanisms are particularly robust, allowing for higher radiotolerance.

14.3.3 Application and Limitation

Among the various methods available, radioisotope cell labeling has become one of the most promising since it permits tracking of cells after injection by different routes to investigate their biodistribution for neurological diseases.

Studies using radiopharmaceuticals to track the fate of transplanted cells have given important clues about the mechanisms of action of cell-based therapies for neurological diseases. In many of these studies, the transplanted cells exerted therapeutic

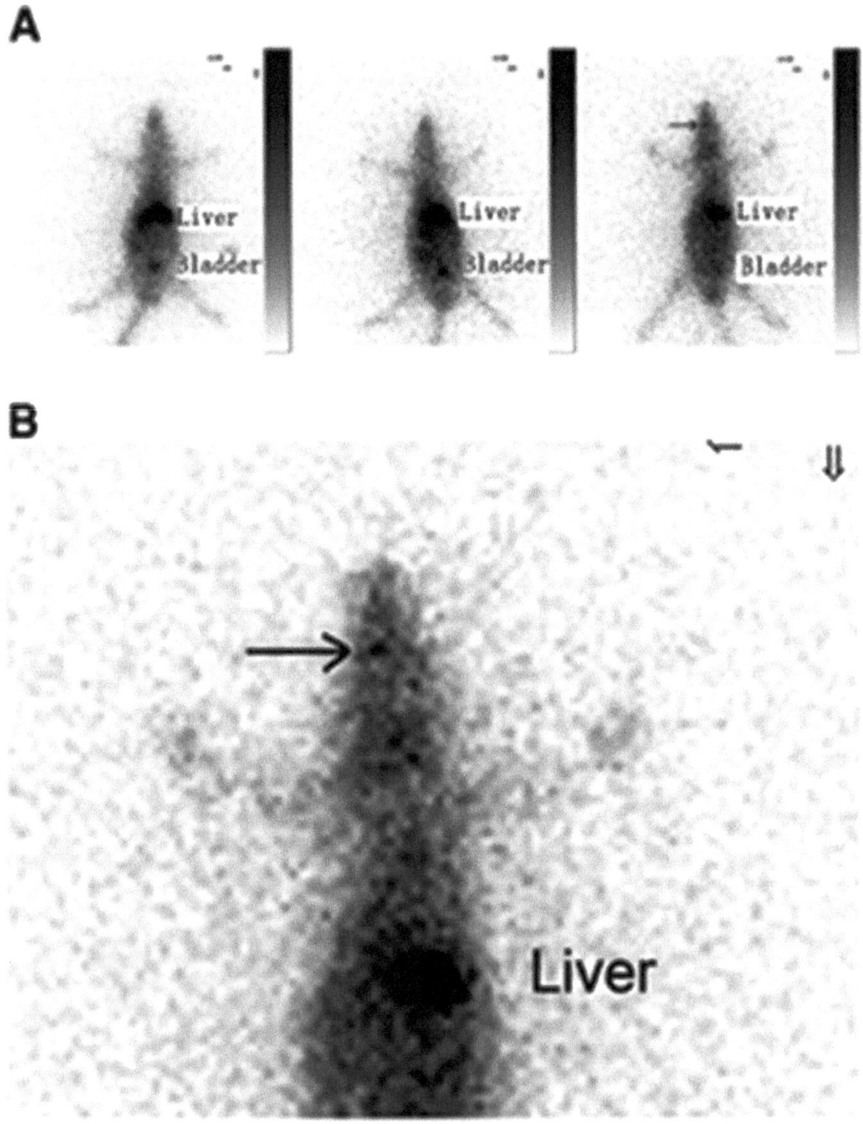

Fig. 14.2 SPECT HSV1-tk-^{131}I-FIAU reporter gene images of experimental middle cerebral artery occlusion rat models with transplanted *BMSCs*. **a** Liver and bladder were clearly imaged. Level of accumulation in brain (*arrow*) was low but increased gradually over time. **b** Local enlargement of A. Greater radioactivity accumulation was seen at cell injection site (*arrow*) (Reprinted and adapted with permission from J Nucl Med. Ref. [44])

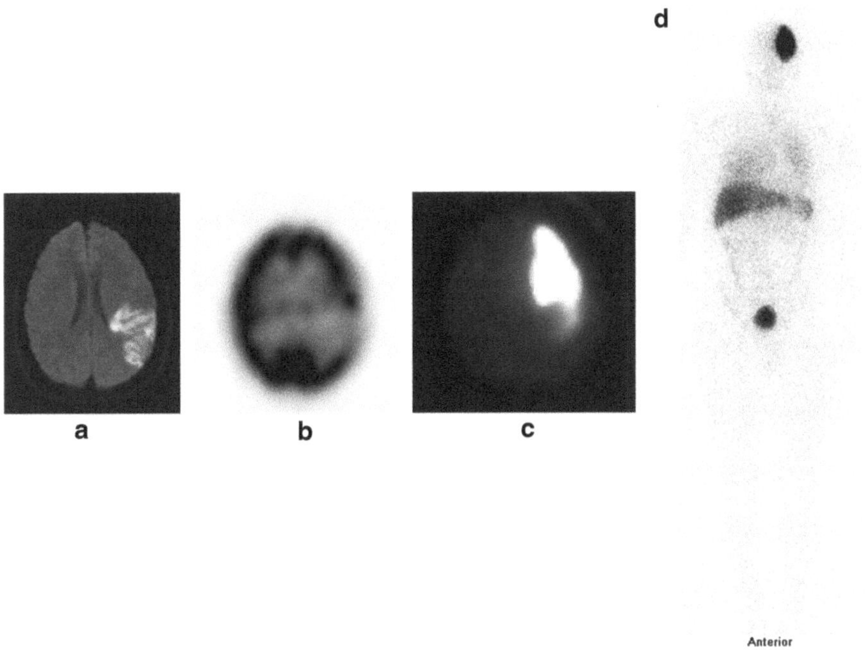

Fig. 14.3 The accumulation of radiopharmaceutically labeled stem cells in stroke. **a** Diffusion MRI showing acute ischemic lesion in the left middle cerebral artery territory. **b** Brain perfusion Tc-99m ECD SPECT showing left temporoparietal hypoperfusion. **c** Tc-99m HMPAO autologous mononuclear bone marrow cell (ABMMN) brain SPECT revealing accumulation of the ABMMN-labeled cells in the left brain hemisphere, more intense in the anterior region. **d** Anterior whole-body scan with Tc-99m HMPAO ABMMN cells revealed left brain, liver, and spleen uptake. There was no lung uptake (Reprinted and adapted with permission from Clin Nucl Med. Ref. [45])

effects by limiting tissue damage and/or by stimulating regeneration and plasticity of the diseased central nervous system. In recent years, bone marrow mononuclear cells (BM-MNCs) remain the primary source of stem cell therapy for treating neurological diseases. ^{111}In-oxine, ^{99m}Tc or ^{99m}Tc-HMPAO, ^{131}I-FIAU, ^{18}F-FHBG, and ^{18}F-FDG are mainly used in several preclinical and clinical trials to track transplantated cells for different neurological diseases, including spinal cord and brain injury, transient cerebral artery occlusion, cerebral ischemia, and ischemic stroke. Research shows that transplantation of radiopharmaceutically labeled stem cells led to accumulation in the site of injection in animal models of spinal cord or brain injury. In contrast, intravenous and intra-arterial injection led to little homing to the brain or spinal cord, and biodistribution was mainly to the liver, spleen, lungs, and kidneys [45] (Fig. 14.3). Although the number of transplanted cells that accumulated at the lesion site was very low in most cases, recent studies have clearly shown that BMSCs may promote functional recovery after various kinds of central nervous system disorders, including ischemic stroke. Using an ^{18}F-FDG, PET/CT apparatus clearly demonstrates that the BMSCs not only enhance functional recovery but also promote the recovery of glucose utilization in the peri-infarct area after ischemic stroke [46] (Fig. 14.4).

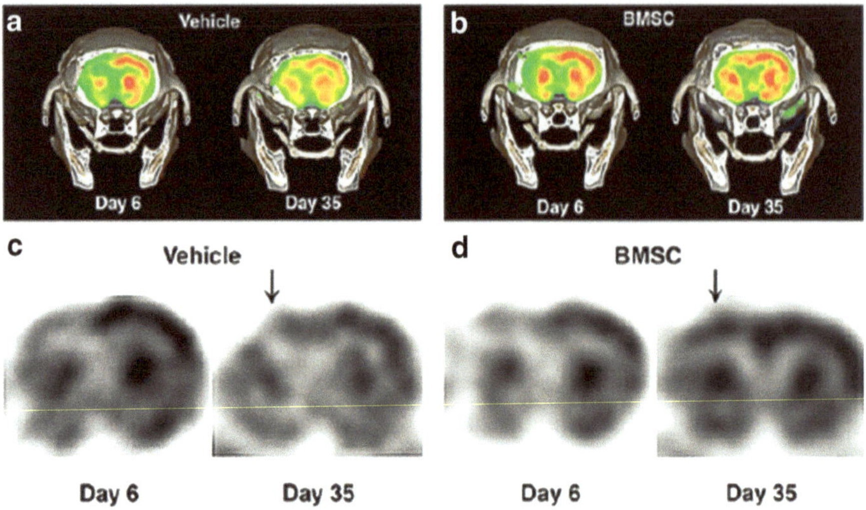

Fig. 14.4 Representative findings of 18F-FDG PET at 6 and 35 days after ischemia. Color (**a**) and *black*-and-*white* (**c**) images of vehicle transplanted animals. Color (**b**) and *black*-and-*white* (**d**) images of *BMSC*-transplanted rats. There was significant increase in local glucose metabolism in the peri-infarct neocortex (*arrows*) (Reprinted and adapted with permission from J Nucl Med. Ref. [46])

Nuclear techniques, including reporter genes and direct cellular radio labeling, afford very good detectability but more limited spatial resolution. Rapid advances in imaging modalities have now resulted in the development of fused imaging systems, such as PET/CT, SPECT/CT, PET/MRI, and even dual-mode systems e.g., MRI/SPECT (PET), improving both sensitivity and specificity significantly. Multimodal imaging is expected to achieve more accurate information of the transplanted cells. In addition to a more precise localization of the homing site, the fusion of nuclear medicine images with CT or MRI allows evaluation of different aspects, such as (1) correlation of cell homing with positive morphological and functional effects, (2) evaluation of adverse reactions including brain hemorrhage or formation of tumors, and (3) the effect of different doses and routes of injection on cell migration and proliferation [47].

Cicchetti et al. combined labeling subventricular zone- (SVZ-) derived neural stem/progenitor cells with SPIOs to evaluate their migration with MRI and simultaneously utilized different radiotracers to analyze physiological aspects with PET [48]. Tang et al. synthesized an MRI/SPECT/fluorescent trimodal probe (125I-fSiO4@SPIOs) for quantitatively tracking MSCs transplanted into stroke rats. They labeled the probe with SPIOs and radioisotope simultaneously, which allowed them to track the labeled MSCs in vivo with high spatial resolution and anatomical localization by MRI and high sensitivity by SPECT. The study demonstrates that 125I-fSiO4@SPIOs are robust probes for long-term tracking of MSCs in the treatment of ischemic stroke [49].

14.3.4 Conclusion and Future Perspectives

Methods for cell tracking with radioisotopes are feasible and efficient, and different studies have used it to monitor migration in cell therapies for neurological diseases. These techniques provide validated quantifications of cell retention in different organs and the dynamics of cell distribution in the whole body. The potential value of the information that radionuclide imaging could provide may justify the use of this technique in the future. The combination of different imaging modalities allows for cell tracking in conjunction with assessing cell metabolism non-invasively, and such methods have the potential of answering important questions in the field of stem cell therapy. PET/MRI will likely emerge as the imaging modality of choice given that it provides exquisite functional and anatomical detail with minimal radiation exposure.

14.4 Optical Imaging

Lan Yuan, Junhao Yan and Yunqian Li

Optical imaging is an easy, fast, and inexpensive tool for monitoring cell transplantation therapy in vivo. Optical imaging encompasses a variety of cell imaging modalities including fluorescence imaging and bioluminescence imaging (BLI). All of these methods involve the detection of photons emitted either by chemical oxidative processes or by external excitation of a fluorophore [17]. The technique is usually based on ex vivo labeling of the donor cells with a fluorescent dye [50]. Compared with other cell imaging technologies, the advantages of fluorescent optical imaging originate from its non-invasiveness, real-time in vivo imaging and high repeatability, sensibility, and security.

14.4.1 Concepts and Principles

Optical imaging of stem cell tracking is a cell imaging technology that introduces an exogenous fluorescence probe; a fluorescence microscope can then be used to trace the implantation, migration, and survival of stem cells and evaluate the clinical effects. Aptamers are single-stranded RNA or DNA oligonucleotides usually 15–60 bases in length that bind specifically to target molecules. Typically, aptamers can be generated from a selection process termed as systematic evolution of ligands by exponential enrichment (SELEX) [51, 52]. Aptamer labeling can alternatively be applied to monitor the differentiation process of stem cells, as shown by Iwagawa et al. [53]. Their selected aptamers showed specific binding affinity to mouse embryonic stem cells (mESCs), with low affinity toward differentiated cell lines. During the course of retinoic acid (RA) differentiation, multiple injections of the aptamer showed declining cell affinity with a progressive decrease in the fluorescent signal. Applying the same principle, stem cells can potentially be incubated and tagged with similar aptamer probes prior to their injection to facilitate colorimetric evaluation of successful cell differentiation [54]. Through external excitation, the fluorescent light group emits the fluorescence after reaching its high energy state [55]. The intensity of fluorescence can reflect the signal strength of stem cells, which is determined by the transplanted stem cells in the body of their implantation, migration, and survival situations. The stem cells labeled with fluorescent dyes were transplanted into the mice and then real-time tracking of the transplanted stem cells was conducted using the fluorescence microscope. BLI is the most well studied of the various optical imaging modalities with regard to stem cell imaging in the brain. Bioluminescence involves transducing a reporter gene, which codes for firefly luciferase or Renilla luciferase, into the stem cells [17]. When the luciferase enzyme reacts with its substrate D-luciferin or coelenterazine, it emits photons, which can then be detected and quantified by a charge-coupled device camera system [56].

14.4.2 Methods and Tracers

Nowadays, much more attention has been paid to the application of using in vivo optical imaging to track the transplanted stem cells in real-time. In fact, there are various fluorescence probes of labeling and tracking stem cells.

Green fluorescent protein (GFP) was first found in *Aequorea victoria* by Shimomura in 1962 and has been widely used to label various kinds of cells through transgenic or transfection [57]. The discovery paved the way for a universal marker for cell structures and cellular processes detectable by fluorescence microscopy. The diversity of FPs has increased tremendously by mutating the original GFP sequence and cloning FPs from distant species like crustaceans. Such FPs can be expressed in mammalian cells, including stem cells and transgenic mice without signs of toxicity. Furthermore, smart multi-label approaches like the Brainbow toolbox have been developed to mark neurons with many different FPs [51]. Under the blue wavelengths of light, GFP can emit green fluorescence. At present, the tracer-labeled cell imaging technology has been extensively used in some animal models. Yang et al. had injected mouse melanoma cells that expressed a high level of GFP into the tail vein or portal vein; the results indicated that the metastatic lesion in the brain, liver, and bone could be clearly visualized using whole-body optical images [58]. In addition, Hideo Shichinohe et al. also transplanted stem cells labeled with GFP into mouse brains to trace the conditions of stem cells [57].

The quantum dot (QD) is a type of near-infrared fluorescent dye with good light stability. Additionally, its fluorescent intensity is 20 times higher than that of traditional organic fluorescence dyes e.g., rhodamine. QDs consist of an inorganic core, a shell of metal, and an outer organic coating with a total diameter of 2–10 nm and are quickly becoming very important for in vivo imaging [17]. Therefore, researchers can observe QD-labeled cells for a long time, and it has been a favorable tool for studying the interaction between biological macromolecules in the cell. Due to its longer wavelengths, the near-infrared (NIR)-emitting QDs can easily penetrate the tissues, including the bone and skin. QDs can be synthesized to the desired specifications, including size, shape, and photon emission energy [17]. They are currently regarded as useful biological probes e.g., labeling the bone mesenchymal stem cells (BMSCs) owing to their nanometer dimensions, attractive optical characteristics, high resistance to bleaching or degradation, and strong fluorescence. QD-labeled BMSCs have been transplanted into the infarcted rat brain and the fluorescence can be tracked for at least 8 weeks after transplantation [59]. Unfortunately, cell labeling with QDs suffer from the same long-term in vivo imaging difficulties as direct labeling in that the signal will be diluted due to cell proliferation. Near-infrared-emitting QDs have a potential for in vivo cell tracking within the brain, as demonstrated by Kawabori et al. [60].

14.4.3 Application and Limitations

Optical imaging has some advantages, such as the lower cost, non-radiative, rapid acquisition, high sensitivity, etc. Therefore, it has been widely applied for in vivo imaging in biological tissues [59]. In recent years, with the in-depth study of stem cells, fluorescence imaging techniques has been extensively used to trace the survival and migration of transplanted stem cells. These aptasensors rely on the highly specific, structure-switching ability of aptamers; they undergo drastic secondary or tertiary folding from their initial conformation upon binding with their target molecules [61]. By labeling aptamers with quencher and fluorophore dyes at their 5′ and 3′ ends, a target-binding event, which causes a displacement of the two dyes, can be transduced to a change in fluorescent signal as a result of Förster resonance energy transfer (FRET) principles [54]. Serial optical fluorescence images in vivo could identify the distribution of cells in the cerebral cortex and demonstrate real-time cell migration in the animal. The optical fluorescence imaging in vivo can provide important information regarding the behaviors of donor cells through non-invasive and serial visualization [50].

Apart from the optical imaging technology, other imaging techniques such as MRI and nuclear imaging also have no limit of penetration depth and can provide brain images in three dimension with better spatial resolution [59]. However, nuclear imaging is unsuitable for long-term monitoring, and the effects of MRI are sometimes impeded due to artifacts e.g., intracranial hemorrhage [62]. Moreover, for visualization purposes with medical imaging modalities (e.g., magnetic resonance imaging), cells can be labeled with contrast agents (e.g., iron oxide nanoparticles), which allows their identification from the surrounding environment. Despite the success of revealing cell biodistribution in vivo, most of the existing agents do not provide information about the status and function of cells following transplantation [54]. Consequently, fluorescence imaging might be an alternative or adjuvant technique under some conditions, such as hemorrhagic infarction and traumatic brain injury.

However, the main problem of optical imaging is the difficulty in locating limited light penetration (from a few millimeters to centimeters); thus, it is not appropriate for large animals. Owing to the absorption and scattering of light, the fluorescence signal in deep tissue is frequently very weak, and the spatially resolved intensity is also low. Additionally, the depth of the light source in the animal model and the minimum observed cell number can vary. Therefore, tracking stem cell using optical imaging technology is only applicable to small animals and cannot be applied to whole body imaging of humans [55]. In spite of BLI that has already been used to study in vivo stem cell migration, viability, immunogenicity, and tumorigenicity in small animal studies, all luciferase emissions are in the visible spectrum, which is prone to scattering and absorption by the tissue. Even firefly luciferase, with a comparatively long peak wavelength of 562 nm, is limited for use in small animals due to a maximum penetration of 3 cm of tissue. Moreover, for BLI, the requirement of engineering cells runs the risk of introducing unwanted mutations. Therefore, BLI may not be feasible for clinical translation [63].

14.4.4 Conclusion and Future Perspectives

Stem cells are highly specialized cells with the ability of self-renewal and multidirectional differentiation potential. In recent years, the application of stem cell transplantation for the treatment of cerebral stroke is utilized increasingly. In order to achieve the perfect therapeutic effect after transplantation, it is necessary to monitor the transplanted cells. Transplanted stem cells can induce and enhance functional recovery in experimental stroke. Invasive analysis has been extensively used to provide detailed cellular and molecular characterization of the stroke pathology and engrafted stem cells. However, postmortem analysis is inappropriate in revealing the time scale of the dynamic interplay between the cell graft, ischemic lesion, and endogenous repair mechanisms [64]. The optical imaging techniques in vivo can trace the survival and migration of transplanted stem cells in real-time and provide a lot more important information. Toshiya Osanai et al. had successfully applied the in vivo fluorescence imaging for monitoring the migration and survival of transplanted stem cells in the brain in real-time [50]. Although this technology is still in its infancy, with the continuous development of medical technologies, these powerful new techniques of using ultrafast lasers, dual-photon imaging, and ballistic photon imaging may see progress in the areas of sensitivity, detection depth, and spatial resolution. In addition, these technologies would allow researchers to non-invasively obtain complete brain images in the future and to monitor transplanted stem cells in the central nervous system [57, 59].

References

1. Azevedo-Pereira RL, Daadi MM. Isolation and purification of self-renewable human neural stem cells for cell therapy in experimental model of ischemic stroke. Methods Mol Biol. 2013;1059:157–67.
2. Roh JK, Jung KH, Chu K. Adult stem cell transplantation in stroke: its limitations and prospects. Curr Stem Cell Res Ther. 2008;3:185–96.
3. Bang OY. Clinical trials of adult stem cell therapy in patients with ischemic stroke. J Clin Neurol. 2016;12:14–20.
4. Kim DE, Schellingerhout D, Ishii K, Shah K, Weissleder R. Imaging of stem cell recruitment to ischemic infarcts in a murine model. Stroke J Cereb Circ. 2004;35:952–7.
5. Wang Y, Xu F, Zhang C, Lei D, Tang Y, Xu H, et al. High MR sensitive fluorescent magnetite nanocluster for stem cell tracking in ischemic mouse brain. Nanomed: Nanotechnol Biol Med. 2011;7:1009–19.
6. Bulte JW, Duncan ID, Frank JA. In vivo magnetic resonance tracking of magnetically labeled cells after transplantation. J Cereb Blood Flow Metab: Off J Int Soc Cereb Blood Flow Metab. 2002;22:899–907.
7. Sykova E, Jendelova P. In vivo tracking of stem cells in brain and spinal cord injury. Prog Brain Res. 2007;161:367–83.
8. Hicks A, Jolkkonen J. Challenges and possibilities of intravascular cell therapy in stroke. Acta Neurobiol Exp. 2009;69:1–11.
9. Bliss TM, Andres RH, Steinberg GK. Optimizing the success of cell transplantation therapy for stroke. Neurobiol Dis. 2010;37:275–83.
10. Walker PA, Harting MT, Shah SK, Day MC, El Khoury R, Savitz SI, et al. Progenitor cell therapy for the treatment of central nervous system injury: a review of the state of current clinical trials. Stem Cells Int. 2010;2010:369578.
11. Arbab AS, Frank JA. Cellular MRI and its role in stem cell therapy. Regen Med. 2008;3:199–215.
12. Sabapathy V, Mentam J, Jacob PM, Kumar S. Noninvasive optical imaging and in vivo cell tracking of indocyanine green labeled human stem cells transplanted at superficial or in-depth tissue of SCID mice. Stem Cells Int. 2015;2015:606415.
13. Tennstaedt A, Mastropietro A, Nelles M, Beyrau A, Hoehn M. In vivo fate imaging of intracerebral stem cell grafts in mouse brain. PLoS One. 2015;10:e0144262.
14. Zhou B, Li D, Qian J, Li Z, Pang P, Shan H. MR tracking of SPIO-labeled mesenchymal stem cells in rats with liver fibrosis could not monitor the cells accurately. Contrast Media Mol Imaging. 2015;10:473–80.
15. Han H, Shi C, Fu Y, Zuo L, Lee K, He Q, et al. A novel MRI tracer-based method for measuring water diffusion in the extracellular space of the rat brain. IEEE J Biomed Health Inform. 2014;18:978–83.
16. Lei Y, Han H, Yuan F, Javeed A, Zhao Y. The brain interstitial system: anatomy, modeling, in vivo measurement, and applications. Prog Neurobiol. 2016. doi:10.1016/j.pneurobio.2015.12.007.
17. Chao F, Shen Y, Zhang H, Tian M. Multimodality molecular imaging of stem cells therapy for stroke. Biomed Res Int. 2013;2013:849819.
18. Srivastava AK, Kadayakkara DK, Bar-Shir A, Gilad AA, McMahon MT, Bulte JW. Advances in using MRI probes and sensors for in vivo cell tracking as applied to regenerative medicine. Dis Model Mech. 2015;8:323–36.
19. Norman AB, Thomas SR, Pratt RG, Lu SY, Norgren RB. Magnetic resonance imaging of neural transplants in rat brain using a superparamagnetic contrast agent. Brain Res. 1992;594:279–83.
20. Andrzejewska A, Nowakowski A, Janowski M, Bulte JW, Gilad AA, Walczak P, et al. Pre- and postmortem imaging of transplanted cells. Int J Nanomedicine. 2015;10:5543–59.

21. Sieber MA, Pietsch H, Walter J, Haider W, Frenzel T, Weinmann HJ. A preclinical study to investigate the development of nephrogenic systemic fibrosis: a possible role for gadolinium-based contrast media. Invest Radiol. 2008;43:65–75.
22. Modo M, Beech JS, Meade TJ, Williams SC, Price J. A chronic 1 year assessment of MRI contrast agent-labelled neural stem cell transplants in stroke. Neuroimage. 2009;47 Suppl 2:T133–42.
23. Ishibashi H, Hirao K, Yamaguchi J, Nabekura J. Inhibition of chloride outward transport by gadolinium in cultured rat spinal cord neurons. Neurotoxicology. 2009;30:155–9.
24. Shi C, Lei Y, Han H, Zuo L, Yan J, He Q, et al. Transportation in the interstitial space of the brain can be regulated by neuronal excitation. Sci Rep. 2015;5:17673.
25. Lee N, Kim H, Choi SH, Park M, Kim D, Kim HC, et al. Magnetosome-like ferrimagnetic iron oxide nanocubes for highly sensitive MRI of single cells and transplanted pancreatic islets. Proc Natl Acad Sci U S A. 2011;108:2662–7.
26. Gavins FN, Smith HK. Cell tracking technologies for acute ischemic brain injury. J Cereb Blood Flow Metab: Off J Int Soc Cereb Blood Flow Metab. 2015;35:1090–9.
27. Cores J, Caranasos TG, Cheng K. Magnetically targeted stem cell delivery for regenerative medicine. J Funct Biomaterials. 2015;6:526–46.
28. Korchinski DJ, Taha M, Yang R, Nathoo N, Dunn JF. Iron oxide as an MRI contrast agent for cell tracking. Magn Reson Insights. 2015;8:15–29.
29. Wang Y, Xu C, Ow H. Commercial nanoparticles for stem cell labeling and tracking. Theranostics. 2013;3:544–60.
30. Heryanto YD, Achmad A, Taketomi-Takahashi A, Tsushima Y. In vivo molecular imaging of cancer stem cells. Am J Nucl Med Mol Imaging. 2015;5:14–26.
31. Shapiro EM. Biodegradable, polymer encapsulated, metal oxide particles for MRI-based cell tracking. Magn Reson Med. 2015;73:376–89.
32. Mathiasen AB, Kastrup J. Non-invasive in-vivo imaging of stem cells after transplantation in cardiovascular tissue. Theranostics. 2013;3:561–72.
33. Genove G, DeMarco U, Xu H, Goins WF, Ahrens ET. A new transgene reporter for in vivo magnetic resonance imaging. Nat Med. 2005;11:450–4.
34. Aghayan HR, Soleimani M, Goodarzi P, Norouzi-Javidan A, Emami-Razavi SH, Larijani B, et al. Magnetic resonance imaging of transplanted stem cell fate in stroke. J Res Med Sci: Off J Isfahan Univ Med Sci. 2014;19:465–71.
35. Savitz SI, Cramer SC, Wechsler L, Consortium S. Stem cells as an emerging paradigm in stroke 3: enhancing the development of clinical trials. Stroke J Cereb Circ. 2014;45:634–9.
36. Dousset V, Brochet B, Deloire MS, Lagoarde L, Barroso B, Caille JM, et al. MR imaging of relapsing multiple sclerosis patients using ultra-small-particle iron oxide and compared with gadolinium. AJNR Am J Neuroradiol. 2006;27:1000–5.
37. Bulte JW. Intracellular endosomal magnetic labeling of cells. Methods Mol Med. 2006;124:419–39.
38. Zhao W, Schafer S, Choi J, Yamanaka YJ, Lombardi ML, Bose S, et al. Cell-surface sensors for real-time probing of cellular environments. Nat Nanotechnol. 2011;6:524–31.
39. Rosado-de-Castro PH, Pimentel-Coelho PM, Gutfilen B, Lopes de Souza SA, de Freitas GR, Mendez-Otero R, et al. Radiopharmaceutical stem cell tracking for neurological diseases. Biomed Res Int. 2014;2014:417091.
40. Roca M, Martin-Comin J, Becker W, Bernardo-Filho M, Gutfilen B, Moisan A, et al. A consensus protocol for white blood cells labelling with technetium-99m hexamethylpropylene amine oxime. International Society of Radiolabeled Blood Elements (ISORBE). Eur J Nucl Med. 1998;25:797–9.
41. Acton PD, Zhou R. Imaging reporter genes for cell tracking with PET and SPECT. Q J Nucl Med Mol Imaging: Off Publ Ital Assoc Nucl Med. 2005;49:349–60.
42. Iyer M, Berenji M, Templeton NS, Gambhir SS. Noninvasive imaging of cationic lipid-mediated delivery of optical and PET reporter genes in living mice. Mol Ther: J Am Soc Gene Ther. 2002;6:555–62.

43. Moreira ML, da Costa MP, de Souza SA, Gutfilen B, Rosado-de-Castro PH. In vivo tracking of cell therapies for cardiac diseases with nuclear medicine. Stem Cells Int. 2016;2016:3140120.
44. Wu T, Lang J, Sun X, Zhang B, Liu Y, An R. Monitoring bone marrow stem cells with a reporter gene system in experimental middle cerebral artery occlusion rat models. J Nucl Med: Off Publ Soc Nucl Med. 2013;54:984–9.
45. Correa PL, Mesquita CT, Felix RM, Azevedo JC, Barbirato GB, Falcao CH, et al. Assessment of intra-arterial injected autologous bone marrow mononuclear cell distribution by radioactive labeling in acute ischemic stroke. Clin Nucl Med. 2007;32:839–41.
46. Miyamoto M, Kuroda S, Zhao S, Magota K, Shichinohe H, Houkin K, et al. Bone marrow stromal cell transplantation enhances recovery of local glucose metabolism after cerebral infarction in rats: a serial 18F-FDG PET study. J Nucl Med: Off Publ Soc Nucl Med. 2013;54:145–50.
47. Daadi MM, Hu S, Klausner J, Li Z, Sofilos M, Sun G, et al. Imaging neural stem cell graft-induced structural repair in stroke. Cell Transplant. 2013;22:881–92.
48. Cicchetti F, Gross RE, Bulte JW, Owen M, Chen I, Saint-Pierre M, et al. Dual-modality in vivo monitoring of subventricular zone stem cell migration and metabolism. Contrast Media Mol Imaging. 2007;2:130–8.
49. Tang Y, Zhang C, Wang J, Lin X, Zhang L, Yang Y, et al. MRI/SPECT/fluorescent tri-modal probe for evaluating the homing and therapeutic efficacy of transplanted mesenchymal stem cells in a rat ischemic stroke model. Adv Funct Mater. 2015;25:1024–34.
50. Osanai T, Kuroda S, Sugiyama T, Kawabori M, Ito M, Shichinohe H, et al. Therapeutic effects of intra-arterial delivery of bone marrow stromal cells in traumatic brain injury of rats – in vivo cell tracking study by near-infrared fluorescence imaging. Neurosurgery. 2012;70:435–44. discussion 444.
51. Tuerk C, Gold L. Systematic evolution of ligands by exponential enrichment: RNA ligands to bacteriophage T4 DNA polymerase. Science. 1990;249:505–10.
52. Ellington AD, Szostak JW. In vitro selection of RNA molecules that bind specific ligands. Nature. 1990;346:818–22.
53. Iwagawa T, Ohuchi SP, Watanabe S, Nakamura Y. Selection of RNA aptamers against mouse embryonic stem cells. Biochimie. 2012;94:250–7.
54. Wiraja C, Yeo D, Lio D, Labanieh L, Lu M, Zhao W, et al. Aptamer technology for tracking cells' status & function. Mol Cell Ther. 2014;2:33.
55. Shen HF, Yao ZF, Xiao GF, Jia JS, Xiao D, Yao KT. Induced Pluripotent Stem cells (iPS Cells): current status and future prospect. Prog Biochem Biophys. 2009;36:950–60.
56. Massoud TF, Singh A, Gambhir SS. Noninvasive molecular neuroimaging using reporter genes: part I, principles revisited. AJNR Am J Neuroradiol. 2008;29:229–34.
57. Shichinohe H, Kuroda S, Lee JB, Nishimura G, Yano S, Seki T, et al. In vivo tracking of bone marrow stromal cells transplanted into mice cerebral infarct by fluorescence optical imaging. Brain Res Brain Res Protoc. 2004;13:166–75.
58. Yang M, Baranov E, Jiang P, Sun FX, Li XM, Li L, et al. Whole-body optical imaging of green fluorescent protein-expressing tumors and metastases. Proc Natl Acad Sci U S A. 2000;97:1206–11.
59. Sugiyama T, Kuroda S, Osanai T, Shichinohe H, Kuge Y, Ito M, et al. Near-infrared fluorescence labeling allows noninvasive tracking of bone marrow stromal cells transplanted into rat infarct brain. Neurosurgery. 2011;68:1036–47. discussion 1047.
60. Kawabori M, Kuroda S, Sugiyama T, Ito M, Shichinohe H, Houkin K, et al. Intracerebral, but not intravenous, transplantation of bone marrow stromal cells enhances functional recovery in rat cerebral infarct: an optical imaging study. Neuropathol: Off J Jpn Soc Neuropathol. 2012;32:217–26.
61. Famulok M, Hartig JS, Mayer G. Functional aptamers and aptazymes in biotechnology, diagnostics, and therapy. Chem Rev. 2007;107:3715–43.
62. Bulte JW. In vivo MRI cell tracking: clinical studies. AJR Am J Roentgenol. 2009;193:314–25.

63. Sutton EJ, Henning TD, Pichler BJ, Bremer C, Daldrup-Link HE. Cell tracking with optical imaging. Eur Radiol. 2008;18:2021–32.
64. Aswendt M, Adamczak J, Tennstaedt A. A review of novel optical imaging strategies of the stroke pathology and stem cell therapy in stroke. Front Cell Neurosci. 2014;8:226.

Kolibri 360 GmbH
Ferdinand-Jühlke-Straße 7
99095 Erfurt, Deutschland
produktsicherheit@kolibri360.de